T3-AKD-600

Mathematical Methods for Social and Management Scientists

T. Marll McDonald

DEAN JUNIOR COLLEGE

HOUGHTON MIFFLIN COMPANY BOSTON

Atlanta Dallas Geneva, Illinois
Hopewell, New Jersey Palo Alto London

Printed in the United States of America

Library of Congress Catalog Card Number: 73-9187

ISBN: 0-395-17089-3

Acknowledgements for Sources

p. 40, Problems 3 and 4; p. 41, Probelm 11—From *Sets and Logic* by Samuel C. Hanna and John C. Saber. © 1971 by Richard D. Irwin, Inc., and reprinted with their permission.

p. 41, Problem 7; p. 344, Problems 18, 19, 20—From *Mathematics for Managerial and Social Sciences* by William L. Hart. © 1970 by Prindle, Weber, & Schmidt and reprinted with their permission.

p. 41, Problem 8; p. 163, Problems 13, 14, 15—From *Basic Concepts of Mathematics* by Charles G. Moore and Charles E. Little. © 1967 by McGraw-Hill Book Company. Used with permission of McGraw-Hill Book Company.

p. 41, Problem 10; p. 298, Problems 5 and 6; p. 342, Problem 10—From *Foundations of Mathematics: With Application to the Social and Management Sciences* by Grace A. Bush and John E. Young. © 1968 by McGraw-Hill Book Company. Used with permission of McGraw-Hill Book Company.

p. 54, Problems 2 and 3; p. 55, Problems 4 and 5; p. 179, Problem 20; p. 291, Problems 2 and 13; p. 297, Problem 4; p. 298, Problems 7, 8, 9—Robert C. Fisher and Allen D. Ziebur, *Integrated Algebra and Trigonometry with Analytic Geometry*, 2nd ed., © 1967, pp. 51, 245, 251, 254, 255, 303. Reprinted by permission of Prentice-Hall, Inc., Englewood Cliffs, New Jersey.

p. 219, Problem 14—From *Basic Mathematics for Management and Economics* by Lyman C. Peck. © 1970 by Scott, Foresman and Company and reprinted with their permission.

pp. 266–267, Example 10—From *Theory and Problems of Statistics* by Murray R. Spiegel. © 1961 by Schaum Publishing Company. Used with permission of McGraw-Hill Book Company.

p. 282, Problems 12 and 13—From *Statistical Analysis for Decision Making* by Morris Hamburg, © 1970 by Harcourt Brace Jovanovich, Inc. and reprinted with their permission.

p. 330, Problem 1—From *Modern Mathematics for Business Students,* by R. E. Wheeler & W. D. Peeples. Copyright © 1969 by Wadsworth Publishing Company, Inc. Reprinted by permission of the publisher, Brooks/Cole Publishing Company, Monterey, California.

p. 341, Problems 2 and 3—From *Applications of College Mathematics* by A. William Gray and Otis M. Ulm. © 1970 by Glencoe Press. Reprinted by permission of Macmillan Publishing Co., Inc.

p. 342, Problem 5—From *Quantitative Approaches to Management,* 2nd ed., by Richard I. Levin and Charles A. Kirkpatrick. © 1971 by McGraw-Hill Book Company. Used with permission of McGraw-Hill Book Company.

p. 342, Problem 6—John E. Freund, *College Mathematics with Business Applications,* © 1969, p. 232. Reprinted by permission of Prentice-Hall, Inc., Englewood Cliffs, New Jersey.

p. 342, Problem 11—J. Ronald Frazer, *Applied Linear Programming,* © 1968, pp. 10–11. Reprinted by permission of Prentice-Hall, Inc., Englewood Cliffs, New Jersey.

pp. 343–344, Problems 15 and 17—John Riner, *Basic Topics in Mathematics,* © 1963, p. 130. Reprinted by permission of Prentice-Hall, Inc., Englewood Cliffs, New Jersey.

p. 402, Problem 48; p. 443, Problems 16, 18, 19—From *Probability and Calculus* by E. R. Mullins, Jr. and David Rosen. © 1971 by Bogden and Quigley, Inc. Used with permission of Prindle, Weber, & Schmidt.

pp. 402–405, Problem 49—The selections are reprinted from *Calculus* by Leonard Gillman and Robert H. McDowell by permission of W. W. Norton & Company, Inc., © 1973.

p. 419, Figure 10.1—Melcher P. Fobes and Ruth B. Smyth, *Calculus and Analytic Geometry,* Volume One, © 1963, p. 274. Reprinted by permission of Prentice-Hall, Inc., Englewood Cliffs, New Jersey.

To Alice

Foreword to the Teacher

The main purpose of many mathematics texts, regardless of their subject matter or grade level, is to *cover* mathematical content. The goal of this text is to *uncover* certain mathematical methods for undergraduate social and management scientists. Students in these fields will not have to set up entire systems of complete problem solutions by themselves. Nor is it likely they will directly encounter problems requiring the type of symbol manipulation often emphasized in a traditional college algebra course. It is essential, however, that they be able to understand mathematical approaches to problem solving and to communicate effectively with specialists in applied mathematics, such as computer programmers.

The methods a student uses to initiate, discover, analyze, abstract, solve, interpret, and check a problem or concept are just as important as the content of the problem or concept itself. Moreover, since formal course work can provide only a fraction of the mathematical knowledge a college student must eventually learn and apply, it is essential mathematics courses stress the methods of learning as much as the specific content of the subject. Learning how to learn topics is just as important as the topics themselves. Most learning takes place by working with specific concrete examples until a pattern reveals itself and leads to a generalization. Consequently, the basic approach of this text to a new topic is the sequence of examples, generalization, and additional applications.

I have known many students who apprehensively enter a mathematics course simply to fulfill an annoying requirement. Nevertheless, I have seen these so-called "weak" students really come alive when they realize that mathematics is formalized common sense, that they can learn significant concepts without formal proofs or extensive computations, that definitions such as the one of standard deviation are arrived at through trial *and* error, and that a working knowledge of mathematics does have significance in their vocational fields. This text embodies many of the oral explanations that have helped my classes to attain the above realizations.

I have kept in mind that many users of this book have studied only one or two years of high school mathematics, which they often never fully understood or have partially forgotten by now. The only prerequisites are the ability to compute with fractions and decimals, to evaluate arithmetic expressions, and to solve equations no more difficult than

$$9.3x + 4(3 - 7x) = 2x - 11.6.$$

The background required for the solution of the problems in the third part of Chap. 1 (Algebra Review) is sufficient algebra for the solution of practically all other examples and problems in the text. Logarithms are used in occasional parts of Chaps. 3 and 4 to provide deeper explanations for students familiar with logs.

Computers have made a significant impact on our lives with their many applications, but the versatility of these machines has seldom been used in mathematics textbooks. Earlier I mentioned that mathematics is best learned through pattern-revealing examples; often the best way to obtain many of these examples is by having the computer produce calculation-laden specific cases. A prime example is the basic acceptance sampling formula in which proportion p of the entire lot is defective, there are n items in the sample to be tested, and the entire lot will be accepted if the sample contains at most r defectives. Then

$$\begin{aligned}&\text{Probability (accepting entire lot)}\\ &\quad= \binom{n}{0} \cdot p^0 \cdot (1-p)^{n-0} + \binom{n}{1} \cdot p^1 \cdot (1-p)^{n-1}\\ &\qquad + \binom{n}{2} \cdot p^2 \cdot (1-p)^{n-2} + \ldots + \binom{n}{r} \cdot p^r \cdot (1-p)^{n-r}.\end{aligned}$$

By changing the values of one or more variables at a time and having the computer do the considerable calculations one can see the cause and effect relationships among the three independent variables and the one dependent variable. Included in the *Instructor's Guide* are the BASIC language programs used to produce the book's examples. Though the book makes no real attempt to teach computer programming, the programs are included in case an instructor who has access to a BASIC-equipped computer wishes to modify or use them as he or she sees fit. A computer was also used in researching one relevant topic not usually found in mathematics texts: the application of probability to law.

This two-semester text is not radically different from the recommendations of the Committee on the Undergraduate Program in Mathematics of the Mathematical Association of America for students in the biological, management, and social sciences. Probability, differential and integral calculus, linear algebra, and work with the results of a computer—all of which are specific recommendations of the CUPM—are included in the text.

The ten chapters are somewhat dependent on one another and the instructor is relatively limited in rearranging their order. Of course, he can and should feel free to add his own explanations, examples, and extra topics.

The author would like to thank the following individuals who reviewed this book in its early stages of development: Michael Bernkopf, Pace University; Charles Burris, University of New Mexico; Carl Cowen, Indiana University (Richmond); Fred Dashiell, University of California (Berkeley); George Dorner, Loyola University of Chicago; Melcher Fobes, College of Wooster; David Hildebrand, University of Pennsylvania; John Matthews, California State College (Fullerton); John A. Moreno, San Diego City College; Gary Nichols, Mount Hood Community College; John Novosel, Southwest College (Chicago); James Ramaley, University of Pittsburgh; and Richard Schwartz, Staten Island Community College.

People credited with the development of this text include the students who made many suggestions in the interests of clarity and readability and who discovered mathematical errors in the interests of accuracy (and for a finder's fee of $1!); the administration at Dean Junior College for providing a nurturant atmosphere where I could experiment with the many teaching methods that have found their way into these pages; Rod Cavedon and Mike Hebert, fellow teachers who constructively criticized preliminary manuscripts; Doug Buchanan and Jim Conlon, former roommates who could not escape the incessant pecking of a typewriter for many months; and the professionals at Houghton Mifflin Company, whose competence immeasurably aided this book.

January 1974 *T. Marll McDonald*

Foreword to the Student

It is the purpose of this text to acquaint you with the mathematics—some of it old, some of it just recently developed—that you will need in future courses, in the actual fields of social science work, or in the business world. The book not only will stress the *content* of this mathematics—its symbolism, formulas, techniques, and applications—but will also heavily emphasize the *methods* of mathematics—the ways we initiate, discover, analyze, set up, solve, interpret, and check a problem or concept. These chapters will show you *how* mathematics is learned, for an essential part of any course is not just the knowledge itself but also the way in which this knowledge is learned. Learning how to learn is of prime importance since you can never obtain in one formal academic course all the mathematics that you must ultimately be able to apply. You should be able to learn mathematics independently and individually since you often will have to do so in the actual practice of your profession.

The least effective way to learn mathematics is to read the sentences and merely agree with what the book is saying. Learning mathematics is an activity, not a spectator sport; you should have a pencil and paper ready as you carefully read—and reread several times if necessary—the text and its examples. Do not hesitate to work through the examples or check the arithmetic calculations because doing them helps you to understand their origins. Perhaps the best way to come to understand a new idea is to work, rework, and constantly come back to it *until it is as familiar to you as if you had invented it yourself*. After all, learning to sew, drive a car, ride a motorcycle, swim, or ski is difficult at first until practice moves the conscious awareness of the basic principles into our subconscious minds. We then may bring the basic principles to immediate recall if needed. An equally effective way to check yourself is to pretend that you are about to explain the new process or idea to your classmates. As you mentally prepare your oral explanation, continually ask yourself: "If I were a fellow student, would I understand my own explanation?"

The text is organized into ten chapters. The first presents the author's philosophy of learning mathematics and a brief description of mathematics itself along with some refresher algebra problems. Chapter 2 deals with sets; though the subject has been sometimes overemphasized in secondary school "modern mathematics," a knowledge of sets from a more advanced viewpoint is quite useful. In Chap. 3 you work with functions which are powerful ways of describing the relation between two variables such as time and population. Chapter 4 is inherently interesting and useful since it concerns itself with money and its behavior when affected by different investment or borrowing policies. You learn the mathematics of chance, or probability, in Chap. 5 along with a few of its commercial applications. Chapter 6 helps you to learn statistics for organizing and interpreting numerical results of an experiment or survey. Chapter 7 deals with solving certain groups of equations. Chapter 8, on linear programming, gives you direct involvement with the new mathematics created during World War II, and you will learn how this powerful tool solves many important problems of peace. Chapters 9 and 10 uncover the ideas behind calculus, which has vast applications to problems in the worlds of science, engineering, social science, and business.

Each section of this text begins with a familiar example, generalizes slowly from it, and then applies the newly obtained principle to other specific problems. Instead of relying solely on formal mathematical proof, the text appeals to common sense as justification for many steps in the development of a formula. The problems at the end of each section are designed to be realistic and meaningful applications of our ideas for you to sink your teeth into.

I hope you will find this book readable, interesting, and different. I especially welcome comments about, or errors found in, this book and will personally answer every letter. Please send them to me at the Math-Science Department, Dean Junior College, Franklin, Massachusetts 02038.

T. Marll McDonald

Contents

Mathematical Methods for Social and Management Scientists

Chapter 1

Introduction

How are problems solved by mathematics?

The emphasis in this text is on using mathematics for solving everyday problems rather than on the study of abstract mathematics for its own sake. For example, geometry originated thousands of years ago for the literal purpose of measuring lengths and areas on the earth (*geo* = earth, *metrein* = to measure). The Greeks, including Euclid (who was the organizer of much of the geometry we studied in high school), developed it into a logical sequence of abstract statements dealing with circles, triangles, line segments, and shapes in a strictly theorem-proof fashion, often without much regard for the ultimate usefulness of the theorems. While many contemporary mathematicians insist upon proving every statement about algebra and applied mathematics as the Greeks did with geometry, we will often appeal to intuition, observation of patterns, and common sense as we develop and use the mathematics in this book.

What is mathematics? There are almost as many definitions of this body of thinking and knowledge as there are mathematicians, but the one we will use is this:

Definition Mathematics is a common sense formalization of the patterns exhibited by sets of numbers and points.

Let us examine the key words in this definition. Mathematics, in its finished form, is obviously quite formalized; its definitions of terms, theorems, and formulas are precisely stated with a high degree of logical reasoning. However, the formal statements in the finished product can almost always be traced back to common sense observations of numbers and points. The commutative property of addition of whole numbers is the formalization generated by the common sense notion that if we add together any two whole numbers the order of addition does not make any difference in the result of the addition operation:

$$3 + 5 = 5 + 3,$$
$$89 + 478 = 478 + 89,$$
$$17{,}416 + 2{,}980{,}123 = 2{,}980{,}123 + 17{,}416, \text{ etc.}$$

In this case we conclude that the commutative property of addition is indeed a valid statement because we have observed a pattern exhibited by the ordinary set of whole numbers. Of course, such a series of consistent observations does not constitute a formal proof, but one of the beautiful things about mathematics is that almost all patterns lead to generalizations that can be formally proven.

Just as it is difficult to understand how a cathedral was built by looking at the finished product after the scaffolding is torn down, it is often difficult to understand a theorem or property without knowing the pattern of specific results that led to the generalized theorem or property itself. Seeing the scaffolding as well as the completed cathedral is almost mandatory in comprehending the building. Likewise, working with the particular pattern of numbers or points is essential in comprehending the formalized and completed theorem or property. In this text we will always balance the finished statements with a perspective on the patterns that give rise to the statements themselves. Of course, we are indebted to other mathematicians for their efforts in actually carrying through the proofs that permit our quick generalizations.

Problem-solving through the four-step cycle

Now that we have examined the interplay between developing and developed mathematics let us see how the latter is used in a problem solving situation. We begin with a simple example.

Example 1 Joe has some dimes, and Mary gives him seven more so that Joe ends up with nineteen dimes. How many did he have to begin with?

Solution Instantaneously we know the answer is twelve dimes, but the problem is well worth a closer look since the general *method* we use to solve *this* problem is the same one we can use to solve almost *any* problem in applied mathematics. A diagram is in order here.

Figure 1.1

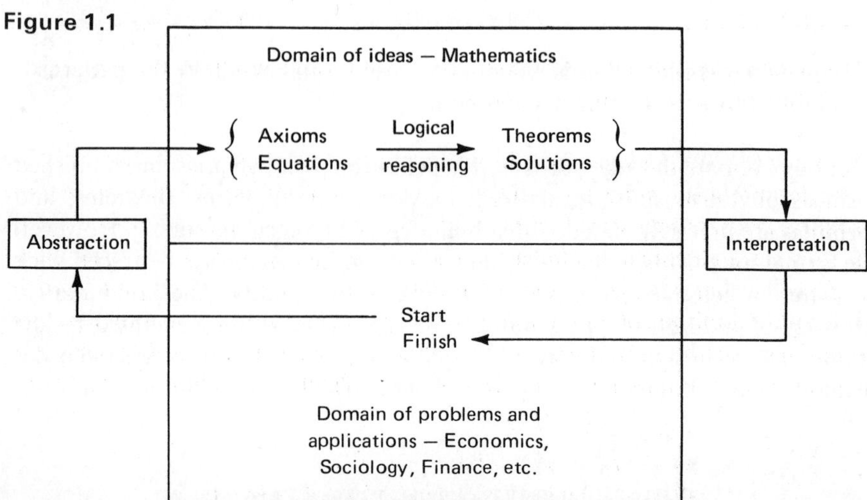

Formal mathematics itself is wholly contained in the upper rectangle since its prime concern is with the logical relationships among statements concerning the ideas of numbers and points. The fact that numbers and points grew out of manipulation of concrete objects and measurements does not always concern the research mathematician, who need not care where the axioms come from or where the logical conclusions from working with them lead to in the real world.

If we go back to the dime problem we can see how it, the diagram, and most word problems are related. Problem solving in mathematics can almost always be analyzed in terms of a four-step cycle: identification, abstraction, equation-solving, and interpretation. The first step is to clearly identify the numbers that the problem involves. After all, if we do not know exactly what we are looking for we will never know if we have found the answer. It is a good idea to clearly specify, in words, the number we are looking for before we identify it with a single letter. In Example 1, then, we are looking for

the number of dimes John had originally.

Then we write

$$x = \text{the number of dimes John had originally.}$$

The next step, abstraction, is the one we remember as "setting up," and it is often the most difficult part. If we wish to use our knowledge of solving equations, we must represent the things and movements of the world of the word problem by a corresponding equation with numbers and operations in the world of mathematics; that is, we must "set up" the unknown number in a mathematical equation that mirrors the conditions of the problem. There is no series of rules on how to automatically accomplish this often difficult translation process, but one rough guide is to write *in words* an intermediate equation that best describes the problem situation before completing the translation process. Here, we will say

(John's original dimes) + (seven dimes) = (nineteen dimes).

When abstracted into the world of mathematics, this word equation yields

$$x + 7 = 19.$$

We then use our knowledge of pure mathematics to solve this mathematical equation, whose solution we easily find to be 12. Obviously, the equation-solving steps can get much more difficult, but the basic principles of identification and abstraction remain the same.

The last step is often overlooked but nonetheless is important. We interpret the solution of the abstract equation which in this case is

$$x = 12$$

in terms of the number we originally sought so that we will have completed the cycle in the diagram. Hence we would say at the finish,

John had twelve dimes originally.

It is also a good idea to check our work by substituting our answer into the original equation to see if the resulting numbers on both sides of the equality sign are really equal.

Numbers and numerals

We should also note in this all-important diagram that numbers and points are *ideas*, not objects. To illustrate, we will prove that

half of eight is zero

and

half of twelve is seven.

We start with eight and twelve in their Hindu-Arabic and Roman forms.

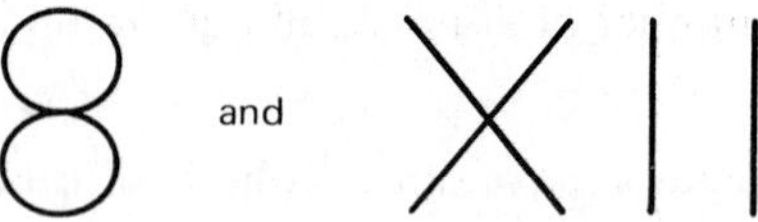

Then if we erase their bottom halves we obtain

respectively. The catch is that while half of one particular *numeral*, or symbol, for the number eight is a *numeral* for the number zero, then half of the *number* or idea symbolized by 8 is the *number* symbolized by 4. Moral of the story: a number is an *idea* which no one has ever seen; a numeral is a *symbol* for a number. In this book we will let the context of a situation determine whether a number (idea) or a numeral (symbol) is being discussed.

We use this cyclical approach of determining the unknown, writing an equation, solving the equation, and interpreting and checking the results again on a more complicated problem.

Example 2 How much 70% acid solution should be added to 30 gallons of a 20% acid solution to obtain a 35% acid solution?

Solution The first reaction—that this is one of those hideous story problems again—is a natural one, especially in light of the actual rarity of such problems in life. Bear in mind, though, that the *method* of setting up and solving relatively simple problems such as this one is what we are learning rather than the memorization of rules for solving these specific problems.

Step 1. The problem is somewhat vague in stating the unknown. The statement "let $x =$ amount of 70% acid" is not precise enough since it does not clearly identify the *number* we are looking for, which is the number of gallons of 70% acid solution that should be added. Thus we have

$x =$ the *number* of gallons of 70% acid solution,

which does specify a number, symbolized by x, that will become part of an equation.

Step 2. But what equation? A diagram is almost always helpful in tracking down the correct equation. We draw and write as much as possible in a diagram, this time showing containers holding the various component solutions.

Figure 1.2

70% solution x gallons	+	20% solution 30 gallons	=	35% solution (x+30) gals.

Here is the key: we are looking for an intermediate equation *in words* that describes the objects in the diagram. The equation springs from the common sense notion that

(number of gallons of *pure* acid in container 1)
+ (number of gallons of *pure* acid in container 2)
= (number of gallons of *pure* acid in container 3).

We refine this information to equation (A):

$$(70\% \text{ of } x) + (20\% \text{ of } 30) = (35\% \text{ of } (x+30)). \qquad \text{(A)}$$

Where did equation (A) come from? For example, 4 gallons of a solution that is 30% acid contains 30% of 4 or 1.2 gallons of pure acid (and 2.8 gallons of pure water). Similarly g gallons of a solution that is 72% pure contains 72% of g or $.72 \cdot g = .72g$ gallons of pure acid. The abstraction process is now complete, and we move on to

Step 3. We now have the equation itself in (A), which becomes, after a slight modification,

$$.70 \cdot x + .20 \cdot 30 = .35 \cdot (x+30).$$

Solving it gives the successive equations

$$\begin{aligned}
&.70x + 6 = .35x + 10.5\\
&.35x + 6 = 10.5\\
&.35x = 4.5\\
&x = 4.5/.35\\
&x = 450/35\\
&x = 90/7\\
&x = 12\,6/7.
\end{aligned}$$

Step 4. All we have to do now is to interpret the final equation

$$x = 12\,6/7,$$

which means we have to add 12 6/7 gallons of 70% acid solution to meet the requirements. We check our result by observing that, where the three measures are in gallons of pure acid, the answer to the following questions is yes.

$$\overbrace{70\%\text{ of }\frac{90}{7}\text{ gal}}^{\text{pure acid}} + \overbrace{20\%\text{ of }30\text{ gal}}^{\text{pure acid}} \overset{?}{=} \overbrace{35\%\text{ of }\left(\frac{90}{7}+30\right)\text{ gal}}^{\text{pure acid}}$$

$$.70 \cdot \frac{90}{7} + .20 \cdot 30 \overset{?}{=} .35 \cdot \left(\frac{90}{7} + \frac{210}{7}\right)$$

$$\overset{.10}{\cancel{.70}} \cdot \frac{90}{\cancel{7}} + .20 \cdot 30 \overset{?}{=} \overset{.05}{\cancel{.35}} \cdot \frac{300}{\cancel{7}}$$

$$9 + 6 \overset{?}{=} 15$$

$$15 \overset{?}{=} 15.$$

Problem set 1–1

The following problem set is designed to give us some practice with applied mathematics. Try to do these problems by the four-step method since it makes many problems easier to understand. We will also make repeated references to Figure 1.1 in the rest of this book.

1. How many ounces of an 80% acid solution should be added to 16 ounces of a 30% acid solution in order to have a 45% solution?
2. How much water should be mixed with 32 pints of a 60% alcohol solution to dilute it to 21%?
3. Mr. Smith must mix gasoline and oil in his snowmobile so that the mixture is 5% oil and 95% gasoline. If 2 1/2 gallons of the mixture is 98% gasoline, then how much oil should he add to bring the oil content up to 5%?
4. Betty's car radiator, with a capacity of 20 quarts, contains a mixture that is 15% alcohol. How much of this solution must be drained off and then replaced by an equal quantity of 95% alcohol solution to bring the strength up to 40%?
5. How much 20% alcohol solution must be mixed with 16 gallons of a 30% solution in order to have a 35% alcohol mixture?
6. How many pounds of grass seed costing 50¢/lb must be mixed with 40 lb of seed costing 90¢/lb to make a mixture worth 80¢/lb? Hint: (*Total value* of 50¢/lb seed) + (*total value* of 90¢/lb seed) = (*total value* of 80¢/lb seed).
7. How many pounds each of candy worth 65¢/lb and $1.25/lb must be mixed to give a 60-lb mixture worth 90¢/lb?
8. Banker Smith has fifty shares of stock of XYZ Corporation, including common and preferred, whose total value is $618. If the preferred stock is worth $15 a share and the common stock is worth $9 a share, then how many of each kind of stock does he have?
9. If eighty-one coins, consisting of dimes and quarters amount to $10.50, how many of each kind are there?
10. If fifty-six coins, which are either dimes or nickels, have a value of $4.85, how many of each kind are there?

11. Refer to the sentence which begins with "The catch is . . . ." Write two similar sentences that explain in what sense we can say that half of twelve is seven and half of eight is three.

Computers

The next part of this chapter provides some basic knowledge of computers along with their capabilities and limitations. All of us have been touched in some way by computers since information about us has been processed by them at one time or another. A few of the widespread applications of computers in our lives include savings and checking accounts at banks, reservation systems at airlines and hotels, student registration and grade reporting, and sending man into space.

Computers can process data at unbelievably fast rates, such as multiplying or dividing hundreds of pairs of seven-digit numbers in one second. Their attached printers can output data at the rate of 2,400 letters per *second*—equivalent to the combined efforts of 480 typists typing at 50 words per minute. This speed alone has made possible the existence of airline reservation systems and their nearly instantaneous response time. Even though the cost of computers is high, often running into hundreds of dollars a day, their speed cuts the cost per unit of doing a business transaction or scientific calculation to well under the cost of doing it by hand.

The advent of computers has had an impact upon the teaching of mathematics, too. Remembering our earlier statement that we often learn general principles by working first with patterns of specific results, let us work with an example to see how a computer can give results which will reveal a pattern to us.

Finding a pattern

Example 3 If a regular polygon has n sides, then how many degrees are in each of its interior angles?

Solution Before we can use a computer to help us answer the question we, as humans, must have a complete understanding of the problem. After obtaining this understanding we will then call upon a computer to do some calculations for our benefit.

The best way to answer such a general question about n sides is for us to answer the question for small, specific values of n that are arranged in an orderly fashion. Thus we should work initially with $n = 3$, then 4, then 5, then 6, and so on. We omit $n = 1$ and $n = 2$ because the smallest number of sides a polygon can have is 3. We will make a table to keep track of our results and carefully examine this table to find the pattern exhibited there.

Suppose we have a regular polygon—that is, a polygon with equal sides and equal interior angles—of three sides. This is merely a fancy way of describing an equilateral triangle, △. Since the sum of the interior angles of a triangle is 180 degrees, this means that each interior angle contains $180 \div 3 =$ 60 degrees. Similarly, a regular polygon of four sides is the familiar square, □, in which each interior angle is a right angle or 90 degrees.

So far our table looks like this:

Table 1.1

Number of sides	Number of degrees in each interior angle
3	60
4	90

What about a regular five-sided polygon, or pentagon? Let us look at one with its center and "spokes" to the corners drawn in.

Figure 1.3

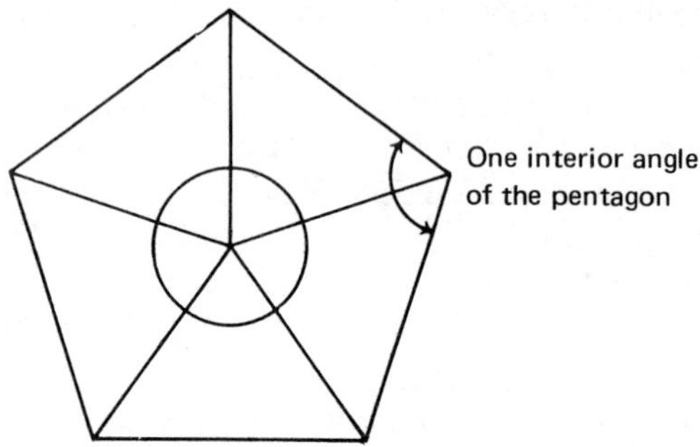

There are five triangles here with a total sum of $5 \cdot 180 = 900$ degrees in their interior angles, but 360 of those degrees are accounted for by the five innermost angles of the triangles which leaves $900 - 360 = 540$ degrees spread equally among the five interior angles of the pentagon itself. Hence each of these five angles measures $540 \div 5$ or 108 degrees.

If we move on to a regular six-sided polygon, or heaxagon, then we dissect it entirely into six congruent triangles, the sum of whose interior angles is $6 \cdot 180$ or 1,080 degrees. The six innermost angles of the triangles comprise 360 of these degrees, so that each interior angle of the hexagon measures

$$(1{,}080 - 360)/6 = 720/6 = 120 \text{ degrees.}$$

We continue in a similar fashion with a regular seven-sided polygon, or heptagon, which we dissect into seven smaller triangles, each having one vertex at the center of the heptagon. Then the sum of the interior angles of all seven triangles is $7 \cdot 180 = 1{,}260$ degrees with 360 of them in the center of the heptagon and $1{,}260 - 360 = 900$ degrees divided equally among the seven interior angles of the polygon. Then each of these seven angles measures $900 \div 7 = 128\,4/7$ degrees.

The pattern for calculating the size of the interior angles should be somewhat clearer by now. If we jump to the case where there is a regular twelve-sided polygon, a regular dodecagon, then we dissect it into twelve triangles whose sum of interior angles is $12 \cdot 180$ degrees. Since 360 of these $12 \cdot 180$ degrees are in the center, $(12 \cdot 180 - 360)$ degrees are in the sum of the interior angles of the dodecagon. Each interior angle thus contains

$$(12 \cdot 180 - 360)/12 \text{ degrees.}$$

Generalizing a formula

The generalization to n sides is an easy one if we look at the expression in the last sentence and recognize that 12 plays the role of the number of sides of the polygon. If there are

$$n$$

sides in a regular polygon, then there are

$$(n \cdot 180 - 360)/n$$

degrees in each interior angle. We bring our table up to date.

Table 1.2

Number of sides	Number of degrees in each interior angle
3	60
4	90
5	108
6	120
7	128 4/7
⋮	⋮
12	$(12 \cdot 180 - 360)/12$
⋮	⋮
n	$(n \cdot 180 - 360)/n$

Now we have a formula for A, the number of degrees in each interior angle of an n-sided polygon:

$$A = (n \cdot 180 - 360)/n.$$

Notice that this formula also holds for values of n less than or equal to 7, as illustrated by the equation

$$(6 \cdot 180 - 360)/6 = 120.$$

We finally call on a computer to calculate the values of A corresponding to larger and larger values of n as we investigate the behaviour of A as n

increases in value. Doing it by hand is out of the question because of the sheer arithmetic; for example, if $n = 873$, then

$$A = (873 \cdot 180 - 360)/873.$$

Remember the state of our problem-solving so far: we want the computer to read a number from a list of increasing values of n, calculate the resulting value of the expression

$$A = (n \cdot 180 - 360)/n,$$

and print the values of n and A. It is to read the next number in the list as the new value of n to work with and continue until the list has been exhausted. Here are the results of the computer's work where the numbers in the first column are values of the variable n (number of sides) and the numbers in the second column are the corresponding values of A (number of degrees in each interior angle).

Table 1.3

3	60
4	90
5	108
6	120
7	128.571
8	135
9	140
10	144
11	147.273
12	150
13	152.308
14	154.286
15	156
20	162
25	165.6
30	168
35	169.714
40	171
45	172
50	172.8
60	174
70	174.857
80	175.5
90	176
100	176.4
150	177.6
200	178.2
300	178.8
350	178.971
400	179.1
500	179.28
600	179.4
700	179.486
800	179.55
873	179.588

```
874          179.588
875          179.589
876          179.589
900          179.6
1000         179.64
1500         179.76
2000         179.82
2500         179.856
3000         179.88
5000         179.928
10000        179.964
15000        179.976
20000        179.982
30000        179.988
50000.       179.993
70000.       179.995
90000.       179.996
99999.       179.996
```

What happens to the corresponding values of A as n becomes larger and larger? They approach 180.000 but never quite reach it. We have reached this conclusion much faster with a computer than without one. In the rest of this text we will often use the power of a computer to illustrate principles and to generate patterns of specific cases that can help to reveal these principles.

BASIC computer language

It is interesting to look at the actual program in the BASIC (*B*eginner's *A*ll-purpose *S*ymbolic *I*nstruction *C*ode, trademark held by Dartmouth College, Hanover, New Hampshire) computer language.

```
10 DATA 3,4,5,6,7,8,9,10,11,12,13,14,15,20,25,30,35,40
20 DATA 45,50,60,70,80,90,100,150,200,300,350,400,500
30 DATA 600,700,800,873,874,875,876,900,1000,1500,2000
40 DATA 2500,3000,5000,10000,15000,20000,30000,50000
50 DATA 70000,90000,99999
60 READ N
70 LET A = (N * 180 - 360)/N
80 PRINT N, A
90 GO TO 60
100 END
```

These instructions to the computer are in an almost-algebraic language with each instruction and list of DATA items assigned a number from 10 through 100, inclusive. Statements 10, 20, 30, 40, and 50 contain the DATA items, or values of N, which we use as input to the formula for A; statement 60 says to READ one value of N from the DATA lists; statement 70 LETs A equal the result of evaluating the expression

$$(N * 180 - 360)/N$$

for the current value of N where * is the multiplication sign; statement 80 PRINTs the current values of N and A on the teletypewriter attached to the computer; statement 90 has the computer GO TO statement 60 to read the *next* number in the DATA list as the new value of N. This read-calculate-print process cycles through until all the numbers in the DATA lists have been read and processed in their order of appearance there. Statement 100 indicates the END of the program to the computer so it can begin to execute the previous nine instructions. Computer programs get far larger and more complex, but the ten-statement instruction set we have just examined is a full-fledged computer program.

Limitations of computers

We have talked about and seen the capabilities of a computer, but what about its limitations? We should realize right away exactly what a computer is—a machine that moves electricity according to human instructions. There is no human genius lurking within a black box that verifies every bit of output from a computer. As long as information can be represented in the form of electrical impulses then the computer can process it in clearly defined operations, such as arithmetic operations or comparisons with other data, at very fast speeds with virtually complete accuracy, regardless of problem complexity. The key to understanding a computer lies in the phrase "as long as information can be represented in the form of electrical impulses" because a computer cannot think. It can only obey explicit instructions specified entirely by people.

The process of thinking is hard even for psychologists to define, but one definite part of it is the marvelous ability of the human mind to associate words in a seemingly endless yet reasonably logical fashion. For example, if we asked three people to associate four immediate and consecutive words with the word "hawk" we might get

"hawk ⟶ bird ⟶ forest ⟶ leaves ⟶ pretty foliage"

from a naturalist;

"hawk ⟶ flight ⟶ airplanes ⟶ trip ⟶ New York"

from a business man; and

"hawk ⟶ Honda ⟶ motorcycle ⟶ riding ⟶ feeling of freedom"

from a motorcyclist. Computers, amidst much fanfare, were programmed at their infancy to translate from one language to another but these expensive and exhaustive efforts were often fruitless because the machines cannot recognize the meaning of words according to their context. One computer, instructed to translate from English to Russian and then from the resulting Russian back to the English, went from "Out of sight, out of mind" to the Russian equivalent back to "Blind, insane." Another computer began with "The spirit is willing, but the flesh is weak" and ended with "The wine is good, but the meat is terrible."

As another example of the machines' inherent limitations, the word "love" to most IBM computers means this consecutive set of thirty-two charged (●) or noncharged (○) miniature doughnut-shaped bits of iron:

Figure 1.4

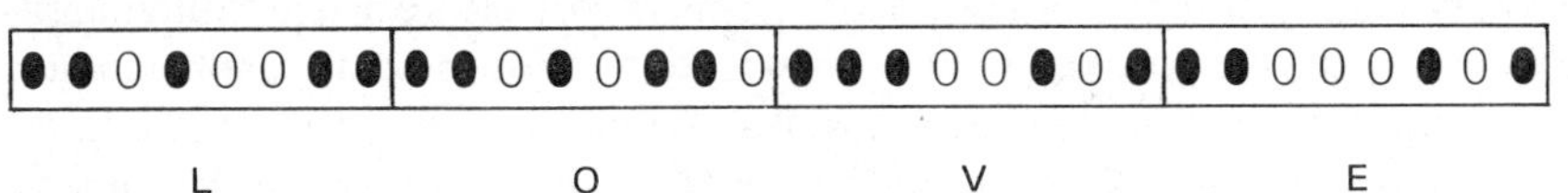

Is anyone really enthusiastic about having a computer select a date for us if our responses to a questionnaire have been reduced to a few hundred charged particles inside a black box?

The next picture shows us the exterior of a small but powerful computer system manufactured by the Data General Corporation of Southboro, Massachusetts. Figure 1.5 shows a computer along with three attached teletypewriters. The computer receives instructions and data from the teletypewriters, processes them internally, and then returns the results to the teletypewriters. A few of the examples, and most of the tables in this book were calculated and printed by a computer system which included a teletypewriter. Table 1.3 is an example of this output.

Throughout the remainder of this book we will concentrate on learning certain mathematical methods and content aided frequently by a computer. Though we must realize their serious and inherent limitations, computers are really quite useful in and even necessary for our highly technological society.

Figure 1.5

Algebra review

The facts learned in high school algebra have become rather hazy to many students, so in this section we will review a few algebra topics. This review includes almost all the topics from algebra needed to understand the upcoming concepts, examples, and homework problems. There are very few explanations of how to do the problems since instructors will want to explain them according to the varying backgrounds of students.

Problem set 1–2

1. Evaluate these expressions.
   a. $3 + 5 \cdot 2$ b. $(3+5) \cdot 2$
   c. $(13.4 - 17.18) - 12$ d. $15 - (6-9)$
   e. $(5+9)/5$ f. $7 + 5 + 9 \div 5$
   g. $5 + 9/5$ h. $16.48 + 23.019 - 15.2$
   i. $60 \div 4 \cdot 3$ j. $(-7) \cdot (-8)$
   k. $60 \div (4 \cdot 3)$ l. $3 \cdot 1.5^2 - 4 \cdot 6 \div 2$
2. Remove parentheses from these algebraic expressions and combine terms whenever possible.
   a. $10 + 3(x-2)$ b. $15 - (7-9)$
   c. $14 - 6(x-5)$ d. $(x^2 - 3x + 2) - (10 - 4x^2)$
   e. $14 - 6(x+5)$ f. $.4(x+15)$
   g. $40\% \cdot (x+15)$ h. 30% of $(20-x)$
   i. $60\% \cdot (90-x)$ j. $y - 5(2y-1)$
   k. $x(x-1) - 2x - (x+3)$ l. $5x - [(3x+1) - 3 - (2x-3)]$
3. If $a = 3$, $b = -2$, $c = 2$, and $d = 0$, then find the value of
   a. $c^2 - 3b$ b. $c^2 - cd$
   c. $5c^2 - 2ab$ d. $2a^3 + b^2$
   e. $(a+c)^2$ f. $a^3b - 6$
   g. $7abc - a^3$ h. $7abc - d^3$
   i. $7b^3 - 5d$ j. $(7b)^3 - 5d$
   k. $(4a^2b) \div (-9c)$ l. $5ab^2 \div (-3b)$
   m. $-b^2$ n. $(-b)^2$
4. Solve these equations for x.
   a. $3x - 5 = 16$ b. $x - 6 = 5x + 14$
   c. $3x + 3 = 10x - 4$ d. $2 - x = 3x + 4x$
   e. $5x = 10x - 15$ f. $20x = 10 \cdot (5-x)$
   g. $9x + 1 = 2(1+x)$ h. $.5x = 10$
   i. $.3x + 2 = 14$ j. $3.8x - 1.36 = 10.6x$
   k. $5x - 1/4 = 3x + 1/2$ l. $(2x+4) = 3x + 8$
   m. $C = 2\pi x$ n. $b^2x = 3b$
   o. $ax - b = c$ p. $.05x = 2h$
   q. $A = p(1+rx)$ r. $s = \frac{1}{2}xt^2$
5. Solve these inequalities for x.
   a. $3x > 6$ b. $x + 9 < 14$

c. $3x + 11 > 26$ d. $3x + 11 \geqslant 26$
e. $6x - 3 < x + 2$ f. $6x - 3 \leqslant x + 2$
g. $-3x > 12$ h. $3x < -12$
i. $3x > -4$ j. $x + 4 \leqslant 5x - 7$
k. $3(2x-7) > 4x + 9$ l. $5(2-x) + 6 \geqslant 7 - (3x+4)$

6. If x is any real number then the absolute value of x is given by this pair of equations:

$$|(x)| = (x) = x \text{ where } x \geqslant 0, \text{ and}$$
$$|(x)| = -(x) \text{ where } x < 0$$

For example,

$$|(3)| = (3) = 3 \text{ since } 3 \geqslant 0 \text{ and}$$
$$|(-5)| = -(-5) = 5 \text{ since } -5 < 0.$$

Taken together these two equations say that if x is any real number then $|x|$ is always nonnegative. Evaluate these expressions.
a. $|17|$ b. $|23.5|$
c. $|-16|$ d. $|0|$
e. $|9-5|$ f. $|5-9|$
g. $|21.3-16.81|$ h. $|16.81-21.3|$
i. $|2 \cdot (5-8)|$ j. $2 \cdot |(5-8)|$
k. $|7+-5|$ l. $|7| + |-5|$

7. Evaluate these square roots to the nearest hundredth. Recall that, for example, $\sqrt{25} = 5$ and *not* ± 5.
a. $\sqrt{64}$ b. $\sqrt{34}$
c. $\sqrt{48}$ d. $\sqrt{16}$
e. $\sqrt{67.8}$ f. $\sqrt{9}$
g. $\sqrt{16+9}$ h. $\sqrt{16+\sqrt{9}}$
i. $\sqrt{251}$ j. $\sqrt{265.1}$
k. $\sqrt{706.23}$ l. $\sqrt{3219}$

8. Remove parentheses and simplify the answer whenever possible.
a. $3a(ab+bc)$ b. $5x(4x-3x^4)$
c. $(x+4)(x-3)$ d. $(2x-7)(x+3)$
e. $(2x-13)(5x+6)$ f. $(3x-4)(5x-7)$
g. $(x+7)^2$ h. $(x-5)^2$
i. $(x+7)^2 - x$ j. $\dfrac{(x+h)^2 - x^2}{h}$
k. $\dfrac{3(x+h)^2 - 4(x+h) + 5 - (3x^2 - 4x + 5)}{h}$
l. $\dfrac{5(x+h)^2 - 7(x+h) - 11 - (5x^2 - 7x - 11)}{h}$

9. Solve these equations by first factoring the expression and then setting each of the resulting factors equal to 0.
a. $x^2 - 10x + 21 = 0$ b. $x^2 + 7x + 10 = 0$

c. $x^2 - 6x - 16 = 0$
d. $x^2 + 6x - 27 = 0$
e. $2x^2 + 5x - 3 = 0$
f. $6x^2 + 15x = 9$
g. $2y^2 + 5 = -11y$
h. $c^2 + 5c = 24$
i. $4c^2 - 36 = 0$
j. $y^2 + 8y = 0$
k. $x(x+11) + 18 = 0$
l. $2c^2 - 20c = -18$

10. Recall the Quadratic Formula, which says that if

$$ax^2 + bx + c = 0 \quad (a \neq 0)$$

then

$$x = \frac{-b \pm \sqrt{b^2 - 4ac}}{2a}$$

Use it to solve these equations for x, simplifying the answer wherever possible.

a. $x^2 - 10x + 21 = 0$
b. $2x^2 + 5x - 3 = 0$
c. $x^2 + 8x = 0$
d. $3x^2 - 2x = 8$
e. $2x^2 = 1 - 3x$
f. $5x^2 + 26x + 5 = 0$
g. $1 - 4x = 8x^2$
h. $x^2 - 81 = 0$
i. $x^2 + 6x = 55$
j. $3x^2 - 4x - 2 = 0$
k. $5x + x(x-7) = 9$
l. $(x-3)(x+4) = 3x - 5$

11. A very powerful form of shorthand notation is the $\sum$ ("sigma") notation, which means to find the sum of a list of numbers. For example, consider the list

4, 7, 8, 13, 20, 23, 24, 26, 27

where the variable x may be replaced by an element of the list. We say that $x_1 = 4$ ("x sub one equals 4") where x_1 represents the first number in the list. Likewise, $x_2 = 7$, $x_3 = 8$, $x_4 = 13$, ..., $x_8 = 26$, and $x_9 = 27$. The symbol

$$\sum_{i=1}^{5} x_i \text{ ("the summation of } x \text{ sub } i \text{ from } i = 1 \text{ to 5")}$$

is evaluated by

1. replacing the subscripts i by positive integers starting at 1 and stopping at 5, inclusive, and then
2. adding the resulting numbers.

Here,

1. $x_1 \quad x_2 \quad x_3 \quad x_4 \quad x_5$
2. $x_1 + x_2 + x_3 + x_4 + x_5 =$
$4 + 7 + 8 + 13 + 20 = 52.$

Thus,

$$\sum_{i=1}^{5} x_i = 52.$$

Convince yourself that

$$\sum_{i=2}^{5} x_i = 48$$

and also that

$$\sum_{i=1}^{6} x_i = 75.$$

If we wish to sum all n values of a list

$$x_1, x_2, x_3, \ldots, x_n$$

we often shorten the symbol

$$\sum_{i=1}^{n} x_i$$

to just

$$\sum x$$

since there is no doubt that we are adding *all* the numbers in the list of possible replacements for x.

If x represents an element from the list

$$-5, -2, 0, 3, 7, 8, 10, 13, 15$$

and if y represents an element from the list

$$-8, -4, -1, 0, 1, 2, 3, 5, 8,$$

then evaluate these symbols.

a. $\sum_{i=1}^{4} x_i$

b. $\sum_{i=1}^{4} y_i$

c. $\sum_{i=1}^{6} x_i$

d. $\sum_{i=3}^{5} y_i$

e. $\sum_{i=1}^{4} x_i$

f. $\sum_{i=1}^{9} x_i$

g. $\sum x$

h. $\sum_{i=6}^{4} y_i$

i. $\sum_{i=4}^{6} x_i \cdot y_i$

j. $\sum_{i=7}^{8} (x_i + 6)$

k. $\sum_{i=1}^{4} (x_i + y_i)$

l. $\sum_{i=1}^{4} x_i + \sum_{i=1}^{4} y_i$

m. $\sum_{i=1}^{6} 7 \cdot y_i$

n. $7 \cdot \sum_{i=1}^{6} y_i$

o. $\sum xy$ p. $\sum x \cdot \sum y$

q. $\sum_{i=4}^{7} (x_i)^2$ r. $\sum x^2$

s. $(\sum x)^2$ t. $\sum y^2$

u. $(\sum y)^2$ v. $\sum x - \sum y$

12. Copy this standard pair of coordinate axes and plot the points corresponding to these pairs of numbers (x, y) in the resulting coordinate plane.

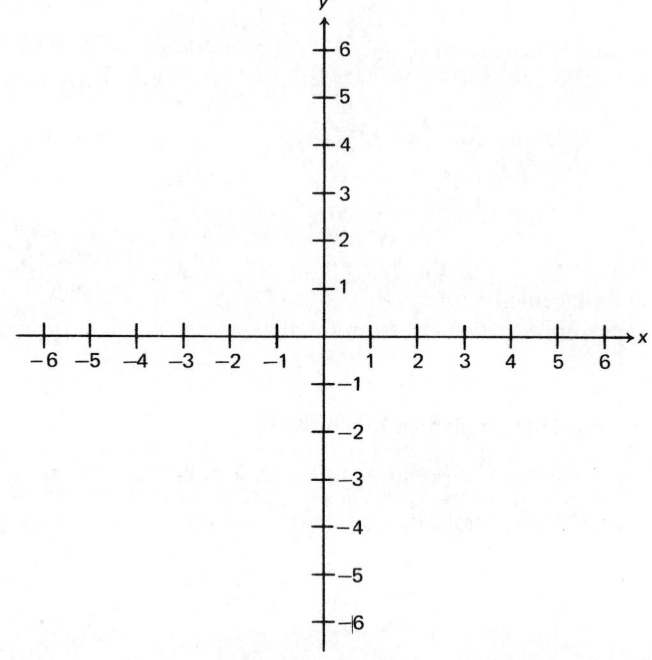

a. $(2, 3)$ b. $(3, 2)$ c. $(4, 1)$
d. $(1, 4)$ e. $(5, 0)$ f. $(-2, 6)$
g. $(-4, 1)$ h. $(-2, 0)$ i. $(-3, -5)$
j. $(-5, -3)$ k. $(0, -4)$ l. $(0, 0)$
m. $(3, -5)$ n. $(-3, 5)$ o. $(3/2, 4)$
p. $(-2.7, 3.2)$ q. $(-4.5, 0)$ r. $(3/4, -13/3)$

13. Solve each of the following systems of equations.

a. $x + y = 3$
$x - y = -2$

b. $x + 7y = 38$
$x + 3y = 22$

c. $x + 2y = -1$
$x + 5y = 17$

d. $2p - 2r = 1$
$p = 7r - 3$

e. $7.5c - 3d = 5$
$4d = 3c - 6$

f. $v + t = 3v - t + 6$
$7 - t = 8 + v$

g. $3x - 2y = 12$
$4x - 5y = 16$

h. $3y - 2 = x$
$4x - 3 = y$

i. $7t + \frac{5}{2}u = 2$
$-2t = u - \frac{17}{2}$

j. $\frac{x+y}{3} = 2$
$\frac{x-y}{2} = -1$

k. $5p + 6q = 2.5p - .28$
$5q = -1.5$

l. $\frac{2x+4y}{-3} = -4y - x$
$-x = 4y + 1$

14. What is wrong with the following "proof" that 3 = 6?

Statement	Reason
$a = b$	Begin with two numbers equal to 3.
$ab = b^2$	Multiply both sides by b.
$ab - a^2 = b^2 - a^2$	Subtract a^2 from both sides.
$a(b-a) = (b+a)\cdot(b-a)$	Factor both sides.
$\frac{a(b-a)}{(b-a)} = \frac{(b+a)\cdot(b-a)}{(b-a)}$	Divide both sides by $(b-a)$.
$\frac{a\cancel{(b-a)}^{1}}{\cancel{(b-a)}_{1}} = \frac{(b+a)\cdot\cancel{(b-a)}^{1}}{\cancel{(b-a)}_{1}}$	Cancellation.
$a = (b+a)$	Properties of 1.
$a = b + a$	Remove parentheses.
$3 = 3 + 3$	First statement: $a = b = 3$.
$3 = 6$	Addition fact.

Chapter 2

Sets

Basic concepts

The last chapter emphasized that mathematics is really a formalized extension of our commonsense knowledge of the measurable world around us. This chapter on sets demonstrates how we formalize our common knowledge of groups of objects and their relationships with one another. Even though part of this chapter may be review for many of us, it is still worthwhile to at least highlight the subject of sets since they do have a real use in explaining future concepts and applications.

Definition of sets

Intuitively, a *set* in mathematics is any collection of objects defined well enough so that we can clearly tell whether or not a specific object is in the collection. The collection of the letters of the alphabet from d through n, inclusive, is a good example of a set because if we are given any object we can clearly determine if it is among these letters or not. On the other hand, the collection of intelligent people who have held elective office in Washington, D.C., is not an example of a set in the mathematical sense of the word. Certainly some citizens would place controversial people such as the current President in that set and others would deny him membership. The point is that intelligence is not a well defined term, and it is open to subjectivity, not objectivity. We have not given a formal definition of a set since it is so basic and innate in our minds that it cannot be explained in more primitive terms.

The notation for sets is reasonably straightforward. They are usually symbolized by capital letters; their objects, *elements*, or *members*, are usually symbolized by numerals or small letters. We use braces, {}, to indicate the existence of a set, and we place the symbols for the elements in the set between the braces. Thus we can symbolize the set mentioned in the previous paragraph by the *listing* or *roster* method.

$$A = \{d, e, f, g, h, i, j, k, l, m, n\}$$

This is read "A is the set whose members (or elements) are d, e, f, g, h, i, j, k, l, m, and n." The Greek letter $\in$ (epsilon) is used as the basic verb to show that a specified object is an element of a set and $\notin$ is the verb to show that a specified object is not an element of a set. The following statements are all true with respect to set A:

$$d \in A,\ k \in A,\ n \in A,\ p \notin A,\ b \notin A$$

We read $d \in A$ as "d is an element of A" or "d belongs to A" or "d is in A," and $b \notin A$ is read "b is not an element of A" or "b does not belong to A" or "b is not in A."

Sometimes it is impractical, if not impossible, to list all the elements in a set. In this case an *ellipsis* or 3 dots (...) is used to indicate that the missing elements continue in the same fashion as those that are present.

$$A = \{d, e, f, \ldots, n\}$$

is a convenient and understandable way to specify set A.

The second way to symbolize a set is by the *set-builder* notation where the common property of the elements of that set is indicated. For example,

$$A = \{x | x \text{ is a whole number between 1 and 30 inclusive}\}$$

It is read "A is the set whose members are all objects x such that (or with the property that) x is a whole number between 1 and 30 inclusive." x is merely a symbol that may represent any one of the specified numbers so that

$$\{t | t \text{ is a whole number between 1 and 30 inclusive}\}$$

specifies exactly the same set.

Whenever we are discussing sets of numbers and their properties we have to specify whether we are discussing positive whole numbers, negative fractions, square roots, or whatever; that is, we must specify a *universal set*, usually symbolized by U, which is the umbrella set that includes all the items of particular interest for a specified situation. A public opinion pollster may choose the registered voters of Los Angeles County or of New Hampshire or of Wichita, Kansas, as his universal set. A quality control manager at an appliance assembly line may choose a specific shipment of transistors in order to determine the quality of their supplier's manufacturing process.

In contrast to the universal set, which is everything under consideration, is the *empty set* or *null set*. It is the set with no members. The set of people currently alive who are over 200 years old, the set of Cadillacs manufactured by the Ford Motor Company, and the set of 20-inch or more snowfalls in Mississippi during this century are all examples of the empty set. The empty or null set is symbolized by $\{\}$ or $\varnothing$ (the Danish letter phi which is pronounced "fee").

We are all familiar with the idea that some sets are totally included in larger encompassing sets. For example, the set of math majors is a part of the set of all students at any college. The set of vowels is totally included in the set of letters of the alphabet. More formally, if A and B are any two sets such that every element of A is also an element of B, then *A is a subset of B* or $A \subseteq B$.

Example 1 Let $A = \{1, 3, 5, 7\}$ and let $B = \{1, 2, 3, 4, 5, 6, 7, 8, 9\}$. Then, since whenever $x \in A$, x is also in B,

$$A \subseteq B$$

is a true statement.

Example 2 If

$$T = \{x|x \text{ is an automobile with two doors}\}$$

and if

$$C = \{x|x \text{ is a convertible currently manufactured in the United States}\},$$

then

$$C \subseteq T.$$

Also,

$$\{1, 3, 5, 7\} \subseteq \{1, 3, 5, 7\}$$

because any element of the first set is indeed an element of the second set.

Classification of sets

We say that two sets A and B are *equal* if they have the same elements. Examples of pairs of equal sets are

$$\{1, 2, 3, 5\} = \{1, 2, 3, 5\}$$
$$\{1, 2, 3, 5\} = \{1, 5, 3, 2\}$$
$$\{1, 2, 3, 5\} = \{1, 5, 3, 2, 3\}$$

Notice that changing the order in which the elements are listed does not change the set itself. We may repeat an element of a set in its listing, but this also does not affect the set. From now on, though, we will assume that the elements we list for a set are all distinct from one another unless a specific situation requires differently.

Sets may also be classified as *finite* or *infinite*. We will call a set *finite* if all its members can be counted. An *infinite* set is a set whose members can not all be counted.

$$A = \{1, 2, 3, 4, \ldots, 999\}$$

is an example of a finite set and

$$B = \{1, 2, 3, 4, \ldots\}$$

is an example of an infinite set since it has no last element to be counted.

One relation which may exist between a pair of sets is that of a *one-to-one correspondence.* We say that two sets A and B are in a one-to-one correspondence if two conditions are met:

1. every element of A has exactly one correspondent in B, and
2. every element of B has exactly one correspondent in A.

In a diagram the corresponding pairs of elements are represented by double-headed arrows ($\longleftrightarrow$).

Example 3 Let $A = \{a, b, c, e, g, h\}$ and let $B = \{x, r, s, u, t, v\}$. Then a one-to-one correspondence may be established as indicated here.

$$A = \{a, b, c, e, g, h\}$$

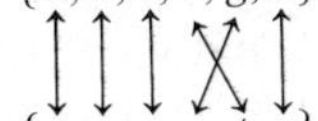

$$B = \{x, r, s, u, t, v\}$$

Sets A and B are also known as *equivalent* sets (see problem 7). It may not seem "natural" to make e correspond with t and g with u, but nothing in our definition prevents "crossing over" among the individual correspondences.

Problem set 2–1

1. Let S be the set of counting numbers between 2 and 15, inclusive. Determine if these statements are true or false.
   a. $3 \in S$ b. $13 \in S$ c. $15 \in S$
   d. $5\,1/2 \in S$ e. $18 \in S$ f. $6 \notin S$
   g. $7\,1/2 \notin S$ h. $2 \notin S$ i. $2 \in S$
2. Rewrite these sets by using the listing or roster notation.
   a. $\{x | x$ is a counting number between 13 and 18, inclusive$\}$
   b. $\{x | x$ is a counting number between 13 and 18, exclusive$\}$
   c. $\{x | x$ is a counting number between 41 and 423, inclusive$\}$
   d. $\{x | x$ is an odd counting number between 7 and 73, inclusive$\}$
   e. $\{t | t$ is an odd counting number between 7 and 73, inclusive$\}$
   f. $\{v | v$ is a fraction between 1 and 3, exclusive, whose denominator is 4$\}$
   g. $\{x | x$ is a positive whole number less than 0$\}$
   h. $\{x | x$ is a counting number greater than 7$\}$
   i. $\{x | x$ is a counting number between 41 and 423, exclusive$\}$
3. Rewrite these sets by using the set-builder notation.
   a. $A = \{5, 6, 7, 8, 9, 10\}$
   b. $B = \{13, 14, 15, 16, \ldots, 78\}$
   c. $C = \{2, 4, 6, 8, \ldots, 64\}$
   d. $D = \{9, 11, 13, 15, \ldots, 117\}$
   e. $E = \{$California, Oregon, Washington$\}$
   f. $F = \{$Richard Nixon, Lyndon Johnson, Hubert Humphrey, Spiro Agnew, Gerald Ford$\}$

g. $G = \{13, 14, 15, 16, \ldots\}$

4. Let $U = \{1, 2, 3, 4, \ldots, 15\}$ with three subsets $A = \{2, 3, 4, 5, 8, 9, 11\}$, $B = \{3, 4, 5, 9\}$, and $C = \{4, 5, 6, 11, 12, 14, 15\}$. Determine if these statements are true or false.
   a. $C \subseteq U$
   b. B and A are in a one-to-one correspondence
   c. $B \subseteq C$
   d. $A \subseteq U$
   e. $A = C$
   f. A and C are in a one-to-one correspondence
   g. $A \subseteq B$
   h. $A = B$
5. Classify each of these sets as finite or infinite.
   a. the odd counting numbers
   b. the residents of Colorado
   c. all fractions with denominators of 13
   d. the counting numbers evenly divisible by 5
   e. the counting numbers evenly divisible by 10
   f. the letters of the English alphabet
   g. the letters of the Chinese alphabet
   h. the set of 30-or-more-inch snowfalls in Miami, Florida, since 1970.
6. Establish a one-to-one correspondence between $C = \{4, 5, 7, 8\}$ and $D = \{13, 15, 11, 16\}$ in two different ways.
7. If A and B are two equal finite sets must they also be *equivalent* sets? Equivalent sets are two sets which are in a one-to-one correspondence. Give an example to illustrate your answer.
8. If A and B are two equivalent finite sets must they also be equal? Give an example to illustrate your answer.
9. a. Explain why $\{2, 7, 6, 5\}$ is not a subset of $\{2, 6, 5\}$.
   b. Explain why $\{14, 16, 17, 18, 19\}$ is not a subset of $\{16, 17\}$.
   c. Explain, using similar reasoning, why we can *not* show that $\{\}$ is not a subset of $\{2, 6, 5\}$. Since we can not prove that $\{\}$ is not a subset of $\{2, 6, 5\}$ we conclude that $\{\}$ is a subset of $\{2, 6, 5\}$. In general, $\{\} \subseteq A$ for any set A because it is impossible to find an element of $\{\}$ which does not belong to A.
   d. A further note about this problem: our conclusion that the empty set is a subset of every set is also an agreement that we make because it is useful. It is part of a larger agreement in mathematics (and elsewhere) that, for a condition to be "fulfilled," really means that it is never violated. For example, suppose a law says that all people born in 1492 must pay income tax in 1992. This law is fulfilled not because any such person pays an income tax but because no one born in 1492 violates this law for obvious reasons.
10. a. List all the subsets of $\{a, b, c\}$. How many are there?
    b. List all the subsets of $\{a, b, c, d\}$. How many are there?
    c. List all the subsets of $\{a, b\}$. How many are there?
    d. List all the subsets of $\{a\}$. How many are there?
    e. Organize your work so far by completing the table on the following page.

Number of elements in a set, n	Number of subsets of that set
1	
2	
3	
4	

f. By looking at the pattern of specific results, how many subsets would a set of five elements have?
g. By looking at the pattern of specific results, how many subsets would a set of six elements have?
h. Review your work so far and recall that $2 = 2^1$, $4 = 2^2$, $8 = 2^3$, and $16 = 2^4$. Complete this statement: if a finite set has n elements then it has ______ subsets.

11. Establish a one-to-one correspondence between the sets $A = \{3, 7, 11, 15, \ldots\}$ and $B = \{30, 70, 110, 150, \ldots\}$.

Operations with sets

One of our earliest experiences with mathematics was learning the four basic operations of addition, subtraction, multiplication, and division. These operations are sometimes called *binary operations* because they take two (hence the name *binary*) numbers and combine them together in a special way (or *operate* on them) so as to obtain a third number. Study this example.

Figure 2.1

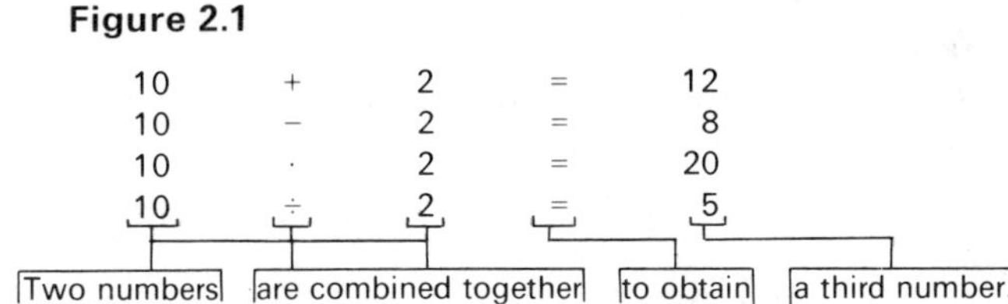

The three numbers are not always different since, for example, $3 + 0 = 3$, $4 \div 1 = 4$, $1 \cdot 1 = 1$. We also have *unary operations* in mathematics where we begin with just one number and operate on it to obtain one number. A common example of a unary operation is the extraction of a square root. The equation $\sqrt{9} = 3$ says that after taking the square root of one number (9) we then have *one* number $(+3)$ as the result. By universal mathematical convention $\sqrt{x}$ is the *positive* number (or perhaps zero) whose square is x. It is therefore wrong to say that $\sqrt{9}$ may also equal -3.

We may also have binary operations on sets where two sets are combined together in a special way to obtain a third set. The two binary operations we will study are *intersection* and *union*.

Intersection and union

Definition The *intersection* of set A and set B is the set whose elements belong both to set A *and* to set B. Symbolically,

$$A \cap B = \{x | x \in A \text{ and } x \in B\}.$$

$A \cap B$ is read "A intersection B."

Example 4 Let $U = \{1, 2, 3, 4, \ldots, 15\}$ (remember that a universal set is always specified for any discussion about sets), $A = \{2, 3, 4, 5, 8, 9, 11\}$, $B = \{3, 9, 4, 5\}$, $C = \{4, 5, 6, 11, 12, 14, 15\}$, and $D = \{1, 2, 7, 10\}$. Then

$$A \cap C = \{4, 5, 11\}$$

because 4, 5, and 11 are the common elements of A and C; that is, they belong both to set A and to set C. Convince yourself that these statements are true by observing that the elements of the intersection do belong to both of the original sets.

$$\begin{aligned} A \cap B &= \{3, 4, 5, 9\} \\ A \cap D &= \{2\} \\ B \cap C &= \{4, 5\} \\ B \cap D &= \{\} = \varnothing \\ C \cap D &= \{\} = \varnothing \end{aligned}$$

Notice that

$$B \cap A = \{3, 9, 4, 5\} = \{3, 4, 5, 9\} = A \cap B.$$

The order of choosing the elements in the intersection is unimportant because the elements in a set can be reordered at will without changing the set membership itself. B and D along with C and D are examples of *disjoint* or *mutually exclusive* sets since they have no elements in common.

Definition The *union* of set A and set B is the set whose elements belong either to set A *or* to set B. Symbolically,

$$A \cup B = \{x | x \in A \text{ or } x \in B\}.$$

$A \cup B$ is read "A union B."

Example 5 Let $U = \{1, 2, 3, 4, \ldots, 15\}$, $A = \{2, 3, 4, 5, 8, 9, 11\}$, $B = \{3, 9, 4, 5\}$, $C = \{4, 5, 6, 11, 12, 14, 15\}$, and $D = \{1,2,7,10\}$. Then

$$\begin{aligned} A \cup C &= \{2, 3, 4, 5, 8, 9, 11, 6, 12, 14, 15\} \\ &= \{2, 3, 4, 5, 6, 8, 9, 11, 12, 14, 15\}. \end{aligned}$$

The word "or" is used in its inclusive sense to indicate "either or both." Though 4 is in both A and C, it still qualifies for membership in the union of the two sets. Likewise 5 and 11 are in the union. Convince yourself that these statements are true by observing that the elements of the union do belong to either or both of the original sets.

$$
\begin{aligned}
A \cup B &= \{2, 3, 4, 5, 8, 9, 11\} \ (= A) \\
A \cup D &= \{1, 2, 3, 4, 5, 7, 8, 9, 10, 11\} \\
&\quad \text{(We customarily reorder the listing of the elements.)} \\
B \cup C &= \{3, 4, 5, 6, 9, 11, 12, 14, 15\} \\
B \cup D &= \{1, 2, 3, 4, 5, 7, 9, 10\} \\
C \cup D &= \{1, 2, 4, 5, 6, 7, 10, 11, 12, 14, 15\}
\end{aligned}
$$

We mentioned that extracting its square root is an example of a unary operation on a single number. We also have a unary operation that begins with a single set and ends with a single set.

Definition The *complement* of a set A is the set whose elements are *not* in A but are in the universal set. Symbolically,

$$A' = \{x | x \notin A \text{ and } x \in U\}.$$

A' is read "A complement" or "A prime."

Example 6 Let $U = \{6, 7, 8, 9, \ldots, 14\}$, $A = \{8, 9, 10, 11, 13\}$, and $B = \{7, 9, 10, 12, 13, 14\}$. Then

$$A' = \{6, 7, 12, 14\}$$

and

$$B' = \{6, 8, 11\}.$$

It is false to say $15 \in B'$ because while 15 is not in B it is not in the universal set.

Hierarchy of operations

In algebra we learned that the operations of multiplication and division are performed before the operations of addition and subtraction, while operations in parentheses are always performed first in the evaluation of expressions. For example,

$$3 + 8 \cdot 16 - 10 \div 5 + \underbrace{(7 + 4)}$$

will be evaluated step by step as indicated below.

$$
\begin{aligned}
&3 + \underbrace{8 \cdot 16} - 10 \div 5 + 11 \\
&3 + 128 - \underbrace{10 \div 5} + 11 \\
&\underbrace{3 + 128} - 2 + 11 \\
&\underbrace{131 - 2} + 11 \\
&\underbrace{129 + 11} \\
&140.
\end{aligned}
$$

A similar hierarchy of operations exists in working with sets. We first perform operations enclosed in parentheses, then complementation, then intersection, then union.

Example 7 Let $U = \{1,2,3,4,\ldots,10\}$, $A = \{3,6,7,8,10\}$, $B = \{1,2,3,4,8,9\}$, and $C = \{7,8,9,10\}$. Then

$$A \cup B' \cap C \cup \underbrace{(A \cup B)} =$$

$$A \cup \underbrace{B'} \cap C \cup \{1,2,3,4,6,7,8,9,10\} =$$

$$A \cup \underbrace{\{5,6,7,10\} \cap C} \cup \{1,2,3,4,6,7,8,9,10\} =$$

$$\underbrace{A \cup \{7,10\}} \cup \{1,2,3,4,6,7,8,9,10\}$$

$$\underbrace{\{3,6,7,8,10\} \cup \{1,2,3,4,6,7,8,9,10\}}$$

$$\{1,2,3,4,6,7,8,9,10\}$$

Venn diagrams

Another useful way of showing the relationships among a universal set and its subsets is a *Venn Diagram* which displays sets as regions of a plane. A universal set U is represented by the points inside a rectangle, a subset A of the universal set is represented by the points inside a circle, and an element of the universal set is represented by a dot inside the rectangle.

Figure 2.2

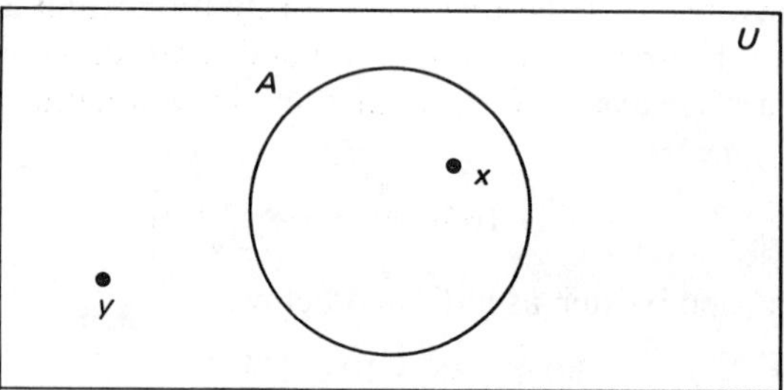

Here we show that $A \subseteq U$, $x \in A$, $x \in U$, $y \notin A$, $y \in A'$, and $y \in U$.

Example 8 If A and B are any two subsets of a universal set U, then find the region of the Venn Diagram representing $A \cap B'$ and shade that region.

Solution By convention a Venn Diagram with two subsets is constructed as

Figure 2.3

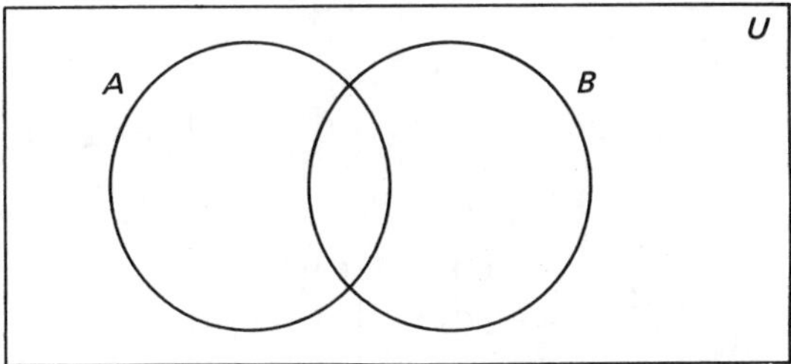

We recall $A \cap B'$ means those points which are both in A and not in (outside) B. The region of those points is shaded below.

Figure 2.4

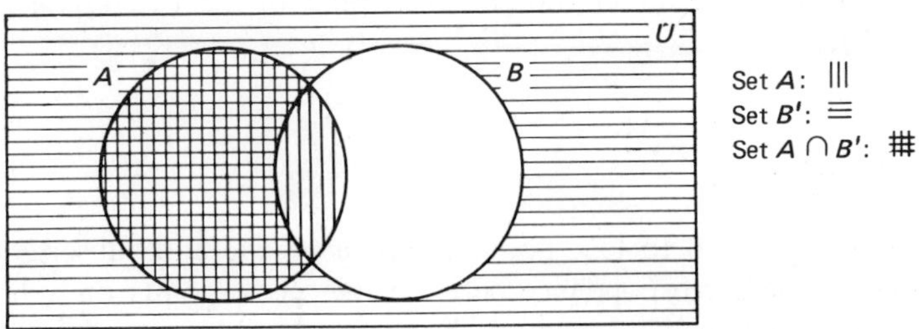

Example 9 If A, B, and C are any three subsets of a universal set U, then find the region of the Venn Diagram representing $A' \cap (B \cup C)$ and shade that region.

Solution By convention a Venn Diagram with three subsets is constructed as

Figure 2.5

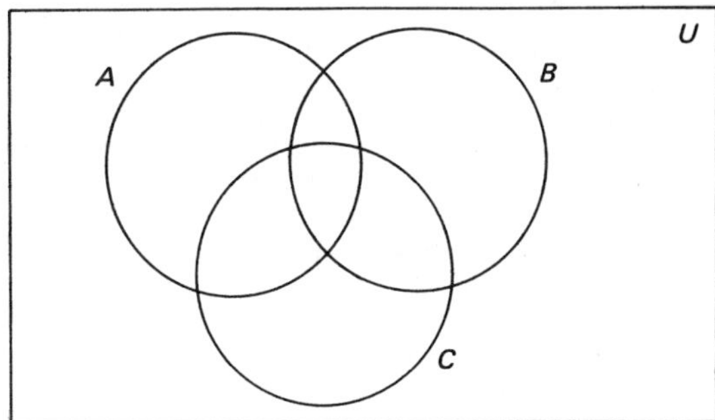

We will shade A' with vertical line segments and $(B \cup C)$ with horizontal line segments so that the final answer (the *intersection* of the two regions) will be the region shaded both vertically *and* horizontally.

Figure 2.6

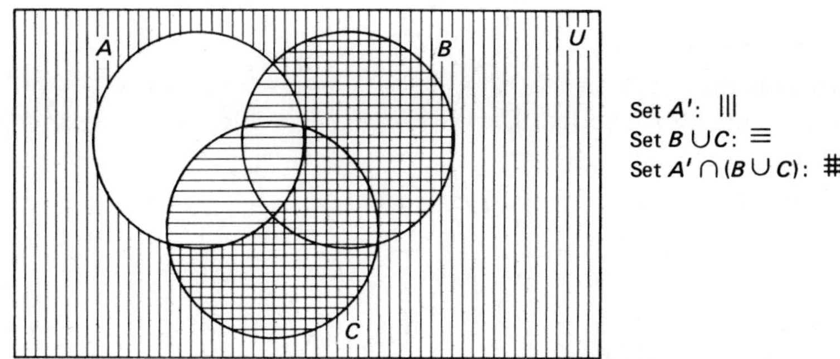

If the original problem had been $A' \cup (B \cup C)$ the final answer would have been the region shaded horizontally only, vertically only, or both ways since we would be finding the union of the two regions A' with $(B \cup C)$.

The number line

Finally, it is fruitful to talk about the *number line* and subsets of it. There is a one-to-one correspondence between the set of points on a line and the set of real numbers. This picture shows the familiar number real line from Algebra I courses.

Figure 2.7

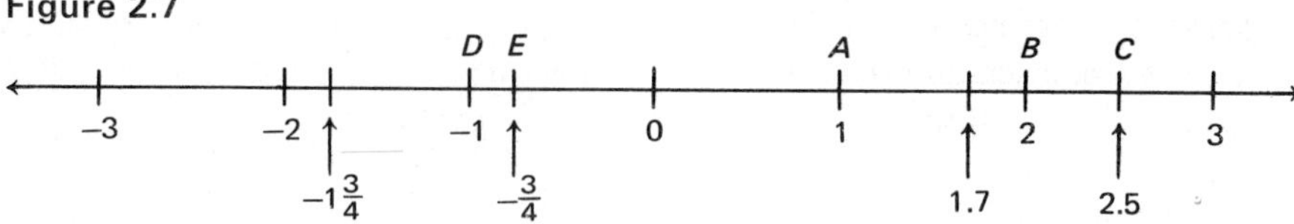

The *coordinate* of a point is the number corresponding to the chosen point. This number line contains, among others, point A with its coordinate 1, point B with its coordinate 2, point C with its coordinate 2.5, point D with its coordinate -1, and point E with its coordinate $-3/4$. The *graph* of a set of numbers is the set of points whose coordinates are elements of that set of numbers.

Example 10 Graph these sets of numbers on a number line:

$$A = \{x | 2 \leqslant x \leqslant 5\}$$
$$B = \{x | -1 \leqslant x < 1.5\}$$
$$C = \{x | x \geqslant 7\}$$
$$D = \{x | x < -3\}$$

Solution Set A is the set of numbers between 2 and 5, inclusive, so its graph is the set of points between the point with coordinate 2 and the point with coordinate 5. To show that these two points are included in the graph we have line segments rising above the two points.

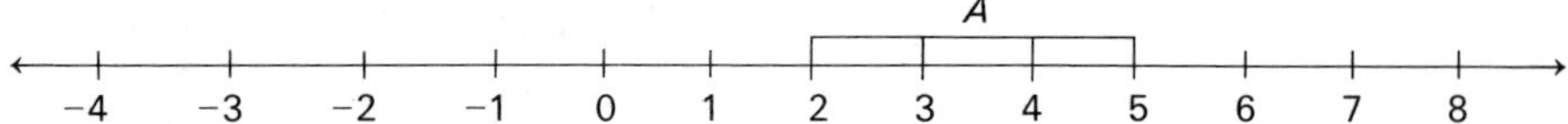

Set B is the set of numbers between -1 and 1.5, including the former and excluding the latter. The exclusion of the point with coordinate 1.5 from the graph is symbolized by a curved line segment rising above the point.

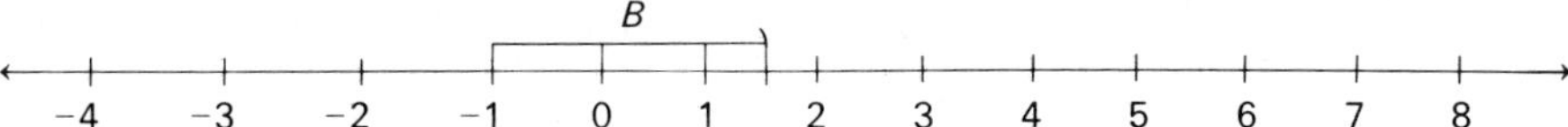

Set C is the set of numbers greater than or equal to 7. This algebraic fact is represented geometrically by points to the right of and including the point with coordinate 7.

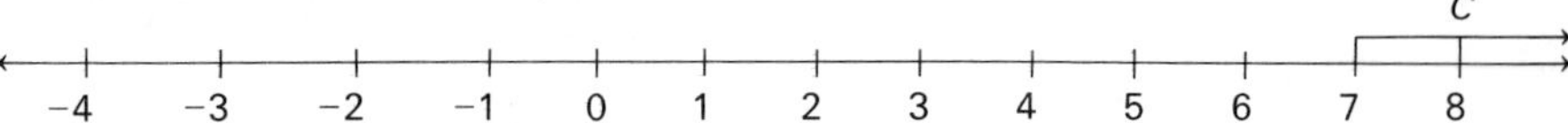

Set D is the set of numbers less than -3. This algebraic fact is represented geometrically by points strictly to the left of the point with coordinate -3.

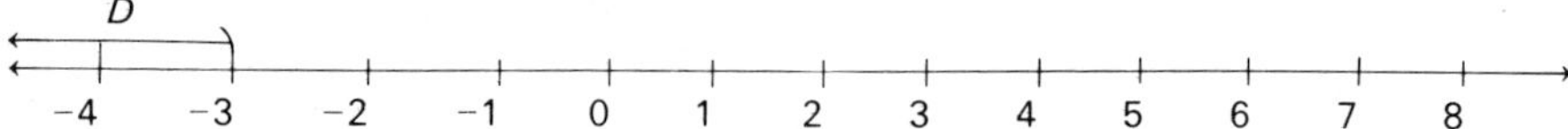

We often use *interval notation* to symbolize sets of numbers or sets of points. A [indicates that a number or point is the lowest in value or left-most of a set, and the number or point is also included in that set. A] indicates that a number or point is the largest in value or right-most of a set, and the number or point is also included in that set. Thus set A may be symbolized by $[2, 5]$. A (indicates that a number or point is the lowest in value or left-most of a set, and the number or point is also excluded from that set. A) indicates that a number or point is the largest in value or right-most of a set, and the number or point is also excluded from that set. Thus set B may be symbolized by $[-1, 1.5)$. Finally, "$-\infty$" is a symbol to show that a set has no lowest element

Table 2.1

Set	Set-builder notation	Interval notation
A	$\{x \mid 2 \leqslant x \leqslant 5\}$	$[2, 5]$
B	$\{x \mid -1 \leqslant x < 1.5\}$	$[-1, 1.5)$
C	$\{x \mid x \geqslant 7\}$	$[7, \infty)$
D	$\{x \mid x < -3\}$	$(-\infty, -3)$
U (the line itself)	$\{x \mid -\infty < x < \infty\}$ or $\{x \mid x \text{ is a real number}\}$	$(-\infty, \infty)$

and "$+\infty$" or "∞" is a symbol to show that a set has no largest element. Therefore, $-\infty$ and $+\infty$, read "minus infinity" and "plus infinity," are *not* symbols for specific numbers. Thus sets C and D may be symbolized by $[7, \infty)$ and $(-\infty, -3)$ respectively. We may summarize this paragraph in Table 2.1.

Problem set 2–2

1. Let $U = \{1, 2, 3, 4, \ldots, 12\}$, $A = \{1, 2, 5, 7, 8\}$, $B = \{3, 6, 7, 9, 11, 12\}$, $C = \{7\}$, $D = \{3, 4, 5, 6, 7, 8, 9\}$, and $E = \{4, 10\}$. Find the following sets.
   a. $A \cap B$ b. $A \cup B$ c. $B \cap C$ d. $B \cap D$
   e. $C \cup B$ f. $D \cup B$ g. $A \cap E$ h. $A \cup E$
   i. $A \cup C$ j. D' k. A' l. C'
   m. B' n. $(A \cup B)'$ o. $A' \cup B'$ p. $A' \cap B'$
   q. $(A \cap B)'$ r. $(C \cup D)'$ s. $C' \cap D'$ t. $A \cup (B \cap D)$
   u. $A \cup B \cap D$ v. $A \cap B \cup D$ w. $A \cap (B \cup D)$ x. $A \cup B \cap D'$
2. Determine whether these statements are true or false with respect to problem 1.
   a. $(A \cup B)' = A' \cap B'$ b. $(A \cup B)' \neq A' \cup B'$
   c. $(A \cap B)' = A' \cap B'$ d. $(A \cap B)' = A' \cup B'$
   e. $B \subseteq B \cup C$ f. $A \subseteq A \cup B$
   g. $A \subseteq A$ h. $A \cap B \subseteq A$
   i. $A \cap B \subseteq B$ j. $B \cap C \subseteq B$
   k. $C \subseteq D$ l. $(B')' = B$
   m. $C \cap E = \varnothing$ n. $D \subseteq D'$
   o. $B \cap D \subseteq B \cup D$ p. $B \cup B' = U$
   q. $C' \cap C = U$ r. $A \cap B \cup C = (A \cap B) \cup C$
   s. $(A \cap B \cap C)' = A' \cup B' \cup C'$ t. $(C \cap D \cap E)' = C' \cup D' \cup E'$
3. Consider this Venn Diagram.

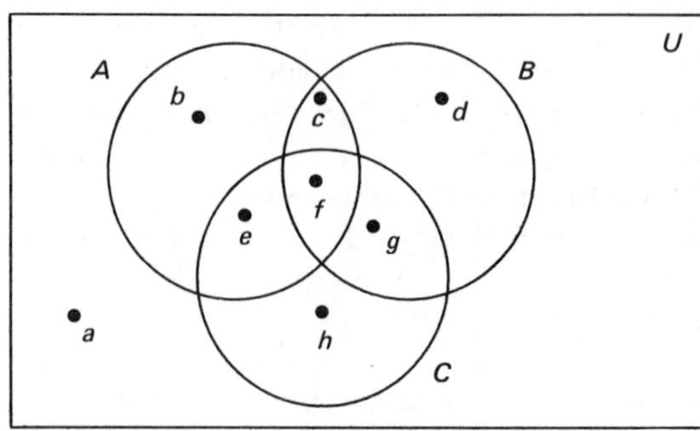

   Determine if these statements are true or false.
   a. $b \in A$ b. $d \notin A$ c. $d \in A'$
   d. $h \in (A \cap C)$ e. $g \in (B \cap C)$ f. $g \in (B \cup C)$
   g. $a \in B'$ h. $a \in (A \cup B \cup C)'$ i. $a \in (A' \cup B' \cup C')$
   j. $d \in (A \cup C)'$ k. $d \in (A \cap C)'$ l. $b \in (A \cup B)$
   m. $b \in (A \cap B)$ n. $d \in (B \cap A' \cap C')$ o. $h \in C'$

4. Shade the region of a Venn Diagram which represents these sets.
   a. $A \cup B$ b. A' c. $A' \cap B$
   d. $(A \cup B)'$ e. B' f. $A' \cap B'$
   g. $B \cap A'$ h. $(A \cap B)'$ i. $A' \cup B'$
5. Shade the region of a Venn Diagram which represents these sets.
   a. $A \cup B \cup C$ b. $A \cup (B \cap C)'$ c. $B \cap (A \cup C)'$
   d. $A \cap B \cap C$ e. $C \cap (A \cup B)'$ f. $(B \cap C) \cap A'$
   g. $(A \cap C) \cap B'$ h. $(A \cup B \cup C)'$ i. $A' \cap B' \cap C'$
6. Let $U = \{x|x \text{ is a car}\}$, $C = \{x|x \text{ is a convertible}\}$, and $R = \{x|x \text{ is red}\}$. Describe these sets in words.
   a. $C \cup R$ b. C' c. $C' \cap R$
   d. $(C \cup R)'$ e. R' f. $C' \cap R'$
   g. $R \cap C'$ h. $(R \cap C)'$ i. $C' \cup R'$
7. Refer to problem 4. Determine which of those pairs of sets are equal by observing whether the shaded regions representing each pair are equal or not. For example, (d) and (f) are equal since the regions representing them in their respective Venn Diagrams are equal.
8. Refer to problem 6. Determine which of those pairs of sets are equal.
9. If set A has five elements and set B has nine elements, then answer these questions about A and B. A Venn Diagram is helpful here.
   a. What is the largest number of elements that $A \cup B$ may have?
   b. When does this largest number occur?
   c. What is the smallest number of elements that $A \cup B$ may have?
   d. When does this smallest number occur?
10. If set C has six elements and set D has eleven elements, then answer these questions about C and D. A Venn Diagram is helpful here.
   a. What is the largest number of elements that $C \cap D$ may have?
   b. When does this largest number occur?
   c. What is the smallest number of elements that $C \cap D$ may have?
   d. When does this smallest number occur?
11. Use a number line to show the graphs of each of these sets. If the set is given in interval notation, then write it in set-builder notation, and vice-versa.
   a. $[1,4]$ b. $[1,4)$ c. $(1,4]$
   d. $(1,4)$ e. $\{x|1 \leqslant x < 4\}$ f. $\{x|x \geqslant -2\}$
   g. $[-2.5, \infty)$ h. $(2.5, \infty)$ i. $[-3.7, 1.8]$
   j. $(-\infty, 4.3]$ k. $\{x|x < 3\}$ l. $[-3,4] \cap [1,6)$
12. A *partition* of a set U refers to a division of U into subsets such that
   1. each pair of subsets is disjoint, and
   2. the union of all the subsets is U.

   For example, if $U = \{1,2,3,4,\ldots,12\}$, $A = \{1,3,5,6,11\}$, $B = \{2,7,8,9\}$, and $C = \{4,10,12\}$, then this division of U into A, B, and C is a partition. Note that

   1. $A \cap B = \varnothing,\quad B \cap C = \varnothing,\quad A \cap C = \varnothing,\quad$ and
   2. $A \cup B \cup C = U.$

Equivalently, we also say that these three subsets are *pairwise disjoint* and *exhaustive* since any *pair* is *disjoint* and the subsets *exhaust* the entire universal set. Determine if these divisions are partitions.

a. $U = \{1, 2, 3, 4, \ldots, 14\}$
$A = \{1, 2, 3, 4\}$, $B = \{5, 6, 8, 9, 10, 11\}$,
$C = \{7, 12, 13, 14\}$

b. $U = \{1, 2, 3, 4, \ldots, 10\}$
$A = \{1, 3\}$, $B = \{2, 5, 7, 9\}$, $C = \{4, 10\}$, $D = \{6\}$

c. $U = \{1, 2, 3, 4, 5, 6, 7\}$
$A = \{1, 5\}$, $B = \{2, 3, 4, 5\}$, $C = \{6, 7\}$

d. $U = \{a, b, c, d, \ldots, z\}$
$A = \{a, e, i, o, u\}$, $B = A'$

Number of elements in a set

In this section we see how we can apply our knowledge of sets and their operations as we determine how many elements are in a specified set. Before we begin a basic definition is in order.

Definition If S is any finite set, then $n(S)$ (read "n of s") is the number of elements in set S.

Example 11 If $S = \{1, 2, 3, 4\}$, then $n(S) = 4$; if $S = \{2, 4, 6, \ldots, 18\}$, then $n(S) = 9$; and if S is the empty set, then $n(S) = n(\varnothing) = 0$.

Example 12 If set A has five elements and set B has nine elements, then what is $n(A \cup B)$?

Solution This is problem 9 of the last section. We saw that $n(A \cup B)$ could be either 9 or 14. Actually, $n(A \cup B)$ can also be any whole number in between. If we are given the additional information that $n(U) = 20$ and $n(A \cap B) = 3$ then we organize all our information into a Venn Diagram.

Figure 2.8

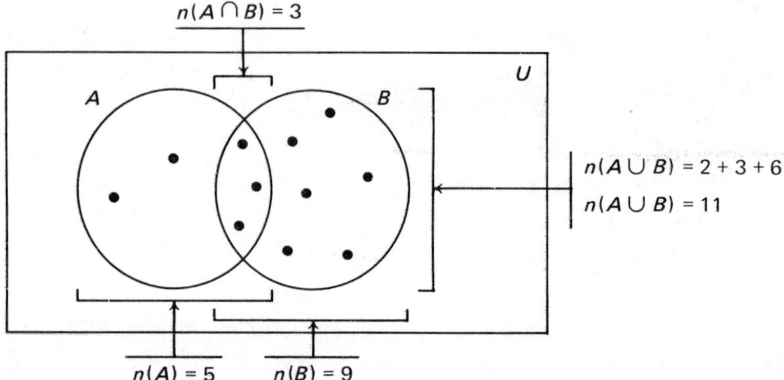

Notice that $n(A) = 5$ does *not* mean that set A has five elements exclusively to itself; these five elements may include some elements also shared with set B. Likewise, $n(B) = 9$ means that some of the nine elements of B may be shared with set A. The facts that there are two elements in A *only* and there are six elements in B *only* are symbolized, respectively, by

$$n(A \cap B') = 2 \quad \text{and} \quad n(B \cap A') = 6.$$

It is tempting to conclude at the very beginning, without a Venn Diagram, that since $n(A) = 5$, $n(B) = 9$, and union means to combine sets,

$$\begin{aligned} n(A \cup B) &= n(A) + n(B) \\ n(A \cup B) &= 5 + 9 \\ n(A \cup B) &= 14. \end{aligned}$$

This temptation is wrong for two reasons. First, inspection of the Venn Diagram shows that $n(A \cup B) = 11$. Second, if we state that

$$n(A \cup B) = n(A) + n(B)$$

or

$$n(A \cup B) = 5 + 9 = 14,$$

we are also saying that

$$n(A \cup B) = (2+3) + (6+3) \tag{A}$$

Equation (A) shows that the three elements in the intersection of A and B are being counted *twice* by this incorrect equation. These three elements are included in

$$n(A) = 2 + 3$$

and in

$$n(B) = 6 + 3.$$

In order to correct equation (A) we must "uncount" these three elements that have been counted twice by subtracting 3.

$$n(A \cup B) = (2 + 3) + (6 + 3) - 3$$

count three elements twice — uncount them

$$\begin{aligned} n(A \cup B) &= 5 + 9 - 3 \\ n(A \cup B) &= n(A) + n(B) - n(A \cap B) \end{aligned} \tag{B}$$

Equation (B) is a very important one. It not only holds true for these two specific sets A and B but also for *any* universal set U with *any* two subsets A and B. Moreover, if there are eleven elements inside the circles there must be $20 - 11 = 9$ elements outside the circles; therefore, $n(A \cup B)'$ must be 9. Notice that if A and B are disjoint sets then

$$n(A \cap B) = n(\varnothing) = 0$$

so that

$$\begin{aligned} n(A \cup B) &= n(A) + n(B) - 0 \\ n(A \cup B) &= n(A) + n(B). \end{aligned} \tag{C}$$

Number of elements in intersecting sets

Example 13 At a downtown newsstand two hundred people were asked if they read one of the city's three newspapers: the *Globe* (G), the *Herald* (H), or the *American* (A). The responses revealed that ninety-two people read the *Globe*, eighty-six read the *Herald*, eighty-three read the *American*, twenty-five read the *Globe* and the *Herald*, twenty-seven read the *Globe* and the *American*, twenty-six read the *Herald* and the *American*, and nine read all three. How many elements are in these sets?

a. G only
b. A only
c. G and H only
d. H and A only
e. $(G \cup A)$
f. $(G \cup A)'$
g. readers of exactly two papers
h. readers of exactly one paper
i. readers of no paper

Solution We organize our information into a Venn Diagram and indicate the partition of these two hundred people into eight subsets.

Figure 2.9

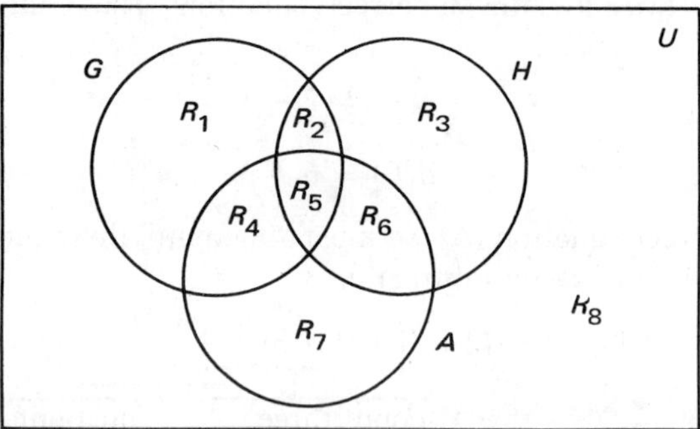

Table 2.2

Region	Set-intersection description	Verbal description of the paper read by a person in the region
R_1	$G \cap H' \cap A'$	*Globe* only
R_2	$G \cap H \cap A'$	*Globe* and *Herald* only
R_3	$G' \cap H \cap A'$	*Herald* only
R_4	$G \cap H' \cap A$	*Globe* and *American* only
R_5	$G \cap H \cap A$	All three papers
R_6	$G' \cap H \cap A$	*Herald* and *American* only
R_7	$G' \cap H' \cap A$	*American* only
R_8	$G' \cap H' \cap A'$	None of the three papers

The key to determining the number of elements in the regions is to work outward from the most inclusive region (R_5) to the least inclusive region (R_8). We will determine these numbers in the order specified.

$$R_5,\ \underbrace{R_2,\ R_6,\ R_4,}_{\substack{\text{can be}\\ \text{interchanged}}}\ \underbrace{R_1,\ R_3,\ R_7,}_{\substack{\text{can be}\\ \text{interchanged}}}\ R_8$$

We are given that $n(R_5) = 9$. Now the twenty-five people who read the *Globe* and the *Herald* also include the nine people who read all three papers; that is,

$$n(R_5 \cup R_2) = 25.$$

Since R_5 and R_2 are disjoint, then by equation (C)

$$n(R_5 \cup R_2) = n(R_5) + n(R_2),$$

or after substitution

$$25 = 9 + n(R_2)$$
$$n(R_2) = 16.$$

This equation says that sixteen people read the *Globe* and the *Herald* only. It is a good idea to fill in these numbers in the appropriate regions of the Venn Diagram as we calculate them. Similar reasoning says that since

$$n(R_5 \cup R_6) = 26$$

then

$$n(R_6) = 17 = \text{number of } \textit{Herald} \text{ and } \textit{American} \text{ only readers.}$$

Also, since

$$n(R_5 \cup R_4) = 27$$

then

$$n(R_4) = 18 = \text{number of } \textit{Globe} \text{ and } \textit{American} \text{ only readers.}$$

Next we have to determine $n(R_1)$, the number of people who read the *Globe* only. Since

$$G = R_1 \cup R_2 \cup R_5 \cup R_4$$

then

$$n(G) = n(R_1 \cup R_2 \cup R_5 \cup R_4),$$

and by an extension of equation (C) applied to four pairwise disjoint sets

$$n(G) = n(R_1) + n(R_2) + n(R_5) + n(R_4).$$

From our previous work and the problem description we know that

$$n(G) = 92,\ n(R_2) = 16,\ n(R_5) = 9,\ \text{and}\ n(R_4) = 18.$$

After substituting these numbers into the previous equation we find that

$$92 = n(R_1) + 16 + 9 + 18$$
$$92 = n(R_1) + 43$$
$$92 - 43 = n(R_1)$$
$$n(R_1) = 49 = \text{number of } \textit{Globe} \text{ only readers.}$$

Before continuing we notice that we had to first work with the inner regions of R_2, R_5, and R_4 in order to have enough information to calculate $n(R_1)$. It is also necessary to be fastidious about the arithmetic because one small error can throw off all the later numbers which depend upon the earlier erroneous value.

Similar reasoning—be sure to follow it through completely—says that since

$$n(H) = n(R_2) + n(R_3) + n(R_6) + n(R_5)$$

then

$$86 = 16 + n(R_3) + 17 + 9$$
$$86 = 42 + n(R_3)$$
$$n(R_3) = 44 = \text{number of } \textit{Herald} \text{ only readers.}$$

Also, since

$$n(A) = n(R_4) + n(R_5) + n(R_6) + n(R_7)$$

then

$$83 = 18 + 9 + 17 + n(R_7)$$
$$83 = 44 + n(R_7)$$
$$n(R_7) = 39 = \text{number of } \textit{American} \text{ only readers.}$$

We organize our work so far into a table.

Table 2.3

Region	Number of elements
R_1	$n(R_1) = 49$
R_2	$n(R_2) = 16$
R_3	$n(R_3) = 44$
R_4	$n(R_4) = 18$
R_5	$n(R_5) = 9$
R_6	$n(R_6) = 17$
R_7	$n(R_7) = 39$
R_8	$n(R_8)$

We only have $n(R_8)$ left to determine. A person interviewed either reads a paper or he does not. There are

$$49 + 16 + 44 + 18 + 9 + 17 + 39 = 192$$

people who read at least one paper and are thus included in the region bounded by

This means that the difference between 192 readers and the 200 people must be the 8 nonreaders or

$$n(R_8) = 8.$$

With this table and the diagram we can immediately find the answers.

a. G only $= R_1$, and $n(R_1) = 49$
b. A only $= R_7$, and $n(R_7) = 39$
c. G and H only $= R_2$, and $n(R_2) = 16$
d. H and A only $= R_6$, and $n(R_6) = 17$
e. $G \cup A = R_1 \cup R_2 \cup R_4 \cup R_5 \cup R_6 \cup R_7$
$n(G \cup A) = n(R_1) + n(R_2) + n(R_4) + n(R_5) + n(R_6) + n(R_7)$
$n(G \cup A) = 49 + 16 + 18 + 9 + 17 + 39$
$n(G \cup A) = 148$
f. $(G \cup A)' = R_3 \cup R_8$
$n(G \cup A)' = n(R_3) + n(R_8)$
$n(G \cup A)' = 44 + 8$
$n(G \cup A)' = 52$
Notice also that $n(G \cup A)' = n(U) - n(G \cup A) = 200 - 148 = 52$.
g. Readers of exactly two papers $= R_2 \cup R_6 \cup R_4$
$n(R_2 \cup R_6 \cup R_4) = n(R_2) + n(R_6) + n(R_4)$
$= 16 + 17 + 18$
$= 51$
h. Readers of exactly one paper $= R_1 \cup R_3 \cup R_7$
$n(R_1 \cup R_3 \cup R_7) = n(R_1) + n(R_3) + n(R_7)$
$= 49 + 44 + 39$
$= 132$
i. Readers of no paper $= R_8$
$n(R_8) = 8$

In spite of the lengthy explanation we have just gone through for this one example, we will find that much of the work can be done mentally. It is not necessary to write out a detailed explanation for all similar problems.

Problem set 2–3

1. Fifty students are interviewed at the "Corral," which is the snack bar at a certain junior college. They are asked about their breakfast habits.

It turns out that thirty-eight have coffee for breakfast, twenty-three have donuts, and seventeen have both. Use a Venn Diagram to determine how many of these interviewees

a. drink coffee only
b. do not drink coffee
c. have neither coffee nor donuts
d. have either coffee or donuts
e. have donuts only
f. do not have donuts

2. Refer to problem 1. Identify each set and its number of elements by using set intersection, union, and complementation where C is the set of coffee drinkers and D is the set of donut eaters. For example, (a) would be

$$n(C \cap D') = 21.$$

3. The public relations officer of Holiday Airlines issued the following statement as evidence of his company's superior departure and arrival record: "On the average, out of every one hundred scheduled flights, ninety-seven depart on time, ninety-four arrive on time, and ninety both depart and arrive on time." The next day this public relations man received a personal call from the president to say he was fired for the statement. Why?
4. The medical files of the seventy-five officers at a certain army post showed that in the past year six officers had seen neither the post physician nor the post dentist, fifty-two had seen the physician at least once, and forty-seven had seen the dentist at least once. How many officers had seen both the physician and the dentist at least once in the past year? How many officers had seen only the dentist in the past year?
5. In a survey of one hundred people it was found that fifty-five were married, fifteen were in college, and ten were both married and in college. Let M be the set of married people and let C be the set of people in college. Fill in these blanks after organizing this information into a Venn Diagram.

a. M' contains ______ people.
b. $M \cap C$ contains ______ people.
c. $M \cup C$ contains ______ people.
d. $M' \cap C$ contains ______ people.
e. $M \cup C'$ contains ______ people.
f. $(M \cup C)'$ contains ______ people.
g. $(M \cap C)'$ contains ______ people.

6. Two hundred five students are at Smithville High School. Five students are studying mathematics, physics, and chemistry; twenty-six are studying mathematics and chemistry; forty-five are studying mathematics and physics; no one is studying physics and chemistry without taking mathematics; fifty-two are studying physics; thirty are studying chemistry; and ninety-six are not studying any of the three subjects.

a. How many are studying chemistry only?
b. How many are studying physics and mathematics only?

c. How many are studying exactly one of the three subjects?
d. How many are studying exactly two of the three subjects?
e. How many are studying physics and chemistry only?
f. How many are studying mathematics only?

7. In a group of two hundred men and women there are fifty women who are Democrats, sixty men who are not Democrats, and forty women who are not Democrats. How many men and how many Democrats are in the group? Use a Venn Diagram to analyze this information.
8. A penthouse party was attended by seventy-one socialites, and during the evening the topic of discussion turned to automobiles. One of the younger socialites, being highly interested in the topic, made some notes and recorded the following information. Of the people at the party, fifty owned Cadillacs, thirty-eight owned Lincolns, and forty-five owned Imperials. As we can already tell, some owned more than one of these makes of cars. In fact, twenty-five owned both Cadillacs and Lincolns, twenty-eight owned both Cadillacs and Imperials, and twenty owned both Lincolns and Imperials. Furthermore, eight of the ultraelite were in attendance, as evidenced by the fact that they owned one of each of the 3 makes. The question is: How many at the party did not own even one of these three makes of automobiles?
9. A survey of 660 workers in a plant indicated that 420 owned their own homes, 440 owned cars, 550 owned televisions, 405 owned cars and televisions, 340 owned cars and houses, 370 owned homes and televisions, and 300 owned all three. How many workers owned a car but neither a television set nor a home? How should you interpret this answer?
10. A student reports that there are thirty-four students in his class. Five have brown hair and blue eyes, seventeen have blond hair and blue eyes, eight have brown hair, twenty-four have blue eyes, and nineteen are blonds. Could he be right?
11. Among a certain group of linguists in French, German, and Spanish, it was learned that twenty-one spoke French, twenty-five spoke German, fourteen spoke French only, sixteen spoke German only, fifteen spoke Spanish only, and three spoke all three, Furthermore, the number who spoke French and German was the same as the number who spoke French and Spanish. Determine the number of linguists who
    a. spoke French and German, but not Spanish
    b. spoke French and Spanish, but not German
    c. spoke German and Spanish, but not French
    d. spoke Spanish
    e. were in the group
12. A young stock broker was asked to use the following procedure for reporting on his success in predicting changes in prices of various stocks. Fifty common stocks with which he had been working were selected. Three intervals of time A, B, and C were selected. Over each interval of time he was to look up in his records the predictions that he had made for the price increase (or possibly decrease) for each stock and compare his

prediction with what actually happened. The young broker reported that his predictions had been correct for all three time intervals on ten stocks, for A and B on twenty-five stocks, for A and C on eighteen stocks, for B and C on thirty-two stocks, and for at least two of the three time intervals for each of the selected stocks. Are his data consistent?

13. Complete this table showing the results of a survey which cross-tabulates the owners of motorcycles according to brand of machine owned ($H =$ the set of Brand H owners, $T =$ the set of Brand T owners, $Y =$ the set of Brand Y owners) and the number of breakdowns the owners have had on their bikes in the past year ($A =$ the set of owners with zero or one breakdowns, $B =$ the set of owners with two or three breakdowns, $C =$ the set of owners with four or five breakdowns, $D =$ the set of owners with six or more breakdowns).

	A	B	C	D	Totals
Y	5	14		4	30
H	9		16	5	
T	2	4	11		
Totals		38		12	

Find the number of elements in these sets.

a. B

b. $C \cup D$

c. H

d. $T \cap D$

e. $(B \cup C)'$

f. $H \cap (A \cup B)$

g. $T \cap (A \cup B)$

h. $Y \cap (A \cup B)$

i. H'

j. $(Y \cup T)'$

What are the percentages that each brand of motorcycle has in the 0 through 3, inclusive, breakdown category?

k. Brand Y

l. Brand H

m. Brand T

14. Refer back to the text's example of the newspapers. If we were asked to find the number of people who read the *Globe* or the *Herald* or the *American*, this number would be symbolized by

$$n(G \cup H \cup A).$$

If we answer

$$n(G \cup H \cup A) = n(G) + n(H) + n(A) \tag{D}$$

the numerical result would be

$$92 + 86 + 83 = 261,$$

which is incorrect. Correct equation (D) by "uncounting" those readers who have been counted a multiple number of times in the right hand side of the equation. Your numeric correction should be equivalent to

$$\begin{aligned} n(G \cup H \cup A) = {} & n(G) + n(H) + n(A) - n(G \cap H) \\ & - n(G \cap A) - n(H \cap A) + n(G \cap H \cap A). \end{aligned}$$

Chapter 3

Functions

Basic concepts and definitions

The concept of a function is a very important one in all areas of mathematics since seemingly diverse topics can be described and analyzed under their common behavior as functions. As usual in our learning of mathematics we will begin with an example of a function, proceed with more precision and generalization, and then go on to further applications.

Example 1 Examine this correspondence between the elements of two sets A and B.

$$\begin{array}{rcccccccccc} A = \{ & 1, & 2, & 3, & 4, & 5, & 6, & 7, & 8, & 9, & 10\} \\ & \downarrow & \downarrow & \downarrow & \downarrow & \downarrow & \downarrow & \downarrow & \downarrow & \downarrow & \downarrow \\ B = \{ & 12, & 17, & 22, & 27, & 32, & 37, & 42, & 47, & 52, & 57\} \end{array}$$

A natural first question to ask is "Where did the elements of set B come from—are they related to those in set A in any logical way?" Observe that if we begin with any number in set A, multiply it by 5, and then add 7 to the product, the resulting sum is the corresponding number in set B. For example, if we choose 3 from set A then $5 \cdot 3$ is 15, and $15+7$ is the 22 in set B. Likewise, 8 in A has as its correspondent in B, $5 \cdot 8+7$ or 47. In general we can express this correspondence between the elements of set A and set B by the equation

$$y = 5 \cdot x + 7$$

where x represents any element of A and y represents its mate in B.

The key word in this discussion so far is "correspondence," since a function is a special kind of correspondence or relationship among the elements in two sets. A working definition of a function is this:

Definition A *function* is a correspondence between any two sets such that for each element in the first set there corresponds exactly one element in the second set.

Other pertinent terminology states that the first of the two sets is the *domain* of the function (set A in our example) and that the second of the two sets is the *range* of the function (in our example it is set B). We will always be dealing with functions whose domains and ranges are sets of real numbers.

Example 2 Consider the function exhibited by this correspondence between sets S and T.

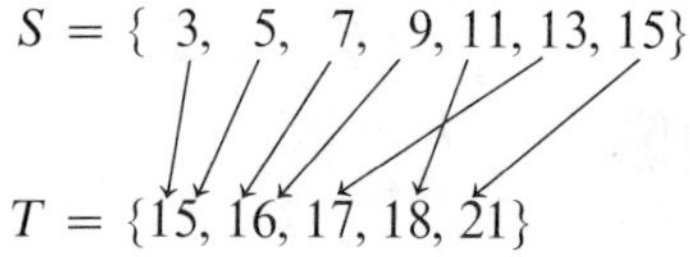

$$S = \{3,\ 5,\ 7,\ 9,\ 11,\ 13,\ 15\}$$

$$T = \{15,\ 16,\ 17,\ 18,\ 21\}$$

The correspondence is still a function because for each element of S there is exactly one element in T, even though 15 in T is the correspondent of both 3 and 5 in S. The correspondences of $11 \longrightarrow 18$ and $13 \longrightarrow 17$ appear to be reversed from a customary left-to-right ordering, but, again, for each element in the domain there is exactly one corresponding element in the range. Finally, this correspondence remains a function in spite of the fact that there is no convenient algebraic formula that determines the elements of set T on the basis of the elements of set S.

Example 3

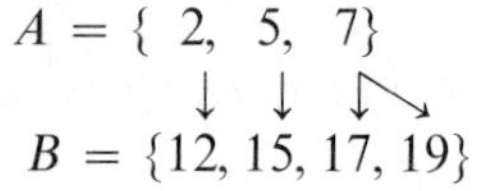

$$\begin{array}{l} A = \{\ 2,\ \ 5,\ \ 7\} \\ \qquad\ \ \ \downarrow\ \ \downarrow\ \ \downarrow\searrow \\ B = \{12,\ 15,\ 17,\ 19\} \end{array}$$

This diagram does not exhibit a function since there is an element of the domain (7) such that corresponding to it in the range there are two numbers (17 and 19). Even though $2 \longrightarrow 12$ and $5 \longrightarrow 15$ are pairs in this correspondence, the one situation of $7 \begin{smallmatrix} \nearrow 17 \\ \searrow 19 \end{smallmatrix}$ prevents the entire correspondence from forming a function.

Examples of functions abound in the world around us. Functions arise in a natural way from the following pairs of sets: gallons of gasoline purchased and their corresponding cost; diameter of a person's waist and the corresponding size of pants he wears; amount of time money is in a savings account and its corresponding accumulated value; time that a car has been moving at a constant rate and the corresponding distance it has traveled; and number of sides in a regular polygon and the corresponding number of degrees in each of its interior angles.

Representing functions

Many functions are given by equations such as $y = 5 \cdot x + 7$ rather than by drawings of direct correspondence between two finite sets. Let us go back to the function in Example 1. If $x \in A$ and $y \in B$, then the equation

$y = 5 \cdot x + 7$ clearly defines this function much more easily and simply than the drawing. Just as we often represent variables by the letter x, we customarily—but not necessarily—represent a function correspondence by the letter f. Mathematicians have agreed that we may also define this function f in Example 1 by the equation

$$f(x) = 5 \cdot x + 7, \quad x \in A$$

where the symbol $f(x)$ is read "f of x" rather than "f times x." The symbol f represents the entire function consisting of the correspondence from set A to set B, while $f(x)$ represents the element in the *range* that is the correspondent of the element x in the *domain*. Thus

$$f(x) = 5 \cdot x + 7$$

means that in function f, if x is any element in the domain, then its corresponding element in the range, symbolized by $f(x)$, equals 5 times x plus 7. In diagrammed format,

$$\underbrace{y}_{\substack{\uparrow \\ \text{range}}} = \underbrace{f(\overset{\substack{\text{domain} \\ \downarrow}}{x})}_{\substack{\uparrow \\ \text{range}}} = \underbrace{5 \cdot \overset{\substack{\text{domain} \\ \downarrow}}{x} + 7}_{\substack{\uparrow \\ \text{range}}}.$$

We may also emphasize the fact that x is a symbol for a domain element by the equation

$$y = f(\text{domain element}) = 5 \cdot (\text{domain element}) + 7.$$

If instead we let s be the variable for the range elements and let t be the variable for the domain elements and symbolize the entire function by g instead of by f, then the equation

$$s = g(t) = 5 \cdot t + 7, \quad t \in A$$

defines exactly the same function as the previous equation. Then

$$g(3) = 5 \cdot 3 + 7$$

or

$$g(3) = 22$$

means that corresponding to 3 in the domain of g is 22 in the range.

Still another way of symbolizing this same function is

$$f\colon x \mapsto 5 \cdot x + 7, \quad x \in A$$

or

$$g\colon t \mapsto 5 \cdot t + 7, \quad t \in A.$$

This symbolism, occasionally used in other texts, emphasizes the key idea of correspondence and shows how we obtain the range element from a specified domain element.

Domain of a function

It is sometimes helpful to think of a function as a machine which inputs values of x (one at a time) from a set called the domain, performs certain mathematical operations on these inputs, and outputs the result of these operations (one at a time) into a set called the range. In Example 1 the machinery multiplies each input value by 5 and adds 7 to the product before outputting this sum into the range.

We move on to the function defined by the equation

$$y = f(x) = 3x^2 + 7x - 14, \quad x \in R.$$

Here the domain is the set of all real numbers and we can find the corresponding range elements merely by "plugging in" the specific value of x into the expression

$$3x^2 + 7x - 14.$$

Some examples of the pairs of numbers in correspondence are in Table 3.1. Be sure to verify each pair.

Table 3.1

Domain x	Range $y, f(x), 3x^2+7x-14$	Symbolism
1	-4	$f(1) = -4$
4	62	$f(4) = 62$
0	-14	$f(0) = -14$
-5	26	$f(-5) = 26$
t	$3t^2 + 7t - 14$	$f(t) = 3t^2 + 7t - 14$
$(t+5)$	$3(t+5)^2 + 7(t+5) - 14$	$f(t+5) = 3t^2 + 37t + 96$
	$= 3(t^2+10t+25) + 7(t+5) - 14$	
	$= 3t + 30t + 75 + 7t + 35 - 14$	
	$= 3t^2 + 37t + 96$	
$(x+h)$	$3(x+h)^2 + 7(x+h) - 14$	$f(x+h)$
	$= 3(x^2+2xh+h^2) + 7x + 7h - 14$	$= 3x^2 + 6xh + 3h^2 + 7x$
	$= 3x^2 + 6xh + 3h^2 + 7x + 7h - 14$	$+ 7h - 14$

Dependent and independent variables

What about the function defined by

$$h(r) = \sqrt{r-3}?$$

We can see that

$$h(3) = \sqrt{3-3} = \sqrt{0} = 0$$

and

$$h(12) = \sqrt{12-3} = \sqrt{9} = 3;$$

but

$$h(2) = \sqrt{2-3} = \sqrt{-1},$$

which does not exist in the real numbers. Thus the domain elements, symbolized by r, cannot include 2 since $h(2) = \sqrt{-1}$ does not exist in the real numbers. The domain is the largest set of real numbers for which the expression $\sqrt{r-3}$ represents a real number, implying that the entire quantity $(r-3)$ must be positive or 0. Hence, the domain is $\{r | r \geqslant 3\}$ and the range is $\{h | h \geqslant 0\}$ since the radical sign calls for the square root which is always nonnegative. In this example of a function we can say, as mathematicians often do, that h is the *dependent variable* and r is the *independent variable.* This is logical enough since the value of h *depends* on the particular value of r, which in turn can be *independently* chosen anywhere from $\{r | r \geqslant 3\}$.

We can have functions with one dependent variable but with several independent variables. An example is given by the familiar equation

$$d = f(r, t) = r \cdot t$$

where the value of d, the distance, depends both upon r, the constant rate, and t, the time. The physical origin of this function limits its domain to $\{(r, t) | r \geqslant 0 \text{ and } t \geqslant 0\}$. Another example of a function with more than one *argument* (The independent variables in a function are sometimes called arguments.) is given by the equation

$$z = f(v, w, x) = v^2 - 3 \cdot w \cdot x + 7 \cdot v \cdot x + 2.43, \qquad \text{(A)}$$

which shows that the corresponding value of z, the dependent variable, depends upon the values of the three argument variables v, w, and x. If R is the set of all real numbers, then the domain here is

$$\{(v, w, x) | v \in R \text{ and } w \in R \text{ and } x \in R\}.$$

Then if $v = 6$, $w = -2$, and $x = .5$,

$$z = f(6, -2, .5) = 6^2 - 3 \cdot -2 \cdot .5 + 7 \cdot 6 \cdot .5 + 2.43$$
$$z = f(6, -2, .5) = 62.43.$$

We close this introductory section on functions by noting that a function can be thought of as a set of ordered pairs of numbers (x, y) such that for each value of x in the domain of the function there is exactly one corresponding value of y or $f(x)$. Thus our first function can be specified by

$$f = \{(x, 5x+7), x \in A\} = \{(1,12), (2,17), (3,22), (4,27),$$
$$(5,32), (6,37), (7,42), (8,47), (9,52), (10,57)\}.$$

The correspondence $3 \longrightarrow 22$ can also be symbolized by $f(3) = 22$ and $(3, 22) \in f$. The ordered pair notation of this set-oriented interpretation emphasizes again that f represents the entire function while $f(x)$ represents an element of the range. Always keep in mind that, regardless of the symbolism used, a function is essentially a certain kind of correspondence between two sets.

Problem set 3–1

1. Which of these tables defines a function? The domain in each table is the set of numbers in the top row.

a.

x	3	4	5	6
y	14	15	16	17

b.

x	3	4	5	6
y	15	14	17	16

c.

x	7	9	11	13
y	8	10	12	10

d.

r	8	10	12	10
s	13	14	15	14

e.

x	8	10	12	10
y	13	14	15	16

f.

x	7	8	9	10
y	16	16	16	16

g.

x	9	25	81	−16
y	9	25	81	−16

h.

x	13.6	−15.8	19.7	13.6
y	21.3	38.6	14.1	23.1

2. If $f(x) = 7x - 3$, then evaluate these expressions.

a. $f(4)$
b. $f(6)$
c. $f(-2)$
d. $f(0)$
e. $f(t)$
f. $f(2r)$
g. $f(x+5)$
h. $f(x+2)$
i. $f(x-8)$
j. $f(x+h)$
k. $\dfrac{f(x+2)-f(x)}{2}$
l. $\dfrac{f(x+h)-f(x)}{h}$

3. If $g(x) = 3$ for all real numbers x in the domain of g, then evaluate these expressions. g is an example of a *constant* function.

a. $g(1)$
b. $g(-7)$
c. $g(5)$
d. $g(x+2)$
e. $g(-78.9204)$
f. $\dfrac{g(x+2)-g(x)}{2}$

4. If $f(x) = 3x^2 + 4x - 6$, then evaluate these expressions.

a. $f(1)$
b. $f(-5)$
c. $f(0)$
d. $f(1/4)$
e. $f(x+3)$
f. $f(x+h)$
g. $\dfrac{f(x+h)-f(x)}{h}$
h. $f(x-5)$

5. A *zero* or *root* of a function f is an element x of the domain such that $f(x) = 0$. For example, $x = 5/3$ is a zero of

$$f(x) = 3x - 5$$

since

$$f(5/3) = 3 \cdot 5/3 - 5 = 0.$$

To find this zero we solved the equation

$$f(x) = 3x - 5 = 0$$

for x. Find the zero of the functions defined by each of these equations.

a. $f(x) = 13x - 26$
b. $f(x) = 4.5x + 8$
c. $g(x) = x^2 - 3x - 10$
d. $h(x) = 3x^2 + 8x - 35$ (Hint: use the quadratic formula.)
e. $F(x) = x^2 + 3$
f. $H(z) = 5$

6. Explain why the set

$$h = \{(x, y) | x^2 + y^2 = 25\}$$

is not a function by producing some specific pairs (x, y) satisfying the equation but with different values of y for the same value of x.

7. Refer to the polygon problem of the first chapter (Example 3) in which we used the expression

$$(n \cdot 180 - 360)/n.$$

Rewrite this expression in function format using A as the dependent variable. What is the domain of the function as far as the nature of the polygon problem is concerned?

8. We will assume that the domain of a function is the largest possible set of real numbers for which range elements exist under the rule of correspondence. We can also limit this largest or "natural" set whenever necessary. For example, in our earlier function

$$h(r) = \sqrt{r - 3}$$

the domain would have to be restricted to the set

$$\{3, 4, 7, 12, 19, 28, \ldots\}$$

in order to produce range elements that are whole numbers. Such a restriction on the domain produces a different function. Find the domain of these functions. The range elements do not have to be whole numbers.

a. $f(t) = 3t + 7$
b. $g(x) = 4$
c. $f(x) = 6x^2 + 5x - 7$
d. $g(z) = \dfrac{5z + 6}{z - 2}$
e. $h(x) = x + 7$
f. $g(t) = 3t - 8$
g. $g(t) = \sqrt{3t - 8}$
h. $h(x) = \dfrac{5x^3 - 19x + 7}{(x + 2) \cdot (3x - 11)}$

9. Consider the function $f(v, w, x)$ as defined in the text by equation (A). Find the value of z corresponding to these triplets of independent variables.
   a. $(2, 4, 5)$ b. $(-1, 3, 0)$
   c. $(1, -2, 5)$ d. $(1, -2, 6)$
   e. $(1, -2, 7)$ f. $(1, -2, 8)$
10. Complete this table for the "piecewise-defined" function, whose domain is the set of all real numbers, given by the equations

$$f(x) = \begin{cases} \sqrt{2-x}, & x < 2 \\ \sqrt{x-2}, & x \geqslant 2 \end{cases}$$

x	−7	−2	1	2	3	11	15
$f(x)$							

11. Problem 8 establishes an alternate interpretation of the domain of a function as the set of real numbers for which there exist corresponding real numbers in the range. It does not specify or ask what the range itself is for the functions there. It is usually much more difficult to actually find the range elements themselves, particularly when we seek the largest or smallest range elements. The next section presents some problems where it is desirable to locate the maximum or minimum values of the range of a function. Chapter 9 (Differential Calculus) is devoted to this problem of locating range elements.

Word problems with functions

So far in this chapter we have worked mostly within the upper half of our diagram on applied mathematics. Now that we are acquainted with some of the formalities of functions we will proceed to their applications. Recall that wherever there is a definite relation between two sets we always hope to describe it with a function. Once the general rule of correspondence is identified we then use it to make further deductions about the situation. We will plunge right into this area of applied mathematics by doing a few examples.

Example 4 A rectangular field whose area is two hundred square yards is bounded on one side by a steep cliff, and a farmer wishes to fence it in. The fence costs \$6 a yard for the three land sides and \$8 a yard for the cliff side. Express the cost of fencing the field, C, as a function of the width, w, of the field.

Solution In order to first understand what the problem is all about we make a picture of the situation, writing in as much data as possible. The goal is to have an expression such as

$$C = f(w) = \text{(some algebraic expression with } w\text{'s and constants).}$$

Figure 3.1

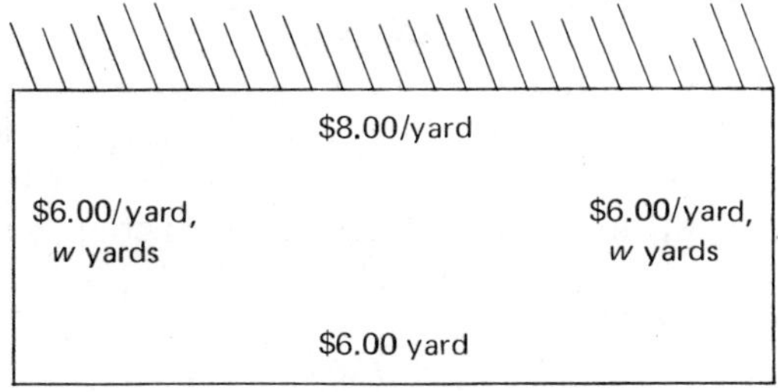

Since common sense tells us that the total cost for the field is the sum of the costs of the four sides, we begin with the front side and proceed clockwise in writing the cost of each side.

$$\begin{aligned} \text{Cost} &= 6 \cdot \text{length} + 6 \cdot w + 8 \cdot \text{length} + 6 \cdot w \\ \text{Cost} &= 14 \cdot \text{length} + 12 \cdot w \qquad \text{(A)} \end{aligned}$$

Thus we have cost expressed as a function of both length and width, $C = f(l, w)$, but we must end up with $C = f(w)$ alone. The key question is: can we express the length in terms of w? The answer is yes, which we can scc by realizing that the area is 200 square yards, or, using an equivalent formula,

$$\begin{aligned} \text{length} \cdot \text{width} &= 200, \\ \text{length} \cdot w &= 200, \\ \text{length} &= 200/w. \end{aligned}$$

We now replace the length by $200/w$ in equation (A).

$$\text{Cost} = 14 \cdot \frac{200}{w} + 12 \cdot w,$$

which we refine to

$$C = f(w) = \frac{2800}{w} + 12w$$

or, in factored form,

$$C = f(w) = 4 \cdot \left(3w + \frac{700}{w}\right).$$

Mathematically speaking, the domain of this function is any number w different from 0—$\{w | w \neq 0\}$—but, as far as the problem itself goes, w must be positive since it is nonsense to talk about a negative width.

Example 5 An apple orchard now has 30 acres and 60 trees per acre with an average yield of 400 apples per tree. For each additional tree that is planted on each acre the average yield per tree is reduced by 10 apples because the soil's fixed nutrients are spread around more thinly. Express the total number of apples in the orchard as a function of the number of trees planted per acre.

Solution We begin by letting t be the independent variable representing the number of *additional* trees planted per acre beyond the starting number of 60, so that $(60+t)$ is then the total number of trees planted per acre. It is not helpful to make any kind of a diagram, so we resort to making a table that will lead us to the desired format of

$$A = f(t) = (\text{expression in } t).$$

In general terms, A will equal

$$(\text{number of acres}) \cdot (\text{number of apples per acre})$$

or

$$A = (\text{number of acres}) \cdot (\text{number of trees per acre}) \cdot (\text{number of apples per tree}) \quad \text{(B)}$$

Table 3.2

Additional trees planted per acre	Total number of trees per acre	Total number of apples per tree
t	$(60+t)$	?

If we can somehow find an expression involving t to replace the question mark in Table 3.2, then we will know the values of the first and second factors in equation (B). We can use Table 3.3 to help us.

Table 3.3

Additional number of trees per acre, t	Total number of apples per tree, or ?
0	400
1	$390 = 400 - 10$
2	$380 = 400 - 20 = 400 - 2 \cdot 10$
3	$370 = 400 - 30 = 400 - 3 \cdot 10$
4	$360 = 400 - 40 = 400 - 4 \cdot 10$
⋮	⋮
t	$400 - t \cdot 10$ or $400 - 10t$

By looking at this pattern, we obtain the expression $(400 - 10 \cdot t)$ for the number of apples per tree as t additional trees are planted per acre.

We can now replace the words in our intermediate equation (B) with the formal mathematical symbols.

$$A = f(t) = (30) \cdot (60 + t) \cdot (400 - 10t)$$
$$A = f(t) = 30 \cdot (24000 - 200t - 10t^2)$$
$$A = 300 \cdot (2400 - 20t - t^2)$$

The domain of this function as it stands is $\{x | x \in R\}$, but we have to limit the domain in light of the physical circumstances of the apples and trees. Obviously, t must be a nonnegative integer and $(400 - 10t)$, the number of apples per tree, must also be a nonnegative integer. If we solve

$$400 - 10t = 0$$

we get

$$-10t = -400$$

or

$$t = -400/-10$$
$$t = 40,$$

meaning that the maximum value of t we can consider is 40 since $(400 - 10t)$ is a negative number for those values of t greater than 40. Thus we formally state the domain as

$$\{t | t = 0, 1, 2, 3, \ldots, 39, 40\}$$

Example 6 A farmer stores his wheat and decides to sell it when he will receive a maximum profit. For each 500 pounds stored 4 pounds a day will be lost because of rats, evaporation, silo leakage, etc. On the other hand, each 100 pounds that is kept will add 8¢ to his profit each day. If he can sell the wheat right now for a profit of 7¢ a pound without any storage, then write the function which expresses his total profit on 6 tons of wheat in terms of the number of days he stores the wheat.

Solution We notice from the outset that several units of weight are involved, so we convert everything to one-pound units. Then every pound of wheat loses $4/500 = 1/125$ lb. a day, adds 8¢ to his profit each day, and sells for 7¢; there are $6 \cdot 2{,}000 = 12{,}000$ lbs. initially involved. We will let t, the independent variable, be the number of days he keeps the wheat and P be the total profit which corresponds to the values of t: $P = f(t)$.

Common sense says that the total profit P equals

[number of pounds sold] · [profit per pound at time of sale].

Further refinement of

[number of pounds sold]

yields

[(initial number of pounds) – (number of pounds lost)].

Further refinement of

[profit per pound at time of sale]

yields

[(initial profit per pound) + (accumulated profit per pound)].

Writing the last two word equations is really the heart of this problem. Now we fill in the four word expressions inside the parentheses with appropriate t's, constants, and operation signs.

$$
\begin{aligned}
P &= [\text{number of pounds sold}] \cdot [\text{profit per pound at time of sale}] \\
&= [(\text{initial number of pounds}) - (\text{number of pounds lost})] \\
&\qquad \cdot [(\text{initial profit per pound}) + (\text{accumulated profit per pound})] \\
&= \left[(12{,}000) - \left(\frac{1}{125} \cdot t \cdot 12{,}000\right)\right] \cdot [(7) + (.08 \cdot t)] \\
&= [12{,}000 - 96t] \cdot [7 + .08t] \\
&= 84{,}000 + 960t - 672t - 7.68t^2 \\
P &= f(t) = -7.68t^2 + 288t + 84{,}000
\end{aligned}
$$

To obtain the domain we realize that if 1/125 of a pound is lost each day then in 125 days it will all be gone, so t is any integer between 0 and 125, inclusive. In practice, of course, the farmer would want to know how long to store the grain in order to have the maximum profit. After we have studied that branch of mathematics known as calculus we will return to this problem to answer his question.

Problem set 3–2

1. A farmers' cooperative association plans to store the wheat raised during the year and sell when they can receive the most for it. They estimate that each original 1,000 pounds they store will lose 5 pounds each day because of evaporation, rodents, etc. For every day they keep it they estimate they will gain 80¢ on each 1,000 pounds sold. They can sell it right now without any storage for \$25 a thousand pounds. Express the future value of each 1,000 pounds of wheat originally stored as a function of the number of days it is kept. What is the domain of the function?
2. A ship is steaming due north at 12 miles per hour. At midnight a lighthouse is sighted at a distance of 3 miles directly west of the ship. If the distance between the ship and the lighthouse t hours later is d miles, find a formula that expresses d in terms of t.
3. An airplane leaves an airport at noon flying due north at 200 miles per hour. At 2:00 p.m. another plane leaves the airport and flies due east at 250 miles per hour. If the distance in miles between the two planes t hours after noon is denoted by d, then find a formula that expresses d as a function of t.

4. An open box is to be made from a rectangular piece of tin 10 inches long and 8 inches wide by cutting pieces x inches square from each corner and bending up the sides. Express the volume, V cubic inches, of the box in terms of x. What is the domain of the resulting function?
5. An open box with a square base is to be made of wood costing 7¢ per square foot for the sides and 9¢ per square foot for the bottom. The volume of the box is to be 10 cubic feet. If the bottom of the box is to be x feet by x feet, then express the total cost, C cents, as a function of x.
6. A man wishes to make a rectangular flower bed against the wall of his house. He has 50 feet of fencing available to use for the other three sides.
   a. Write the growing area of the flower bed as a function of its width and find the domain of this function.
   b. If the flower bed has dimensions of 10′ by 30′ and he covers the bed with topsoil which costs $4.00 a cubic yard to a depth of 6 inches, then what would his total cost be?
   c. If the flower bed is w feet wide and he covers it with topsoil which costs c dollars a cubic yard to a depth of d inches, then what would his total cost be?
7. A senior class decides to take a trip to Washington, D.C., by bus. The bus company will rent a vehicle whose capacity is eighty people so long as at least thirty people sign up. The fare will be $40 a person if the minimum requirement of thirty people is met, and it will decrease by 50¢ per person for everyone for each additional person who goes. For example, if thirty-two people go then each of them will pay $39.
   a. Write the bus company's total revenue, R, as a function of x, the number of additional people who go, and give the domain of the function.
   b. What is the total revenue if thirty people go? Forty? Fifty? Eighty?
   c. Write the bus company's total revenue, R, as a function of x, the total number of people who go, and give the domain of the function.
8. A layout editor wishes to have 50 square inches of printed matter on each page of a book. The printed matter must have margins of 2 inches on the top and bottom and margins of 1 inch along the two sides. Write the total area of the paper as a function of the overall width of the paper. If the pages may have a maximum width of 9 inches, then give the domain of this function.
9. A retailer wants to price a toy car which costs him $2 so as to maximize his total profit. He knows from experience that if he charges x dollars apiece, he will be able to sell $300-100x$ of the cars.
   a. What is his profit per car when he charges x dollars apiece?
   b. What is his total profit if he charges $2.20 apiece? $2.30? $2.40? $2.50? $2.60? $2.70? $2.80?
   c. From looking at your results in tabular form from part (b), estimate how much he should charge per car in order to maximize his total profit. We will determine the exact value of x after we have studied differential calculus in Chapter Nine.

10. Rework Example 5 by letting the independent variable t represent the *total* number of trees planted per acre.

Graphs of functions

Now that we are familiar with functions in terms of their algebraic definition and some applications we will next analyze them from a geometric viewpoint. The specific function we will work with is defined by an equation.

Example 7 Consider the function defined by

$$y = f(x) = 1.5 \cdot x + 2$$

where $x \in R$. Remember that we may consider a function to be a set of pairs of corresponding numbers (x, y) which satisfy the equation defining the function. One such pair in this function is $(6, 11)$. There is also a one-to-one correspondence between the set of all ordered pairs of numbers and the set of all points in the coordinate plane we studied in high school algebra. Constructing the graph of this function is thus a matter of finding those points in the plane whose coordinates satisfy the equation.

Plotting coordinates of a function

We proceed by first finding algebraically some pairs of numbers (x, y) that satisfy the equation, organizing them into a tabular format, and locating the corresponding points in the coordinate plane. We choose as our values of x in the domain

$$-3, -1, 0, 1, \text{ and } 5,$$

which give us the corresponding values of y in the range as indicated next in Table 3.4.

Table 3.4

Domain x	Range y
−3	−2.5
−1	.5
0	2
1	3.5
3	6.5

Then plotting the points with coordinates

$$(-3, -2.5), (-1, .5), (0, 2), (1, 3.5), \text{ and } (3, 6.5)$$

gives us

Figure 3.2

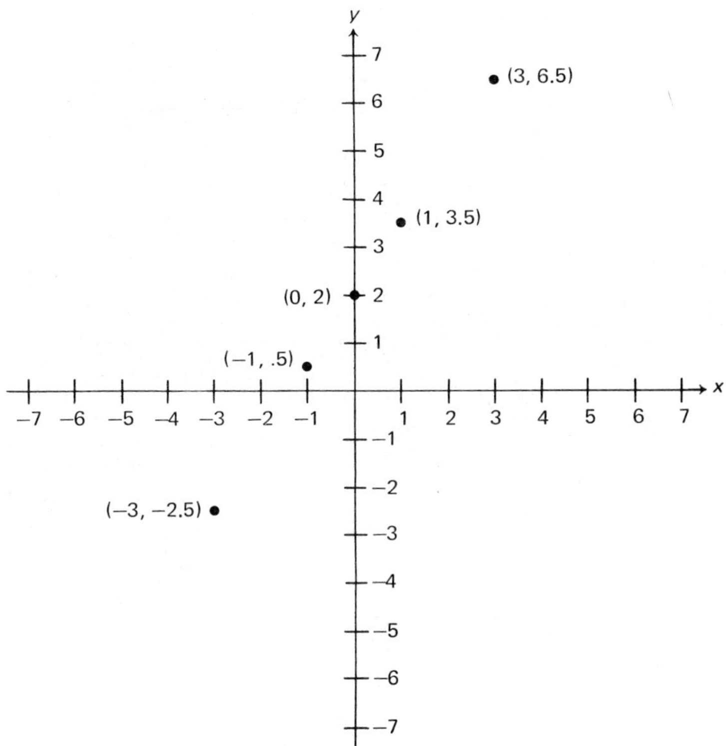

It appears that the dots are part of a line, so we connect them by constructing the line passing through each dot. (See Fig. 3.3 overleaf.)

If we look more closely at the equation defining our function,

$$y = 1.5 \cdot x + 2,$$

we can generalize from it by noting that it is of the general form

$$y = (\text{a constant}) \cdot x + (\text{a constant})$$

or, in an abbreviated form,

$$y = m \cdot x + b.$$

Such a general equation defines what is called a *linear function.* The terminology is appropriate for the following two reasons.

1. The equation does define a *function*—for any value of x in the domain the value of the corresponding expression $m \cdot x + b$ represents exactly one number y in the range.

Figure 3.3

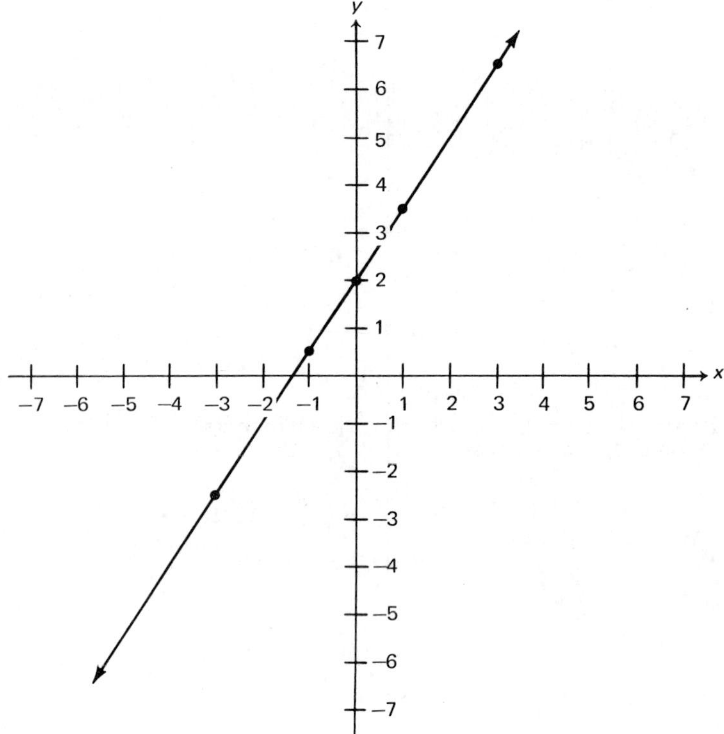

2. We could also show (though we have not actually done so) that the graph of the equation $y = m \cdot x + b$ will always be a *line*, regardless of the specific values of m and b.

Example 8 What do we know about the set of points whose coordinates satisfy the equation

$$3x - 5y = 7?$$

Solution If we solve this equation for y, we have

$$\begin{aligned} &3x - 5y = 7 \\ &-5y = 7 - 3x \\ &-5y = -3x + 7 \\ &y = -3x/-5 + 7/-5 \\ &y = (3/5) \cdot x + -(7/5) \qquad \text{(A)} \end{aligned}$$

Slope of a line

Equation (A) is therefore of the form $y = m \cdot x + b$, and so the original equation also defines a function.

An important property of lines is their *slope* or "steepness" or direction. Look at lines A, B, C, D, and E in this picture.

Figure 3.4

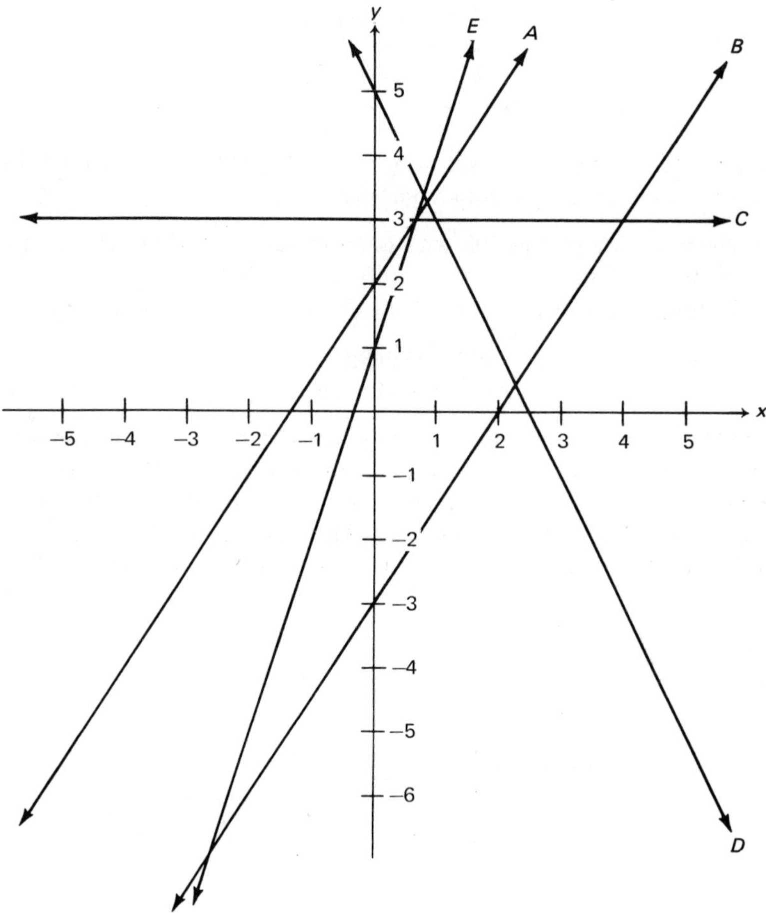

It is geometrically apparent that

1. A and B are parallel and have the same steepness and direction;
2. A, B and E are "increasing" lines, meaning that if a point moves along the line from left to right it also increases in a vertical direction;
3. D is a "decreasing" line, meaning that if a point moves along the line from left to right it also decreases in a vertical direction;
4. E is a steeper line than A;
5. C is parallel to the x axis.

The equations of the lines from Figure 3.4 in $y = mx + b$ form are given in Table 3.5.

Table 3.5

Line	Equation
A	$y=1.5\cdot x+2$
B	$y=1.5\cdot x+-3$
C	$y=0\cdot x+3$
D	$y=-2\cdot x+5$
E	$y=3\cdot x+1$

Next is the all-important definition of slope, which lets us quantify our qualitative observations 1 through 5 above.

Definition If *any* two points with coordinates (x_1, y_1) and (x_2, y_2) (read "x sub one, y sub one" and "x sub two, y sub two") are on a line that is not parallel to the y axis, then the *slope* of that line is the number

$$m = \frac{\text{vertical change}}{\text{horizontal change}} = \frac{y_2 - y_1}{x_2 - x_1}.$$

Rephrasing slightly, we may say that the slope of the line through any two points on it is the ratio of the rise (change in a vertical direction) to the run (change in a horizontal direction) or, more important for future, the change in y divided by the change in x. The ratio $(y_2 - y_1)/(x_2 - x_1)$ does *not* depend on which points (x_1, y_1) and (x_2, y_2) are chosen on the line (see

Figure 3.5

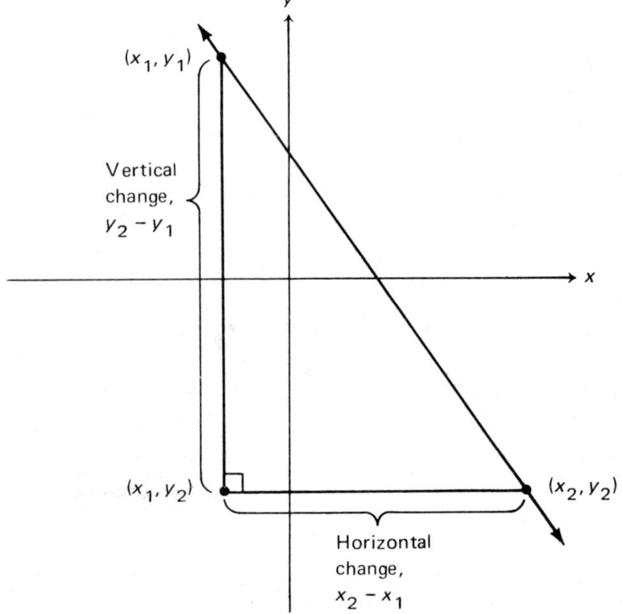

problem 13). We summarize this paragraph and the definition in a diagram of a point moving along the line from (x_1, y_1) to (x_2, y_2). (See Fig. 3.5.)

Example 9 Consider the line that is the graph of the function defined by the equation

$$4x - 7y = -14.$$

If we solve this equation we get

$$y = \frac{4}{7} \cdot x + 2.$$

Next we find two pairs of numbers satisfying the previous equation so that these two pairs of numbers will be the coordinates of two points on the line. Since two points uniquely determine a line, we will then be able to draw the graph of the entire line. If we let x be 0 and 7 then the corresponding values of y are 2 and 6; therefore two points on the line have coordinates $(0, 2)$ and $(7, 6)$. We let these coordinates be the (x_1, y_1) and the (x_2, y_2), respectively, in the general slope formula. Then substitution gives us

$$m = \frac{y_2 - y_1}{x_2 - x_1} = \frac{6 - 2}{7 - 0} = \frac{4}{7}.$$

Figure 3.6 gives us a geometric interpretation of the significance of the number 4/7 as the slope of this line.

Figure 3.6

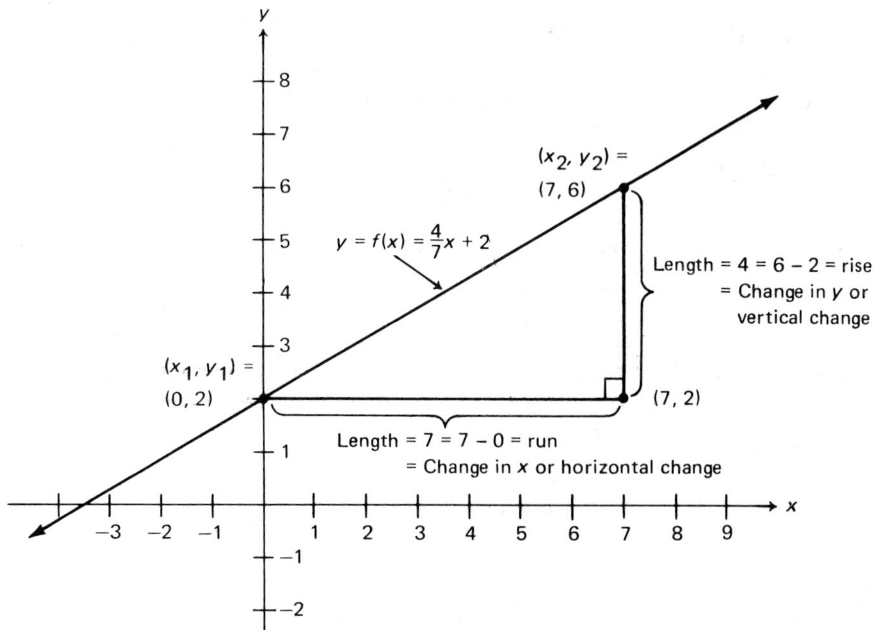

It is interesting to note that the slope 4/7, obtained by using the slope formula, is precisely the same number as the coefficient of x in the equation

$$y = \frac{4}{7}x + 2.$$

This statement can be shown to hold true in general; that is, if a line is given by the equation $y = m \cdot x + b$, then the slope of the line is always the numerical value of m.

We refer to Table 3.5 and add new information to it. The new third column of this table has as its elements the slopes of the lines which we obtain by merely inspecting the coefficients of x in the equations.

Table 3.6

Line	Equation	Slope
A	$y = 1.5 \cdot x + 2$	1.5
B	$y = 1.5 \cdot x + -3$	1.5
C	$y = 0 \cdot x + 3$	0
D	$y = -2 \cdot x + 5$	−2
E	$y = 3 \cdot x + 1$	3

Using slopes for algebraic interpretations

Now we list some algebraic interpretations, by using slopes, of our five earlier geometric observations of the lines in Figure 3.4.

1. A and B are parallel, and their slopes are equal.
2. A, B, and E are increasing lines, and their slopes are positive.
3. D is a decreasing line, and its slope is negative.
4. E is a steeper line than A, and the slope of E, m_E ("m sub e") = 3, is greater than the slope of A, $m_A = 1.5$.
5. C is parallel to the x-axis, and its slope is 0: $m_C = 0$.

Example 10 Graph the function defined by the equation

$$y = f(x) = x^2 - 2x - 3.$$

Solution We do this by finding several—seven will usually suffice for our purposes—pairs of numbers satisfying the equation, plotting their corresponding points, and connecting them in a follow-the-dots fashion.

Table 3.7

x	−2	−1	0	1	2	3	4
y	5	0	−3	−4	−3	0	5

Figure 3.7

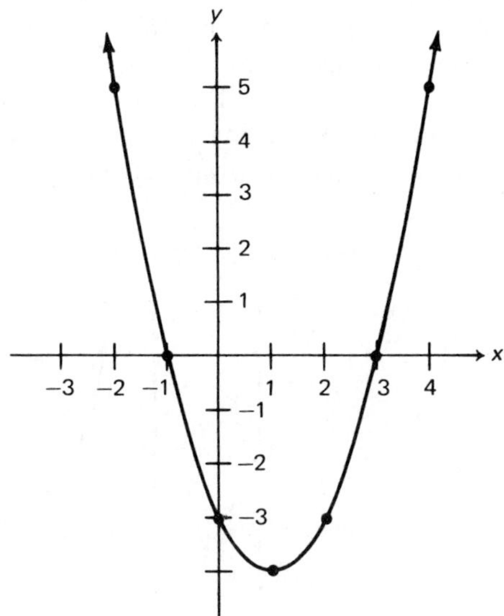

The graph in Example 10 is an example of a *parabola.* In fact, the graph of any function of the form

$$y = f(x) = ax^2 + bx + c$$

where $a \neq 0$ is a parabola.

Example 11 Graph the function defined by the equation

$$y = f(x) = \frac{5}{(x-2)}$$

Solution The algebraic preliminaries yield Table 3.8.

Table 3.8

x	−3	−2	−1	0	1	2	3	4	5
y	−1	−1.25	−1.67	−2.5	−5	$\frac{5}{0}$: does not exist	5	2.5	1.67

Something unusual happens to the values of y corresponding to those of x between 1 and 3, especially since 2 as a candidate for the domain has no correspondent in the range. We examine the values of y corresponding to additional values of x chosen from the interval $(1, 3)$.

Table 3.9

x	1.5	1.7	1.9	1.95	1.99	1.999	2.001	2.01	2.05	2.1	2.3	2.5
y	−10	−16.67	−50	−100	−500	−5000	5000	500	100	50	16.67	10

We see that for values of x just below 2 the corresponding values of y are large negative numbers. For values of x just above 2 the corresponding values of y are large positive numbers. Whenever this algebraic situation of dividing a nonzero number by zero occurs in general in evaluating $f(x)$, its geometrical indication is a vertical line through $(x, 0)$ where x is the number producing a zero divisor. This vertical line is an example of an *asymptote*. The graph of the function appears next in Figure 3.8.

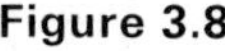

Figure 3.8

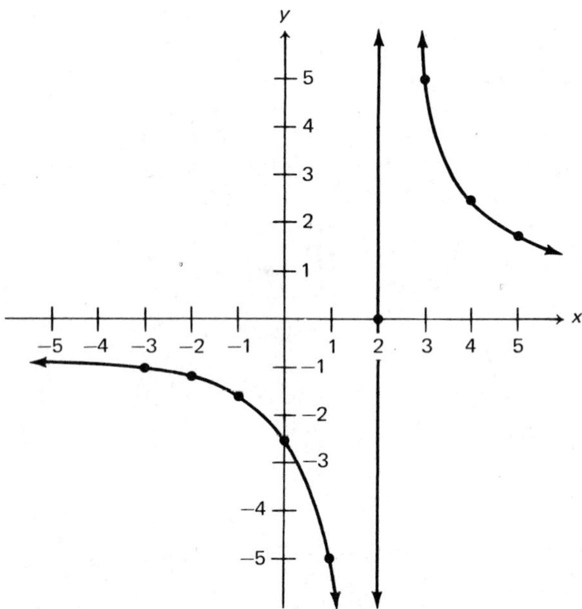

The asymptote itself is *not* part of the graph of the function, and $x = 2$ is *not* in the domain of the function. It is a line which the two curves approach but never quite reach. For example, if x is 2.000001 then y is 5,000,000 so that $(2.000001, 5{,}000{,}000)$ is a point on the curve just barely to the right of the asymptote and far above the x-axis.

We close by noting that so far we have been given the equation defining a function and have proceeded to find its graph. Occasionally we are given information about the graph of a function and must find the equation itself. Problem 15 is an example of this situation.

Problem set 3–3

1. Rewrite, where possible, each of these equations in standard $y = m \cdot x + b$ form. Then draw the graph of each equation after algebraically finding two pairs of numbers satisfying the equation and plotting the two corresponding points. You may wish to plot a third point to check on the

accuracy of the arithmetic. Find the slope of each line by using the slope formula and compare the numerical result with the value of m in the equation $y = m \cdot x + b$.

a. $y = 2x + 5$ b. $3y = 12x + 1$
c. $5x + 4y = 7$ d. $7x - 3y - 5 = 0$
e. $3y - 7x + 2 = 0$ f. $3x + 7y - 4 = 0$
g. $y + 3 = 0$ h. $5 - y = 0$
i. $2x + 0y = 6$ (careful!) j. $4x + 7 = 0$

2. Draw the graphs of the equations of parts d, e, and f from the previous problem on the same set of axes.
   a. How do the pairs of lines from parts d and f appear to be related?
   b. What is the product of their slopes?
   c. What is the product of the slopes of the lines from parts e and f?
   d. What generalization do the last two questions suggest?
3. Draw the line through these points with the given slope by using the interpretation of slope as

$$\frac{\text{vertical change}}{\text{horizontal change}} = \frac{\text{change in } y}{\text{change in } x}.$$

a. $(2, 1/2)$; $m = 3 = \frac{3}{1}$ b. $(-2, 4)$; $m = \frac{-3}{5}$

c. $(-2, 4)$; $m = \frac{3}{-5}$ d. $(4, -3)$; $m = \frac{1}{2}$

e. $(2, 5)$; $m = 0$ f. $(-5, -3)$; $m = -2$
g. $(5, 3)$; $m = 3.5$ h. $(-2, -6)$; $m = 4$

4. Let points A, B, C, and D be the four vertices of a quadrilateral. Draw the graph of each quadrilateral, and then determine *algebraically* which of the following quadrilaterals are parallelograms (i.e., have two pairs of parallel sides).
   a. $A(4, 4)$, $B(0, 0)$, $C(5, 0)$, $D(5, 4)$
   b. $A(8, 2)$, $B(-1, 3)$, $C(-2, -6)$, $D(5, -4)$
   c. $A(3, -5)$, $B(4, -6)$, $C(-1, 6)$, $D(-2, 5)$
   d. $A(2, 4)$, $B(3, 8)$, $C(5, 1)$, $D(4, -3)$
   e. $A(3, 8)$, $B(6, 4)$, $C(-2, 7)$, $D(-4, -3)$
5. Draw the graph of each of the functions defined by these equations. Using seven pairs of points, including some negative choices for x, should suffice to indicate the general pattern in each case, but be prepared to plot more points in order to indicate the complete behavior of the graph.
   a. $y = f(x) = 3x^2 - 4x + 1$
   b. $y = g(x) = -2x^2 + 3x - 2$
   c. $s = h(t) = 1.5t^2 - 2t$
   d. $y = f(x) = x^3 - 4x^2 + x + 6$
   e. $y = f(x) = x^2 + 5$
6. Inspect carefully these equations, each of which defines a function, and then draw the graph of the function. Using seven pairs of points, including

some negative choices for x, should suffice to indicate the general pattern in each case.

a. $y = f(x) = \dfrac{2}{x+1}$ b. $y = f(x) = \dfrac{4}{x-3}$

c. $s = f(t) = \dfrac{x-2}{x+3}$ d. $r = h(t) = \sqrt{t-3}$

e. $s = f(x) = x + 5$ f. $s = g(x) = \dfrac{x-5}{(x+2)\cdot(x-6)}$

7. In a function each value of x in the domain has exactly one corresponding element y in the range. The geometrical equivalent of this statement says that the graph of a function is a set of points in the coordinate plane which is met exactly once by any vertical line through the domain portion of the x-axis. Which of these graphs are the graphs of a function in the indicated domain $[a, b]$?

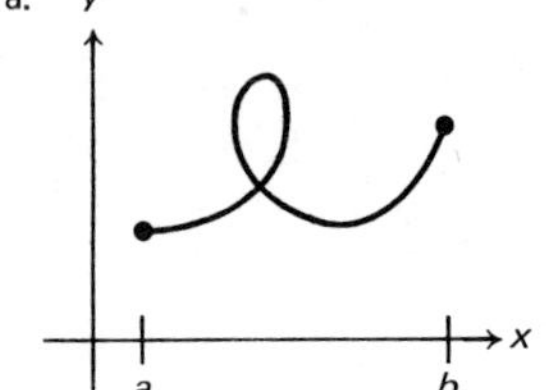

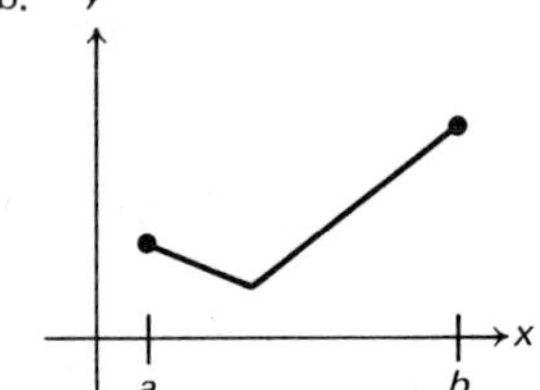

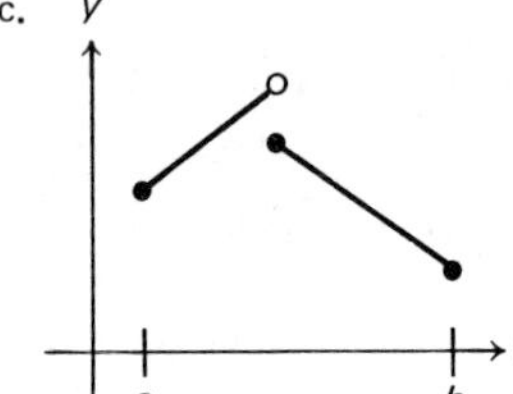

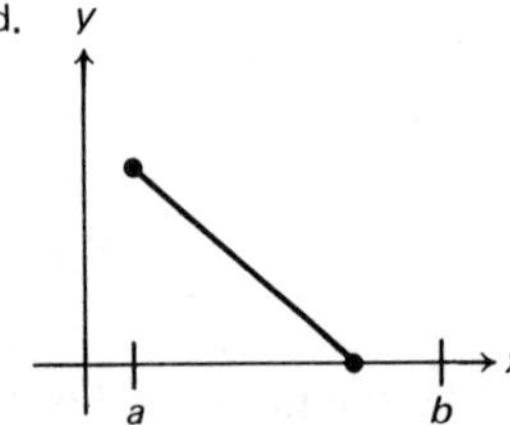

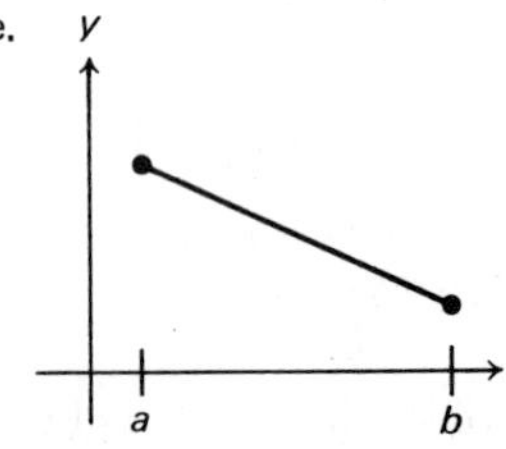

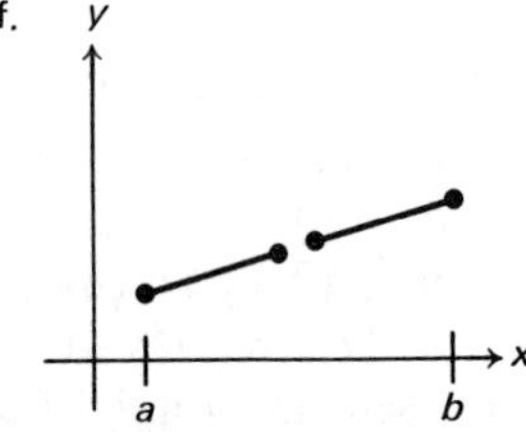

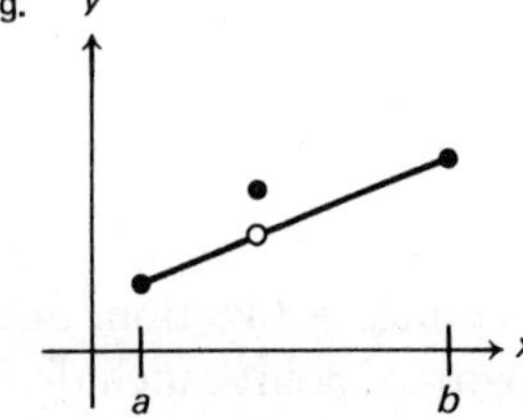

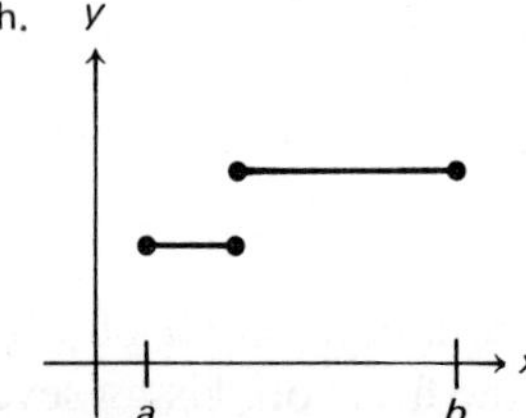

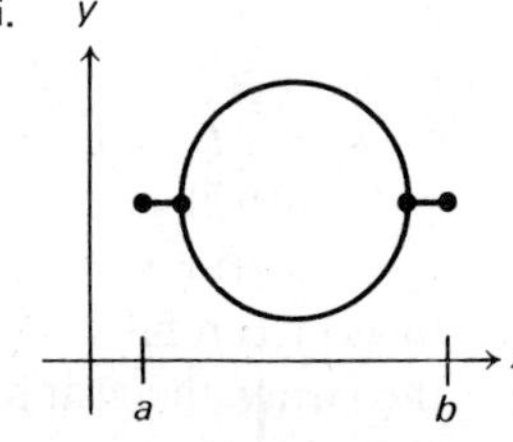

8. In Example 9 we saw that the graph of the function given by an equation of the form

$$y = f(x) = ax^2 + bx + c, \quad a \neq 0$$

is a parabola. Why is the restriction $a \neq 0$ necessary?

9. In the definition of slope, the text says that the line under consideration is not parallel to the y-axis. Why does the definition of slope fail to hold for a line parallel to the y-axis?

10. A projectile fired straight up attains a velocity of v feet per second after t seconds of flight, and the relation between the numbers v and t is a linear one. If the projectile is fired at a velocity of 1,000 feet per second and reaches a velocity of 350 feet per second after 2 seconds of flight, then find the formula for v in terms of t. At what time does the projectile reach its highest point?

11. We can assume that the temperature, $t°$ Fahrenheit, at a point h feet above the surface of the earth is a linear function of h for limited altitudes under certain atmospheric conditions. Suppose the temperature on the surface of the earth is 70° and the temperature at 3,000 feet is 61°. Find the formula for t in terms of h, and sketch the graph for those values of h between 0 and 20,000.

12. The *y-intercept* of a line is the point in the coordinate plane where the line crosses the y-axis. At this point the x-coordinate is 0, and we use this fact to calculate the y-coordinate of the y-intercepts. For example, to find the y-intercept of the graph of $3x - 11y = 6$, we set x equal to 0 and solve for y, which gives us the pair of numbers $(0, y)$ that we are looking for. It is $(0, -6/11)$ in this case. Find the y-intercept of each of the lines in Problem 1. What generalization does the pattern of your answers suggest?

13. The points $(3, 26/7)$ and $(-1, 10/7)$ are on the line which is the graph of

$$y = \frac{4}{7} \cdot x + 2,$$

which we studied in Example 8. Using these two points, calculate the slope of the line. Does your answer for the slope depend on which particular pair of points on the line you use in calculating it?

14. The sales manager of a local motorcycle shop claims that the shop's sales of motorcycles will increase \$6 for each additional dollar spent on advertising over the local radio station. If the bike shop has averaged monthly sales of \$20,000 while spending \$150 a month on the commercials, then find the linear equation relating the shop's expected sales to the amount spent on the radio commercials. What is the expected sales income if \$170 of air time is bought?

15. It is sometimes necessary to find the equation of the line which passes through a given point with a given slope. As an example, we may know that a line with slope 5/8 passes through the point $(3, 4)$, and we are

looking for the equation that generates that line. In general, a line with slope m which passes through a specific point (x_1, y_1) has as its equation

$$y - y_1 = m \cdot (x - x_1).$$

Thus in the previous example the equation of the line is

$$y - 4 = (5/8) \cdot (x - 3)$$
$$y - 4 = (5/8)x - 15/8$$
$$y = (5/8)x + 17/8$$
$$y = .625x + 2.125.$$

Write the equation of the lines through point (x_1, y_1) with slope m.

a. $(1, 4)$; 2
b. $(3, 2)$; -4
c. $(-2, 4)$; 5/6
d. $(1, 21)$; 3
e. $(-8, -6)$; 3
f. $(-7.43, 2.786)$; 0
g. $(-2, 1)$; $-9/7$

(What fact do (d) and (e) tell you?)

16. Construct the graph of the function in Problem 10 of Problem set 3–1. Restrict the domain to the interval $[-7, 18]$.

Sequences

So far the functions we have studied have had the real numbers as their domain, except for, on occasion, a few specified intervals or numbers. There are particular kinds of functions, called *sequences*, with domains consisting only of the positive integers. Sequences play an important role in later work, so we will learn their basic properties now and some of their many applications in the next chapter. Sequences are also interesting from a purely mathematical viewpoint because of the way they very neatly yield patterns which lead to general formulas.

Example 12 Consider the sequence a which is the infinite set

$$\{3, 7, 11, 15, 19, 23, 27, \ldots\}$$

The equation $a_3 = 11$ (read "a sub 3 equals 11") is true, and it is a compact way of expressing the idea that the third element of sequence a is the number 11. We can also observe that

$$a_5 = 19, \quad a_1 = 3, \text{ and } a_8 = 31.$$

The subscripts of the symbols a_1, a_2, a_3, etc., are sometimes called *index numbers* or *indices* because they serve as labels to locate the numbers in the sequence by their positions. For example, the index of 15 is 4 because 15 is the fourth element in the sequence. Also, the elements in a sequence are often called its *terms*; a_3 is the third term in sequence a, b_7 is the value of the seventh term of a particular sequence b, and d_{15}

is the fifteenth term of sequence d. Technically, since a is a sequence it is by definition a function and consists of the ordered pairs

$$(1, 3), (2, 7), (3, 11), (4, 15), (5, 19), \ldots$$

However, since the domain is the same in any sequence, we usually represent it by the range elements only and often omit the braces when symbolizing these range elements.

Example 13 Examine the sequence

$$b = 3, 2\,1/2, 2\,1/3, 2\,1/4, 2\,1/5, \ldots$$

by looking for a general formula that will give the value of the nth term in the sequence b_n by merely plugging the particular value of n into this to-be-determined formula.

Solution We organize our search by making a table and slightly rewriting the original numbers.

Table 3.10

n	1	2	3	4	5	...	n
b_n	$2 + 1/1$	$2 + 1/2$	$2 + 1/3$	$2 + 1/4$	$2 + 1/5$	...	?

By looking at the pattern exhibited by the pairs of numbers in this table, it is evident that the correspondent of n is

$$2 + \frac{1}{n}.$$

Thus we can immediately conclude, without writing out all the intermediate numbers, that

$$2\,1/17$$

is the seventeenth term of sequence b:

$$b_{17} = 2\,1/17.$$

Sequence b can also be identified by

$$b = \left\{2 + \frac{1}{n}\right\}$$

since this notation indicates how the elements of b are determined.

We return to sequence a (from Example 12) to look for a formula that generates the set

$$3, 7, 11, 15, 19, \ldots$$

from its understood domain of

$$1, 2, 3, 4, 5, \ldots.$$

We organize these two sets into a table as before, which gives us

Table 3.11

n	1	2	3	4	...	n
a_n	3	7	11	15	...	?
a_n, rewritten	$3+0\cdot4$	$3+1\cdot4$	$3+2\cdot4$	$3+3\cdot4$	...	?

The bottom row of the table holds the key to the search. We created it by noticing that the terms of the sequence begin with 3 and regularly increase from there in steps of 4. Thus the second term equals 3 plus one 4, the third term equals 3 plus two 4's, the fourth term equals 3 plus three 4's, etc. If we continued in such a pattern we could conclude that the thirteenth term equals 3 plus $(13-1)\cdot4$, the twenty-first term equals 3 plus $(21-1)\cdot4$, and in general that the nth term equals 3 plus $(n-1)\cdot4$:

$$a_n = 3 + (n-1)\cdot 4.$$

Arithmetic progressions

Sequence a is an example of a very important type of sequence known as an *arithmetic progression.* By definition an arithmetic progression (often abbreviated to AP) is a sequence in which each term after the first equals the one before it plus a constant—in a, for example, the third term of 11 equals $7+4$, the sixth term of 23 equals $19+4$, and in general

$$a_n = a_{n-1} + 4.$$

In words this equation says that the nth term of sequence a equals its predecessor plus 4 (where $n \geqslant 2$). We would like to have a formula for the nth term of *any* AP, and we can obtain it by generalizing from our earlier AP of

$$a_n = 3 + (n-1)\cdot 4.$$

This specific AP has 3 as its first term, which is symbolized in general by a_1. It has 4 as its constant difference, which is symbolized in general by d. Then, by using these letters to replace 3 and 4, respectively, we conclude that in any AP

$$a_n = a_1 + (n-1)\cdot d.$$

Any AP is completely determined by its first term a_1 and constant difference d.

Example 14 Let c be the AP whose first term is 25 and whose constant difference is -3 so that

$$c = \{25, 22, 19, 16, 13, \ldots\}.$$

Then

$$c_9 = 25 + (9-1) \cdot -3 = 1,$$
$$c_{13} = 25 + (13-1) \cdot -3 = -11,$$

and

$$c_n = 25 + (n-1) \cdot -3.$$

It is often necessary to find the sum of a certain number of terms of an AP, and we use sequence a again as an example of the summing process. The symbol S_n means the sum of the first n terms of the progression as illustrated here.

$$S_1 = a_1 = 3$$
$$S_2 = a_1 + a_2 = 3 + 7 = 10$$
$$S_3 = a_1 + a_2 + a_3 = 3 + 7 + 11 = 21$$
$$S_4 = a_1 + a_2 + a_3 + a_4 = 3 + 7 + 11 + 15 = 36$$
$$S_5 = a_1 + a_2 + a_3 + a_4 + a_5 = 3 + 7 + 11 + 15 + 19 = 55$$
$$\cdots\cdots\cdots\cdots\cdots\cdots\cdots\cdots\cdots\cdots\cdots\cdots$$
$$S_n = a_1 + a_2 + a_3 + a_4 + a_5 + \cdots + a_n = ?$$

The question we wish to answer is, "By what particular expression involving n can we replace the question mark for this particular arithmetic progression?" A logical way to search for this expression is to create a table showing a summary of our work so far.

Table 3.12

n	1	2	3	4	5	...	n
S_n	3	10	21	36	55	...	?

Unfortunately, the pattern is not revealed by the table so we must look elsewhere for the formula. We do so by working with, for example, S_7, which is the sum of the first seven terms. We know that

$$S_7 = 3 + 7 + 11 + 15 + 19 + 23 + 27,$$

but the key here is to rewrite the terms in reverse order so that

$$S_7 = 27 + 23 + 19 + 15 + 11 + 7 + 3.$$

Now we rewrite these two equations again, add them together (equals added to equals give equals), and then solve the resulting equation for S_7.

$$S_7 = 3 + 7 + 11 + 15 + 19 + 23 + 27$$
$$S_7 = 27 + 23 + 19 + 15 + 11 + 7 + 3$$

$$2 \cdot S_7 = (3+27) + (7+23) + (11+19) + \cdots + (27+3)$$
$$2 \cdot S_7 = 7 \cdot (3+27) \text{ (since all the numbers in the seven pairs of parentheses equal } 3+27\text{)}$$

$$S_7 = \frac{7}{2} \cdot (3+27) \qquad \text{(A)}$$

$$S_7 = 105$$

Of course, we could have found that S_7 equaled 105 much more easily by direct addition in the first place, but doing it this admittedly drawn-out way lets us generalize to S_n for any AP rather quickly, especially if we look at equation (A),

$$S_7 = \frac{7}{2} \cdot (3 + 27).$$

The first term (a_1) is 3, and 27 is the seventh term (a_7) of our particular AP. We can therefore rewrite the equation as

$$S_7 = \frac{7}{2} \cdot (a_1 + a_7).$$

By looking at the correspondence among the 7's it seems reasonable to conclude that the sum of the first eight terms of this AP would be

$$S_8 = \frac{8}{2} \cdot (a_1 + a_8),$$

that the sum of the first thirteen terms is

$$S_{13} = \frac{13}{2} \cdot (a_1 + a_{13}),$$

and that, in general, the sum of the first n terms is

$$S_n = \frac{n}{2} \cdot (a_1 + a_n).$$

It is also true that we can conclude that for *any* AP with n initial terms, first term a_1, and nth term a_n,

$$S_n = \frac{n}{2} \cdot (a_1 + a_n).$$

As an example of using this formula consider the AP

$$a = \{12, 10, 8, 6, \ldots\}.$$

Then the sum of the first fifteen terms is

$$S_{15} = \frac{15}{2} \cdot (a_1 + a_{15}).$$

a_1 is 12, and a_{15} is given by our earlier formula

$$a_n = a_1 + (n-1) \cdot d,$$

which becomes, in this specific case,

$$\begin{aligned} a_{15} &= 12 + (15-1) \cdot -2 \\ &= -16. \end{aligned}$$

Then through substitution we obtain

$$S_{15} = \frac{15}{2} \cdot (12 + -16)$$

$$S_{15} = -30.$$

Notice that in this problem we first had to find a_{15} by substituting into the formula for a_n and then substituting that result into the formula for S_n. Is there any way we can do similar summation problems by substitution into just one formula? Fortunately, the answer is yes, and we develop another formula to find S_n by combining the formulas for a_n and S_n according to this diagram:

Figure 3.9

Substitution

$$S_n = \frac{n}{2} \cdot [a_1 + a_n] \qquad a_n = a_1 + (n-1) \cdot d$$

$$S_n = \frac{n}{2} \cdot [a_1 + a_1 + (n-1) \cdot d]$$

$$S_n = \frac{n}{2} \cdot [2 \cdot a_1 + (n-1) \cdot d]$$

This last formula is the one we usually use for S_n. Notice that all we have to know in order to find S_n for any AP are the values of n, the number of terms, the first term (a_1); and the constant difference between any two terms (d).

Example 15 Consider the sequence

$$b = \{7, 10, 13, 16, \ldots\}.$$

Find the sum of the first twenty-one terms.

Solution First of all, b is an AP so we can proceed by using the last formula where $n = 21$, $a_1 = 7$, and $d = 3$.

$$S_{21} = \frac{21}{2} \cdot [2 \cdot 7 + (21-1) \cdot 3]$$

$$S_{21} = \frac{21}{2} \cdot [14 + 60]$$

$$S_{21} = \frac{21}{2} \cdot 74$$

$$S_{21} = 777$$

Truly, we can say that the formula for S_n lets us add without adding!

Example 16 Find the AP whose fifth term is 32 and whose eleventh term is 74.

Solution The phrase "find the AP" means we have to find the formula for a_n, which then determines the progression itself:

$$a_n = a_1 + (n-1) \cdot d$$

Thus we are looking for those values a_1 and d of the one AP with $a_5 = 32$ and $a_{11} = 74$. Now we know the values of

$$a_5 \text{ and } a_{11}$$

of any AP, which are

$$a_1 + (5-1) \cdot d \quad \text{and} \quad a_1 + (11-1) \cdot d,$$

respectively. Then the equations $a_5 = 32$ and $a_{11} = 74$ in our case become

$$a_1 + 4d = 32$$

and

$$a_1 + 10d = 74.$$

It is now a matter of solving the two equations in two unknowns whose solution is $a_1 = 4$ and $d = 7$ so that the final answer is

$$a_n = 4 + (n-1) \cdot 7.$$

If we wish to simplify the formula, we write

$$a_n = 7n - 3.$$

We could also symbolize the sequence by writing $\{7n-3\}$ or, as other authors sometimes do, $\langle 7n-3 \rangle$.

Geometric progressions

Another important type of sequence is a *geometric progression*, which is by definition a sequence in which each term after the first equals the one before it times a constant. As an example, the sequence

$$a = 2, 2/3, 2/9, 2/27, 2/81, \ldots$$

has 2 as its first term, a_1, and 1/3 as its constant multiplier, symbolized by r. Notice that the fourth term equals the third term times (1/3) $(2/27 = 2/9 \cdot 1/3)$. The letter r is used because the ratio of any two successive terms, $a_n \div a_{n-1}$, is always 1/3—for example, $2/27 \div 2/9 = 1/3$.

We will analyze geometric progressions in very much the same way we analyzed arithmetic progressions, by searching for formulas for a_n and S_n. We begin the search for a_n with the sequence mentioned in the last paragraph. We choose to rewrite the terms in such a fashion that the generating pattern reveals itself as shown in these equations.

$$\begin{aligned}
a_1 &= 2 \\
a_2 &= 2 \cdot (1/3) \\
a_3 &= [2 \cdot (1/3)] \cdot (1/3) = 2 \cdot (1/3)^2 \\
a_4 &= [2 \cdot (1/3)^2] \cdot (1/3) = 2 \cdot (1/3)^3 \\
a_5 &= [2 \cdot (1/3)^3] \cdot (1/3) = 2 \cdot (1/3)^4 \\
a_6 &= [2 \cdot (1/3)^4] \cdot (1/3) = 2 \cdot (1/3)^5
\end{aligned}$$

(Notice the pattern in the last four equations.)

$$\begin{aligned}
a_7 &= 2 \cdot (1/3)^6 \\
a_8 &= 2 \cdot (1/3)^7 \\
&\cdots\cdots \\
a_n &= 2 \cdot (1/3)^{n-1}
\end{aligned}$$

We can see that the same pattern holds true for a_1 and a_2 as well since

$$a_1 = 2 \cdot (1/3)^{1-1} = 2 \cdot (1/3)^0 = 2 \cdot 1 = 2$$

and

$$a_2 = 2 \cdot (1/3)^{2-1} = 2 \cdot (1/3)^1 = 2 \cdot (1/3).$$

We will see why $(1/3)^0 = 1$ in the next section.

Now that we have found a formula that lets us find the value of a_n for any n in this one particular GP, we wish to generalize to a formula for a_n in *any* GP. We look again at our particular GP,

$$a_n = 2 \cdot (1/3)^{n-1}.$$

If we realize that 2 plays the role of a_1 and 1/3 plays the role of r, then the formula for the nth term of any GP with a first term of a_1 and constant multiplier r is

$$a_n = a_1 \cdot r^{n-1}.$$

Example 17 Consider the geometric progression b with $b_1 = 2$ and $r = 6$ so that

$$b = \{2, 12, 72, 432, \ldots\}.$$

Then

$$b_4 = 2 \cdot 6^{4-1} = 2 \cdot 6^3 = 2 \cdot 216 = 432.$$

The final formula we wish to derive is that of S_n, the sum of the first n terms of any GP. We derive it by working with S_5 in the previous example and then generalizing to any GP.

$$\begin{aligned}
S_5 &= a_1 + a_2 + a_3 + a_4 + a_5 \\
S_5 &= 2 + 2 \cdot 6 + 2 \cdot 6^2 + 2 \cdot 6^3 + 2 \cdot 6^4
\end{aligned}$$

If we multiply both sides of the last equation by 6 and then subtract the resulting equation term by term we obtain

$$\begin{array}{rl} S_5 = & 2 + 2\cdot 6 + 2\cdot 6^2 + 2\cdot 6^3 + 2\cdot 6^4 \\ 6\cdot S_5 = & \phantom{2 +{}} 2\cdot 6 + 2\cdot 6^2 + 2\cdot 6^3 + 2\cdot 6^4 + 2\cdot 6^5 \\ \hline S_5 - 6\cdot S_5 = & 2 + \quad 0 + \quad 0 + \quad 0 + \quad 0 - 2\cdot 6^5 \end{array}$$

$$S_5\cdot 1 - S_5\cdot 6 = 2 - 2\cdot 6^5$$
$$S_5\cdot(1-6) = 2 - 2\cdot 6^5$$
$$S_5\cdot(1-6) = 2\cdot(1-6^5)$$
$$S_5 = \frac{2\cdot(1-6^5)}{1-6}$$

Since $6^5 = 7776$, the answer S_5 is

$$\frac{2\cdot(1-7776)}{-5} = \frac{2\cdot -7775}{-5} = 2\cdot 1555 = 3110.$$

If

$$S_5 = \frac{2\cdot(1-6^5)}{1-6}$$

then it seems reasonable to conclude for this sequence b that

$$S_7 = \frac{2\cdot(1-6^7)}{1-6},$$

$$S_{12} = \frac{2\cdot(1-6^{12})}{1-6}, \quad \text{and finally}$$

$$S_n = \frac{2\cdot(1-6^n)}{1-6}.$$

To generalize, we realize that 2 plays the role of a_1 and 6 plays the role of r so that the formula for the sum of the first n terms of any GP with a first term of a_1 and constant multiplier r is

$$S_n = \frac{a_1\cdot(1-r^n)}{1-r} \qquad (r \neq 1).$$

If $r = 1$, as in $\{3, 3, 3, 3, 3, \ldots\}$, then S_n is n 3's added together or $n\cdot 3 = 3n$. Note here that the formula for S_n lets us add without adding. Sometimes the numerator and denominator are multiplied by -1 so that the resulting equivalent formula is

$$S_n = \frac{a_1\cdot(r^n-1)}{r-1} \qquad (r \neq 1).$$

We will use this equivalent formula in the next chapter.

This concludes the section on sequences. We shall see them again in the next chapter on the mathematics of finance. With reference to our diagram in the first chapter about problem solving, what we have done so far is to work solely within the upper half of the rectangle, in the world of ideas. We now know the mathematics needed to solve certain kinds of problems which will arise from the world of applications. Keep in mind, too, that we have made quite hasty generalizations from specific cases to the formulas without any formal proof. Such reasoning can lead to serious errors in mathematics, but this text will point out those very few occasions when this is the case in our areas of study. We also acknowledge our indebtedness to other mathematicians. Their work with the rather difficult proofs of the formulas has made our generalizations valid ones.

Problem set 3–4

1. Find the next three terms in these sequences.
   a. $2, 5, 8, 11, 14, \ldots$
   b. $15, 9, 3, -3, -9, \ldots$
   c. $-6, -1, 3, 6, 8, \ldots$
   d. $5/6, 8/9, 11/12, 14/15, \ldots$
   e. the sequence with $a_1 = 2$ and $a_n = a_{n-1} + 3$
   f. the sequence with $a_1 = 3$ and $a_n = 4 \cdot a_{n-1}$
   g. $5, 5/2, 5/4, 5/8, \ldots$
   h. $1/8, 3/8, 9/8, 27/8, \ldots$
2. Write out the first seven terms of the GP with $a_1 = 2$ and $r = 6$. Find their sum. Then evaluate the expression

$$2 \cdot (1 - 6^7)/(1 - 6)$$

   and compare your answers.
3. Find a_n for these progressions.
   a. $n = 5,\ a = -6, -4, -2, \ldots$
   b. $n = 13,\ a = 5, 12, 19, 26, \ldots$
   c. $n = 10,\ a = 1/6, 1/3, 1/2, 2/3, \ldots$
   d. $n = 7,\ a = 1/5, 2/5, 4/5, 8/5, \ldots$
   e. $n = 8,\ a = 243, 81, 27, 9, \ldots$
   f. $n = 6,\ a = 50, 10, 2, \ldots$
4. Find d in the AP such that $a_1 = -5$ and $a_{11} = 65$.
5. Find S_n for these progressions.
   a. $n = 14,\ a = 7, 12, 17, 22, \ldots$
   b. $n = 9,\ a = 3, 6, 12, 24, \ldots$
   c. $n = 23,\ a = -17, -11, -5, 1, \ldots$
   d. $n = 7,\ a = 1/16, 1/4, 1, 4, \ldots$
6. The 3rd and 20th term of an AP are 15 and 117, respectively. Find a_n and a_7.
7. Find a formula for a_n if an AP is involved with $a_{15} = 37$ and $a_{20} = 117$.

   In the next four problems involving AP's three of the numbers a_1, a_n, d, n, and S_n are given. Find the remaining two.
8. $a_1 = 3,\ d = 6,\ n = 15$
9. $a_1 = -2,\ n = 8,\ S_n = 68$
10. $n = 6,\ a_n = -10,\ S_n = -15$
11. $d = 3,\ n = 15,\ a_n = 31$

    In the next four problems involving GP's three of the numbers a_1, a_n, r, n, and S_n are given. Find the remaining two.

12. $a_1 = 24,\ r = 1/2,\ n = 10$
13. $r = -5,\ n = 3,\ a_n = 1/5$
14. $a_1 = 9,\ n = 2,\ S_n = 12$
15. $a_1 = 2/3,\ r = -2/3,\ n = 4$
16. Find the formula for a_n in a GP with $a_3 = 45$ and $a_7 = 3{,}645$.
17. Find the formula for a_n in a GP with $a_2 = 40$ and $a_5 = 2{,}560$.
18. Alice Smith borrows $1,200 now and will pay it back with 12 end-of-month payments of $100 each plus monthly interest at the rate of 1/2% of the unpaid balance. What is the total amount of interest she pays? (Hint: make a table showing her payments for the first few months and note the pattern of results.)
19. Find the sum of the first 50 odd positive integers.
20. Find the sum of the first 100 positive integers.
21. Karl Brown agrees to pay off a $15,000 mortgage on his house by paying $120 toward the principal at the end of each month plus interest on the unpaid balance. The interest is charged at the rate of 6% a year (1/2% a month).
    a. Find a formula that gives his total monthly payment.
    b. How many payments does he make?
    c. How much interest does he pay altogether?
    d. How much does he pay altogether to the lender?

    (Hint: make a table showing his payments of interest and principal for the first few months and note the pattern of results.)
22. Ajax, Inc., loses $1,500 on its first wrench (initial plant costs) and then obtains a profit of 30¢ on all subsequent wrenches. How many wrenches will Ajax have to make before it has broken even?
23. How many times a day (from 12:00:00 A.M. to 11:59:59 P.M.) does a grandfather clock strike if it strikes only the hours?
24. The number of bacteria in a culture triples every four hours. If there are B bacteria present now, how many will there be in twenty-four hours?
25. Derive the formula for the sum of the first five terms of any GP by using a generalization of the method we used for $n = 5$ and $a_n = 2 \cdot 6^{n-1}$. Can you derive the formula for the sum of the first n terms of any GP in a similar way?
26. Give the algebraic reasons for each step in the derivation of S_5 in Example 16.
27. Find the AP with $a_1 = 8$, $a_6 = 38$, and $S_3 = 34$. Check your formula by making sure all three of these specific equations hold true. What is your conclusion?

Exponential functions

We now turn our attention to the type of function which includes the concept of exponents. Functions involving exponents have many widespread and important applications in psychological learning theory, growth of money, radioactive decay, and, as we shall see in the next section, population growth.

A review of some elementary algebra is in order before we can study exponential functions themselves. We are familiar with symbols of the form b^m where b is any real number and where m is a positive integer:

$$b^m = b \cdot b \cdot b \cdot \cdots \cdot b \quad (m\ b\text{'s multiplied together}).$$

As we recall, 3^4 means $3 \cdot 3 \cdot 3 \cdot 3$, which yields 81 after multiplication. The term 3^4 is a *power*, 3 is its *base*, and 4 is its *exponent*. Multiplication of powers with identical bases is accomplished by adding exponents:

$$3^4 \cdot 3^2 = 3^{4+2}\ (=3^6)$$

We can verify this equation by placing it under the magnifying glass of a step-by-step algebraic analysis.

$$\begin{aligned} 3^4 \cdot 3^2 &= (3 \cdot 3 \cdot 3 \cdot 3) \cdot (3 \cdot 3) && \text{definition of a power} \\ &= 3 \cdot 3 \cdot 3 \cdot 3 \cdot 3 \cdot 3 && \text{removing parentheses} \\ &= 3^6 && \text{definition of a power} \\ &= 3^{4+2} && \text{addition fact} \end{aligned}$$

In general, $b^m \cdot b^n = b^{m+n}$.

What about b^0, where b itself is different from 0? What is the value, for example, of 5^0? We are immediately tempted to say 0, since 5^0 appears to mean 0 5's multiplied together which would be 0. However, we want to look at this situation from another point of view. We want to be able to multiply 5^0 once it is evaluated, and we insist on multiplying this new power with a zero exponent in exactly the same way that we multiplied old powers with positive exponents. This last sentence (be sure to note the tenses of the verbs) illustrates a most important idea of mathematics—the Preservation of Principle idea, which essentially states that we want as few laws as possible to hold true in as many places as possible. Thus, while we do not know the value of 5^0 we do know how we want to be able to multiply with it.

$$\begin{aligned} 5^2 \cdot 5^0 &= 5^{2+0} && \text{Preservation of Principle} \\ 5^2 \cdot 5^0 &= 5^2 && \text{addition fact} \\ \frac{5^2 \cdot 5^0}{5^2} &= \frac{5^2}{5^2} && \text{division of both sides of the equation by } 5^2 \\ \frac{25 \cdot 5^0}{25} &= \frac{25}{25} && \text{definition of a power} \\ 5^0 &= 1 && \text{result of division} \end{aligned}$$

Generalizing, if b is any nonzero real number, we must define

$$b^0 = 1$$

if we want to preserve the law of multiplication of powers.

What about 5^{-2}—what is its value? The original definition of a power does not give the symbol a meaning, since it makes no sense at all to discuss the product of -2 fives. We rely on the Preservation of Principle idea to tell us how to make a useful definition of 5^{-2}, as illustrated by the next series of equations and their justifications.

$5^2 \cdot 5^{-2} + 5^{2+-2}$	Preservation of Principle: we want to multiply powers with negative exponents in exactly the same way that we multiplied powers with nonnegative exponents
$5^2 \cdot 5^{-2} = 5^0$	addition fact
$\dfrac{5^2 \cdot 5^{-2}}{5^2} = \dfrac{1}{5^2}$	division by 5^2 and previous work with 5^0
$5^{-2} = \dfrac{1}{5^2}$	cancellation

We could go on to show that we should define $5^{-3} = 1/5^3$, $5^{-4} = 1/5^4$, $5^{-7} = 1/5^7$, etc., in much the same manner. In general, if b is any positive real number and m is any integer, then

$$b^{-m} = \frac{1}{b^m}$$

if we want to preserve our original law of multiplication of powers. Also, b^m is always a positive number. For example, $5^{-1000} = 1/5^{1000}$, which is almost 0 but still positive.

Fractional exponents

We will not go into a detailed explanation of roots, which is needed for a full development of powers with fractional exponents, but we will have a general definition and then several examples to refresh our memories. In the equations b is a positive real number, p is an integer, and q is a positive integer.

$$b^{p/q} = \sqrt[q]{b^p}$$

or equivalently

$$b^{p/q} = (\sqrt[q]{b})^p,$$

provided that each expression represents a real number. We use the first or second equation depending upon ease of computation. The second equation is generally the better one to start with. Logarithms, which we have not studied, will always let us evaluate $b^{p/q}$ where the above two equations do not give us a "nice" answer.

$$\begin{aligned}
9^{1/2} &= (\sqrt[2]{9})^1 = \sqrt{9} = 3\\
8^{2/3} &= (\sqrt[3]{8})^2 = (2)^2 = 4\\
25^{3/2} &= (\sqrt[2]{25})^3 = (5)^3 = 125\\
64^{-5/6} &= (\sqrt[6]{64})^{-5} = (2)^{-5} = 1/2^5 = 1/32\\
49^{-3/2} &= (\sqrt[2]{49})^{-3} = (7)^{-3} = 1/7^3 = 1/343\\
(-49)^{1/2} &= (\sqrt[2]{-49})^1 \text{ is not a real number. In general, even roots of negative numbers are never real numbers.}
\end{aligned}$$

We are almost ready to discuss b^x for $b > 0$ and x any real number, but we still lack a knowledge of irrational exponents such as $\sqrt{2}$ and π. We will assume that $4^{\sqrt{2}}$ and 4^{π} behave in a "natural" way. That is, since

$$1.4 < \sqrt{2} < 1.5 \quad \text{and} \quad 3.14 < \pi < 3.15$$

then

$$4^{1.4} < 4^{\sqrt{2}} < 4^{1.5} \quad \text{and} \quad 4^{3.14} < 4^{\pi} < 4^{3.15}.$$

In other words, $4^{\sqrt{2}}$ is close to $4^{1.4}$ because $\sqrt{2}$ is close to 1.4. A detailed explanation of irrational exponents belongs in more advanced mathematics. Now we are ready for the definition of an exponential function.

Definition If b is any positive real number and x is any real number, then the equation

$$y = f(x) = b^x$$

defines an *exponential function* with base b.

We use the functions defined by $y = 2^x$ and $y = (1/5)^x$ as examples to investigate the properties of exponential functions in general. To graph the function defined by

$$y = f(x) = 2^x$$

we make a table as illustrated below, plot the corresponding points, and join them with a smooth curve. Be sure to check the calculations of y. The fifth table entry states that

$$\sqrt[2]{2} \doteq 1.4.$$

The symbol $\doteq$ means "is approximately equal to." Since $\sqrt[2]{2}$ is not exactly equal to any finite decimal number, the 1.4 says that the nearest approximation ending in tenths is 1.4.

Table 3.13

x	y
−3	1/8
−2	1/4
−1	1/2
0	1
.5	$\sqrt[2]{2} \doteq 1.4$
1	2
1.5	$\sqrt[2]{8} \doteq 2.8$
2	4
3	8

Figure 3.10

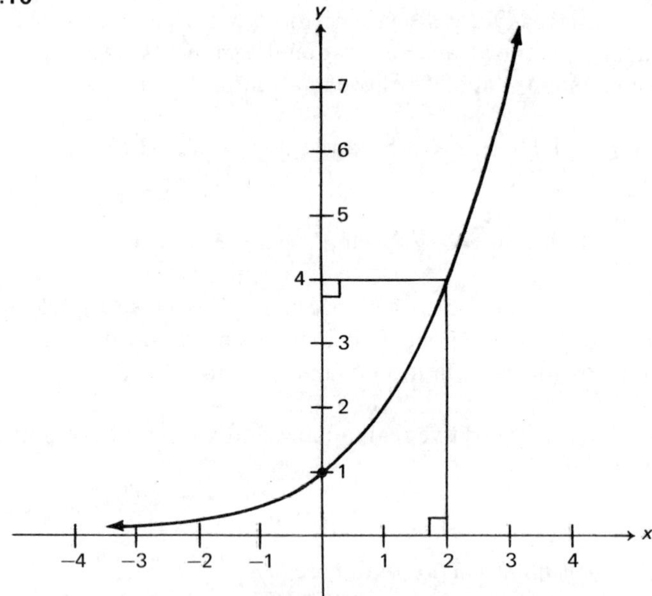

We find the graph of the function $y = (1/5)^x$ by proceeding in a similar fashion. Check the calculations here.

Table 3.14

x	y
−3	125
−2	25
−1	5
0	1
1	1/5
2	.04
3	.008

Figure 3.11

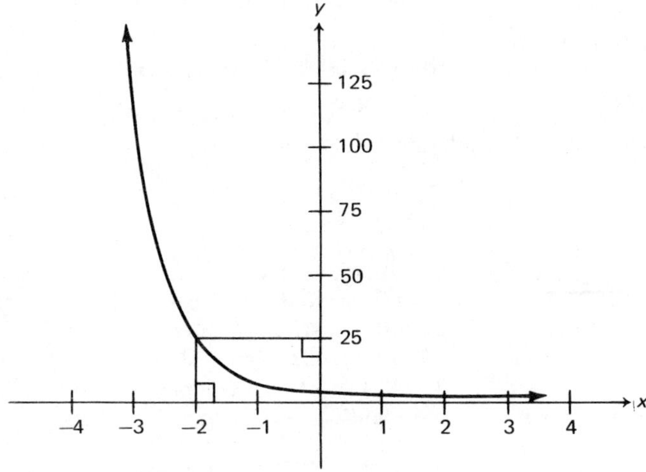

Notice that we have different scales on the axes.

The graph of $y = (1/5)^x$ is a decreasing curve, and that of $y = 2^x$ is an increasing curve. We could generalize to prove the statements in this table for the graph of $y = b^x$, $b > 0$. The base of the exponential function is b, y is a range element corresponding to a domain element x, and r and s are specific elements of the domain with corresponding range elements b^r and b^s.

Table 3.15

If $r < s$ and	then
$b < 1$	$b^r > b^s$ and the graph is decreasing
$b = 1$	$b^r = b^s$ and the graph is parallel to the x-axis
$b > 1$	$b^r < b^s$ and the graph is increasing

The irrational number *e*

A very important number found in theoretical and applied mathematics is the irrational number e (named for Leonhard Euler, the Swiss mathematician who discovered it), which is approximately equal to 2.718. An example of its use is in the general learning curve

$$y = k - a \cdot e^{-cx}$$

with positive constants k, a, and c. It shows the improvement in a skill (y) as a function of the independent variable time (x).

Example 18 A typing teacher finds that the relationship between words typed per minute (y) and the number of weeks in class (x, $0 \leqslant x \leqslant 10$) for beginning students is

$$y = 50 - 50 \cdot e^{-1x} = 50 \cdot (1 - e^{-x}).$$

Table I in the appendix gives the values of e^x and e^{-x} for various values of x. Then making a table of values and plotting them gives us

Table 3.16

x	$50 \cdot (1 - e^{-x})$	
0	$50 \cdot (1-1) = 0$	
1	$50 \cdot (1-.368) = 31.6$	
2	$50 \cdot (1-.135) = 43.25$	
3	$50 \cdot (1-.050) = 47.5$	
4	49.1	Double
5	49.65	check
6	49.85	these
7	49.95	calculations
8	$50 \cdot (1-.000+) \doteq 50$	
9	$\doteq 50$	
10	$\doteq 50$	

Figure 3.12

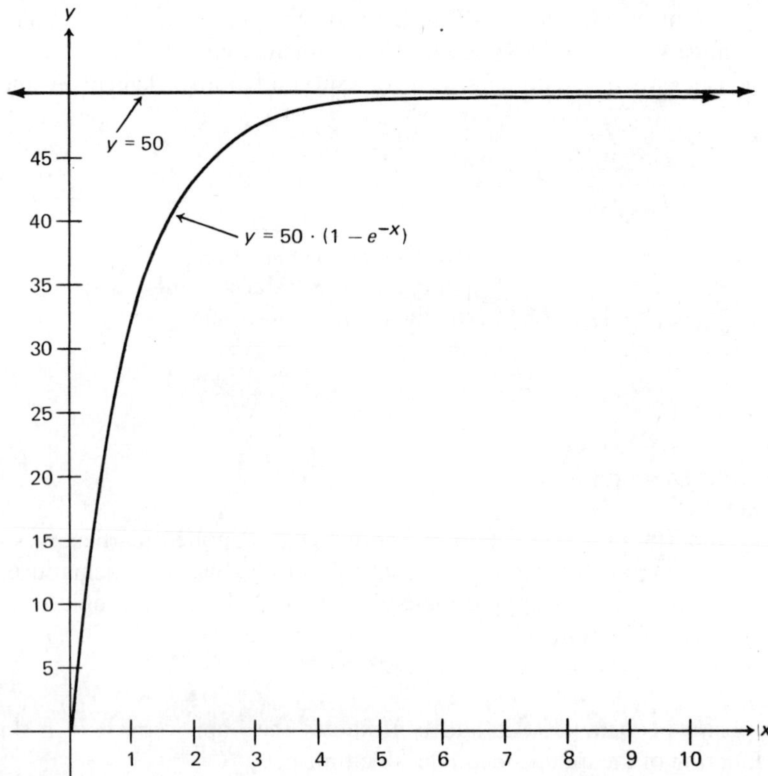

As our intuition might suggest, a student improves rapidly at first and then his weekly rate of improvement tapers off. His typing ability always increases each week. The line $y = 50$, which the graph never really intersects, is an example of a *horizontal asymptote.*

Problem set 3–5

1. Find the value of these numbers.
   a. 3^2 b. 3^{-2} c. -3^2
   d. $(-3)^2$ e. -3^{-2} f. $(-3)^{-2}$
   g. $\dfrac{3}{5^{-2}}$ h. $\left(\dfrac{3}{5}\right)^{-2}$ i. $\left(\dfrac{5}{3}\right)^{2}$
   j. $32^{-3/5}$ k. $.04^{-1/2}$ l. $1024^{.3}$
   m. $16^{3/2}$ n. $64^{4/3}$ o. $32^{4/5}$
2. We have insisted all along that the base of an exponential function must be a positive number. Evaluate $(-8)^{1/3}$ and $(-8)^{2/6}$. By comparing results we have an example of the difficulties that can arise with using a negative number as the base of an exponential function.
3. Which is larger, $7x^0$ or $(7x)^0$?
4. Which is larger, $-(-4^{-2})$ or 4^{-2}?
5. For each point below find the base, b, of the exponential function $y = b^x$ whose graph passes through the given point.
   a. $(2, 9)$ b. $(4, 81)$ c. $(-2, .01)$
   d. $(2, 13)$ e. $(0, 1)$ f. $(0, 5)$
6. Use Figure 3.10 to carry out the following instructions.
   a. Evaluate $2^{3/4}$. b. Evaluate $2^{-1/3}$.
   c. Evaluate $\sqrt[5]{2}$. d. Solve $2^x = 6$.
   e. Solve $2^x = 1/4$. f. Solve $2^x = -1/4$.
7. The Quotient Law of Powers states that if b^m is divided by b^n, then the quotient is b^{m-n}. If b^0 were defined to be 0, then show that this definition would give a result that contradicts this law.
8. Construct the graphs of these functions for the domain $[-4, 4]$.
   a. $y = 1.5^x$ b. $y = 1.5^{-x}$ c. $y = (1/3)^x$
   d. $y = (1/3)^{-x}$ e. $y = 1.5^{2x}$ f. $y = (1/3)^{2x}$
   g. $y = -2^x$ (Hint: use Table 3.13 to help you plot points.)
9. Suppose that the percentage of television viewers of a quiz show who remember the emcee's name is given by

$$y = f(x) = 100 - 90 \cdot e^{-.3x}$$

   where x is the number of times they have viewed the program. Use Table I in the appendix to calculate the percentages corresponding to
   a. $x = 1$. b. $x = 2$. c. $x = 3$.
   d. $x = 4$. e. $x = 5$. f. $x = 7$.
   g. $x = 10$. h. $x = 13$. i. $x = 16$.
   j. $x = 19$. k. Graph the results of a through j.

10. Assume that a book publisher's annual profit from the sales of a certain book is given by the equation

$$y = f(x) = 7{,}000 + 10{,}000 \cdot e^{-.4x}$$

where y is the annual profit in dollars and x is the number of years the text is on the market. Find the annual profit at the end of one, two, three, four, five, seven, and ten years. Graph the results.

11. A balloon full of 4,000 cubic feet of helium gas is leaking so that at the end of any minute the volume is only 90% of what it was at the beginning of that minute. If the leak starts now, find a formula that relates the volume of gas, V, to t, the number of minutes in the future. Work out the specific results for the first few minutes and then note the pattern which should occur in the results.

12. The number of ants in a colony continually increases by 4% every week. If there are two hundred ants now, then how many will there be one week from now, two weeks from now, three weeks from now, four weeks from now, nine weeks from now, t weeks from now? Hint: think of an increase of 4% of an amount A as resulting in an ending quantity of 104% of A or $(1.04) \cdot A$ or $A \cdot (1.04)$.

13. The *half-life* of a decaying substance is the amount of time necessary for one-half of any amount of it to decay. If there are G grams of Radium 214 present now, and if the substance loses 20% of itself every minute, then carefully draw a graph for $0 \leqslant t \leqslant 4$ minutes. Using the graph, estimate the half-life of Radium 214.

14. Assume that the market research department of a frozen food company estimates that y, the monthly demand (in hundreds of packages) for frozen lobsters, is a function of x, the number of months the product is advertised, according to the formula

$$y = f(x) = 80 - 70 \cdot e^{-.8x}.$$

Find the predicted monthly sales after zero, one, two, three, four, five, seven, nine, and twelve months of the advertising campaign. Graph the results.

15. Show, by using the same method of the text, that we should define 6^0 as 1 and 6^{-3} as $1/6^3$.

Population growth

One of the more interesting and important applications of exponential functions is the study of population growth.

Example 19 Consider problem 12 of the last problem section.

Solution We worked with an ant population initially equal to two hundred and continually increasing at a constant rate of 4% each week. Our answer to the problem should have been

$$P = 200 \cdot (1 + .04)^n$$

where P is the final population and n is the number of weeks the colony grew. If the derivation of this formula is not clear, then let us analyze it on a week-by-week basis.

At the end of the first week the population increases by 4% of 200, or 8 ants, so that the amount at the end of the first week, which equals the amount at the beginning of the second week, is 208 ants. We recall that if the population of 200 ants increases by 4% then the new number of ants is also equal to

$$104\% \text{ of } 200 = 1.04 \cdot 200 = (1+.04) \cdot 200$$

or

$$[200 \cdot (1+.04)].$$

Then if the second week's initial amount of

$$[200 \cdot (1+.04)]$$

increases by 4% or .04, its terminal amount is

$$[200 \cdot (1+.04)] \cdot (1+.04)$$

or

$$[200 \cdot (1+.04)^2].$$

Notice that *all 208* ants increase by 4% during the second week instead of just the original 200 ants. This expression is also the number of ants present at the beginning of the third week; if it again increases by 4% then the number present at the end of the third week is

$$[200 \cdot (1+.04)^2] \cdot (1+.04)$$

or

$$200 \cdot (1+.04)^3.$$

We organize our results so far into a table to look for a pattern.

Table 3.17

End of week number	Number of ants present, P
1	$200 \cdot (1+.04)^1$
2	$200 \cdot (1+.04)^2$
3	$200 \cdot (1+.04)^3$
⋮	⋮
n	$200 \cdot (1+.04)^n$

The last pair of table entries is the generalization of the specific results from the first three pairs.

We wish to generalize to a formula which helps us calculate the value of the dependent variable P which results from the interaction of P_0 ("P sub zero"), the initial number in any specified population; r, the constant rate of increase per given time period, usually years; and n, the number of such time periods. It should be clear that in our specific formula,

$$P = 200 \cdot (1+.04)^n.$$

We know that 200 is the initial number P_0 in the population, .04 or

4% is the constant rate r of increase per week, and n is the number of weeks. The generalized formula then becomes

$$P = P_0 \cdot (1+r)^n.$$

Table V in the appendix gives the values of $(1+r)^n$ for various values of n and r. For example, in 10 weeks the number of ants present is

$$200 \cdot (1+.04)^{10} = 200 \cdot 1.480,$$

which is 296 after rounding to the nearest whole number. Notice that P is a function of P_0, r, and n according to the equation

$$P = f(P_0, r, n) = P_0 \cdot (1+r)^n.$$

Sometimes we are asked to calculate the average annual rate of increase, r, on the basis of given values of P, P_0, and n. It is then necessary to solve the equation

$$P = P_0 \cdot (1+r)^n$$

for r. This solution is carried out as follows.

$$\frac{P}{P_0} = (1+r)^n \qquad \text{Divide both sides by } P_0.$$

$$\sqrt[n]{\frac{P}{P_0}} = (1+r) \qquad \text{Take the } n\text{th root of both sides.}$$

$$\sqrt[n]{\frac{P}{P_0}} - 1 = r \qquad \text{Subtract 1 from both sides.}$$

(A) $$r = \sqrt[n]{\frac{P}{P_0}} - 1 \qquad \text{Reverse the previous equation.}$$

In the application of equation (A) it is often necessary to use logarithms to calculate r. Tables II and III in the appendix are tables of common logarithms. As an example, if $P_0 = 2{,}000$, $P = 3{,}800$, and $n = 20$, then $r = .0326$ or 3.26% (Check this answer if you have studied logarithms.).

We are also sometimes asked to find the doubling time of a population with P_0 members initially present, that is, how long it takes P_0 to grow to $2 \cdot P_0$. We solve the appropriate equation

$$2 \cdot P_0 = P_0 \cdot (1+r)^n$$

for n.

$$2 = (1+r)^n \qquad \text{Divide both sides by } P_0.$$

$$(1+r)^n = 2 \qquad \text{Reverse the previous equation.}$$

$$\log_{10}(1+r)^n = \log_{10} 2 \qquad \text{Take the } \log_{10} \text{ of both sides.}$$

$$n \cdot \log_{10}(1+r) = \log_{10} 2 \qquad \text{Power law of logarithms.}$$

$$n = \frac{\log_{10} 2}{\log_{10}(1+r)} \qquad \text{Divide by } \log_{10}(1+r).$$

(B) $$n = \frac{.3010}{\log_{10}(1+r)} \qquad \text{Log}_{10} 2 = .3010$$

For example, it would take

$$n = \frac{.3010}{\log_{10}(1+.04)}$$

$$= \frac{.3010}{.0170}$$

$$= 17.7$$

or eighteen weeks to double the ant population described in Example 19.

Solving demographic problems

Now that we have worked within the upper half of our problem-solving diagram by learning the necessary mathematics, we apply it to *demography.* Demography is the study of human population growth. The next two figures appear by permission of the magazine *Scientific American,* which published Kingsley Davis' article "Population" in its September 1963 issue. The first figure shows the world's population, including 1960 to 2000 projections, as a function of time.

Figure 3.13

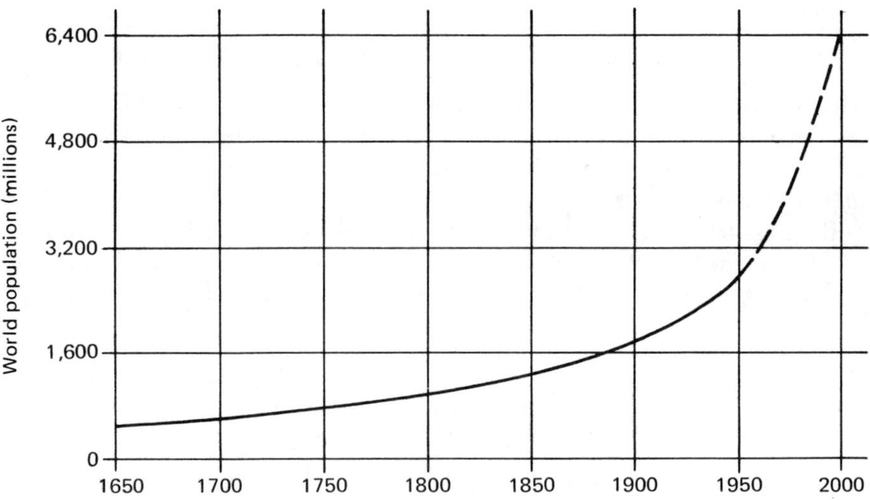

There is no one equation of the form

$$P = P_0 \cdot (1+r)^n$$

with the curve as its graph since, as the next table shows, the population has not grown at a constant rate.

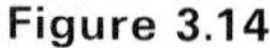

Figure 3.14

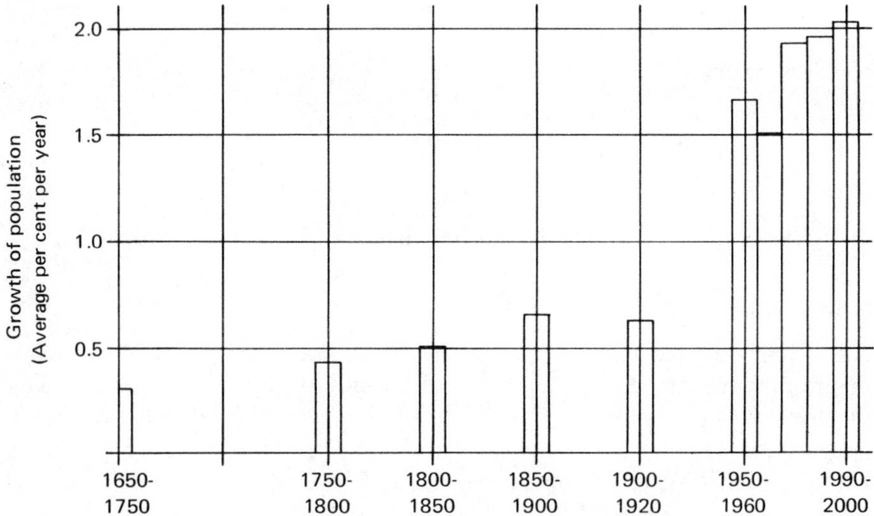

Notice that the rate of increase itself also is increasing with the passage of time.

$$\text{net annual rate of increase} = (\text{annual birth rate}) - (\text{annual death rate}).$$

The world has generally had a high and constant birth rate, but with improvements in medicine and health services the death rate, especially in infancy, has dramatically decreased to account for the increase in the net annual rate of increase.

If we assume that a present world growth rate of 1.7%—perhaps a slightly low figure—will hold in the future, then applying the rate to our population function and using a 1964 base figure of 3.22 billion people yields Table 3.18.

Table 3.18

Year	Population (billions)	People per square mile
1964	3.22	61.4
1975	3.88	74.0
2000	5.94	113.3
2025	9.08	178.3
2050	13.89	265.0
2075	21.25	405.3
2100	32.51	620.1
2200	177.93	3,394.0
2300	973.99	18,578.8
2400	5,330.39	101,677.2

> This last figure may be compared with certain current population densities. In 1960 Manhattan Island (New York County) had a density of 77,194 persons per square mile, the highest for any political unit in the United States. For the city of New York the population density then was 24,697, and for New York City and its surrounding suburbs (the New York urbanized area) it was 7,462. By comparison, the population density in the well-populated state of Massachusetts in 1960 was only 650 per square mile, and in the United States as a whole, a low 50.5.[1]

The figure for the year 2400 becomes even more dramatic when we realize that with 3,097,600 square feet in a square mile and 101,677.2 persons there, each person will have 30.46 square feet to himself or a square plot of earth (which could include the Sahara desert, South American jungle, Siberian wasteland, or other nonarable location) about 5 1/2 feet on a side.

If the annual growth rate of the world continues at its present 1.88% annual rate, the population increase would be even more marked; by 2025, for example, there would be 10.03 billion instead of the 9.08 billion based on a 1.70% growth rate. It is useful to inquire about the doubling time of the world's population, and we do this by using our previous equation (B),

$$n = \frac{.3010}{\log_{10}(1+r)},$$

which upon substitution yields

$$n = \frac{.3010}{\log_{10}(1+.0188)}$$

$$n = \frac{.3010}{.008086}$$

$$n = 37.2 \text{ years.}$$

We can conclude from this number that by the time some of us are still alive the world might have four times as many people as there were when we were born.

Analyzing population trends

So far we have seen the ominous effects of the present growth curve of the world's population. We now turn our attention to the United States where the picture is somewhat brighter. We use Bureau of the Census figures to analyze recent United States population trends.

We see that the United States' growth rate is declining in contrast to the increasing growth rate of the world. One very likely explanation is the widespread use of the oral contraceptive pill which began around 1961.

It is one thing to note that the present American population growth rate is 1.0% per year, but another to see how this figure is derived from the

[1] David M. Heer, *Society and Population* © 1968. Reprinted by permission of Prentice-Hall, Inc., Englewood Cliffs, New Jersey.

Table 3.19

Years	Average annual rate of change
1930–1939	0.8%
1940–1944	1.1%
1945–1949	1.6%
1950–1959	1.7%
1960–1961	1.6%
1962–1962	1.5%
1963–1965	1.3%
1966–1966	1.1%
1967–1971	1.0%

number of children in a family. For example, if every thousand American couples has 2,360 children, then we wish to convert this number into a specific value of r, the resulting annual rate of increase of the United States population.

In order to seek out this conversion formula we must first understand what a *birth cohort* is.

> A birth cohort is a group of women born in the 12-month period centering on January 1 of the year by which the group is identified. For example, the cohort of 1900 was born between July 1, 1899, and June 30, 1900. The reason for defining birth cohorts on the basis of 12-month periods centering on January 1 rather than on the basis of conventional calendar years is that it makes more convenient the analysis of fertility trends by age and calendar year.[2]

We turn again to Bureau of the Census figures for the results of population prediction based on analysis of cohort fertility. Cohort fertility here means the number of children born to each birth cohort of 1,000 in its lifetime, and the Bureau assumed that the fertility rate is the same for all cohorts in the population for a given series. The figures are summarized under the Bureau's heading.

> In millions. As of July 1. Includes Armed Forces abroad. Projections are consistent with the April 1, 1970 census. These sets of projections were computed using the "cohort-component" technique and assume that annual net immigration will be 400,000 and that completed fertility of all women (i.e., the average number of children per 1,000 women at end of childbearing) will move gradually toward the following levels: Series B, 3,100; Series C, 2,775; Series D, 2,450; and Series E, 2,110.[3]

The estimated United States population on 1 July 1970 was 205 million.[4]

[2] Clyde V. Kiser, Wilson H. Grabill, and Arthur A. Campbell, *Trends and Variations in Fertility in the United States* (Cambridge, Massachusetts: Harvard University Press, 1968), p. 293.
[3] U.S. Bureau of the Census, *Statistical Abstract of the United States* (1972), Tables 7 and 6.
[4] Ibid.

Table 3.20

Date	Series B-3,100	C-2,775	D-2,450	E-2,110
1975	218	217	217	216
1980	237	234	231	228
1985	258	252	246	240
1990	279	270	261	251
1995	299	287	274	261
2000	322	305	288	271
2005	350	327	304	281
2010	381	350	320	291
2015	413	373	336	300
2020	447	397	351	307

Source: U.S. Bureau of the Census, *Statistical Abstract of the United States* (1972), Tables 7 and 6.

We can convert the numerical changes from 2015 to 2020 to average annual growth rates, as shown in the next table with the Series letters in reversed order. We investigate the changes from 2015 to 2020 to see what the average annual growth rates would eventually become if all women who have not yet borne children would so according to the given Series rates.

Table 3.21

Series and cohort birth rate per thousand, x	Average annual growth rate, r
E—2,110	0.46%
D—2,450	0.88%
C—2,775	1.25%
B—3,100	1.59%

These average annual growth rates were calculated by using the previous formula (Equation (A)) that we developed for r. For example, if each group of 1,000 women has 2,775 children (Series C), then the population will increase from $P_0 = 373{,}000{,}000$ to $P = 397{,}000{,}000$ in the $n = 5$ years from 2015 to 2020. According to equation (A),

$$r = \sqrt[5]{\frac{397{,}000{,}000}{373{,}000{,}000}} - 1$$

$$r = 1.25\%.$$

The relationship between x, the cohort birth rate per thousand, and r, the average annual growth rate, becomes more meaningful if we construct a graph from the entries of Table 3.21.

Figure 3.15

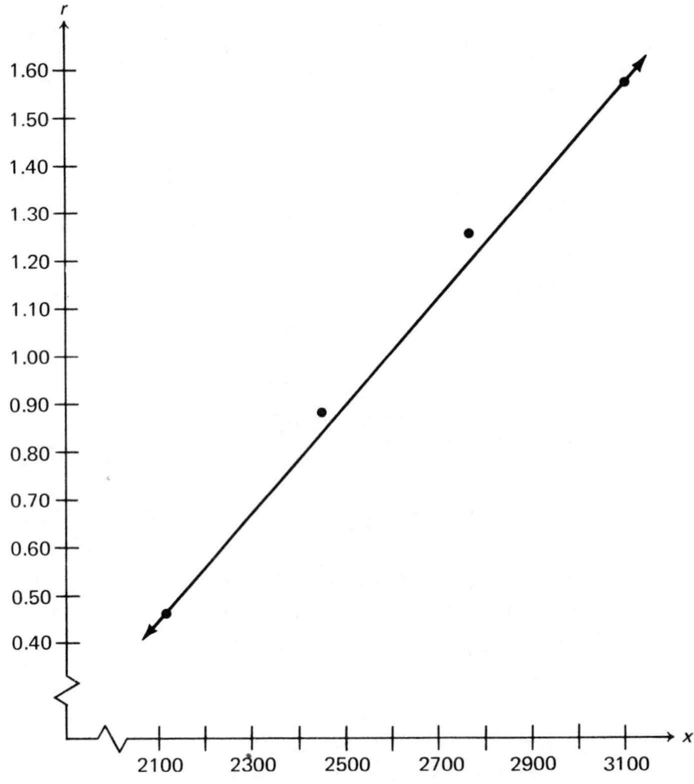

We connect the two extreme dots by a line even though all four of them do not exactly lie on the line. We now search for the equation of this line, which relates r to x, by first noting that the line passes through (2110, 0.46) and (3100, 1.59) so that its slope is

$$\begin{aligned} m &= \frac{r_2 - r_1}{x_2 - x_1} \\ &= \frac{1.59 - 0.46}{3100 - 2110} \\ &= .00114. \end{aligned}$$

Since one of the points (x_1, r_1) on the line is (3100, 1.59), then by the results of Problem 15 of Problem set 3–3 its equation is

$$r - 1.59 = .00114 \cdot (x - 3100)$$

or

$$r = .00114x - 1.944 \quad \text{(right?).}$$

Now we can directly see the effects of varying the cohort fertility rate. For example, if each birth cohort of 1,000 has $x = 2{,}360$ children, then the average annual population growth rate, r, will approach approximately

$$r = .00114 \cdot 2360 - 1.944$$

or

$$r = 0.75\%,$$

which means that the population will double itself in about

$$n = \frac{.3010}{\log_{10}(1+0.75\%)}$$

$$n = \frac{.3010}{.003245} \qquad \text{(from Table A.3)}$$

$$n = 93 \text{ years.}$$

This section is just the beginning of the mathematics of demography. The output of projected population from the input of raw births and deaths, age distributions, economic conditions, family styles, availability of birth control methods, etc., emerges after much analysis and work with the data. Still, in a world with finite resources and land area the equation expressed by the Zero Population Growth (ZPG) organization of

$$2 + 2 = 0$$

(two adults plus two children equals zero population growth) is certainly to be considered.

Problem set 3–6

1. If the population of the world was 3.55 billion in 1969[5] and if it increases by 1.9% a year, then find the population in
   a. 1970 b. 1980 c. 1990
   d. 2000 e. 2030 f. 2060
   Table IV in the appendix contains the values of $(1.019)^n = (1+1.9\%)^n$.
2. Draw the graph of
   $$P = (3 \text{ billion}) \cdot (1+1.9\%)^n$$
   for $0 \leqslant n \leqslant 100$. The population of the world was 3 billion in 1960.
3. If the population of the United States was 200 million in 1968 then find its population in 2025 and 2050 if the average annual growth rate remains constant at
   a. 1% b. 1.1% c. 1.3%
   d. 1.7% e. 2% f. 2.5%

[5] *United Nations Demographic Yearbook*, 1969 (New York: United Nations, 1970).

This table of values of $(1+r)^n$ is necessary for Problem 3.

Time	Average annual growth rate 1.0%	1.1%	1.3%	1.7%	2.0%	2.5%
57	1.7632	1.8656	2.0880	2.6139	3.0917	4.0856
82	2.2612	2.4524	2.8838	3.9839	5.0724	7.5746

4. If every cohort has x children, then what is the resulting annual net rate of change in the current American population for these values of x?
   a. 2300 b. 2400 c. 2500
   d. 2800 e. 3100 f. 4000
   g. In order for the population growth rate to approach .8% a year in America, what should the cohort fertility rate be?
   h. to approach 1.1%? i. to approach 1.7%?
   j. to approach 2.0%? k. to approach 2.5%?
5. If there are 300 ants in a colony now and 500 ants two weeks later, then find the average weekly rate of growth.

Problems 6, 7, 8, and 9 require a knowledge of logarithms.

6. The text has mentioned that the present world population growth rate is 1.88%. Show how this figure was derived if the population was 3.176 billion in 1963 and 3.552 billion in 1969.
7. Complete this table.

Average annual growth rate, %	.5	1.0	1.5	1.7	2.0	2.5
Years for P_0 to double at above rate	138.7					

8. If the population of the world was three billion in 1960 and if it had been growing at an average annual rate of 1%, then when was the population
   a. 3 million? b. 300,000? c. 30,000?
   d. 3,000? e. 300? f. 30?
9. Calculate the average annual growth rate corresponding to the Bureau of the Census predictions for 2010 to 2015 for
   a. Series E b. Series D
   c. Series C d. Series B
10. So that we will not become overwhelmed by the sheer numbers of population prediction, we should pay close attention to the words of a competent demographer, Philip Hauser.

> Projections of future populations are admittedly fictions. No one can actually predict future population and anyone who claims he can is either a fool or a charlatan. Yet the projections of the demographers are more than exercises in

> arithmetic; they make it possible for us to see the implications of observed rates of growth. The fact that man is able to consider these implications is one reason why the projected numbers will never be reached, for recognition of the problems posed by his birth rate will move man to modify it. Such a modification, however, will not be automatic. It requires policy decisions and implementation of policy. . . .[6]

TIME magazine, on pages 58 and 59 of its 13 September 1971, issue, goes well beyond arithmetic possibilities in an essay entitled "Population Explosion: Is Man Really Doomed?" The May 19, 1972, issue of *LIFE* magazine provides an excellent description of the population problem.

[6] Philip Hauser, *Population Perspectives* (New Brunswick, New Jersey: Rutgers University Press, 1960).

Chapter 4

Mathematics of Finance

Short-term installment buying

Buying merchandise on credit, or borrowing money for any number of reasons, is very much a part of American life. Very few people—and certainly not the federal government—have any guilt feelings about going into debt. Since nothing is free, we pay money for the convenience of later payments and for the use of other people's money. This charge is called *interest*. The basic formula used to calculate this dollar amount of interest (i) involves three other related variables: the amount borrowed or *principal* of the loan (p); the annual rate of interest (r); and the number of years for which the loan is made (t). These variables are related by the formula

$$i = p \cdot r \cdot t.$$

Therefore, if we know any three of the four variables we can find the fourth. We illustrate the meaning of this formula by studying a few examples.

Example 1 A man borrows \$400 at an annual rate of 6% and pays it back 8 months later. What interest does he pay?

Solution After converting the percentage to a decimal and the months to years we obtain, by substitution into

$$i = p \cdot r \cdot t,$$

the successive equations

$$\begin{aligned} i &= 400 \cdot .06 \cdot 8/12 \\ &= 24 \cdot 2/3 \\ &= \$16.00. \end{aligned}$$

Example 2 Mary Smith borrows \$600 and pays back \$624 nine months later. What interest rate is she being charged?

Solution Only two of the four variables are directly given to us ($p = 600$ and $t = 9/12$), but if she pays back a total of \$624 and only borrows \$600, this means the difference of (\$624 – \$600) = \$24 is the interest. Now our familiar formula gives us

$$\begin{aligned} i &= p \cdot r \cdot t \\ 24 &= 600 \cdot r \cdot 9/12 \\ 24 &= 600 \cdot r \cdot 3/4 \\ 24 &= 450 \cdot r \\ .0533 &\doteq r \\ r &\doteq 5.33\%. \end{aligned}$$

Example 3 Tom Brown borrows \$900 from the First National Bank and agrees to pay it back in twelve monthly payments of \$81, each of which covers both principal and interest. What annual rate of interest is he paying?

Solution This problem appears similar to Example 2. Making twelve payments of \$81 at the end of each month means that he pays back $12 \cdot \$81 = \$972$, of which \$900 is the principal and \$72 is the interest. Then direct substitution yields

$$\begin{aligned} i &= p \cdot r \cdot t \\ 72 &= 900 \cdot r \cdot 1 \\ 72/900 &= r \\ r &= 8/100 \\ r &= 8\%. \end{aligned}$$

Right? No, wrong! The formula

$$\text{interest} = \text{principal} \cdot \text{annual rate} \cdot \text{time in years}$$

is applicable when, and only when, the borrower has the *full use* of the principal for the *full term* of the loan. Remember, Tom pays back \$81 one month after he borrows the \$900, so he does *not* have a principal of \$900 during the second month, as the amount owed is than \$900 – \$81 = \$819.[1] Thus he has the use of the full \$900 *only* during the first month of the twelve-month loan period.

If the amount Tom owes is *not* \$900 for each of the twelve months, then what *is* the amount owed? To answer this question we make a table (4.1) showing the decline of the amount owed in successive months.

We now find the *average* amount owed by adding all the individual amounts owed and dividing their sum by 12. Before we grind through the addition

$$900 + 819 + 738 + \cdots,$$

[1] The principal of a loan or investment is the money *originally* invested in the stated transaction. Hence, the *principal* is \$900 during the first month and all succeeding months. The *amount owed* (*amount available*, *balance*, or *debt*) is \$819 during the second month, \$738 during the third month, etc.

Table 4.1

During month	Amount owed	Total end of month payment	Payment to principal	Payment to interest
1	900	81	81	0
2	819	81	81	0
3	738	81	81	0
4	657	81	81	0
5	576	81	81	0
6	495	81	81	0
7	414	81	81	0
8	333	81	81	0
9	252	81	81	0
10	171	81	81	0
11	90	81	81	0
12	9	81	9	72
		972	900	72

we should notice that these numbers form an arithmetic progression with first term $a_1 = 900$, $n = 12$ terms, and a common difference of $d = -81$. We use the formula

$$S_n = \frac{n}{2} \cdot [2 \cdot a_1 + (n-1) \cdot d],$$

which was derived in Figure 3.9, to add these terms. It becomes, after substitution,

$$\begin{aligned} S_{12} &= \frac{12}{2} \cdot [2 \cdot 900 + (12-1) \cdot (-81)] \\ &= 6 \cdot [1800 + -891] \\ &- 6 \cdot [909] \\ &= 5454. \end{aligned}$$

Thus the average amount Tom owes, or has the use of, is

$$\frac{5454}{12} = \$454.50.$$

It is *this* value of p that we substitute into the interest formula.

$$\begin{aligned} i &= p \cdot r \cdot t \\ 72 &= 454.50 \cdot r \cdot 1 \\ r &\doteq 15.84\% \quad \text{(Right?)} \end{aligned}$$

Consequently, Tom's annual interest rate is not 8% but a figure almost twice that much, since we have based the calculations for his rate not on the full \$900 but on the average amount of the debt. Notice that the bank would be charging him interest on a figure of \$900 each month of the loan, regardless of the fact that reduction of the debt is occurring, if it claimed the rate to be 8%.

We should note that we assumed (as we always will in these problems unless specified otherwise) that the \$81 sum of each payment reduced the outstanding principal so that the last \$81 eliminated the remaining \$9 of outstanding principal and covered the full \$72 of interest. If we had assumed that the interest was paid back entirely in the first payment or that each payment covered \$6 in interest and \$75 in payment toward the principal, we would have obtained slightly different rates. Problems 2, 3, and 4 at the end of this section are examples of these varying rates.

If all this seems a little difficult to comprehend, then we should look at Example 3 from the banker's viewpoint. Tom has the entire \$900 for the first month which, we will agree, goes from January 1 to January 31. When Tom gives him back \$81 on January 31, what does the banker do with it—does he just set it aside in a special compartment and leave it there until all \$900 are back on December 31? Of course not! He loans out the first \$81 repayment immediately, so that it earns interest for him from somebody else, and after it comes back on February 28 with its earned interest it goes out on March 1 to a third borrower; this cycle continues for the succeeding nine months so that eventually the first \$81 has earned interest from twelve different borrowers! Similarly, Tom's February 28 repayment is loaned out immediately, and this \$81 has earned interest from ten other borrowers by December 31. The pattern should be apparent now. If you still do not believe the way in which banks really earn almost 16% on a nominal rate of 8% for installment loans, try borrowing \$900 for twelve months at 8% to be repaid in one lump sum of \$972 twelve months later; almost no bank in the country will agree to these terms.

On the other hand, the 4% interest the bank pays us as depositors is a true 4% annual rate if the interest is calculated just once at the end of the year. If we were to put \$800 in the bank for a one-year period, we would certainly end up with our \$800 plus 4% of it or

$$\begin{aligned} &800 + 4\% \text{ of } 800 \\ &= 800 + 32 \\ &= \$832. \end{aligned}$$

In the meantime the bank will loan the \$800 out at, say, a nominal 5% rate. Then the borrower would pay back

$$\begin{aligned} &800 + 5\% \text{ of } 800 \\ &= 800 + 40 \\ &= \$840. \end{aligned}$$

in monthly installments of $\$840/12 = \$70$ each. If we analyze the situation of this borrower of our money, we see that his outstanding amounts owed of

$$800, 730, 660, 590, \ldots$$

form an AP with $a_1 = 800$, $n = 12$, and $d = -70$. Then the sum of these monthly principals is

$$\begin{aligned} S_n &= \frac{n}{2} \cdot [2 \cdot a_1 + (n-1) \cdot d] \\ S_{12} &= \frac{12}{2} \cdot [2 \cdot 800 + (12-1) \cdot -70] \\ &= 6 \cdot [1600 + -770] \\ &= 6 \cdot [830] \\ &= 4980. \end{aligned}$$

The average amount owed is therefore

$$4980/12 = 415.$$

Upon solving the basic formula for r we have

$$\begin{aligned} i &= p \cdot r \cdot t \\ 40 &= 415 \cdot r \cdot 1 \\ 40/415 &= r \\ .0964 &\doteq r \\ r &\doteq 9.64\%. \end{aligned}$$

This means that by the time the year is over the bank has turned our \$800 into

$$800 + 9.64\% \text{ of } 800 = 800 + 77.12 = \$877.12.$$

The bank gives us \$32 of the interest it has accumulated and keeps the other \$45.12 for itself. Alternatively, the bank's profit margin is not

$$5\% - 4\% = 1\%$$

but

$$9.64\% - 4\% = 5.64\%.$$

Conclusion: Banks do not really earn as little profit as mere subtraction of percents might indicate.

The direct ratio formula

We have noticed that there is a considerable amount of arithmetic involved in calculating the annual interest rate. We hopefully ask whether there is a formula we can plug numbers into to obtain the rate as a result, and the answer is "yes." The derivation of the formula is rather complex,[2] but it does take into account the declining amounts owed on the loan and we accept it as valid. The formula, called the *direct ratio formula*, is

$$r = \frac{6 \cdot m \cdot I}{3 \cdot P \cdot (n+1) - I \cdot (n-1)}$$

[2] H. E. Stelson, "The Rate of Interest in Installment Payment Plans," *American Mathematical Monthly* (April, 1949), pp. 257–261.

where

r = annual interest rate
m = number of payments in a year (12 for monthly payments)
I = total amount of interest
P = principal of the loan, or the unpaid balance if there is a down payment applied to the cash price of an article
n = number of repayment periods in the loan

If we use this formula for our first problem in installment buying where $m = 12$, $I = 72$, $P = 900$, and $n = 12$ we have

$$r = \frac{6 \cdot 12 \cdot 72}{3 \cdot 900 \cdot 13 - 72 \cdot 11}$$

$$r = \frac{5184}{34308}$$

$$r \doteq .1511$$

$$r \doteq 15.11\%.$$

The answer is a little lower than our previously obtained 15.84%. The direct ratio formula is easier to use than our first method of calculating the average amount owed (assuming the interest is paid at the end of the period) and using it as the value of p in the basic formula $i = p \cdot r \cdot t$. There are several other formulas available. These, along with our two methods, give different values for r (see problems 2, 3, and 4) because they differ in their assumptions about how each payment is split into payment to be applied to the principal and interest payment. (We will account for these discrepancies at the end of this chapter in problem 10 of Problem set 4–4.) All of these formulas for calculating r give higher rates than the immediate solution of $i = p \cdot r \cdot t$ for r.

The Federal Truth-in-Lending Act, which became effective in 1969, forced many reluctant banks, finance companies, and merchandisers to state several factors involved in installment buying other than the monthly payment alone. A finance agreement with a store such as Sears, Roebuck must state the cash price, the dollar amount of the finance charge, the amount and numbers of monthly payments, and the annual percentage rate. Banks and credit unions generally have the lowest interest rates, followed by department stores (But watch out for the "revolving" charge plans—if a person isn't careful he can go around and around almost indefinitely!), finance companies, and loan sharks.

Problem set 4–1

1. A person borrows $700 and makes twelve monthly payments of $63 each. Calculate the interest; explain how an erroneous rate of 8% is obtained; calculate the annual interest rate by the method of averaging the amounts owed.

2. The cash price of a washer is \$230. It can be purchased for \$14 down and \$30 a month for eight months. If the monthly payments are first applied to the principal and then to the interest, calculate the annual rate of interest.
3. Rework problem 2, but this time assume that the monthly payments are first applied to the interest and then to the unpaid balance. Use the average-amount-owed method as it applies to this situation.
4. In problems 2 and 3 the interest charge is \$24. In this problem assume that the same \$24 is spread equally over the eight months. We know that $24 \div 8 = 3$, and therefore each monthly payment is split into \$3 toward the interest and \$27 toward the principal. Find the average amount owed under this assumption and the resulting annual interest rate.
5. A friend offers to tide you over when a hot-tip horse comes in last at the race track. You borrow \$20 from him and make twenty-two weekly repayments of \$1 each. What annual rate are you paying according to the direct ratio formula?
6. Acme Motor Rebuilders sells a rebuilt engine whose cash price is \$399 for \$39 down and \$35 monthly for eleven months. What annual interest rate is the company charging for the privilege of deferred payments according to the average-amount-owed method?
7. A man borrows \$100 from a finance company and makes eight monthly repayments of \$18 each. Find his annual rate of interest by using the average-amount-owed method after writing out all eight outstanding balances. (These amounts owed can never be negative.) Then find it by using the direct ratio formula.
8. Joe Smith buys \$600 of furniture on the "4% add on plan," meaning that 4% of the purchase price will be added on as an interest charge for each year of the loan. Joe pays back

$$600 + 4\% \text{ of } 600 = \$624$$

in twelve monthly payments of

$$624/12 = \$52 \text{ each.}$$

According to the direct ratio formula, what rate is he paying? What does the average-amount-owed method yield for the interest rate?
9. It is tempting to believe that a finance charge of \$5 per \$100 of original principal per year means an annual rate of 5%. If we borrow \$1,000 under these circumstances, how much interest would we pay in three years? If we pay the interest and principal in thirty-six equal monthly payments, what are these payments? What is the annual rate of interest according to the direct ratio formula?
10. If Alice Jones borrows \$400 on the "6% plan" for twelve months, then her interest charge is 6% of \$400 or \$24. Suppose Alice makes six payments of \$35.33 and unexpectedly inherits \$500. She decides to pay off the balance of the loan immediately in order to receive a reduction of the interest. Alice reasons, "The interest charge is \$24 spread over the twelve

payments. Since I am paying the loan back in 6/12 of the allotted time, I should pay 6/12 of $24 or $12 in interest and hence obtain a

$$24.00 - 12.00 = \$12.00$$

reduction." The banker calculates her interest charge differently by the commonly used Rule of 78's:

$$12 + 11 + 10 + \cdots + 2 + 1 = 78.$$

The Rule divides the interest charge for one year into seventy-eight equal parts. He reasons, "During the first part of the loan she is using more of the bank's money than during the last part of the loan. She should be charged more for having more money at the earlier time. Therefore, the fraction of 24.00 that is charged in the first month is 12·(1/78) or 12/78, so that 12/78 of 24.00 (= $3.69) is the interest charged then." By a continuation of this reasoning, 11/78 of 24.00 (= $3.38) is charged for the second month, 10/78 of 24.00 (= $3.08) is charged for the third month, etc. At the end of the six months

$$\begin{aligned} &\frac{12}{78}\cdot 24.00 + \frac{11}{78}\cdot 24.00 + \frac{10}{78}\cdot 24.00 + \cdots + \frac{7}{78}\cdot 24.00 \\ &= 24.00 \cdot (12/78 + 11/78 + 10/78 + \cdots + 7/78) \\ &= 24.00 \cdot 54/78 \\ &= \$17.15. \end{aligned}$$

is charged, so the interest reduction is only

$$24.00 - 17.15 = \$6.85$$

instead of the expected $12.00.

Complete this table.

At the end of month	Fraction of 24.00 that is charged as interest	Amount of 24.00 that is charged as interest	Cumulative percent in column 2	Percent of time of loan
1	12/78 ≐ 15.4%	12/78 of 24.00 = $3.69	15.4%	1/12 = 8.3%
2	11/78 ≐ 14.1%	11/78 of 24.00 = $3.38	29.5%	2/12 = 16.7%
3	10/78 ≐ 12.8%	10/78 of 24.00 = $3.08	42.3%	3/12 =25%
4				
5				
6				
7				
8				
9				
10				
11				
12				

If the money were borrowed for two years, then the fact that

$$(24 + 23 + 22 + \cdots + 1) = 300$$

would mean that 24/300 of the total interest is charged during the first month, 23/300 during the second month, etc.

Compound interest and present value

In the previous section we noticed how banks and finance companies make a reasonably profitable living by lending money. It is now a natural question to ask, "But what about the money I put in the bank—what happens to it?" We investigate with a particular example.

Example 4 Susie Smith opens a savings account with $400 on 1 July 1973, and the bank pays her interest at the rate of 6% compounded annually. What amount does she have in her account five years later?

Solution We solve this problem by calculating the amount she has on each July 1 during the five-year period. On June 30, 1974, the bank adds 6% of $400, or $24, to the original principal, yielding the new balance of $400 + $24 or $424. From July 1, 1974, to June 30, 1975, all $424—not just the original $400—are earning interest for Susie. On June 30, 1975, the bank credits her account with interest equal to 6% of $424, or $25.44, so that her new balance is $449.44. The third year runs from July 1, 1975, to June 30, 1976, during which time the $449.44 increases itself by 6%, or $26.97, to a new amount of $476.41. When the fourth year ends on June 30, 1977, the account is increased again by interest equal to $28.58, yielding the updated amount $504.99. During the last year this balance grows, and when the bank credits her account with 6% of $504.99 or $30.30 on June 30, 1978, the final amount is $535.29.

Interest calculated in this manner is called *compound interest* because the entire amount in the savings account—not just the original principal of $400—increases by a certain fixed percentage at the end of each interest-compounding period. In other words, compound interest means that interest is earned on previously earned interest and the original principal, and not just on the original principal alone. If interest is earned only by the original principal, then it is called *simple interest*. For example, if the 6% figure had been applied to the original principal only, then each of the five yearly interest payments on June 30 from 1974 through 1978, inclusive, would have been only $24. Hence

$$400 + 5 \cdot 24 = 400 + 120 = \$520.00.$$

and the interest earned on the accumulated interest is

$$535.29 - 520.00 = \$15.29.$$

If we wanted to find the value of Susie's savings after twenty years, we could, of course, continue this same process year by year fifteen more times to

Table 4.2

At the end of year	The amount of savings is
1	400 increased by 6% = 400 + 6% of 400 = $400 \cdot 1 + 400 \cdot 6\%$ = $400 \cdot (1 + 6\%) = 400 \cdot (1 + 6\%)^1$. Notice that increasing 400 by 6% gives $400 \cdot (1 + 6\%)^1$.
2	Amount at beginning of year 2 + 6% of that amount, or the result of increasing $[400 \cdot (1 + 6\%)^1]$ by 6%. This result is $[400 \cdot (1 + 6\%)^1] \cdot (1 + 6\%) = [400 \cdot (1 + 6\%)^2]$.
3	Amount at beginning of year 3 + 6% of that amount, or the result of increasing $[400 \cdot (1 + 6\%)^2]$ by 6%. This result is $[400 \cdot (1 + 6\%)^2] \cdot (1 + 6\%) = [400 \cdot (1 + 6\%)^3]$.
4	Amount at beginning of year 4 + 6% of that amount, or the result of increasing $[400 \cdot (1 + 6\%)^3]$ by 6%. This result is $[400 \cdot (1 + 6\%)^3] \cdot (1 + 6\%) = [400 \cdot (1 + 6\%)^4]$.
5	Amount at beginning of year 5 + 6% of that amount, or the result of increasing $[400 \cdot (1 + 6\%)^4]$ by 6%. This result is $[400 \cdot (1 + 6\%)^4] \cdot (1 + 6\%) = [400 \cdot (1 + 6\%)^5]$.

end up, after considerable arithmetic, with the answer. Instead, we make a table of our results to see if a pattern emerges that we can express by a formula. If so, we could then just substitute numbers into that formula to obtain the desired amount.

The pattern in Table 4.2 has emerged fairly well, for at the end of the fifth year Susie's amount is

$$400 \cdot (1 + 6\%)^5.$$

It seems reasonable to conclude that at the end of year twenty her amount would be

$$400 \cdot (1 + 6\%)^{20}.$$

Fine, but how do we evaluate expressions such as $(1 + 6\%)^5$ and $(1 + 6\%)^{20}$ without doing all of the direct multiplication implied by the expressions? Fortunately, Table V in the appendix gives the values of $(1 + r)^n$ for various combinations of values of r and n, and since we are concerned here with $r = 6\%$ and $n = 20$, we see that

$$(1 + 6\%)^{20} = 3.20713.$$

To obtain her actual amount we multiply by 400 and round off to the nearest penny, which gives us $1,282.85. If we worked with $r = 6\%$ and $n = 5$, then

$$(1 + 6\%)^5 = 1.33822$$

so that $400 \cdot 1.33822$ is $535.29. We agree that the table entries are easier than the year-by-year method of analysis and direct multiplication. Finally, we note that if Susie's \$400 had earned simple interest at 6% each year for twenty years instead of compound interest it would have grown to

$$\begin{aligned} &= 400 + 20 \text{ increases of } (6\% \text{ of } 400) \\ &= 400 + 20 \cdot 24 \\ &= 400 + 480 \\ &= \$880.00. \end{aligned}$$

The interest on the previous interest thus amounts to

$$1{,}282.85 - 880.00 = \$402.85.$$

Almost all banks pay compound interest on their depositors' savings accounts rather than simple interest.

General formula for compound growth

We now wish to generalize our method of calculating compound interest into a formula, so let us look at Example 4 from a more general perspective. We obtained the value \$535.29 (S, for the sum of original principal plus interest earned) by evaluating the expression

$$400 \cdot (1+6\%)^5.$$

The original amount of principal (P) is \$400, 6% is the constant percentage by which the previous amount is increased at the end of each compounding period (or, .06 is the interest rate (r) per compounding period), n is the number of compounding periods, and 1 is a constant. We use these symbols S, P, r, n, and 1 as we write the general formula for the compound growth of money:

$$S = P \cdot (1+r)^n \tag{A}$$

Notice the similarity between this formula and the formula arising from problem 12 of Problem set 3–5 and from Example 18 of Chapter 3. Only the letters are different since in one situation we are describing the growth of a population at a constant rate and in the other we are describing the growth of an initial amount of money, also at a constant rate.

Example 5 The amount of \$700 is invested at an annual rate of 6% where the interest is paid quarterly (four times a year at the end of each three-month period). How much money will be in the account ten years later?

Solution We are looking for the value of S and know that P is \$700. In addition, we must find r and n in order to use our basic formula (A). Recall that r represents the interest rate *per compounding period*, which means here *not* that the amount is increased by a full 6% every three months but that it is increased by 6%/4 or 1 1/2% every three months. Likewise, there are *not* ten compounding periods in the ten years but forty, since four periods per year for ten years means forty compounding

Figure 4.1

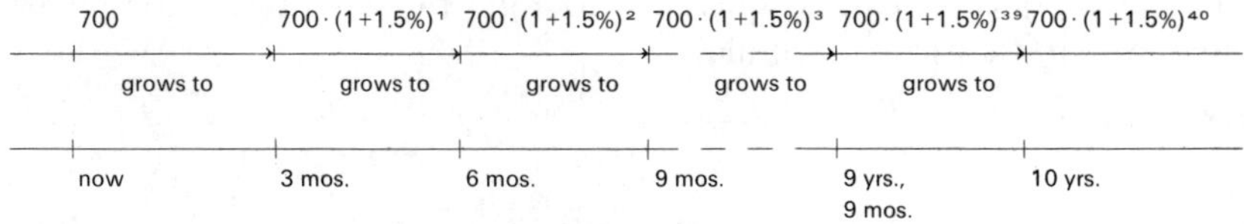

periods in that time. We can also examine our problem by portraying it on a *time line* as shown in Figure 4.1.

Substitution into our basic formula

$$S = P \cdot (1+r)^n$$

now yields

$$\begin{aligned} S &= 700 \cdot (1+1\,1/2\%)^{40} \\ S &= 700 \cdot (1+1.5\%)^{40} \\ S &= 700 \cdot 1.81403 \qquad \text{(from Table V)} \\ S &= \$1{,}269.82. \end{aligned}$$

Incidentally, if the interest were compounded *annually* at 6% the $700 would grow to

$$\begin{aligned} 700 \cdot (1+6\%)^{10} &= 700 \cdot 1.79085 \\ &= \$1{,}253.60 \end{aligned}$$

for an amount $16.22 less than if the interest were compounded *quarterly.* Why? Because the interest is earning additional interest less frequently than if it is compounded four times a year.

Example 6 An interesting variation of the compound interest formula is found in the following type of question: If $3,000 is invested at 6% compounded quarterly for one year, then what interest rate compounded once a year would yield an equal amount?

Solution We realize that the answer must be a little over 6% because 6% compounded quarterly, from the previous problem, yields a little more than a straight 6% increase once a year. Translating from words into a mathematical equation we have

$$\begin{aligned} &(\$3{,}000 \text{ at } r\% \text{ compounded annually}) \\ &\qquad = (\$3{,}000 \text{ at } 6\% \text{ compounded quarterly}) \\ &3{,}000 \cdot (1+r)^1 = 3{,}000 \cdot (1+1.5\%)^4 \\ &(1+r) = (1+1.5\%)^4 \\ &1+r = 1.06136 \qquad \text{(from Table V)} \\ &r = 1.06136 - 1 \\ &r \doteq .0614 \\ &r \doteq 6.14\% \end{aligned}$$

Generalizing, if we want to know (for a one-year period) what annual rate r', compounded once a year, is equivalent to a given annual rate r compounded n times a year, we use the formula

$$r' = (1+r/n)^n - 1.$$

Present value

The last topic of this section is that of *present value*, which is closely related to compound interest. We introduce it with an example.

Example 7 If a man wants to have \$700 in five years, how much money should he invest now if it will earn 5% compounded semiannually?

Solution Before attempting to work with any formula we should note that the answer will be less than \$700. After all, we wish to invest P dollars of principal now that will grow to \$700 in five years. We also notice that the variables involved—the future amount ($S = 700$), the present principal ($P =$ the unknown), the interest rate per compounding period ($r = 2.5\%$), and the number of compounding periods ($n = 10$)—are precisely those of the compound interest formula

$$S = P \cdot (1+r)^n.$$

Since we know three of these variables we are now able to solve for the fourth by substituting the known variables into the formula and solving for P.

$$\begin{aligned} S &= P \cdot (1+r)^n \\ 700 &= P \cdot (1+2.5\%)^{10} \\ 700 &= P \cdot 1.28009 && \text{(from Table V)} \\ P &= 700 \div 1.28009 && \text{(solving for } P\text{)} \\ P &= \$546.84 \end{aligned}$$

We could check our answer by verifying that

$$700 = 546.84 \cdot (1+2.5\%)^{10}.$$

We can generalize from this particular result by noticing that, in order to solve for P in the formula

$$S = P \cdot (1+r)^n,$$

we divide both sides of the equation by $(1+r)^n$ to obtain

$$S/(1+r)^n = P$$

or

$$P = S/(1+r)^n.$$

Unfortunately, the use of this equation usually involves an awkward division, but we can partially overcome this arithmetic by rewriting $S/(1+r)^n$ in this manner:

$$\begin{aligned} S/(1+r)^n &= \frac{S}{(1+r)^n} \\ &= \frac{S \cdot 1}{1 \cdot (1+r)^n} && \text{(property of 1)} \\ &= \frac{S}{1} \cdot \frac{1}{1 \cdot (1+r)^n} && \text{(multiplication of fractions)} \\ &= \frac{S}{1} \cdot \frac{1 \cdot (1+r)^{-n}}{1} && \text{(negative exponents)} \\ &= S \cdot (1+r)^{-n} && \text{(properties of 1).} \end{aligned}$$

P is the *present value* of a savings goal which must be set aside now to grow to S dollars in the future, while in the meantime P dollars earn $r\%$ interest for each of n compounding periods. This is given by the formula

$$P = S \cdot (1+r)^{-n}. \tag{B}$$

Table VI in the appendix lists the values of $(1+r)^{-n}$ for various values of r and n.

Example 8 Karen Jones wishes to have \$600 in eighteen months for a trip to Hawaii. If the money she sets aside now earns 5% compounded monthly, how much does she have to deposit in the bank now to accomplish her goal?

Solution The present value of the future \$600 is given by the formula

$$P = S \cdot (1+r)^{-n},$$

and here $S = 600$, $r = 5/12\%$, and $n = 18$. Then

$$\begin{aligned} P &= S \cdot (1+r)^{-n} \\ &= 600 \cdot (1 + 5/12\%)^{-18} \\ &= 600 \cdot .927889 && \text{(from Table VI)} \\ &= \$556.73. \end{aligned}$$

We close by noting that we really have not learned a formal proof of the derivations of the equivalent formulas

$$S = P \cdot (1+r)^n \qquad \text{(future sum of principal and interest)}$$

and

$$P = S \cdot (1+r)^{-n} \qquad \text{(present value of future sum of principal and interest).}$$

We can be assured, however, that other mathematicians have proven these formulas to be valid.

Problem set 4–2

1. Find the resulting compound amount under these circumstances.

Principal	Annual rate, %	Compounded	Years
200	4	annually	7
1400	5	semiannually	20
1000	6	annually	8
1000	6	semiannually	8
1000	6	quarterly	8
1000	6	monthly	8

2. a. What is the difference between investing \$100 at 3% compounded semiannually for five years and investing \$100 at 6% compounded semiannually for five years?
   b. Is the second amount at least twice the first amount?
   c. Is the interest earned the second way at least twice that earned the first way?
3. Find the present value of each amount under these circumstances.

Amount	Annual rate, %	Compounded	Years
300	4	quarterly	15
1600	5	semiannually	4
1000	2	semiannually	10
1000	3	semiannually	10
1000	4	semiannually	10
1000	5	semiannually	10
1000	6	semiannually	10
1000	8	semiannually	10
1000	10	semiannually	10

4. Often a bank will advertise that it pays an annual rate of 5% compounded monthly. What annual rate compounded once a year is this equivalent to if the period of comparison is one year?
5. Is it more profitable to invest \$100 at 4% compounded monthly or at 4 1/2% compounded annually, and how much is the difference in one year?
6. Big Daddy, the 320-pound president of the Angle City Roamers motorcycle club, decides to purchase a new machine whose present purchase price is \$2,000. If inflation raises the price 3% compounded annually, how much should Big Daddy put in the bank now, earning interest at 6% compounded quarterly, in order to have enough to buy the bike in five years?
7. At what annual interest rate will \$4,500 grow to \$5,713.79 in six years if the compounding of interest occurs quarterly?
8. If \$1,000 has grown to \$1,200 where the annual rate of growth is 5.5% compounded monthly, then how long did this take?

9. If you have studied logarithms, then use them where necessary to complete this table.

Principal	Number of compounding periods per year at an annual rate of 100%	Amount of the end of one year
1000	1 (annually)	2000.00
1000	2 (semiannually)	2250.00
1000	4 (quarterly)	
1000	12 (monthly)	
1000	365 (daily)	

(Hint: log (1 +1/365) = .00119)

10. If \$300 is invested at 4 1/2% compounded annually for ten years in Bank A, while simultaneously \$300 is invested under identical circumstances in Bank B for twenty years, will B's amount be at least twice A's amount?
11. An easily remembered and useful principle about investments is the "Rule of 72": If 72 is divided by the percentage annual compound interest rate, then the quotient fairly well approximates the number of years it takes for the principal to double itself. Using the Rule of 72 and the nearest entries from Table V in the appendix, complete this table.

Principal	Annual rate, %	Number of yearly compounding periods	Estimated years to double	Actual years to double
100	4	1	72÷4 = 18	18
100	4	2	72÷4 = 18	35 6-monthly periods
100	6	1	72 ÷ 6 = 12	12
100	6	2	72 ÷ 6 = 12	
100	6	4		
100	5	1		
100	5	4		
100	8	2		
100	8	4		

12. Refer to the end of the explanation after Example 3 of this chapter. If our bank pays us depositors at an annual rate of 4% compounded monthly, then what is its profit (in dollars) on the loan it has made with our \$800?

Annuities

We begin this section by going back to the problem from the last section about Karen Jones and her trip to Hawaii. Karen had to deposit a lump sum of \$556.73 to have it accumulate to \$600 eighteen months later under the condition of 5% (annually) compounded monthly, but it is often impractical to make such a lump sum deposit. Normally Karen would make eighteen equal monthly deposits and have them collectively grow to \$600 while earning interest. This series of equal periodic payments is an example of an *annuity*. What should these deposits be? We calculate that $600 \div 18 = 33.33$, but this is not the answer since, for example, the first \$33.33 deposit grows to more than \$33.33 in the first eighteen months. Hence our answer will be a little less than \$33.33.

We will guess that \$30.00 is the correct answer and then immediately see if the sum of the eighteen *end*-of-month deposits plus their accumulated interest gives us \$600. If this sum is less than \$600, we will know that the monthly deposits will have to be increased slightly from \$30. If this sum is greater than \$600, we will know that the monthly deposits will have to be decreased slightly. Besides, perhaps this trial-and-error approach will reveal a pattern to us that will be useful in generalizing to a formula about the future amount after depositing R dollars monthly for n months at an interest rate i per month. Our example is realistic[3] because in almost all commercial applications the payments are made at the *end* of each month.

We study a *time line* to analyze the problem.

Figure 4.2

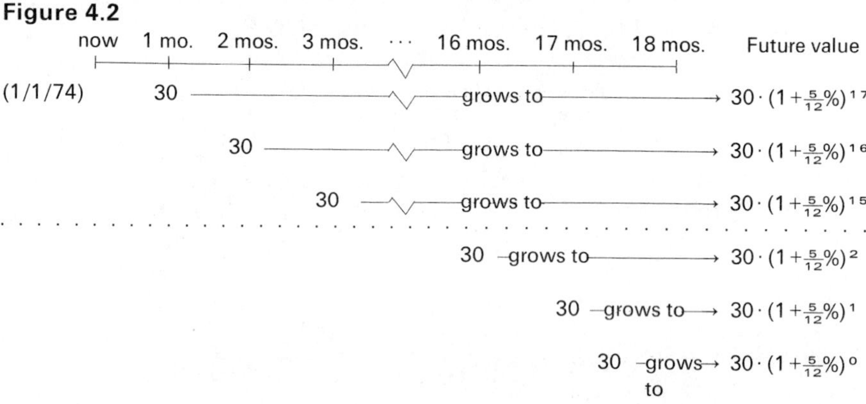

If the eighteen-month period begins on 1 January 1974, then the first deposit is made on January 31 so it earns interest for seventeen months (February 1974 to June 1975, inclusive) compounded at 5/12% per month so the future value of this \$30 is $30 \cdot (1+5/12\%)^{17}$. The second \$30 deposit, made on February 28, earns interest for sixteen months (March 1974 to June 1975, inclusive) compounded at 5/12% per month. Therefore, the future value of this \$30 is $30 \cdot (1+5/12\%)^{16}$. The third \$30 deposit, made on March 31, by similar reasoning grows to $30 \cdot (1+5/12\%)^{15}$. This process continues through the sixteenth deposit of \$30 made at the end of the sixteenth month,

[3] There is regrettable disagreement about symbolism among authors of texts on the mathematics of finance. From now on we will use i to represent the interest rate per compounding period and I as the total amount of interest involved. The symbol r will no longer be used, but R will be used to symbolize the amount of the periodic payment of an annuity.

so that its value two months later is $30 \cdot (1+5/12\%)^2$; the 31 May 1975 deposit is worth $30 \cdot (1+5/12\%)^1$ on July 1; and the June 30 deposit of \$30 is worth exactly \$30 when it is withdrawn the next day. We rewrite 30 as $30 \cdot 1$ or $30 \cdot (1+5/12\%)^0$ to be consistent with the previous future values. Now, we could use Table V to conclude that

$$30 \cdot \left(1+\frac{5}{12}\%\right)^{17} = 30 \cdot 1.07324 = 32.1972,$$

$$30 \cdot \left(1+\frac{5}{12}\%\right)^{16} = 30 \cdot 1.06879 = 32.0637,$$

$$30 \cdot \left(1+\frac{5}{12}\%\right)^{15} = 30 \cdot 1.06435 = 31.9305,$$

$$\cdots\cdots\cdots\cdots\cdots\cdots\cdots\cdots\cdots\cdots\cdots\cdots,$$

$$30 \cdot \left(1+\frac{5}{12}\%\right)^{2} = 30 \cdot 1.00835 = 30.2505,$$

$$30 \cdot \left(1+\frac{5}{12}\%\right)^{1} = 30 \cdot 1.00417 = 30.1251,$$

$$30 \cdot \left(1+\frac{5}{12}\%\right)^{0} = 30 \cdot 1.00000 = 30.0000.$$

We could also evaluate the remaining twelve future values represented by the ...'s and add them to the six future values already listed above. This addition would reflect the common sense fact that the total future value of all the monthly deposits (which we want to be \$600) is the sum of the individual future values, which include their earned interest. This fact is why we are adding

$$32.1972 + 32.0637 + 31.9305 + \cdots + 30.2505 + 30.1251 + 30.0000.$$

Such an approach, while entirely correct, is downright messy with all the multiplications and additions involved. If we look again at our time line diagram (Figure 4.2) with its right-most column of numbers we want to add, then we see that this sum can be expressed, adding from the bottom to the top, as

$$\begin{aligned} \text{Sum} &= 30 \cdot \left(1+\frac{5}{12}\%\right)^{0} + 30 \cdot \left(1+\frac{5}{12}\%\right)^{1} + 30 \cdot \left(1+\frac{5}{12}\%\right)^{2} + \cdots \\ &\quad + 30 \cdot \left(1+\frac{5}{12}\%\right)^{15} + 30 \cdot \left(1+\frac{5}{12}\%\right)^{16} + 30 \cdot \left(1+\frac{5}{12}\%\right)^{17} \end{aligned}$$

$$\begin{aligned} \text{Sum} &= 30 \cdot \left[\left(1+\frac{5}{12}\%\right)^{0} + \left(1+\frac{5}{12}\%\right)^{1} + \left(1+\frac{5}{12}\%\right)^{2} + \cdots \right. \\ &\quad \left. + \left(1+\frac{5}{12}\%\right)^{15} + \left(1+\frac{5}{12}\%\right)^{16} + \left(1+\frac{5}{12}\%\right)^{17}\right]. \end{aligned} \tag{A}$$

If we examine the expression inside the brackets, we see that it is a geometric progression with

$$\text{first term } a_1 = \left(1+\frac{5}{12}\%\right)^0 = 1,$$

$$\text{constant multiplier } r = \left(1+\frac{5}{12}\%\right),$$

and

$$\text{number of terms } n = 18.$$

From the previous chapter we know that in any GP

$$S_n = \frac{a_1 \cdot (1-r^n)}{1-r}.$$

It is also true that multiplying both numerator and denominator of this fraction by -1 yields the equivalent equation

$$S_n = \frac{a_1 \cdot (r^n-1)}{r-1}.$$

By the substitution of our specific numerical values into this equivalent equation we have

$$[\,] = \frac{1 \cdot \left(\left(1+\frac{5}{12}\%\right)^{18} - 1\right)}{\left(1+\frac{5}{12}\%\right) - 1}$$

$$[\,] = \frac{\left(1+\frac{5}{12}\%\right)^{18} - 1}{\frac{5}{12}\%}. \tag{B}$$

From equation (A) we know that the sum we are looking for has just reduced itself to

$$\text{Sum} = 30 \cdot [\,]$$

$$\text{Sum} = 30 \cdot \frac{\left(1+\frac{5}{12}\%\right)^{18} - 1}{\frac{5}{12}\%} \tag{C}$$

so that by Table V

$$\text{Sum} = 30 \cdot \frac{1.07771 - 1}{.004167}$$

$$\text{Sum} = 30 \cdot \frac{.07771}{.004167}$$

$$\text{Sum} = 30 \cdot 18.65 \tag{D}$$

$$\text{Sum} = \$559.50.$$

Unfortunately, monthly deposits of \$30 do not quite accomplish the desired goal. We could work with \$31 in the same lengthy fashion, but an important result of our incorrect guess of \$30 lies in equation (D). In words, equation (D) says that \$30.00 times 18.65 yields \$559.50 as the value of the eighteen \$30 end-of-month deposits earning 5/12% interest per month. This means that we are looking for an amount R which, when deposited under identical circumstances, will yield \$600 as the future value of the deposits and interest. The fact that the monthly payments were \$30 did not enter in the figure 18.65 at all.

$$559.62 = 30.00 \cdot 18.65 \quad \text{(wrong value of } R\text{)}$$
$$600.00 = R \cdot 18.65 \quad \text{(correct value of } R\text{)}$$
$$\frac{600.00}{18.65} = R$$
$$R = \$32.17$$

General formula for the future sum

It is really equation (B) that is the most meaningful one to us as we make a very important generalization. We think of the eighteen as a specific case of n, the number of end-of-month deposits into a savings account, and 5/12% as a specific case of i, the interest rate that is applied to each of the n deposits at the end of each month. The expression [ ] on the left hand side of the equation is symbolized by $s_{\overline{n}|i}$ ("s angle n at i"), and it represents the future sum of n one-dollar deposits and their earned interest. To summarize,

$$s_{\overline{n}|i} = \frac{(1+i)^n - 1}{i} \tag{E}$$

Table VII in the appendix lists values of $s_{\overline{n}|i}$ for various common values of n and i.

Instead of having the deposits equal to one dollar, they may be any amount R, so that the last equation in its most general and useful format becomes

$$S = R \cdot s_{\overline{n}|i} \left(= R \cdot \frac{(1+i)^n - 1}{i} \right). \tag{F}$$

Example 9 Joe Smith plans to set aside \$70 at the end of every month for the next six years. If it earns interest at 5 1/2% compounded monthly, then how much will he have at the end of that time?

Solution This is a straightforward application of the general formula (equation (F)) for the future sum of n deposits where $R = 70$, $i = 5.5\%/12$, and $n = 72$ since there are seventy-two months in six years.

$$S = R \cdot s_{\overline{n}|i}$$
$$S = 70 \cdot s_{\overline{72}|5.5\%/12}$$
$$S = 70 \cdot 85.0734$$
$$S = \$5{,}955.14$$

The answer is reasonable, since if no interest were involved seventy-two deposits of \$70 each would be worth \$5,040.00. Notice that in Table VII 5.5%/12 appears as (5.5/12)%.

We also notice in passing that equation (F), which is the general formula for the future accumulated sum of deposits and interest, can be expressed as a function with multiple independent variables R, i, and n:

$$S = f(R, i, n) = R \cdot s_{\overline{n}|i}$$
$$= R \cdot \frac{(1+i)^n - 1}{i}.$$

Example 10 Karen Jones changes her mind and decides to visit Europe in two years for three weeks of skiing. She estimates that the cost will be \$700, and she wishes to make monthly deposits in her savings account that pays 4 1/2% compounded monthly. What should these deposits be?

Solution First of all, equation (F) does apply because her equal periodic deposits form an annuity with $n = 24$ (twenty-four months in two years) and $i = 4.5\%/12$. We have the desired sum of deposits and earned interest, S, given as \$700. After substitution equation (F) becomes

$$S = R \cdot s_{\overline{n}|i}$$
$$700 = R \cdot s_{\overline{24}|\,4.5\%/12}$$
$$700 = R \cdot 25.0647$$
$$\frac{700}{25.0647} = R$$
$$R = 27.9277$$
$$R = \$27.93.$$

The answer is reasonable, since if no interest were involved twenty-four deposits of \$27.93 each would be worth $24 \cdot \$27.93 = \$670.32$. This means that the deposits will have earned \$29.68 in interest.

General formula for *R*

Notice that using equation (F) to solve for R when S, n, and i are known leads to an awkward division. Watch what happens if we solve this equation for R.

$$S = R \cdot s_{\overline{n}|i}$$
$$R \cdot s_{\overline{n}|i} = S$$
$$R = \frac{S}{s_{\overline{n}|i}}$$
$$R = \frac{S \cdot 1}{1 \cdot s_{\overline{n}|i}}$$

$$R = \frac{S}{1} \cdot \frac{1}{s_{\overline{n}|i}}$$

$$R = S \cdot 1/s_{\overline{n}|i} \qquad \text{(G)}$$

Table VIII in the appendix gives the values of $1/s_{\overline{n}|i}$ for various common values of n and i. Now we can find the value of R by direct multiplication instead of by division.

Example 11 Susie Thayer wishes to have $2,000 for an addition to her house four years from now. If she makes monthly deposits paying 5% compounded monthly into her savings account to have the money available, then what should these deposits be?

Solution This problem requires the direct use of equation (G) with $S = 2000$, $n = 48$, and $i = (5/12)\%$.

$$R = S \cdot 1/s_{\overline{n}|i}$$

$$R = 2000 \cdot 1/s_{\overline{48}|(5/12)\%}$$

$$R = 2000 \cdot .018863 \qquad \text{(from Table VIII)}$$

$$R = \$37.73$$

If no interest were earned she would have to make deposits of $2000 \div 48 =$ $41.67, so our answer is reasonable.

Whenever we know the desired sum of an annuity, the number of payments in that annuity, and the interest rate per payment/interest-compounding period, we use equation (G) to find the value of R, the required periodic payment. Whenever we know the periodic payment into an annuity, the number of payments in it, and the interest rate per payment/interest-compounding period, we use the equation (F) to find the value of S, the future sum of the annuity. There is no real difference between equation (F) and equation (G)—one is just easier to use under certain circumstances than the other. We insisted that the deposits occur at the end of each month so that the accumulations of principal and interest would form a geometric progression.

Example 12 Joe Smith wishes to withdraw $100 at the end of each month from his savings account for the next three years while the balance is earning interest at the rate of 6% compounded monthly. What should his one lump sum deposit be now in order to permit the future periodic $100 monthly withdrawals?

Solution First of all, this situation is an annuity since there are a series of equal payments made to Joe. Instead of having many small equal deposits grow to one large future amount, we are beginning with one large amount of money now and will deplete it by many small equal withdrawals. We are trying to find the *amount* or *present value* of an annuity (symbolized by A) in a situation such as Joe's in contrast to finding the *sum* (symbolized by S) of an annuity in a situation such as Karen's.

Thirty-six withdrawals of \$100 each means that \$3,600 in all will be withdrawn. Joe does *not* have to have \$3,600 in the bank because if he deposits, say, \$3,400 in the bank now the interest earned by the amount on deposit may well add up to \$200 before all of the original principal and earned interest is eventually withdrawn.

A time line is useful in analyzing the problem.

Figure 4.3

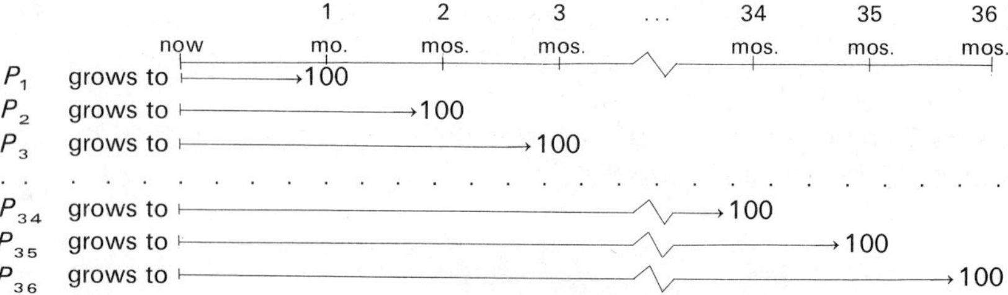

P_1 represents the amount of *principal* that must be present in the account now to permit a withdrawal of \$100 one month from now; it is really the *present value* of that \$100. P_2 represents the amount of *principal* that must be present in the account now to permit a withdrawal of \$100 two months from now; it is really the *present value* of that \$100. Finally, P_{36} represents the amount of *principal* that must be present in the account now to permit a withdrawal of \$100 thirty-six months from now; it is really the *present value* of that \$100. Here is the common sense key to the solution: the total amount that must be deposited now to meet all the future withdrawals is the sum of the individual present values of these future withdrawals. In symbols,

$$A = P_1 + P_2 + P_3 + \cdots + P_{34} + P_{35} + P_{36}.$$

Since

$$P = S \cdot (1+i)^{-n}$$

for any present value situation and $S = 100$ and $i = 6\%/12 = .5\%$ for all of the future withdrawals then, with the use of Table VI in the appendix,

$$P_1 = 100 \cdot (1+.5\%)^{-1} = 100 \cdot .995025 = \$99.50$$
$$P_2 = 100 \cdot (1+.5\%)^{-2} = 100 \cdot .990074 = \$99.01$$
$$P_3 = 100 \cdot (1+.5\%)^{-3} = 100 \cdot .985148 = \$98.52$$

. . . .

$$P_{34} = 100 \cdot (1+.5\%)^{-34} = 100 \cdot .844018 = \$84.40$$
$$P_{35} = 100 \cdot (1+.5\%)^{-35} = 100 \cdot .839819 = \$83.98$$
$$P_{36} = 100 \cdot (1+.5\%)^{-36} = 100 \cdot .835640 = \$83.56$$

We could calculate the middle thirty present values and add them to the six already on hand to obtain the sum which is the final answer, but there is a great deal of drudgery involved in such a task.

General formula for A

Let us find the sum another way. We proceed as follows.

$$A = P_1 + P_2 + P_3 + \cdots + P_{34} + P_{35} + P_{36}$$
$$A = P_{36} + P_{35} + P_{34} + \cdots + P_3 + P_2 + P_1$$
$$A = 100 \cdot (1+.5\%)^{-36} + 100 \cdot (1+.5\%)^{-35} + 100 \cdot (1+.5\%)^{-34} + \cdots + 100 \cdot (1+.5\%)^{-3} + 100 \cdot (1+.5\%)^{-2} + 100 \cdot (1+.5\%)^{-1}$$
$$A = 100 \cdot [(1+.5\%)^{-36} + (1+.5\%)^{-35} + (1+.5\%)^{-34} + \cdots + (1+.5\%)^{-3} + (1+.5\%)^{-2} + (1+.5\%)^{-1}].$$

Inspection of the expression in the brackets shows that it is a geometric progression with $n = 36$ terms, first term $a_1 = (1+.5\%)^{-36}$, and constant multiplier $r = (1+.5\%)^1$. We know that in any GP

$$S_n = \frac{a_1 \cdot (r^n - 1)}{r - 1}.$$

Therefore,

$$A = 100 \cdot [\,],$$

and after substitution

$$[\,] = \frac{(1+.5\%)^{-36} \cdot ((1+.5\%)^{36} - 1)}{(1+.5\%)^1 - 1}$$

$$[\,] = \frac{(1+.5\%)^{-36} \cdot (1+.5\%)^{36} - (1+.5\%)^{-36} \cdot 1}{1 + .5\% - 1}$$

$$[\,] = \frac{(1+.5\%)^{-36+36} - (1+.5\%)^{-36}}{.5\%}$$

$$[\,] = \frac{(1+.5\%)^{0} - (1+.5\%)^{-36}}{.5\%}$$

$$[\,] = \frac{1 - (1+.5\%)^{-36}}{.5\%}$$

This means that

$$A = 100 \cdot [\,]$$
$$A = 100 \cdot \frac{1 - (1+.5\%)^{-36}}{.5\%}. \qquad \text{(H)}$$

We could look up $(1+.5\%)^{-36}$ in Table VI in the appendix and then substitute this number into equation (H) to obtain the answer of $3287.20 (right?), but it is easier if we first obtain a generalization of equation (H) so we can make use of a certain table. The symbol A played the role of the amount needed to establish an annuity (A), 100 played the role of the future equal periodic withdrawals (R), 36 played the role of the number of withdrawal/

interest-compounding periods (n), and .5% played the role of the interest rate per withdrawal/interest-compounding period (i). By replacing these numbers by their respective letters we have the all-important, hasty, but valid generalization that

$$A = R \cdot \frac{1-(1+i)^{-n}}{i}. \tag{I}$$

The above fractional expression is usually symbolized by $a_{\overline{n}|i}$ ("a angle n at i"); it represents the present value of an annuity of \$1 for n periods.

$$A = R \cdot a_{\overline{n}|i} \tag{J}$$

represents the present value of an annuity of \$$R$ for n periods. Table IX in the appendix contains the values of

$$a_{\overline{n}|i} = \frac{1-(1+i)^{-n}}{i}$$

for various common values of n and i.

If we return to our example in equation (H) we see it is an application of equation (J), so that

$$\begin{aligned} A &= 100 \cdot a_{\overline{36}|.5\%} \\ A &= 100 \cdot a_{\overline{36}|(6/12\%)} \qquad \text{(to conform to Table IX's format)} \\ A &= 100 \cdot 32.872 \\ A &= \$3{,}287.20, \end{aligned}$$

which, as expected, is less than \$3,600.

Finally, we are often given the values of A, n, and i and are asked to find R. Using equation (J) often results in an awkward division after substitution, so when we solve it in general for R we obtain

$$R = A \cdot 1/a_{\overline{n}|i}. \tag{K}$$

Table X in the appendix contains the values of $1/a_{\overline{n}|i}$ for various common values of n and i.

Example 13 Pete Frawley is the beneficiary of a life insurance policy of \$10,000. If he wishes to deplete it by making equal monthly withdrawals for the next four years while money is worth 5% compounded monthly, then what is the amount of each of these withdrawals?

Solution This problem involves a direct application of equation (K) with

$$A = 10{,}000, \quad n = 48, \quad \text{and} \quad i = (5/12)\%.$$

Therefore,

$$\begin{aligned} R &= 10{,}000 \cdot 1/a_{\overline{48}|(5/12)\%} \\ R &= 10{,}000 \cdot .023034 \qquad \text{(from Table X)} \\ R &= \$230.34. \end{aligned}$$

Future sum and present value

Equations (F) and (J) are really the summary of all our work in this section. Equation (F) lets us find the future sum of many small equal future deposits, while equation (J) lets us find the present value needed to permit many small equal future withdrawals. We may construct the general graph of each of these equations.

Figure 4.4

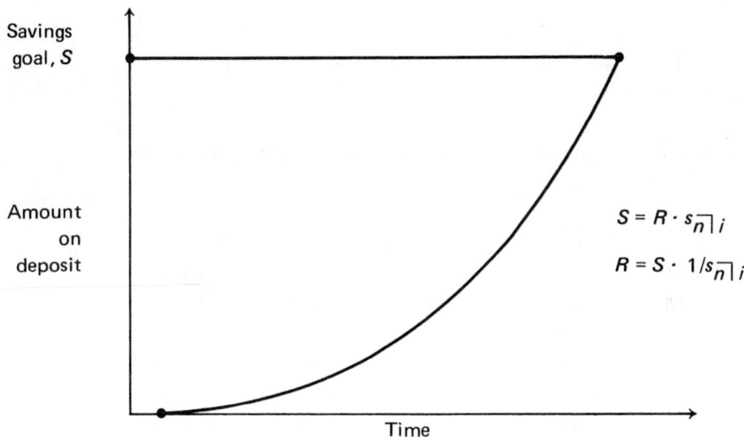

Figure 4.5

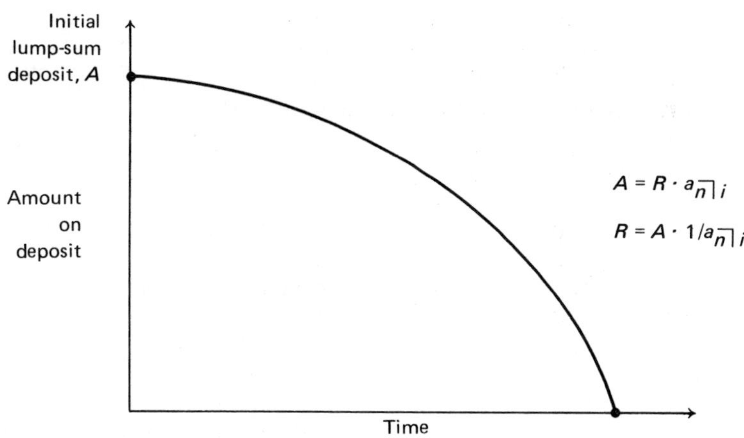

Notice in Figure 4.4 that there is no money ($0.00) on deposit until the end of the first payment period.

We close this section of the chapter by remarking that in all these problems the withdrawal/deposit periods and the interest-compounding periods coincided. Though this is not always so in practice, for our purposes we pass over the more complicated situations. The ideas there are basically the same, but the details of the calculations are different. Likewise, we assumed in the problems and formulas that the annuities consisted of *equal* periodic payments, an assumption which does not always hold. When these so-called *variable annuities* do exist the already complicated formulas become even more complicated. A very specialized branch of applied mathematics, called *actuarial science*, concerns itself with the mathematics of finance and its applications to life expectancy, insurance premiums, benefits, policies, etc. A great deal of probability theory, which we will study later, is also involved in the work of an actuary.

Problem set 4–3

1. Find the future amount, or sum, of each of these annuities

Payment	Annual rate, %	Paid	Term
100	4	monthly	3 years
100	5	monthly	3 years
100	6	monthly	3 years
100	7	monthly	3 years
50	7	monthly	3 years
100	4	monthly	15 years
100	5	monthly	15 years
100	6	monthly	15 years
100	7	monthly	15 years
50	7	monthly	15 years

2. Find the present value of each of these annuities.

Payment	Annual rate, %	Paid	Term
100	4	monthly	8 years
100	5	monthly	8 years
100	6	monthly	8 years
100	7	monthly	8 years
50	7	monthly	8 years
100	4	monthly	16 years
100	5	monthly	16 years
100	6	monthly	16 years
100	7	monthly	16 years
50	7	monthly	16 years

3. If \$30 is deposited at the end of each month in a savings account paying interest at an annual rate of 5% compounded monthly for four years, then how much interest has it earned at the end of that time?

4. If \$30 is withdrawn at the end of each month from a savings account paying interest at an annual rate of 5% compounded monthly for four years, then how much interest has the present value of that annuity earned by the end of that time (when the original present value has been completely withdrawn)?
5. Complete this table where the payments are deposited or withdrawn at the end of each month.

Payment	Annual rate, %	Term	Present value	Sum
500	6	10 years		
20	5	2 years		
20	5	2 1/2 years		
20	5	3 years		
	5	2 years	1000	
	5	4 years	1000	
	5	4 years		600
	5	8 years		600

6. Refer to Problem 6 of the last section. Since Big Daddy most likely can not set aside such a large lump sum, assume that he can set aside the necessary amount by equal periodic bank deposits. What should each of these deposits be, assuming they are made at the end of each quarter?
7. Hugh H. Carson wishes to retire at age thirty and withdraw \$1,000 a month at the end of each month for the next forty years.
   a. If his lump sum deposited earns interest at 6% compounded monthly, what should that sum be?
   b. How much money will he withdraw altogether?
   c. How much interest will his initial deposit earn at the end of the first month?
8. Rework Problem 7, changing the interest rate to 5%.
9. Miss Valerie W. Compact wishes to have \$2,400 to buy a Vega two years from now. If she deposits \$90 a month in a special savings account that earns 4% compounded monthly, will she have enough? If not, what should the deposits be?
10. A *sinking fund* is an account established for the purpose of accumulating money to meet future obligations. If the town of Newbury is establishing a sinking fund for the purpose of replacing its incinerator in twenty years, what should its semiannual payments into the fund be if the incinerator, which costs \$81,970 now, will have its price raised 1 1/4% per year because of inflation? The sinking fund earns interest at 6% compounded semiannually. What is the *unfunded principal* (difference between the desired end amount and amount in the sinking fund) after ten years?
11. If \$15,520.90 is deposited in Hugh H. Brown's account and he withdraws \$100 at the end of each month while the balance earns 6% compounded monthly, then how long will the money last? How much is withdrawn altogether? How much money is in his account during the second month?

12. Art Merrick bought a pick-up truck whose cash price was \$2,800 with a \$500 down payment and with the balance financed over twenty-four months with equal payments at the end of each month. If interest is charged at 7% compounded monthly on the unpaid principal, then what should Art's monthly payment be to pay off the balance? Hint: look at this problem from the finance company's point of view. It is really depositing \$2,300 with Art and will withdraw R dollars a month from him while the balance of its deposit (which is the unpaid principal from Art's point of view) is earning interest for the finance company at 7/12% per month. Thus \$2,300 is the amount of an annuity established with the finance company as the beneficiary so the problem actually is solvable by a direct application of equation (K).
13. The Commonwealth of Massachusetts, like several other states, conducts a weekly lottery. Several million tickets are sold for 50¢ over several weeks, and after a two-step selection process, one person is awarded \$1,000,000 for his ticket. He does not receive the money in one lump sum. Instead, he receives \$50,000 on the day of his winning and a check for \$50,000 at the *beginning* of each year for the next nineteen years.
   a. Complete this time line diagram which shows the winner's receipts. We assume he wins the lottery and receives his first check on January 1, 1974.

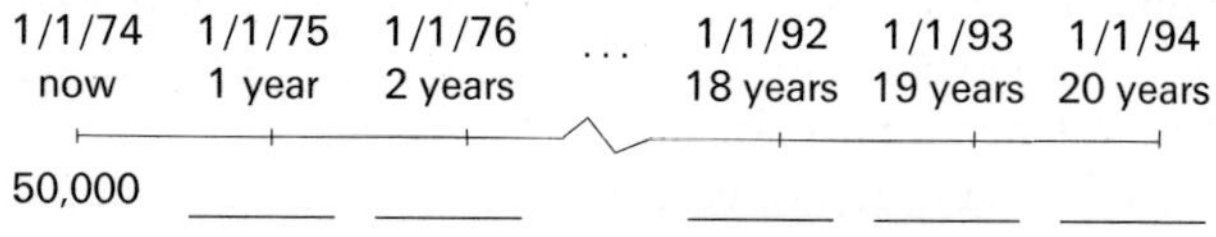

1/1/74	1/1/75	1/1/76	...	1/1/92	1/1/93	1/1/94
now	1 year	2 years		18 years	19 years	20 years
50,000	______	______		______	______	______

   b. How much money must the state set aside on the day of the million dollar win in order to permit the nineteen future payments of \$50,000 each if it receives 6% compounded annually on its lump-sum deposit in a bank? Hint: think of the 1 January 1975 payment as occurring on 31 December 1974, the 1 January 1976 payment as occurring on 31 December 1975, etc. Then the Commonwealth really must provide for 19 future *end*-of-year withdrawals.
   c. How much will the Commonwealth have to pay altogether to each winner?
   d. If Massachusetts withholds an income tax on the winnings at a flat rate of 5%, then what is the net amount of money it pays to each million-dollar winner?
   e. Rework part (c) if the state's deposit earns 7% compounded annually. $(1+7\%)^{-19} = .276508$.
   f. Rework part (c) if the state's deposit earns 8% compounded annually. $(1+8\%)^{-19} = .231712$.

Mortgages

This section on the mathematics of finance deals with an area of present or future concern to almost all of us—mortgages, or the repayment of a long-term debt in regular payments that include both principal and interest. Whenever we buy a house or a major piece of real estate we generally must borrow some money to be repaid over a long period of time with, of course, interest. It is the aim of this section to tie together the six variables involved in the lending of money under mortgage circumstances. In this text they will always be

P = principal borrowed,
i = monthly interest rate,
n = number of equal monthly payments,
R = amount of these monthly payments,
I = dollar amount of interest,
S = total amount repaid (sum of principal and interest).

We will work with a particular example to understand the principles involved and then make a generalization of our method that will apply to all monthly mortgage calculations.

Determining the monthly payment

Example 14 Tom Brown borrows $1,000 to add a patio onto his house and pays back the money over a one-year period by making equal monthly payments. These end-of-month payments include interest at the constant rate of 6% annually or (6/12)% (=1/2%) monthly on the *unpaid balance.* What should these monthly payments be, and how much does he pay back?

Solution To begin—the amount R of the monthly payment, if no interest were charged, would be 1000/12 = $83.33, but since interest is charged we know from the beginning that R is somewhat greater than $83.33. The key to this problem is to realize that an annuity is involved since Tom is making a series of equal periodic payments to the banker. If we look at the situation from the banker's viewpoint we see this fact: the banker has invested $1,000 in Tom, and will receive R dollars a month from him to reduce this investment *and* pay the interest on the unpaid balance. So, the banker has an annuity established in his benefit that begins with an initial present value of $1,000 and pays him R dollars a month, while the remaining amount owed earns interest that goes to him at the rate of 6/12% or 1/2% a month. We now have enough data to solve the annuity equation

$$R = A \cdot 1/a_{\overline{n}|i}$$

of the previous section for the required monthly payment R, since A = the principal $P = 1000$, $i = (6/12)\%$, and $n = 12$ monthly payment/compounding periods.

$$R = 1000 \cdot 1/a_{\overline{12}|(6/12)\%}$$
$$R = 1000 \cdot .086068$$
$$R = \$86.07$$

Since

$$86.07 \cdot 12 = \text{a total repayment of } \$1{,}032.84$$

this means that

$$1{,}032.84 - 1000.00 = \$32.84$$

is the interest charge.

Wait a minute—$32.84 seems a little low for an interest charge. After all, with an annual interest rate of 6%, principal of $1000, and time of one year, the interest should be

$$I = 1000 \cdot .06 \cdot 1$$

or $60.00, not $32.84.

The discrepancy arises because of the way interest is calculated on a mortgage. The rate of 6% a year is, of course, 1/2% a month, and in mortgate payments the interest is calculated by multiplying the *unpaid balance*—not the full original principal—by the monthly percentage rate. If, in contrast, the interest were calculated by multiplying 1,000 by 1/2% each month then the total interest amount would be $60.

We continue with our example by building a table which shows a month-by-month accounting of the money.

Table 4.3

At the end of month	Balance outstanding during that month	Interest on the unpaid balance	Payment toward principal	Total payment
1	$1,000.00	$5.00	$81.07	$86.07
2	918.93	4.59	81.48	86.07
3	837.45	4.19	81.88	86.07
4	755.57	3.78	82.29	86.07
5	673.28	3.37	82.70	86.07
6	590.58	2.95	83.12	86.07
7	507.46	2.54	83.53	86.07
8	423.93	2.12	83.95	86.07
9	339.98	1.70	84.37	86.07
10	255.61	1.28	84.79	86.07
11	170.82	.85	85.22	86.07
12	85.60	.43	85.64	86.07
		$32.80	$1,000.04	$1,032.84

During the first month the unpaid principal is $1,000, so the interest on that amount for that time is 1/2% of $1,000.00 or $5.00. Since the monthly payment (which covers interest and principal) is $86.07, $5.00 covers the interest. Thus the difference of $81.07 reduces the unpaid balance of \$1,000.00 to \$918.93. Consequently, the balance outstanding during the second month is \$918.93, which earns interest for the bank equal to 1/2% of \$918.93 or \$4.59. Subtracting \$4.59 from the monthly payment of \$86.07 means that \$81.48 of it reduces the balance, or amount owed, to \$837.45. Then this amount earns interest during the third month, and the process cycles through in the same fashion until the last payment reduces the balance to zero. The sum of the monthly amounts paid on the principal is \$1,000.04 instead of the expected \$1,000.00 because of rounding off in the calculations, but in practice the final payment is usually adjusted by the pennies necessary to make this sum exactly equal to the principal itself.

Comparing the two methods we have studied so far for repaying debts on an installment plan—short-term borrowing and mortgages—leads to an interesting conclusion. If we borrowed \$1,000 at 6% for one year under the so-called "add on" conditions we would pay back \$1,000 + 6% of \$1,000 = \$1,000 + \$60 = \$1,060 in twelve monthly payments of \$88.33 each (right?). Calculating the interest rate by the average principal method gives us an annual interest rate of 11.67%, since the average amount owed is \$514.18 (right on both these figures?). To find the interest rate under mortgage conditions we sum the second column in the table (1,000.00 + 918.93 + 837.45 + ⋯ + 85.60) to get 6,559.21, divide by 12 for an average principal borrowed of \$546.60, solve the interest equation

$$32.84 = 546.60 \cdot r \cdot 1$$

for r so that $r = 6.01\%$, and conclude that under these circumstances "6%" does mean a true rate of 6%.

The generalization is now relatively easy. If we know how much we must borrow (P), the monthly interest rate, (i), and the number of months of the loan (n), then solving the annuity equation

$$R = P \cdot 1/a_{\overline{n}|i}$$

for R tells us the monthly payment amount. Again, the lending institution's loan to the borrower is an investment of P dollars from which the institution will be withdrawing R dollars monthly from the borrower as an annuity while the unwithdrawn and declining balance is still earning interest at $i\%$ for each of the n months. Then making n monthly payments of R dollars each means we have paid back a total of

$$S = n \cdot R$$

dollars, and subtracting the principal from the total amount repaid gives the dollar amount of the interest (I).

$$I = S - P$$

or

$$I = n \cdot R - P.$$

Changing interest rates and repayment periods

There are many paths we could explore about mortgage payments but time unfortunately does not permit this. We will, though, go through two of them as we investigate a realistic mortgage situation to inquire about the effects of (1) changing the interest rate and (2) changing the number of repayment periods.

Example 15 Suppose \$20,000 is borrowed at an annual rate of 6% for thirty years. What are R, S, and I?

Solution The general annuity equation we use is

$$R = P \cdot 1/a_{\overline{n}|i},$$

which in our case becomes

$$R = 20{,}000 \cdot 1/a_{\overline{360}|(6/12)\%}$$
$$R = 20{,}000 \cdot .0059954 \qquad \text{(to seven decimal places)}$$
$$R = \$119.91.$$

This means that 360 monthly payments of \$119.91 each will pay back the original principal of \$20,000 plus all the interest that the unpaid principal has earned for the bank. How much is this interest? We calculate that 360 monthly payments of \$119.91 each means that a total of

$$360 \cdot \$119.91 = \$43{,}167.60$$

is paid to the bank so

$$\$43{,}167.60 - \$20{,}000.00 = \$23{,}167.60$$

is the interest—which is more than the principal! For example, the interest charged for the first month is 6/12% of \$20,000 or \$100, which means that the principal is only reduced by \$19.91 at the end of the first month. This situation of interest exceeding principal is quite common with high interest rates and long repayment periods.

Let us return to the twin paths of changing interest rates and repayment periods mentioned several paragraphs ago. The next two pages of computer-generated calculations for a principal of \$20,000 show the corresponding values of R, I, and S as the annual interest rate and repayment time both change. Notice that i is given as an annual rate and that it is printed as a capital letter because the teletypewriter does not have lower case letters.

IF THE LIFE OF THE MORTGAGE IS 30 YEARS, THEN:

I (%)	R	INTEREST	TOTAL REPAID
4	95.48	14372.8	34372.8
4.5	101.34	16482.4	36482.4
5	107.37	18653.2	38653.2
5.5	113.56	20881.6	40881.6
6	119.91	23167.6	43167.6
6.5	126.41	25507.6	45507.6
7	133.06	27901.6	47901.6
7.5	139.84	30342.4	50342.4
8	146.75	32830.	52830.
8.5	153.78	35360.8	55360.8
9	160.92	37931.2	57931.2

IF THE LIFE OF THE MORTGAGE IS 29 YEARS, THEN:

I (%)	R	INTEREST	TOTAL REPAID
4	97.19	13822.1	33822.1
4.5	103	15844	35844.
5	108.97	17921.6	37921.6
5.5	115.11	20058.3	40058.3
6	121.4	22247.2	42247.2
6.5	127.84	24488.3	44488.3
7	134.43	26781.6	46781.6
7.5	141.14	29116.7	49116.7
8	147.99	31500.5	51500.5
8.5	154.95	33922.6	53922.6
9	162.03	36386.4	56386.4

IF THE LIFE OF THE MORTGAGE IS 28 YEARS, THEN:

I (%)	R	INTEREST	TOTAL REPAID
4	99.04	13277.4	33277.4
4.5	104.79	15209.4	35209.4
5	110.72	17201.9	37201.9
5.5	116.79	19241.4	39241.4
6	123.02	21334.7	41334.7
6.5	129.4	23478.4	43478.4
7	135.92	25669.1	45669.1
7.5	142.57	27903.5	47903.5
8	149.35	30181.6	50181.6
8.5	156.25	32500	52500.
9	163.26	34855.4	54855.4

IF THE LIFE OF THE MORTGAGE IS 27 YEARS, THEN:

I (%)	R	INTEREST	TOTAL REPAID
4	101.04	12737	32737
4.5	106.74	14583.8	34583.8
5	112.61	16485.6	36485.6
5.5	118.63	18436.1	38436.1
6	124.8	20435.2	40435.2
6.5	131.11	22479.6	42479.6
7	137.56	24569.4	44569.4
7.5	144.15	26704.6	46704.6
8	150.86	28878.6	48878.6
8.5	157.68	31088.3	51088.3
9	164.63	33340.1	53340.1

IF THE LIFE OF THE MORTGAGE IS 26 YEARS, THEN:

I (%)	R	INTEREST	TOTAL REPAID
4	103.21	12201.5	32201.5
4.5	108.86	13964.3	33964.3
5	114.67	15777	35777.
5.5	120.63	17636.6	37636.6
6	126.73	19539.8	39539.8
6.5	132.98	21489.8	41489.8
7	139.37	23483.4	43483.4
7.5	145.88	25514.6	45514.6
8	152.52	27586.2	47586.2
8.5	159.28	29695.4	49695.4
9	166.14	31835.7	51835.7

```
IF THE LIFE OF THE MORTGAGE IS 25   YEARS, THEN:
I (%)          R           INTEREST        TOTAL REPAID
4              105.57      11671           31671
4.5            111.17      13351           33351.
5              116.92      15076           35076.
5.5            122.82      16846           36846.
6              128.94      18682           38682.
6.5            135.04      20512           40512.
7              141.36      22408           42408.
7.5            147.8       24340           44340.
8              154.36      26308           46308.
8.5            161.05      28315           48315.
9              167.84      30352           50352.
```

First of all, a small change in the interest rate can have a ballooning effect on the monthly payment and interest. For example, if the time is held constant at thirty years and the annual interest rate changes from 6% to 6 1/2%, then the corresponding dollar change in the interest itself is over $2,300. Changing the annual rate from 7% to 8% with $20,000 principal and twenty-five years affects the monthly payment by $13 and the interest amount by a little less than $4,000. These two examples illustrate why prospective home owners should pay close attention to seemingly insignificant rate changes.

It is often quite advantageous to reduce the repayment period as much as possible. As an example, borrowing $20,000 at 5% for thirty years means a monthly repayment of $107.37 and total repayment of $38,653.20, while condensing the time to twenty-five years increases the monthly repayments by less than $10 to $116.92 and decreases the total repaid by almost $3600—certainly a considerable saving!

We can summarize this section and this chapter by noting that the mathematics of finance is fairly complicated and can get much more involved than our work. Still, as the last two paragraphs illustrate, a little knowledge of fundamentals can make a substantial difference in the amounts we pay to our creditors or collect from our debtors.

Problem set 4–4

1. Complete this table about borrowing money under mortgage circumstances.

Amount borrowed (P)	Number of monthly payments (n)	Amount of monthly rate, % (R)	Annual interest rate, % ($12 \cdot i$)	Total amount repaid (S)	Total interest paid (I)
$20,000	300	$128.94	6		
20,000	360		6		
20,000	420		6		
20,000	480		6		
20,000	300		7		
20,000	360		7		
20,000	420		7		
20,000	480		7		
40,000	480		7		

2. According to the computer-generated tables in the text, \$20,000 borrowed at 7% for thirty years results in a monthly repayment of \$133.06, and \$20,000 borrowed at 6% results in a monthly repayment of \$119.91. If we were given only this information we would guess that \$20,000 borrowed at 6.5% for thirty years would result in a monthly repayment exactly halfway between \$119.91 and \$133.06, or \$126.49. What is the difference between this guess and the true answer from the table? We often use this method of *interpolation* when a given annual percentage rate does not appear in a table.
3. Complete this table.

Amount borrowed	Time, years	Annual rate	Monthly payment	Total repaid	Interest paid
\$20,000	25	6%			
19,000	25	6%			
18,000	25	6%			
17,000	25	6%			
16,000	25	6%			
15,000	25	6%			

When completed, this table should show you how expensive it is to borrow, say, \$1,000 additional on a mortgage. Thus it is worthwhile to increase the down payment and decrease the principal that is borrowed as much as possible.

4. If \$30,000 is borrowed at 6 3/4% for twenty-five years, then the monthly repayment is \$207.30. If the rate is increased to 7% then find the new monthly repayment. Calculate the increase in total interest paid that is caused by the increase in the interest rate from 6 3/4% to 7%.
5. If we generalized from the text's example of \$1,000 borrowed at 6% for twelve months (Example 14) by working in exactly the same manner with \$10,000 for twenty years, then the monthly repayment is \$71.65 and our repayment schedule would look like this:

At the end of month	Principal outstanding during this month	Interest outstanding on principal	Payment to reduce outstanding principal	Total payment
1	\$10,000.00	\$50.00	\$21.65	\$71.65
2	9,978.35	49.89	21.76	71.65
3	9,956.59			
4		49.67	21.98	
5	9,912.74		22.09	
6				
7				
8	9,846.14	49.23	22.42	

a. Complete the missing table entries.

If we continued in the same fashion for four more months and added the last three columns, we would see that the total interest paid in this first year of the loan is \$592.72. The total payment to reduce the outstanding principal would be \$267.08, and the total repaid would be $12 \cdot \$71.65 = \$859.80$. This example shows that most of the early repayments just pay interest and relatively little outstanding principal. If we continued in the same fashion we would build a *loan progress chart* such as this:

At the end of *year* no.	Percent of time of loan	Cumulative principal repaid	Cumulative percent of principal repaid
1	5	\$267.08	2.7
2	10	550.60	5.5
3	15	851.63	8.5
4	20	1,171.22	
5	25	1,510.52	
6	30	1,870.75	
7	35	2,253.20	
8	40	2,659.24	
9	45	3,090.32	
10	50	3,547.99	
11	55	4,033.88	
12	60	4,549.75	
13	65	5,097.43	
14	70	5,678.89	
15	75	6,296.22	
16	80	6,951.62	
17	85	7,647.45	
18	90	8,386.19	
19	95	9,170.50	
20	100	10,000.00	

b. Complete the missing table entries.

6. Continue the same table of the last problem by completing the missing table entries.

At the end of year	Percent of time of loan	Outstanding principal repaid during the year	Total amount repaid	Interest paid during the year	% of total repaid that is interest	% of total repaid that reduces principal
1	5	\$267.08	\$859.80	\$592.72	68.9	31.1
2	10	283.54	859.80	576.26	67.0	33.0
3	15	301.03	859.80	558.77		
4	20	319.59	859.80	540.21		
5	25	339.30	859.80	520.50		
6	30	360.23	859.80	499.57		
7	35	382.45	859.80	477.35		

At the end of year	Percent of time of loan	Outstanding principal repaid during the year	Total amount repaid	Interest paid during the year	% of total repaid that is interest	% of total repaid that reduces principal
8	40	406.04	859.80	453.76		
9	45	431.08	859.80	428.72		
10	50	457.67	859.80	402.13		
11	55	485.89	859.80	373.91		
12	60	515.87	859.80	343.93		
13	65	547.68	859.80	312.12		
14	70	581.46	859.80	278.34		
15	75	617.33	859.80	242.47		
16	80	655.40	859.80	204.40		
17	85	695.83	859.80	163.97		
18	90	738.74	859.80	121.06		
19	95	784.31	859.80	75.49		
20	100	829.48	856.62	27.14		
Cumulative Totals		\$10,000.00	\$17,192.82	\$7,192.82		

7. If \$17,000 is borrowed for twenty-eight years at an annual rate of 6 1/2%, find the amount of the monthly payment. Also determine the composition of the first three payments in terms of interest and principal repaid, the total amount repaid, and the total interest paid. What percent of the total amount repaid is the total interest paid at the end of the three months?
8. Refer to the problem in the text (Example 14) where \$1,000 was borrowed for one year at 6%. If this were repaid under short term installment conditions (where 6% of \$1,000 or \$60.00 is charged as interest), then calculate the annual interest rate using the direct ratio formula.
9. The second table of Problem 5 was constructed by letting a computer continue the first table amounts by working until 240 months had elapsed. The \$1,870.75 figure representing the principal repaid at the end of the sixth year was found by subtracting the unpaid principal from \$10,000. The \$1,870.75 is the borrower's *equity* or amount of ownership he has in the house. Suppose we did not have a computer available to do the monthly calculations one by one, but we still had to find the equity in a mortgage of \$25,000 for thirty years at 6 1/2%, where Bill Wheeler is the borrower, and ten years have elapsed. Here $R = \$158.03$ (right?). Think of this situation from the banker's point of view. He no longer has \$25,000 invested in Bill, but he still has enough "deposited" to permit twenty years of withdrawals of \$158.03 per month from his investment. The amount needed to establish this annuity is

$$A = R \cdot a_{\overline{n}|i}$$
$$A = 158.03 \cdot a_{\overline{240}|(6.5/12)\%}$$
$$A = 158.03 \cdot 134.125 \qquad \text{(from Table IX)}$$
$$A = \$21{,}195.77.$$

The amount of ownership the banker has in \$21,195.77, so Bill's equity is

$$25{,}000.00 - 21{,}195.77 = \$3{,}804.23.$$

a. Use this method to obtain a number which is close to the \$1,870.75 listed in the second table of Problem 5.
b. Find the equity that John Armstrong has in his house whose purchase price is \$30,000 ten years later on a twenty-five year mortgage at 7.5%.
c. Find the equity that Bill Drake has in his ranch whose purchase price is \$70,000 after twenty-five years on a forty-year mortgage at 7%.
d. Karen Evans has the chance to buy a home by taking over the mortgage of \$18,000 at 5 1/2% for twenty-five years after eight years have elapsed on this mortgage. How much does she have to pay the current owner in order to buy out his equity? How much altogether has the current owner paid to the bank before Karen takes over? How much of this is interest?

10. In this problem we finally knot the loose ends in determining the "true" annual interest rate on a monthly installment loan. Toward the end of the first section of this chapter and in Problems 2, 3, and 4 for that section we noted that there are several different ways of computing the interest rate depending on the "... assumptions about how each payment is split into payment to be applied to the principal and interest payment." The key to understanding this problem is to notice that *the series of equal payments form an annuity.* Consider Example 3, in which Tom Brown repays his \$900 loan in twelve installments of \$81 each. Let us look at it from the bank's point of view as it determines the interest rate it is really charging. The bank has "deposited" \$900 in Tom and will be withdrawing \$81 at the end of each month, while the principal "on deposit" for that month earns interest at i% per month. The formula

$$A = R \cdot a_{\overline{n}|i}$$

applies here since the deposit of \$900 ($=A$) permits 12 ($=n$) withdrawals of \$81 ($=R$) each. We are therefore looking for that value of i for which

$$900 = 81 \cdot a_{\overline{12}|i}$$

becomes a true statement. Consequently, we are looking for that value of i for which

$$900 = 81 \cdot \frac{(1-(1+i)^{-12})}{i} \qquad \text{(A)}$$

becomes a true statement. Now, equation (A) cannot be solved for i in terms of R, A, and n, so we will let i take on different values until a value of i is found such that equation (A) is satisfied.

Suppose we guess that the annual rate is 6%, meaning that the monthly rate is .5%. Then

$$81 \cdot \frac{(1-(1+.5\%)^{-12})}{i} = 941.16.$$

This means that the bank would have to "deposit" \$941.16 in Tom (the present value of the annuity or the amount of the loan) to permit twelve withdrawals of \$81 each at a true annual rate of 6% for the unwithdrawn balance. We next guess that the annual rate is 12%, meaning that the monthly rate is 1%. Then

$$81 \cdot \frac{(1-(1+1\%)^{-12})}{i} = 911.66,$$

which is still too large a present value. If we guess at the annual rate of 18% or a monthly rate of 1.5%, then

$$81 \cdot (1-(1+1.5\%)^{-12})/1.5\% = 883.52.$$

The rate of 18% is too high because if it were the rate then only \$883.52 would have to be loaned out to bring in twelve payments of \$81 each.

Since \$900 is between \$911.66 and \$883.52, we reason that the corresponding monthly rate i is between 1% and 1.5% or that the yearly rate is between 12% and 18%. What we do next is to start over and have a computer evaluate

$$A = 81 \cdot \frac{(1-(1+i)^{-12})}{i}$$

in steps of 1 for the annual interest rate until the value of A is just under 900. Then we will have bracketed the true annual rate, which we call r for the purpose of identification in this problem, between two whole numbers.

```
INPUT THE PRINCIPAL, MONTHLY PAYMENT, NUMBER OF MONTHLY
  PAYMENTS, INITIAL GUESS, AND STEP VALUE.
? 900,81,12,1,1


PRINCIPAL        TRIAL ANNUAL RATE, %          PRESENT VALUE

   900                   1                        966.739
   900                   2                        961.646
   900                   3                        956.395
   900                   4                        951.273
   900                   5                        946.164
   900                   6                        941.162
   900                   7                        936.14
   900                   8                        931.156
   900                   9                        926.222
   900                   10                       921.344
   900                   11                       916.482
   900                   12                       911.659
   900                   13.                      906.871
   900                   14                       902.14
   900                   15                       897.424

TRUE ANNUAL RATE IS BETWEEN 14    AND 15    %.
```

We then search the interval [14, 15] for r in steps of .1.

```
INPUT THE PRINCIPAL, MONTHLY PAYMENT, NUMBER OF MONTHLY
  PAYMENTS, INITIAL GUESS, AND STEP VALUE.
? 900,81,12,14,.1

PRINCIPAL        TRIAL ANNUAL RATE, %         PRESENT VALUE

   900                 14                        902.14
   900                 14.1                      901.661
   900                 14.2                      901.192
   900                 14.3                      900.718
   900                 14.4                      900.252
   900                 14.5                      899.763

TRUE ANNUAL RATE IS BETWEEN 14.4          AND 14.5         %.
```

We can go even finer in our search by examining the interval [14.4, 14.5] in steps of .01.

```
INPUT THE PRINCIPAL, MONTHLY PAYMENT, NUMBER OF MONTHLY
  PAYMENTS, INITIAL GUESS, AND STEP VALUE.
? 900,81,12,14.4,.01

PRINCIPAL        TRIAL ANNUAL RATE, %         PRESENT VALUE

   900                 14.4                      900.252
   900                 14.41                     900.204
   900                 14.42                     900.162
   900                 14.43                     900.107
   900                 14.44                     900.049
   900                 14.45                     900.001
   900                 14.46                     899.955

TRUE ANNUAL RATE IS BETWEEN 14.45         AND 14.46        %.
```

To the nearest hundredth, $r = 14.45\%$ is the *true* annual rate.

The difficulties of this method of obtaining the true annual rate are considerable. In fact, calculating r without a computer or knowledge of logarithms is virtually impossible. The main advantages of the direct ratio formula as an approximation of r are its accuracy and relative ease of application.

In closing this problem we return to Problems 2, 3, and 4 of Problem set 4–1 and find the true annual rate with the aid of a computer.

```
INPUT THE PRINCIPAL, MONTHLY PAYMENT, NUMBER OF MONTHLY
  PAYMENTS, INITIAL GUESS, AND STEP VALUE.
? 216,30,8,1,2

PRINCIPAL        TRIAL ANNUAL RATE, %         PRESENT VALUE

   216                 1                         239.099
   216                 3                         237.326
```

```
   216                    5                      235.558
   216                    7                      233.824
   216                    9                      232.Ø97
   216                    11                     23Ø.395
   216                    13                     228.7Ø8
   216                    15                     227.Ø44
   216                    17                     225.394
   216                    19                     223.765
   216                    21                     222.15
   216                    23                     22Ø.556
   216                    25                     218.976
   216                    27                     217.416
   216                    29                     215.87

TRUE ANNUAL RATE IS BETWEEN 27    AND 29    %.



INPUT THE PRINCIPAL, MONTHLY PAYMENT, NUMBER OF MONTHLY
  PAYMENTS, INITIAL GUESS, AND STEP VALUE.
? 216,3Ø,8,28,.1

PRINCIPAL        TRIAL ANNUAL RATE, %        PRESENT VALUE

   216                    28                     216.641
   216                    28.1                   216.564
   216                    28.2                   216.486
   216                    28.3                   216.41
   216                    28.4                   216.332
   216                    28.5                   216.256
   216                    28.6                   216.177
   216                    28.7                   216.1Ø2
   216                    28.8                   216.Ø24
   216                    28.9                   215.948

TRUE ANNUAL RATE IS BETWEEN 28.8        AND 28.9        %.



INPUT THE PRINCIPAL, MONTHLY PAYMENT, NUMBER OF MONTHLY
  PAYMENTS, INITIAL GUESS, AND STEP VALUE.
? 216,3Ø,8,28.8,.Ø1

PRINCIPAL        TRIAL ANNUAL RATE, %        PRESENT VALUE

   216                    28.8                   216.Ø24
   216                    28.81                  216.Ø16
   216                    28.82                  216.ØØ8
   216                    28.83                  216.ØØ1
   216                    28.84                  215.994

TRUE ANNUAL RATE IS BETWEEN 28.83       AND 28.84       %.
```

a. Complete this table. The unpaid balance at the end of eight months should be within a few cents of \$0.00 when this loan is analyzed as an annuity. Here

$$i = 28.83\%/12 \doteq 2.4\%.$$

At the end of month	Principal outstanding during that month	Interest on the outstanding principal	Payment toward outstanding principal	Total payment
1	\$216.00	2.4% of \$216.00 = \$5.18	\$24.82	\$30.00
2	191.18	2.4% of 191.18 = 4.59	25.41	30.00
3	165.77	2.4% of 165.77 = 3.98	26.02	30.00
4				
5				
6				
7				
8				

b. Use the direct ratio formula to find an approximation to the true annual interest rate.

Chapter 5

Probability

Basic concepts

We live in a world of uncertainties about many aspects of life, except, of course, death and taxes. We are interested in predicting future occurrences of the weather, life expectancy, sporting events, economic conditions, quality of output from industrial processes, and so forth. That branch of mathematics known as *probability* devotes itself to the analysis of occurrences and their degree of certainty.

The phrase "degree of certainty" needs further explanation. It represents the same undefined concept as likelihood or chance. Thus we assume that with respect to a football game between A and B we may speak of the probability of A's winning, the degree of certainty of A's winning, the likelihood that A will win, and A's chances of winning. This last sentence suggests that we will try to give a quantitative description of the likelihood that A wins by assigning a number to this event indicative of its degree of certainty, and this assignment is precisely what we do later in this section.

Before getting into the details of the formal study of probability, it is revealing to look more closely at the foundations of mathematics and the logic of true and false. All statements such as "If $2x + 19 = 15$, then $x = 3$" and "$7 \cdot 2 = 9$" are either true or false and nothing in between. The mathematics of probability deals with statements such as "On July 1, 2001, the temperature at 2:00 p.m. in Northfield, Minnesota, will be between 62.5° and 72.5°." We cannot unquestionably say whether this statement is true or false, but we still would like to measure in some way the degree of certainty of the event described by the statement. We will also see that a basic knowledge of probability is often useful to social scientists and to managers as we examine some interesting applications of probability to law, quality control, and statistics.

Before we can study these applications it is necessary to work within the upper half of the diagram of mathematics and its applications. Basically, our task in this chapter is to create a meaningful measure of probabilities of occurrences. We begin with an example of a six-sided die, and we want to know the probability of having the face with two dots showing on top after the die is rolled once. One reasonable way to solve this problem is to actually roll the die many times, say one hundred, and observe the number of times the two-dot side appears. Thus if the two-dot side occurs nineteen times we form the ratio 19/100 and claim that since two dots turned up 19% of the time in the past, the chances of two dots turning up on the next roll should be 19%. Assigning a probability of 19% or .19 to the future possible occurrence of two dots (a 2), based on past experience, is an example of *empirical* probability. In general, if such an event has occurred m times in n trials of an experiment ($0 \leqslant m \leqslant n$, of course, where m and n are integers) then we say the degree of certainty, or the probability, of that event's occurring on the next trial is m/n. The number n must be reasonably large before we can assign an accurate empirical probability to the event. There are rather involved tests to see if n is "large enough." We will not study these tests.

Not all occurrences can have their probabilities calculated on the basis of performing an experiment or locating historical data. For example, suppose an engineer wishes to produce a rocket that will fire 95% of the time, and further suppose that he knows the empirical probabilities of the various components' reliabilities. Since it is impossible to test the rocket without destroying it, the engineer looks for a mathematical model that will tell him how to weave together the various input probabilities into one output probability. Calculating the probability of an occurrence based on certain theoretical mathematical principles prior to or instead of experimentation is called *a priori* probability. Unless otherwise stated, all the probabilities we work with will be *a priori* probabilities.

Three basic definitions

The following three definitions help us identify the essential concepts of probability.

> Definition The *sample space* of an experiment, symbolized by S, is the set of all possible outcomes of an experiment.

Consider the experiments of (a) flipping a coin, (b) rolling a die, (c) rolling a pair of dice, and (d) drawing a card from a standard deck. Then the respective sample spaces are:

a. $S = \{H, T\}$, and $n(S)$[1] $= 2$;
b. $S = \{1, 2, 3, 4, 5, 6\}$, and $n(S) = 6$;

[1] Recall from Chapter 2 that $n(S)$ is the number of elements in set S.

c. assuming that the dice are of different colors, so that (2, 5) symbolizes a 2 on the red die and a 5 on the green die while (5, 2) symbolizes a 5 on the red die and a 2 on the green die—

$$\begin{aligned} S = \{&(1,1), (1,2), (1,3), (1,4), (1,5), (1,6),\\ &(2,1), (2,2), (2,3), (2,4), (2,5), (2,6),\\ &(3,1), (3,2), (3,3), (3,4), (3,5), (3,6),\\ &(4,1), (4,2), (4,3), (4,4), (4,5), (4,6),\\ &(5,1), (5,2), (5,3), (5,4), (5,5), (5,6),\\ &(6,1), (6,2), (6,3), (6,4), (6,5), (6,6)\}, \text{ and } n(S) = 36; \end{aligned}$$

d. $$\begin{aligned} S = \{&2C, 3C, 4C, 5C, 6C, 7C, 8C, 9C, 10C, JC, QC, KC, AC,\\ &2D, 3D, 4D, 5D, 6D, 7D, 8D, 9D, 10D, JD, QD, KD, AD,\\ &2H, 3H, 4H, 5H, 6H, 7H, 8H, 9H, 10H, JH, QH, KH, AH,\\ &2S, 3S, 4S, 5S, 6S, 7S, 8S, 9S, 10S, JS, QS, KS, AS\}, \end{aligned}$$

(C, D, H, and S represent the suits Clubs, Diamonds, Hearts, and Spades, respectively), and $n(S) = 52$.

Sometimes we wish to organize a sample space in a different way. For example, a sample space for the third experiment could be the different set

$$S = \{2, 3, 4, 5, 6, 7, 8, 9, 10, 11, 12\}$$

if we are interested only in the sum of the two face-up sides. Most problems require that we use as detailed or "fine" a sample space as possible. We will always do so unless context or instructions clearly state otherwise; see part (g) of Problem 5 for such an example.

Definition An *event* is any subset of a sample space.

We examine as examples of events four respective subsets of the previous sample spaces.

In (a), E = getting a head = $\{H\}$, and $n(E) = 1$;

in (b), E = getting a number between 2 and 5, inclusive = $\{2, 3, 4, 5\}$, and $n(E) = 4$ (Event E is said to *occur* if the outcome is a 2, a 3, a 4, or a 5.);

in (c), E = having the sum greater than 9 = $\{(4, 6), (5, 5), (5, 6), (6, 4), (6, 5), (6, 6)\}$, and $n(E) = 6$;

in (d), E = drawing a queen or a red face card (an ace is not considered to be a face card) = $\{QC, QD, QH, QS, JD, JH, KD, KH\}$, and $n(E) = 8$.

Now we are ready to state the all-important basic formula by which we calculate the number which measures the intuitive but undefined idea of probability.

Definition If a sample space S has $n(S)$ equally likely elements and if E is any associated event with $n(E)$ elements, then the *probability that event E occurs* (symbolized by $P(E)$) is calculated by the equation

$$P(E) = \frac{n(E)}{n(S)} = \frac{\text{the number of ways the event can occur}}{\text{the number of ways the experiment can occur}}.$$

This equation is certainly reasonable since it conforms to our intuitive notion of chance or likelihood. After all, if an urn contains three red marbles, five blue ones, and two white ones and if one marble is chosen blindly from the well-shaken urn, then it has three chances out of a total of ten of being red, which gives us 30% as the probability of occurrence of a red marble. Note that our formula also gives 3/10 or 30% as the answer here. On the other hand, the undefined phrase "equally likely" is part of the definition. Intuitively—and this is the best we can do—we think of an equally likely element in the sample space S as one which is not favored or unfavored when an element is chosen from S. All the elements of S must be equally likely before we may use our equation to calculate probability.

Calculating probabilities

With these important preliminary concepts of sample space and event in mind, we proceed to calculate the probabilities of the previous four events with respect to their sample spaces, whose elements we assume to be equally likely.

a. $$P(\text{getting a head}) = \frac{n(E)}{n(S)} = \frac{\text{number of ways of getting a head}}{\text{number of ways a coin can land}}$$

$$= \frac{n(\{H\})}{n(\{H, T\})} = \frac{1}{2} = .50.$$

b. $$P(\text{getting a number between 2 and 5 inclusive}) = \frac{n(E)}{n(S)}$$

$$= \frac{\text{number of ways of getting a number between 2 and 5 inclusive}}{\text{number of ways a die can land}}$$

$$= \frac{n(\{2, 3, 4, 5\})}{n(\{1, 2, 3, 4, 5, 6\})} = \frac{4}{6} \doteq .67.$$

We are *not* requiring *all four* outcomes in the event to occur; we are calculating the likelihood that *any one* outcome in the event occurs.

c. $$P(\text{having the sum greater than 9}) = \frac{n(E)}{n(S)}$$

$$= \frac{\text{number of ways the sum can be greater than 9}}{\text{number of ways the two dice can land}}$$

$$= \frac{n(\{(4, 6), (5, 5), (5, 6), (6, 4), (6, 5), (6, 6)\})}{n(\{(1, 1), (1, 2), \ldots, (1, 6), (2, 1), \ldots, (6, 5), (6, 6)\})}$$

$$= 6/36 \doteq .17.$$

d. $$P(\text{drawing a queen or a red face card}) = \frac{n(E)}{n(S)}$$

$$= \frac{\text{number of ways of drawing a queen or a red face card}}{\text{number of ways of drawing a card from a standard deck}}$$

$$= \frac{n(\{QC, QD, QH, QS, JD, JH, KD, KH\})}{n(\{2C, 3C, 4C, \ldots, AC, 2D, \ldots, AD, 2H, \ldots, AS\})}$$

$8/52 \doteq .15.$

In practice it is usually not necessary to list the elements of the sample space as we have done in the fourth step of each of these examples. The important point is to determine accurately the number of elements in the event and in the equally likely sample space. Notice in example (c) that if we used the sample space

$$\{2, 3, 4, 5, 6, 7, 8, 9, 10, 11, 12\},$$

an incorrect application of our basic formula for $P(E)$ would say

$$P(\text{sum greater than 9}) = 3/11 \neq 6/36.$$

This application is incorrect because the elements in this alternate sample space are *not* equally likely. A sum of exactly 2, for example, can occur in only one way—(1, 1)—but a sum of exactly 3 can occur in two ways—(1, 2) and (2, 1).

Example 1 An urn contains seven black, four green, six white, and three blue marbles all of equal size, shape, weight, etc. The urn is shaken thoroughly and a marble is drawn by a blindfolded person. Find the probability that it is

a. green,
b. white or blue,
c. yellow,
d. not purple,
e. not black.

Solution

a. $$P(\text{green}) = \frac{\text{number of green marbles}}{\text{number of marbles}} = \frac{4}{20} = .20$$

b. $$P(\text{white or blue}) = \frac{\text{number of white or blue marbles}}{\text{number of marbles}}$$

$$= \frac{6+3}{20} = \frac{9}{20} = .45.$$

c. $$P(\text{yellow}) = \frac{\text{number of yellow marbles}}{\text{number of marbles}} = \frac{0}{20} = .00.$$

d. $$P(\text{not purple}) = \frac{\text{number of nonpurple marbles}}{\text{number of marbles}} = \frac{20}{20} = 1.00.$$

e. $$P(\text{not black}) = \frac{\text{number of nonblack marbles}}{\text{number of marbles}} = \frac{4+6+3}{20} = \frac{13}{20} = .65.$$

Example 1 shows some important implications of the basic definition of probability computation. In (c) the occurrence of a yellow marble is an impossible event and its probability is 0; in (d) the occurrence of a nonpurple marble is a certain event, and its probability is 1; and in (a), (b), and (e) the events may occur but do not have to. In general, (c) illustrates that when E is impossible, $P(E) = 0$; (d) illustrates that when E is certain, $P(E) = 1$; and (a), (b), and (e) illustrate that otherwise $0 < P(E) < 1$. We can organize these past few sentences into a diagram.

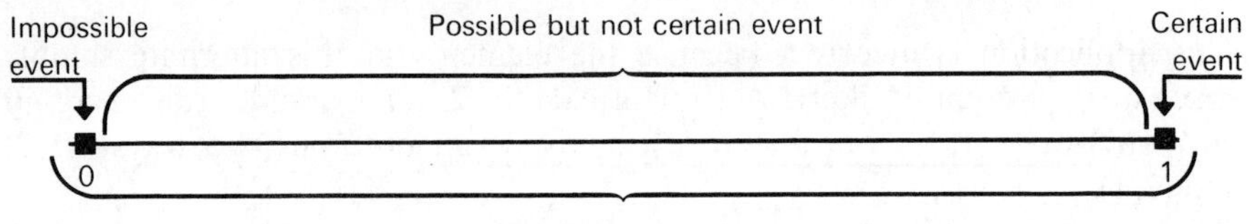

If E is any event, then $0 \leqslant P(E) \leqslant 1$.

So far we can summarize and generalize our work about probability with these general statements about a finite sample space S of n elements

$$S = \{s_1, s_2, s_3, s_4, \ldots, s_n\}$$

in which each separate element s is assigned a unique number $P(s)$, called its probability, with these properties:

1. $$0 \leqslant P(s) \leqslant 1 \text{ if } s \in \{s_1, s_2, s_3, s_4, \ldots, s_n\}.$$
2. $$P(s_1) + P(s_2) + P(s_3) + P(s_4) + \cdots + P(s_n) = 1.$$

   By way of analogy, we may liken probability on a sample space to butter that is spread over pieces of toast. The total amount of butter spread on all the separate pieces is 1 unit and the sum of all the probabilities of the separate individual sample space elements is 1. The butter may not be spread evenly, and the individual probabilities may not all be equal to one another.
3. If we agree that the n outcomes in the sample space are equally likely, then

   $$P(s_1) = P(s_2) = P(s_3) = P(s_4) = \cdots = P(s_n).$$

   Since the sum of these n equal numbers is 1 then

   $$n \quad P(s) = 1$$

or

$$P(s) = \frac{1}{n}.$$

4. Often an event E of interest is the union of several separate sample space elements. If these elements or *points*, as they are sometimes called, are equally likely, then

$$P(E) = \frac{n(E)}{n(S)}.$$

Part (*b*) of Example 1 illustrates this case.

Let us look back at part (e) from Example 1, which states that $P(\text{not black}) = .65$. We can further refine this equation to

$$P(\text{not black}) = .65 = 1 - .35 = 1 - P(\text{black})$$

since .35 is the probability that the marble drawn will be black.

Because the two expressions at the ends of this chain of equalities are equal we now have

$$P(\text{not black}) = 1 - P(\text{black}).$$

We can generalize this idea to any event of any sample space; that is, if E is any event, then

$$P(E') \text{ or } P(\text{not } E) \text{ or } P(E \text{ does not occur}) = 1 - P(E) \text{ or } 1 - P(E \text{ occurs}).$$

In the previous equation E' represents the complement of set E as mentioned in Example 6 of Chapter 2. We also note that solving

$$P(\text{not } E) = 1 - P(E)$$

for $P(E)$ yields

$$P(E) = 1 - P(\text{not } E)$$

or

$$P(E \text{ occurs}) = 1 - P(E \text{ does not occur}). \tag{A}$$

This method of calculating $P(E)$ is occasionally a useful one, as we will see later.

Using a tree diagram

Example 2 An experiment consists of the two-step process of rolling a fair die and then flipping a fair coin. We are interested in the sample space and its number of elements.

Solution We introduce the aid of a *tree diagram* here, which is a way of visualizing the outcomes of the experiment. We first picture those outcomes of rolling a fair die or, for that matter, any die.

Figure 5.1

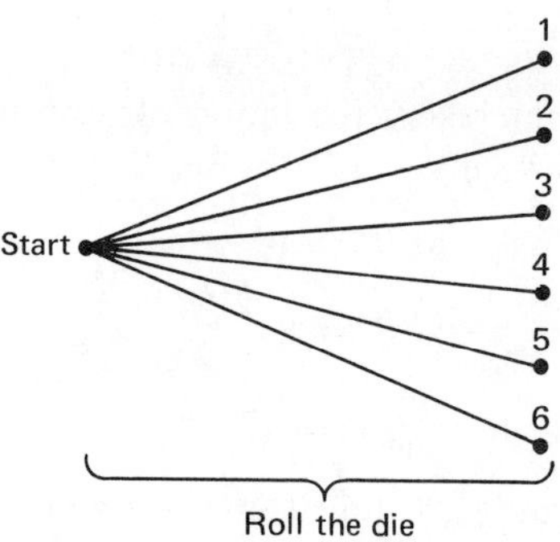

Since any outcome of the die may be followed by any outcome of flipping our coin, we add these branches to the tree:

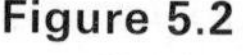
Figure 5.2

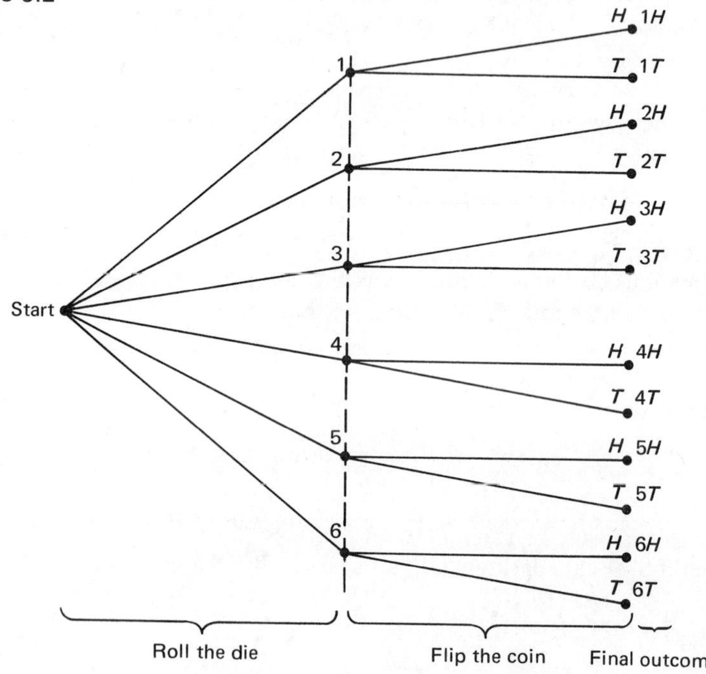

Direct counting of all the outcomes of the composite operations which are at the end of each complete path through the tree yields twelve as the number of elements in the overall sample space. We now notice that the answer is twelve in a problem where 6 and 2 were original input data. Could we have obtained 12 as the answer by direct multiplication without first making the diagram and then counting? The answer is a definite yes, and this problem generalizes to the following important concept.

Fundamental principle of counting If a job can be done by successively completing a first task in t_1 ways and then, for each of these t_1 ways, by next completing a second task in t_2 ways, then the entire job can be done in $t_1 \cdot t_2$ different ways.

We may take a separate look at this problem for its relation to probability. Since the sum of the probabilities of the 12 elements in the sample space

$$\{1H, 1T, 2H, 2T, \ldots, 6H, 6T\}$$

is 1 and we assume that each element is equally likely to occur, we conclude that the probability of any outcome's occurring is 1/12. Equivalently, we can also say that the probability of any path through the tree is 1/12. Then, for example,

$$P(\text{odd number on a die})$$

$$= \frac{\text{number of outcomes with an odd number on the die}}{\text{number of outcomes}} = \frac{6}{12} = .50.$$

Problem set 5–1

1. A fair die is tossed. Write each event in set notation and then find its probability.
   a. 3
   b. an odd number
   c. greater than 4
   d. less than 1
   e. a prime number (1 is not a prime number)
   f. greater than 0
   g. not greater than 3
   h. evenly divisible by 3
   i. between 3 and 5, exclusive of both 3 and 5
   j. between 3 and 5, inclusive
2. A loaded die is tossed so that the probability of an outcome is proportional to its number of dots. Since the total number of dots is

   $$1 + 2 + 3 + 4 + 5 + 6 = 21,$$

   then $P(4 \text{ dots}) = 4/21$. Now find the probabilities of events (a) through (j) in the previous problem. Incidentally, even though the events in the sample space $\{1, 2, 3, 4, 5, 6\}$ are not equally likely the first two general properties of probability mentioned earlier in our summary still apply.
3. A card is drawn at random from a well-shuffled standard deck. Find the probability that it is
   a. red
   b. an honor card (A ten is not an honor card here, but an ace is.)
   c. a black seven

d. black or a seven
e. the five of hearts
f. a diamond
g. red or a three or a jack or the three of hearts or the five of clubs
h. not a diamond
i. not a seven
j. not a 4 or a spade
k. greater than a 5

4. A pair of dice, colored red and green (in that order) is thrown. Find the probabilities of these events.
   a. the sum is 5
   b. the sum is 7 or 11
   c. a red 5
   d. a green 3 or a green 4
   e. the sum is evenly divisible by 3
   f. the sum is an even number
   g. the sum is an odd number
   h. a red 2 and a green prime number
   i. a green 3 or a sum of 4
   j. a green 3 or a 4 on either die
5. Which of these assertions is true?
   a. A card drawn from a standard, well-shuffled deck is either a diamond or not a diamond. Therefore, $P(\text{diamond}) = 1/2 = .50$.
   b. A card drawn from a standard, well-shuffled deck is either a red card or not a red card. Therefore, $P(\text{red}) = 1/2 = .50$.
   c. A fair coin is tossed and the outcome is either a tail or not a tail. Therefore, $P(\text{tail}) = 1/2 = .50$.
   d. A coin is tossed and the outcome is either a tail or not a tail. Therefore, $P(\text{tail}) = 1/2 = .50$.
   e. There are eight states whose first names begin with "M" and eight states whose first names begin with "N." Therefore, if a person is chosen at random from the population of these sixteen states, P(first name of his state begins with "M") $= 8/16 = 1/2$.
   f. The sentence "Now is the time for all good men to come to the aid of their party" contains sixteen words. The words are written on sixteen identical-sized pieces of paper and placed in a hat where a piece is drawn out at random. Then P(drawing "is") $= P$(drawing "their") $=$ 1/16.
   g. If a pair of fair coins is tossed, then the sample space for the possible number of heads is $\{0, 1, 2\}$. Therefore, $P(\text{at least one head}) = 2/3$.
6. Empirical probability is widely used in the insurance industry, where mathematicians estimate the probability of a person's death, along with other factors (occupational hazards, medical records, etc.), in order to determine the premium for his policy. The American Experience Table of Mortality, listed here, shows the death rates per every 100,000 people in the United States from age ten through the ages of their deaths, based upon

American experience table of mortality

Age	Number living	Number dying	Yearly probability of dying	Yearly probability of living	Age	Number living	Number dying	Yearly probability of dying	Yearly probability of living
10	100 000	749	0.007 490	0.992 510	53	66 797	1 091	0.016 333	0.983 667
11	99 251	746	0.007 516	0.992 484	54	65 706	1 143	0.017 396	0.982 604
12	98 505	743	0.007 543	0.992 457	55	64 563	1 199	0.018 571	0.981 429
13	97 762	740	0.007 569	0.992 431	56	63 364	1 260	0.019 885	0.980 115
14	97 022	737	0.007 596	0.992 404	57	62 104	1 325	0.021 335	0.978 665
15	96 285	735	0.007 634	0.992 366	58	60 779	1 394	0.022 936	0.997 064
16	95 550	732	0.007 661	0.992 339	59	59 385	1 468	0.024 720	0.975 280
17	94 818	729	0.007 688	0.992 312	60	57 917	1 546	0.026 693	0.973 307
18	94 089	727	0.007 727	0.992 273	61	56 371	1 628	0.028 880	0.971 120
19	93 362	725	0.007 765	0.992 235	62	54 743	1 713	0.031 292	0.968 708
20	92 637	723	0.007 805	0.992 195	63	53 030	1 800	0.033 943	0.966 057
21	91 914	722	0.007 855	0.992 145	64	51 230	1 889	0.036 873	0.963 127
22	91 192	721	0.007 906	0.992 094	65	49 341	1 980	0.040 129	0.959 871
23	90 471	720	0.007 958	0.992 042	66	47 361	2 070	0.043 707	0.956 293
24	89 751	719	0.008 011	0.991 989	67	45 291	2 158	0.047 647	0.952 353
25	89 032	718	0.008 065	0.991 935	68	43 133	2 243	0.052 002	0.947 998
26	88 314	718	0.008 130	0.991 870	69	40 890	2 321	0.056 762	0.943 238
27	87 596	718	0.008 197	0.991 803	70	38 569	2 391	0.061 993	0.938 007
28	86 878	718	0.008 264	0.991 736	71	36 178	2 448	0.067 665	0.932 335
29	86 160	719	0.008 345	0.991 655	72	33 730	2 487	0.073 733	0.926 267
30	85 441	720	0.008 427	0.991 573	73	31 243	2 505	0.080 178	0.919 822
31	84 721	721	0.008 510	0.991 490	74	28 738	2 501	0.087 028	0.912 972
32	84 000	723	0.008 607	0.991 393	75	26 237	2 476	0.094 371	0.905 629
33	83 277	726	0.008 718	0.991 282	76	23 761	2 431	0.102 311	0.897 689
34	82 551	729	0.008 831	0.991 169	77	21 330	2 369	0.111 064	0.888 936
35	81 822	732	0.008 946	0.991 054	78	18 961	2 291	0.120 827	0.879 173
36	81 090	737	0.009 089	0.990 911	79	16 670	2 196	0.131 734	0.868 266
37	80 353	742	0.009 234	0.990 766	80	14 474	2 091	0.144 466	0.855 534
38	79 611	749	0.009 408	0.990 592	81	12 383	1 964	0.158 605	0.841 395
39	78 862	756	0.009 586	0.990 414	82	10 419	1 816	0.174 297	0.825 703
40	78 106	765	0.009 794	0.990 206	83	8 603	1 648	0.191 561	0.808 439
41	77 341	774	0.010 008	0.989 992	84	6 955	1 470	0.211 359	0.788 641
42	76 567	785	0.010 252	0.989 748	85	5 485	1 292	0.235 552	0.764 448
43	75 782	797	0.010 517	0.989 483	86	4 193	1 114	0.265 681	0.734 319
44	74 985	812	0.010 829	0.989 171	87	3 079	933	0.303 020	0.696 980
45	74 173	828	0.011 163	0.988 837	88	2 146	744	0.346 692	0.653 308
46	73 345	848	0.011 562	0.988 438	89	1 402	555	0.395 863	0.604 137
47	72 497	870	0.012 000	0.988 000	90	847	385	0.454 545	0.545 455
48	71 627	896	0.012 509	0.987 491	91	462	246	0.532 468	0.467 532
49	70 731	927	0.013 106	0.986 894	92	216	137	0.634 259	0.365 741
50	69 804	962	0.013 781	0.986 219	93	79	58	0.734 177	0.265 823
51	68 842	1 001	0.014 541	0.985 459	94	21	18	0.857 143	0.142 857
52	67 841	1 044	0.015 389	0.984 611	95	3	3	1.000 000	0.000 000

very careful records that are kept over a period of many years. The symbol ${}_{x}l_{y}$ represents the probability that a person x years old will live to be y years old. For example, ${}_{23}l_{27}$ is the probability that a person aged twenty-three will live to be at least twenty-seven. This means that our sample space is the 90,471 people who are alive at age twenty-three. The table on previous page tells us that 87,596 are still alive at age twenty-seven, so

P(living from age 23 to age 27)

$$= \frac{\text{number of people alive at 23 who lived to 27}}{\text{number of people who were alive at 23}} = \frac{87{,}596}{90{,}471} \doteq .968.$$

Also, ${}_{x}d_{y}$ is defined as the probability that a person x years old will die by the time he is y years old, and since dying is the opposite of living, ${}_{x}d_{y} = 1 - {}_{x}l_{y}$.

a. Find ${}_{18}l_{24}$.
b. Find ${}_{10}l_{40}$.
c. Find ${}_{35}l_{55}$.
d. Find the probability that you will live to be seventy-two years old.
e. What is the half-life of a person twenty years old, i.e., in how many years will half the people aliv⁻ at twenty be dead?
f. What is your half-life?
g. Between what two consecutive ages does the probability of death become greater than .100?
h. Find ${}_{18}d_{24}$.
i. Find ${}_{10}d_{40}$.
j. Find ${}_{20}d_{65}$.
k. According to the table ${}_{95}d_{96} = 3/3 = 1$. Here is an event whose probability is 1, and yet the event is not a certain one. Is the statement ${}_{94}d_{96} = 1$ a true one? Is every person alive at 94 dead at 96?

7. Draw a tree diagram for the experiment of flipping a fair coin three times. How many elements are there in the sample space?
8. If the sample space S is the result of the previous problem, then find the probabilities of these events:
   a. exactly 1 head
   b. exactly 2 heads
   c. all tails
   d. no heads
   e. at least 1 head
   f. more tails than heads
   g. more heads than tails
9. If a coin is flipped these numbers of times, then how many different outcomes are possible in each case?
   a. 1 b. 2 c. 3
   d. 4 e. 5 f. n
10. Jeff's wardrobe consists partly of white, blue, yellow, red, and gold shirts and accompanying these are ties whose colors are black/white squares,

brown circles, cerise, and green. Aesthetics aside, how many shirt–tie selections can Jeff wear?

11. A commuter train between Westover and Urbania has seven intermediate stations. How many one-way tickets between any two stations are possible?
12. Let $S = \{\text{Sue, Bob, Madeleine, Jim, Alice, Clark, Frank}\}$. A committee of one man and one woman is to be chosen from these people. Draw a tree diagram to calculate the number of ways this can be done.
13. The senior class of eighty at Brandon High School has thirty-five girls. Howard is a senior there.
    a. If a senior is chosen at random, find the probability that it is Howard.
    b. If a senior boy is chosen at random, find the probability that it is Howard.
14. An urn contains four red, one blue, and three white marbles. We make two draws with replacement (replace the first marble and shake the urn before the second is drawn).
    a. Draw a tree diagram for this experiment, and then determine the number of possible different color selections after examining the tree diagram.
    b. Could you have used the Fundamental Principle of Counting before answering (a)? If not, why not?
15. The question is the same as above, only this time the first marble is *not* replaced before the second one is drawn.
16. An urn contains three externally identical envelopes with one containing a penny, another a nickel, and the third a dime. We make three successive drawings with replacement and shaking.
    a. Draw the tree diagram for this experiment.
    b. How many paths are there in it?
    c. Are the outcomes equally likely?
    d. Write the sample space in terms of the total cent value of the three envelopes drawn.
    e. Are the outcomes equally likely now? If not, why not?
17. Which of the following statements is a correct interpretation of the forecast "the probability for rain tomorrow is 20%."?
    a. It will rain one minute out of every five.
    b. During the twenty-four hour interval, it will rain approximately 20% (four hours forty-eight minutes) of the time.
    c. Historically, for such weather conditions, it has rained one time out of every five.
    d. Weather conditions seldom duplicate exactly those previously recorded. Some conditions appear at some times; others appear at other times. From these past performances a compromise "percentage" of rain is reached.
    e. History has nothing to do with the forecast; the weatherman considers that he will be correct in forecasting rain 20% of the time.
18. A friend offers to sell you a ticket in a drawing for a radio, and a total of fifty tickets have been printed. After you buy the forty-fifth and last sold ticket, your friend says that if you buy another one you double your

chances of winning. Is he correct, assuming that the last four tickets would not otherwise be sold?

19. The question is the same as above, only this time with the concluding assumption that the last tickets will be sold.
20. You are at a carnival and the barker, amidst much fanfare and fast talking, asks you to bet him even money on a game of drawing a card. This means your 50¢ entrance fee is lost if the card's outcome is as he predicts, or you get back your 50¢ plus his 50¢ if the card's outcome is different from his prediction. He has three cards in an unseen tray under his counter. The cards are one with red on both sides, one with white on both sides, and one with red on one side and white on the other side. You draw a card at random from the tray without either of you seeing the reverse side. If it is white he says, "That is not the red–red card. Therefore, it is either the white–red card or the white–white card, and I predict that its other side is white." If it is red he makes a corresponding prediction.
    a. In the first instance what is the probability that the other side is white?
    b. Should you accept his offer?
    c. What does he say if the first card drawn is red?

Counting techniques

In the last section we measured the probability of an event with the ratio of the number of ways the event can occur to the number of equally likely ways the event's related experiment can occur. The problems we have done so far have given us no real difficulty in determining the two numbers we had to divide, but there are many applications of probability that require a knowledge of more advanced counting techniques. Thus we turn our attention to these techniques before studying more about probability itself.

Let us review Problem 10 of the last section where the answer, by the Fundamental Principle of Counting, is $5 \cdot 4 = 20$. Suppose that Jeff's selection of socks contains black, brown, white, green, red, and blue pairs, and we want to know the total number of different ways he can select the three accessories as he dresses. The Fundamental Principle does not quite apply since it only works with jobs that are subdivided into just two tasks. We can generalize it to n tasks, so its application here says that Jeff's job of choosing three accessories is broken down into the successive completion of three tasks. These three tasks are choosing one shirt, choosing one tie, and choosing one pair of socks which can be done in, respectively, five, four, and six ways, so that the total number of ways of doing the entire job is

$$5 \cdot 4 \cdot 6 = 120.$$

After a generalization to the situation of n tasks, the Fundamental Principle of Counting (abbreviation: FPC) now becomes

If a job can be done by successively completing n tasks such that the first one can be done in t_1 ways, and then for each of these t_1 ways the second task can be done in t_2 ways, and then for each of these t_2

ways the third task can be done in t_3 ways and then . . . and the nth task can be done in t_n ways, then the entire job can be done in $t_1 \cdot t_2 \cdot t_3 \cdot \cdots \cdot t_n$ different ways.

Example 3 In Ohio license plates are often of the form $LLDDDD$ (where L represents a letter and D represents a digit, such as in AC1843). The two letters can be any except I or O and the first digit cannot be 0 because of possible confusion of these letters and digits. How many different license plates are possible?

Solution We think of the overall job as selecting the six characters of the license plate, and we can do this by successively completing the six tasks of selecting the first character, then the second character, then the third character, than the fourth character, then the fifth character, and then the sixth character. These tasks can be done in respectively $t_1 = 24$ ways (the letters I and O are excluded), $t_2 = 24$ ways, $t_3 = 9$ ways (0 cannot be the first digit), $t_4 = 10$ ways, $t_5 = 10$ ways, and $t_6 = 10$ ways. Then the Fundamental Principle of Counting tells us the total number of different ways of doing the entire job is

$$24 \cdot 24 \cdot 9 \cdot 10 \cdot 10 \cdot 10 = 5{,}184{,}000.$$

We may certainly conclude that the Fundamental Principle of Counting lets us count without counting, since we do not have to list all the separate outcomes of a job and count them one by one in order to determine the total number of outcomes.

We now introduce a symbol which will prove quite useful later on. Some of us may remember the special symbol for the product of all the positive integers from a given one, n, down to the positive integer 1:

$$n! = n \cdot (n-1) \cdot (n-2) \cdot \cdots \cdot 3 \cdot 2 \cdot 1.$$

The symbol $n!$ is read "n factorial." Then, for example,

$$5! = 5 \cdot 4 \cdot 3 \cdot 2 \cdot 1 = 120,$$
$$7! = 7 \cdot 6 \cdot 5 \cdot 4 \cdot 3 \cdot 2 \cdot 1 = 5040,$$
$$2! = 2 \cdot 1 = 2,$$

and

$$1! = 1.$$

For a reason we will see later, $0!$ is defined to be 1. Symbols such as $(-7)!$, $(5/8)!$, and $(\sqrt{3})!$ are not defined at our level of mathematics.

Permutations

Example 4 How many three-letter words can be formed from the letters of the word MARCH if the words do not have to make sense and no letter may be repeated in the word?

Solution We can form a typical three-letter word by successively choosing three letters to fill in these blanks with:

The first letter in a word can be any one of the five, but then the second one can only be chosen in four ways since it must be different from its predecessor. The third letter must be different from its predecessors, and hence can be any one of the remaining three unchosen ones. We may place these numbers in the blanks to represent the number of ways of doing each successive task

$$\underline{5}\ \underline{4}\ \underline{3}$$

and applying the Fundamental Principle of Counting gives

$$5 \cdot 4 \cdot 3 = 60$$

as the answer.

In the previous example we actually chose three distinct objects out of five distinct objects where the order of selection was important. That is, even though RAM and MAR are selections with identical elements, they are still different from each other because the order of appearance of the letters is different. RAM and MAR are examples of *permutations*, or ordered selections of distinct objects from a specific set.

Definition An *r-permutation* of n objects is an ordered selection of r distinct objects chosen from a set of n distinct objects where $r \leqslant n$.

In our instance (Example 4) $n = 5$, $r = 3$, and the number of 3-permutations that exist is sixty. We symbolize this fact by the equation

$${}_5P_3 = 60$$

which is pronounced "5 p 3 equals 60," meaning that the number of 3-permutations we can form from five distinct objects is sixty, or, in other words, the number of different ways of choosing three distinct objects out of five distinct objects is sixty. In other books ${}_5P_3$ also may appear as P_3^5 or $P(5, 3)$.

Is there a formula we could have used to obtain the output of 60 directly from the inputs of 5 and 3 without using the Fundamental Principle of Counting as an intermediate step? Yes, and this chain of equations shows how.

$${}_5P_3 = 60 \qquad \text{previous result}$$

$${}_5P_3 = 5 \cdot 4 \cdot 3 \qquad \text{FPC—Fundamental Principle of Counting}$$

$${}_5P_3 = \frac{5 \cdot 4 \cdot 3}{1} \qquad \text{division by 1 leaves a number unchanged}$$

$${}_5P_3 = \frac{5 \cdot 4 \cdot 3 \cdot 2 \cdot 1}{2 \cdot 1} \qquad \text{multiply the numerator and denominator of the previous fraction by } 2 \cdot 1$$

$${}_5P_3 = \frac{5!}{(2)!} \qquad \text{definition of } n!$$

$${}_5P_3 = \frac{5!}{(5-3)!} \qquad 2 = 5 - 3$$

Since we have the answer of $5!/(5-3)!$ expressed directly in terms of the original numbers 5 and 3, we ask these questions:

$$_6P_3 \stackrel{?}{=} 6!/(6-3)!,$$
$$_7P_3 \stackrel{?}{=} 7!/(7-3)!,$$
$$_7P_4 \stackrel{?}{=} 7!/(7-4)!,$$
$$_7P_5 \stackrel{?}{=} 7!/(7-5)!, \quad \text{and in general}$$
$$_nP_r \stackrel{?}{=} \frac{n!}{(n-r)!}$$

The answer to all these questions is yes, which we could prove in each case by using the FPC. The last question-turned-equation says that the number of ways of choosing r distinct objects out of n distinct objects—that is, the number of r-permutations that exist—where order is important is

$$_nP_r = \frac{n!}{(n-r)!}.$$

What about $_{26}P_3$, which represents the number of three-letter words that we can form from our alphabet? Here $n = 26$ and $r = 3$, so we evaluate

$$_{26}P_3 = 26!/(26-3)! = 26!/23!$$

in this fashion:

$$\frac{26 \cdot 25 \cdot 24 \cdot \cancel{23!}}{\cancel{23!}} = 26 \cdot 25 \cdot 24 = 15{,}600.$$

Notice that we did not completely write out 26! as

$$26 \cdot 25 \cdot 24 \cdot 23 \cdot 22 \cdot \cdots 3 \cdot 2 \cdot 1$$

and 23! as

$$23 \cdot 22 \cdot 21 \cdot 20 \cdot 19 \cdot \cdots 3 \cdot 2 \cdot 1$$

since writing the numerator as $26 \cdot 25 \cdot 24 \cdot 23!$ allowed us to directly cancel its 23! factor with the entire denominator.

We saw that the number of 3-permutations of five distinct objects was sixty where selections such as MAR and RAM were considered to be different from each other. What happens if we want to know the number of possible selections of three distinct objects out of five distinct objects where the order of selection is not important? Then the six 3-permutations of, say, MAR (MAR, MRA, AMR, ARM, RAM, RMA) count as just one selection since they all have the same letters.

Combinations

A definition is in order before we answer the question posed in the last paragraph.

Definition An *r-combination* of n objects is an unordered selection of r distinct objects chosen from a set of n distinct objects where $r \leqslant n$.

The only difference between this definition of a combination and the previous one of a permutation is that the prefix "un" was added to "ordered." It means, for example, that while MAR and RAM are different permutations they are the same combination, or order makes a difference in counting permutations, but it does not make a difference in counting combinations. Of course, MAR and HAM are both different permutations and different combinations.

We will symbolize the number of 3-combinations that are possible by choosing them from the letters MARCH by ${}_5C_3$ (read "5 c 3" or, occasionally, "5 choose 3"). The only way we can evaluate ${}_5C_3$ at present is to list directly all the 3-combinations and then count them. Proceeding, we have

MAR MAC MAH MRC MRH MCH ARC ARH ACH RCH

so that

$${}_5C_3 = 10.$$

We naturally search for a way that we could obtain the output answer of 10 directly from the input numbers of 5 and 3 without having to first list and then count. We begin this search by writing all the 3-permutations and 3-combinations in a rectangular array with six horizontal rows and ten vertical columns. Notice that the top row is exactly the same as our previous list used to evaluate ${}_5C_3$, and that any one column contains only one combination although it has six permutations.

MAR	MAC	MAH	MRC	MRH	MCH	ARC	ARH	ACH	RCH
MRA	MCA	MHA	MCR	MHR	MHC	ACR	AHR	AHC	RHC
ARM	ACM	AHM	RCM	RHM	CHM	RCA	RHA	CHA	CHR
AMR	AMC	AMH	RMC	RMH	CMH	RAC	RAH	CAH	CRH
RMA	CMA	HMA	CMR	HMR	HMC	CAR	HAR	HAC	HRC
RAM	CAM	HAM	CRM	HRM	HCM	CRA	HRA	HCA	HCR

Since this array has six rows and ten columns, it has $6 \cdot 10 = 60$ elements. Now—and here is the key to our discovering a formula to evaluate ${}_5C_3$—we take the equation

$$6 \cdot 10 = 60$$

and legally rewrite 6 as 3!, 10 as ${}_5C_3$, and 60 as ${}_5P_3$ in order to introduce the numbers 5 and 3 into our discussion. After substitution the resulting true equation is

$$3! \cdot {}_5C_3 = {}_5P_3 \qquad \text{(A)}$$

or

$$3! \cdot {}_5C_3 = \frac{5!}{(5-3)!}$$

by the formula for ${}_5P_3$. Then we multiply both sides by 1/3! (or, equivalently, divide both sides by 3!) to obtain, after cancellation,

$${}_5C_3 = \frac{5!}{3! \cdot (5-3)!},$$

which does indeed have the answer 10 expressed solely in terms of the inputs 5 and 3.

We can now presume that since

$${}_5C_3 = \frac{5!}{3! \cdot (5-3)!}$$

then ${}_6C_3$, the number of ways of selecting a subset of three elements from a universal set of six distinct elements, should equal

$$\frac{6!}{3! \cdot (6-3)!},$$

${}_7C_3$ should equal

$$\frac{7!}{3! \cdot (7-3)!},$$

${}_7C_4$ should equal

$$\frac{7!}{4! \cdot (7-4)!},$$

${}_7C_5$ should equal

$$\frac{7!}{5! \cdot (7-5)!},$$

and in general ${}_nC_r$, where r and n are positive integers (although r may be 0) with $r \leqslant n$, should equal

$$\frac{n!}{r! \cdot (n-r)!}.$$

The complete proof necessary to change the verbs in the last sentence from "should" to "does" is omitted in this text. It is based on a generalization from equation (A). Instead of having 3! rows with ${}_5C_3$ elements in each row, we have $r!$ rows with ${}_nC_r$ elements in each row. There are

$$r! \cdot {}_nC_r = {}_nP_r$$

elements in the entire array, so

$${}_nC_r = \frac{{}_nP_r}{r!}$$

or, after some work with the right-hand side of this last equation,

$$_nC_r = \frac{n!}{r! \cdot (n-r)!}.$$

Example 5 Evaluate $_{30}C_4$.
Solution By substituting 30 for n and 4 for r in the general formula we have

$$_{30}C_4 = \frac{30!}{4! \cdot 26!}.$$

Notice how we rewrite the numerator and smaller of the two numbers in the denominator

$$_{30}C_4 = \frac{30 \cdot 29 \cdot 28 \cdot 27 \cdot 26!}{4 \cdot 3 \cdot 2 \cdot 1 \cdot 26!},$$

so that after cancellation we have

$$_{30}C_4 = \frac{\overset{15}{\cancel{30}} \cdot 29 \cdot \overset{7}{\cancel{28}} \cdot \overset{9}{\cancel{27}} \cdot \cancel{26!}}{\cancel{4} \cdot \cancel{3} \cdot \cancel{2} \cdot 1 \cdot \cancel{26!}} = 27{,}405.$$

Permutations or combinations?

Students are often confused about whether a particular problem is concerned with permutations or with combinations. Remember, from the definitions, that we must first be selecting r distinct objects from n distinct objects where $r \leq n$. If the problem implies that two selections of r objects each with the same elements but with a different ordering of those elements are different from each other, then we use $_nP_r$; otherwise, $_nC_r$ is the formula to use. For example, suppose a basketball coach must form a team of center, left forward, right forward, left guard, and right guard from a squad of twelve boys. If he chooses Jones, Smith, Brown, Hastings, and White in that order, then the team is:

Table 5.1

Position	Player
center	Jones
left forward	Smith
right forward	Brown
left guard	Hastings
right guard	White

Choosing Hastings, Smith, White, Jones, and Brown in that order yields this team:

Table 5.2

Position	Player
center	Hastings
left forward	Smith
right forward	White
left guard	Jones
right guard	Brown

It is a different team, particularly if Hastings is a 5′ 6″ dribbling and ball-handling expert! Then, since order is important and does make a difference, the answer is ${}_{12}P_5 = 95{,}040$. On the other hand, if any player can play any position, the order of selection of the team is not important and there are only ${}_{12}C_5 = 792$ different teams (right on both these numbers?). We note again that the FPC, permutation symbol, and combination symbol really do let us count without counting.

Often one problem requires us to use more than one counting principle.

Example 6 How many seven-member committees can be formed from 9 men and 14 women if there is to be a majority of women?

Solution The total number of such committees is the sum of the four-woman, five-woman, six-woman, and seven-woman committee possibilities. Let us look first at the number of four-woman committees, each of which also has three men on it. We can take the job of forming a four-woman committee and break it down into *two* tasks: selecting the four women from the pool of fourteen and then selecting the three men from the pool of nine that are available. The women can be selected in ${}_{14}C_4$ ways since the selection of Helen, Bess, Julie, and Nancy, for example, is the same as the selection of Bess, Nancy, Julie, and Helen. Then the three men can be chosen in ${}_9C_3$ different ways, so that by the FPC the number of these four-woman, three-man committees is ${}_{14}C_4 \cdot {}_9C_3$. Similar reasoning says the number of five-woman committees is ${}_{14}C_5 \cdot {}_9C_2$, the number of six-woman committees is ${}_{14}C_6 \cdot {}_9C_1$, and the number of seven-woman committees is ${}_{14}C_7 \cdot {}_9C_0$. Then the total number of woman-majority committees is the sum of these four products, which is

$$84{,}084 + 72{,}072 + 27{,}027 + 3{,}423 = 186{,}615.$$

We close this section by noting that ${}_5C_5$ is 1 since the only way to choose five objects out of five is to take all of them at once. The reason 0! is defined to be 1 is so that replacing n by 5 and r by 5 in the formula for ${}_nC_r$ will give us 1 as the answer. Other symbols for ${}_5C_3$ are $C(5, 3)$, C_3^5, and $\binom{5}{3}$. This last form is common among authors, and we will frequently interchange ${}_nC_r$ and $\binom{n}{r}$ in the rest of this book.

Problem set 5–2

1. Evaluate these expressions.
   a. ${}_5P_4$ b. ${}_5P_3$ c. ${}_5P_2$ d. ${}_5P_1$
   e. ${}_5P_0$ f. ${}_5P_5$ g. ${}_4P_4$ h. ${}_3P_3$
   i. ${}_6P_6$ j. What generalization do the last four results suggest?
   k. ${}_9P_3$ l. ${}_{31}P_3$ m. ${}_{1001}P_2$ n. ${}_{15}P_6$
2. Evaluate these expressions.
   a. ${}_8C_3$ b. ${}_8C_2$ c. ${}_8C_1$ d. ${}_9C_1$
   e. ${}_{10}C_1$ f. ${}_{11}C_1$ g. ${}_{385}C_1$
   h. What generalization do the last five results suggest?
   i. ${}_{26}C_3$ j. ${}_5C_5$ k. ${}_6C_6$ l. ${}_7C_7$
   m. ${}_{40}C_{40}$
   n. What generalization do the last four results suggest?
   o. ${}_{52}C_5$ p. What set contains ${}_{52}C_5$ elements? q. ${}_5C_0$
3. How many five-card poker hands are possible if they are drawn from a standard deck?
4. Explain why a combination lock is incorrectly named in light of our definition of a combination. What should it be called?
5. The father of twelve children must decide which subset of four children will ride with him in the family's sedan. How many ways can he do this?
6. The father of twelve children must decide which subset of eight children will ride with his wife in the family's Ford Econoline van. How many ways can he do this?
7. What generalization do the previous two problems suggest? Can you prove it?
8. The next-to-the-last sentence of the next-to-the-last paragraph of this section of the text starts with "Similar reasoning shows . . .". Explain in a paragraph why the number of five-women committees is ${}_{14}C_5 \cdot {}_9C_2$.
9. How many three-digit numbers can be formed from our ten digits if
   a. repetition of digits is not allowed?
   b. repetition of digits is not allowed and the number is even?
   c. repetition of digits is allowed?
   d. repetition of digits is allowed and the number is even?
10. Refer back in the text to Example 3 regarding license plates. If we accept its original statements with these independent modifications, then how many different license plates are possible if
    a. they end in an even digit?
    b. they end in 00?
    c. plates with RE as the first two letters are not allowed (REplacement plates for those that are lost or stolen)?
    d. repetition of letters is not allowed?
    e. repetition of letters is not allowed, and neither is repetition of digits?
    f. repetition of letters is not allowed, and the order of digits does not make a difference (e.g. AC1843 = AC3418 and BA1964 ≠ AB1964)?

g. repetition of letters is allowed, and the order of digits does not make a difference?

h. repetition of letters is not allowed, the order of the letters does not make a difference (e.g. AC1843 = CA1843), and the order of digits does not make a difference?

i. repetition of letters is not allowed, the order of the letters does not make a difference, the order of digits does not make a difference, and the number may end in an odd digit?

11. How many seven-person committees can be formed from nine men and fourteen women?

12. How many different management teams of President, Vice-President, Secretary, and Treasurer can be formed from fifteen recent college graduates if

    a. any graduate can do any job?

    b. four persons are qualified to be only the president, five are qualified to be only the treasurer, and the remainder equally qualified for the remaining jobs?

    c. three persons are qualified only for the presidency, six only for the vice-presidency, and four only for the secretariatship?

13. In how many different ways can three officers and five enlisted men line up for chow if only the first person must be an officer?

14. In how many different ways can three officers and five enlisted men line up for chow if the first three persons must be officers?

15. In how many different ways can three officers and five enlisted men line up for chow if the first and last person must be an officer?

16. How many different ways can five couples be seated in a row if

    a. a boy has the first seat and the couples sit next to their dates with no people of the same sex sitting next to each other?

    b. the couples sit next to their dates?

    c. the girls sit together?

    d. the girls sit together in the five left-most seats?

    e. the girls and boys are seated alternately, but not necessarily next to their dates?

    f. there are no restrictions on how the ten people may sit?

17. A grievance committee of six is to be selected from ten labor and seven management men.

    a. How many committees are possible?

    b. How many unipartisan committees are possible?

    c. How many committees favoring the union are possible?

    d. How many committees favoring management are possible?

    e. How many committees favoring neither are possible?

    f. How many bipartisan committees are possible?

    g. Compare your answer from part (a) with the sum of your answers from parts (c), (d), and (e).

    h. How many committees are possible if they contain four management men and the first person selected is a management chairman while the second person selected is a union recorder?

i. How many committees are possible if they favor the union and the first person selected is a management chairman while the second person selected is a union recorder?

18. Part (j) of Problem 1 indicated that ${}_nP_n = n!$ or, in words, the number of ways of choosing n distinct objects out of n where order is important is $n!$. Then the number of nine-letter words we can form from CHAMPIONS is 9! or 362,880 (right?). What about the word SASSAFRAS—how many different nine-letter words can we form from it? The answer is not ${}_9P_9 = 9!$ since the original nine objects are not distinct, which eliminates our use of the permutation symbol. However, we will analyze it in this fashion: our job is to form a nine-letter word by filling in these nine blanks.

— — — — — — — — —

The key here is to break the job down into the *four* successive tasks of selecting three blanks to fill with an A, then selecting any one of the remaining six blanks to fill with an F, then selecting any one of the remaining five blanks to fill with an R, and then finally selecting any four of the remaining four blanks to fill with an S. The number of ways of doing these four selections, respectively, is ${}_9C_3$, ${}_6C_1$, ${}_5C_1$, and ${}_4C_4$.

a. How many ways may the entire job be done?
b. How many rearrangements of the word MISSISSIPPI are possible?
c. How many rearrangements of the word ALFALFA are possible?
d. How many rearrangements of the word MATHEMATICS are possible?
e. How many rearrangements of the word PITCHFORKED are possible?
f. How many ways may four red, two blue, five white, and one green marble be placed in a row?
g. How many ways may four red, two blue, five white, and onc green marble be placed in a row if the end marbles must be of the same color?
h. How many rearrangements of the word DIVIDED are possible?
i. How many rearrangements of the word DIVIDEND are possible?
j. Four men, three women, and six children stand in a row to pose for a picture. How many pictures are possible?
k. Suppose the photographer in part (j) jerks the camera so that when the pictures are developed he can identify an individual only as a man, woman, or child. How many possible pictures are there now?

19. A football coach has a squad of twenty-five boys with eighteen capable of playing any of the seven line positions and the remainder capable of playing any of the four backfield positions. How many different teams can he field? Do not evaluate your answer.

20. How many baseball teams are possible if a coach can form them from seven outfielders, eight infielders, six pitchers, and three catchers? There are three outfielders, four infielders, one pitcher, and one catcher on a team.

21. How many subsets with four elements in each subset does a set of n elements have where $n \geqslant 4$? Hint: answer the question for small, successive values of $n \geqslant 4$ and look for a pattern in the results which lets you write s, the number of subsets, as a function of n.

22. There is just one way of placing two distinguishable balls into one cell if the order of the balls in a cell is not important. How many ways may one place these two balls
    a. into two numbered cells?
    b. into three numbered cells?
    c. into four numbered cells?
    d. into r numbered cells?

    How many ways may one place three distinguishable balls
    e. into one numbered cell?
    f. into two numbered cells?
    g. into r numbered cells?
    h. Construct an appropriate table which summarizes your work so far.

    How many ways may one place four distinguishable balls
    i. into one numbered cell?
    j. into two numbered cells?
    k. into r numbered cells?
    l. Expand your table from part (h) to summarize your entire work so far.

    How many ways may one place n distinguishable balls
    m. into one numbered cell?
    n. into two numbered cells?
    o. into r numbered cells?
23. Solve the following equations for n.
    a. ${}_{n+1}C_{n-1} = 21$
    b. ${}_nP_{n-1} = 56$
    c. ${}_{n+1}C_3 = 2 \cdot {}_nC_2$
24. Show that $\dfrac{{}_nP_r}{r!} = \dfrac{n!}{r! \cdot (n-r)!} = {}_nC_r$.
25. A committee of five is to be chosen from eight Democrats and six Republicans by drawing five names out of a hat where each of the pool of fourteen has his name written on an identical piece of paper. Find the number of different ways of choosing this committee such that there are:
    a. three Republicans on the committee.
    b. four Democrats on the committee.
    c. more Democrats than Republicans on the committee.
    d. more Republicans than Democrats on the committee.
    e. same as (d), only there are now eight Democrats and eight Republicans in the pool.
    f. same as (d), only there are now eight Democrats and ten Republicans.
    g. same as (d), only there are now eight Democrats and twelve Republicans.
    h. same as (d), only there are now eight Democrats and fourteen Republicans.
    i. same as (d), only there are now eight Democrats and sixteen Republicans.
    j. same as (d), only there are now eight Democrats and r Republicans, where $r \geqslant 3$.

k. same as (d), only there are now d Democrats and r Republicans where $d \geqslant 3$ and $r \geqslant 3$.
l. same as (d), only there are now d Democrats and r Republicans where $d \geqslant 4$ and $r \geqslant 4$. There are seven people on the committee.
m. same as (d), only assume there are now d Democrats and r Republicans with c people on the committee where c is an odd number greater than 1 with $d \geqslant (c+1)/2$ and $r \geqslant (c+1)/2$.

26. The New York Yankees have twenty-five players on the team. Assume that each player can play any position, including pitcher and catcher. How many ways can the manager choose a nine-man team to play in a game?
27. Use the information in problem 26. How many ways can the manager choose his batting order to hand to the umpire at the beginning of the game.
28. The New York Yankees have twenty-five players on the team. Eight of them are pitchers, three are catchers, seven are infielders, and seven are outfielders. How many ways can the manager choose a nine-man team to play in a game?
29. Use the information in problem 28. How many ways can the manager choose his batting order to hand to the umpire at the beginning of the game?
30. The owner of an appliance store receives a shipment of thirty radios. He knows this shipment contains four defective radios. He decides to select five radio sets at random without replacement, and to reject the entire shipment if the sample contains at least one defective radio. In how many ways can he select a sample of five radios which contain at least one defective radio?

Probabilities of compound events

Now that we have studied certain counting techniques we will return to the study of probabilities of more complicated events.

Example 7 A student questions fifty other students about their breakfast habits, and the fifty respondents indicate that thirty-three have coffee, eighteen have donuts, and fourteen have both. If a student from this sample is selected at random, find the probability that he has coffee or donuts for breakfast, which is symbolized appropriately by $P(C \text{ or } D)$.

Solution We organize the information into a Venn Diagram.

Figure 5.3

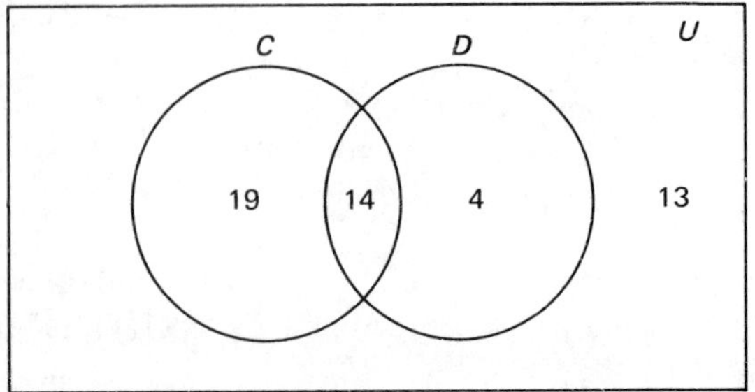

Since there are $19+14+4 = 37$ people who had either coffee or donuts (or both, since we use the word "or" in its inclusive sense, as explained in Example 5 of Chapter 2 which follows the definition of the union of two sets), then, by the definition of probability,

$$P(C \text{ or } D) = \frac{n(C \text{ or } D)}{n(S)} = \frac{37}{50} = .74.$$

This answer of .74 is the end of the search for the specific number asked for but let us look at it a little more closely. The number of people who drink coffee or eat donuts is 37, and it is obtained by adding 19, 14, and 4. Notice that 37 could also have been obtained by adding the number of coffee drinkers (33) and the number of donut eaters (18) and then immediately subtracting 14 from the sum to "uncount" those 14 people who were counted twice by the preceding addition operation. This subtraction is also mandated by our formula from Chapter 2 (equation (B) of the third section) about the number of elements in the union of two sets:

$$n(C \cup D) = n(C) + n(D) - n(C \cap D). \qquad \text{(A)}$$

In our case,

$$\begin{aligned} n(C \text{ or } D) &= n(C) + n(D) - n(C \text{ and } D) \\ &= 33 + 18 - 14 \\ &= 37. \end{aligned} \qquad \text{(B)}$$

Notice that equations (A) and (B) above differ slightly in their left-hand members. We agree here that the number of elements in the union of set C with set D, $n(C \cup D)$, is the same as the number of people who have coffee or donuts, $n(C \text{ or } D)$. We justify this agreement by noting that the union symbol, $\cup$, is defined in terms of the English word "or," so we often freely interchange the two symbols whenever it is convenient. Furthermore and similarly, the equation

$$n(C \cap D) = n(C \text{ and } D) \quad (= 14 \text{ in this specific example})$$

is justified by our earlier definition of "$\cap$" in terms of the English word "and."

We return to the original problem solution where we saw that

$$P(C \text{ or } D) = \frac{n(C \text{ or } D)}{n(S)} = \frac{37}{50}.$$

We now choose to expand 37/50 so that

$$\begin{aligned} P(C \text{ or } D) &= \frac{n(C \text{ or } D)}{n(S)} = \frac{37}{50} \\ &= \frac{33+18-14}{50} \\ &= \frac{33}{50} + \frac{18}{50} - \frac{14}{50} && \text{addition of fractions} \\ P(C \text{ or } D) &= P(C) + P(D) - P(C \text{ and } D) && \text{calculation of probability} \end{aligned}$$

General formula for $P(E \text{ or } F)$

We can now make the hasty but easily verified generalization (Problem 2) to the situation involving any two events E and F.

$$P(E \text{ or } F) = P(E) + P(F) - P(E \text{ and } F)$$

If we consider the situation of two disjoint events, such as the one E of drawing a diamond from a standard deck and the other F of drawing a black card, then

$$P(E \text{ or } F) = P(E) + P(F)$$

since

$$P(E \text{ and } F) = n(\{\})/n(S) = 0/n(S) = 0,$$

and thus

$$\begin{aligned} P(E) + P(F) - P(E \text{ and } F) &= P(E) + P(F) - 0 \\ &= P(E) + P(F). \end{aligned}$$

In fact, if $E_1, E_2, E_3, \ldots, E_n$ are n pairwise disjoint events then

$$P(E_1 \text{ or } E_2 \text{ or } E_3 \text{ or } \ldots \text{ or } E_n) = P(E_1) + P(E_2) + P(E_3) + \cdots + P(E_n).$$

Example 8 From a box which contains five red, six white, and nine black marbles, two marbles are drawn without replacement. That is, the first marble is *not* replaced before drawing the second marble. Find the probability that both marbles drawn are the same color.
Solution We analyze the event's description that "both marbles drawn are the same color," which means that

(both are red or both are white or both are black).

Since this one event is the union of the three pairwise disjoint events

both are red, both are white, both are black,

we can now say

$$\begin{aligned} &P(\text{both marbles are the same color}) \\ &\quad = P(\text{both are red } \textit{or} \text{ both are white } \textit{or} \text{ both are black}) \\ &\quad = P(\text{both are red}) + P(\text{both are white}) + P(\text{both are black}). \quad \text{(C)} \end{aligned}$$

We can set up the equation

$$P(\text{both are red}) = \frac{\text{number of ways of choosing two red marbles}}{\text{number of ways of choosing two marbles}}.$$

The numerator is ${}_5C_2$ since we want to know the number of ways of choosing two things out of five where the order of selection is not important and, similarly, the denominator is ${}_{20}C_2$. Hence

$$
\begin{aligned}
P(\text{both are red}) &= \frac{{}_5C_2}{{}_{20}C_2} \\
&= \frac{\dfrac{5!}{2!\cdot 3!}}{\dfrac{20!}{2!\cdot 18!}} \\
&= \frac{\dfrac{5\cdot \overset{2}{\cancel{4}}\cdot \cancel{3!}}{\cancel{2}\cdot \cancel{3!}}}{\dfrac{\overset{10}{\cancel{20}}\cdot 19\cdot \cancel{18!}}{\cancel{2}\cdot \cancel{18!}}} \\
&= \frac{10}{190} \\
&\doteq .053.
\end{aligned}
$$

Similarly,

$$
\begin{aligned}
P(\text{both are white}) &= \frac{{}_6C_2}{{}_{20}C_2} \\
&= \frac{15}{190},
\end{aligned}
$$

and

$$
\begin{aligned}
P(\text{both are black}) &= \frac{{}_9C_2}{{}_{20}C_2} \\
&= \frac{36}{190}.
\end{aligned}
$$

If we now continue directly from equation (C) above we have

$$
\frac{10}{190} + \frac{15}{190} + \frac{36}{190} = \frac{61}{190} \doteq .321
$$

as the final answer.

It is important to realize that we took the event as originally described in Example 8 and rewrote it as a statement containing several *or*'s before we could substitute into our formula

$$
\begin{aligned}
&P(E_1 \text{ or } E_2 \text{ or } E_3 \text{ or } \ldots \text{ or } E_n) \\
&= P(E_1) + P(E_2) + P(E_3) + \cdots + P(E_n).
\end{aligned}
$$

This successive rewriting of a verbal description is often necessary in order to abstract a word problem into a formula. Then we can use our knowledge of the formal statements of the mathematics of probability. We also note that

$$\begin{aligned} &P(\text{both marbles are different colors}) \\ &\quad = 1 - P(\text{both marbles are the same color}) \\ &\quad = 1 - \frac{61}{190} = \frac{129}{190} \doteq .679. \end{aligned}$$

The justification for this last chain of equations is found in equation (A) of the first section of this chapter.

Conditional probability

So far we have ways of determining $P(E)$, $P(E')$, and $P(E \text{ or } F)$. We now turn our attention to seeking a general formula for $P(E \text{ and } F)$, which will let us calculate such probabilities as P(red on first draw and black on second draw) with reference to the last example. We begin our search by reviewing the coffee and donut example, number 7.

Example 9 If a person is selected at random, what is the probability that he drinks coffee?
Solution This is straightforward.

$$P(C) = n(C)/n(S) = 33/50 \text{ or } .66.$$

Example 10 If a person who eats donuts is selected at random, what is the probability that he also drinks coffee?
Solution This is not as straightforward. Here the sample space for the experiment of selecting a student is not the original universal set of the fifty students but just the eighteen donut eaters. We symbolize the required probability by

$$P(C \mid D),$$

which is read "the probability of C given D" and means the probability that a person who eats donuts (D) is also a coffee drinker (C). Then the Venn Diagram shows that fourteen of these eighteen donut eaters also have coffee, so that

$$P(C \mid D) = 14/18 \doteq .78.$$

Since 14 is $n(C \text{ and } D)$ and 18 is $n(D)$ we can also say

$$P(C \mid D) = \frac{n(C \text{ and } D)}{n(D)},$$

or if we divide both numerator and denominator of the last equation by $n(S) = 50$, then

$$P(C \mid D) = \frac{n(C \text{ and } D)/n(S)}{n(D)/n(S)}.$$

If we now apply the basic formula for calculating probability to this fraction it yields

$$P(C \mid D) = \frac{P(C \text{ and } D)}{P(D)}.$$

$P(C \mid D)$ is an example of *conditional* probability, since we ask for the probability that a person drinks coffee given the *condition* that he is a donut eater.

We may safely generalize from the two specific events C and D in this sample space to any two events E and F in any equally likely sample space:

$$P(E \mid F) = \frac{P(E \text{ and } F)}{P(F)} = \frac{n(E \text{ and } F)}{n(F)} \qquad \text{(D)}$$

Example 11 Find the probability that a coffee drinker also eats donuts.

Solution We know that a person will be chosen at random from the coffee drinkers, and we want to know the conditional probability that he eats donuts. This is $P(D \mid C)$, which, after replacing E by D and F by C in the previous equation, equals

$$\begin{aligned} &P(D \text{ and } C)/P(C) \\ &\quad = (14/50)/(33/50) \\ &\quad = 14/33. \end{aligned}$$

Basic multiplication principle

Now we will take the basic equation for conditional probability, which is (D) above, and rewrite it slightly:

$$\frac{P(E \mid F)}{1} = \frac{P(E \text{ and } F)}{P(F)}$$

$$\frac{P(E \mid F)}{1} = \frac{P(F \text{ and } E)}{P(F)}.$$

If we now multiply both sides of the second equation by $P(F)$, the result is

$$P(F) \cdot P(E \mid F) = P(F \text{ and } E),$$

which in reversed format becomes

$$P(F \text{ and } E) = P(F) \cdot P(E \mid F).$$

This important equation says that the probability that events F and E both occur is the probability that F occurs times the conditional probability that E occurs, given that F also occurs or has occurred. Since remembering formulas is easier if the letters are in alphabetical order, we replace F by E and E by F so that in final form the very important equation which allows us to calculate the probability of the joint occurrence of two events is

$$P(E \text{ and } F) = P(E) \cdot P(F \mid E). \tag{E}$$

Equation (E) expresses the basic *multiplication principle* of probability.

Example 12 With reference to the earlier sample space of five red, six white, and nine black marbles in Example 8 calculate the probability that the first marble drawn is red and the second one drawn is black.

Solution By the previous general equation (E),

$$P(\text{red and black}) = P(\text{red}) \cdot P(\text{black} \mid \text{red}).$$

$P(\text{red on first draw})$ is 5/20 and the probability of a black on the second draw, given that the first draw is red, is 9/19; we know that there are nine black marbles and only nineteen left to choose from because the first ball, which has to be red, is not replaced. Consequently

$$P(\text{red and black}) = \frac{5}{20} \cdot \frac{9}{19} = \frac{1}{4} \cdot \frac{9}{19} = \frac{9}{76} \doteq .12.$$

Similarly, $P(\text{black and white})$ would be

$$P(\text{black}) \cdot P(\text{white} \mid \text{black}) = 54/380 \quad (\text{right?})$$

and $P(\text{black and black})$ would be

$$P(\text{black on first draw}) \cdot P(\text{black on second} \mid \text{first was black})$$

$$= \frac{9}{20} \cdot \frac{8}{19} = \frac{72}{380} = \frac{18}{95} \doteq .19.$$

Notice that the last event is the same as simultaneously or successively selecting two black marbles, which can be done in ${}_9C_2$ ways out of a sample space of ${}_{20}C_2$ ways, so that its probability is

$${}_9C_2/{}_{20}C_2 = 36/190 = 18/95 \doteq .19.$$

We can often look at a probability problem in two different ways and still come up with the same answer as illustrated by the previous two equations.

Independent events

Example 13 The situation is the same as in Example 12, with two marbles being chosen from five red, six white, and nine black ones; but here the first marble is replaced and shaken with the rest before the second one is drawn. Find P(red on first draw and black on second draw) under these modified circumstances.

Solution

$$P(\text{red and black}) = P(\text{red}) \cdot P(\text{black} \mid \text{red}) = \frac{5}{20} \cdot \frac{9}{20} = \frac{9}{80} \doteq .11$$

Notice that $P(\text{black} \mid \text{red})$ is 9/20 here (instead of 9/19 as in the previous example) because there are twenty marbles to choose from since the first marble is replaced. Consequently, it is possible to draw the same marble twice.

If we asked for the probability in the previous example that the second marble is black, P(black on second draw), the answer is 9/20. $P(\text{black} \mid \text{red})$ was also 9/20, and thus

$$P(\text{black} \mid \text{red}) = P(\text{black}).$$

The events of "black on second draw, given that the first draw is red" and "black on the second draw" are *independent* events. This means that the probability of getting a black marble on the second draw is the same, regardless of what happens on the first draw. It also means that the occurrence of one event in no way influences the occurrence of the other.

In general, we say two events E and F are *independent* of each other if

$$P(E \mid F) = P(E) \quad \text{and} \quad P(F \mid E) = P(F).$$

Consequently,

$$P(E \text{ and } F) = P(E) \cdot P(F \mid E) = P(E) \cdot P(F).$$

In general, if n events $E_1, E_2, E_3, \ldots, E_n$ are all independent of one another, then

$$P(E_1 \text{ and } E_2 \text{ and } E_3 \text{ and } \ldots \text{ and } E_n) = P(E_1) \cdot P(E_2) \cdot P(E_3) \cdot \cdots \cdot P(E_n).$$

If the n events are not known to be independent of each other, then

$$P(E_1 \text{ and } E_2 \text{ and } E_3 \text{ and } \ldots \text{ and } E_n) = P(E_1) \cdot P(E_2 \mid E_1) \cdot P(E_3 \mid E_1 \text{ and } E_2) \cdot P(E_4 \mid E_1 \text{ and } E_2 \text{ and } E_3) \cdot \cdots \cdot P(E_n \mid E_1 \text{ and } E_2 \text{ and } E_3 \text{ and } \ldots \text{ and } E_{n-1}).$$

Often a tree diagram is useful in calculating probabilities.

Example 14 Urn I contains three red, five blue, and seven white marbles, while Urn II contains six red, four blue, and six white marbles. An urn is chosen at random, and then two marbles are chosen from that urn without replacement. Find the probability that the marbles are different in color.

Solution We make a tree diagram of the first step of the three-step process and show the probabilities.

Figure 5.4

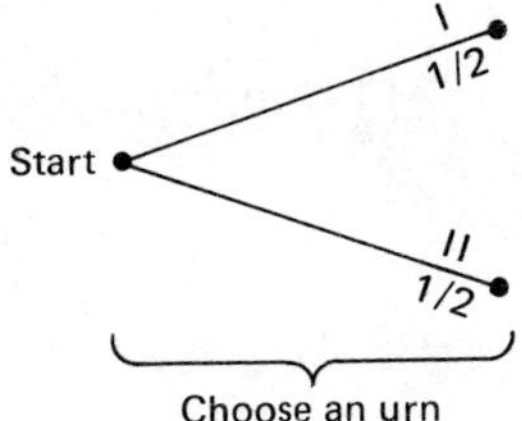

Now we write the possible ways the two remaining steps may occur and their corresponding probabilities. We also calculate the probability of each complete path through the tree by multiplying the probabilities of each component branch. For example,

$$P(\text{Urn I and blue on first draw and white on second draw} = P(\text{Urn I}) \cdot P(\text{blue on first draw} \mid \text{Urn I is chosen})$$
$$\cdot\, P(\text{white on second draw} \mid \text{Urn I is chosen and blue on first draw})$$
$$= \frac{1}{2} \cdot \frac{5}{15} \cdot \frac{7}{14} = \frac{1}{2} \cdot \frac{1}{3} \cdot \frac{1}{2} = \frac{1}{12}.$$

See Figure 5.5 opposite.

The answer is the sum of the probabilities of those complete paths through the tree where the marbles are of different colors:

$$\begin{aligned} P(\text{different colors}) &= 15/420 + 21/420 + 15/420 + 35/420 + 21/420 \\ &\quad + 35/420 + 24/480 + 36/480 + 24/480 \\ &\quad + 24/480 + 36/480 + 24/480 \\ &= 142/420 + 168/480 \\ &\doteq .69. \end{aligned}$$

We close this section by emphasizing that the best initial approach to take to an applied probability problem often is to analyze the description of the event and determine whether it is a simple quotient $n(E)/n(S)$, a *not* statement, an *or* statement, an *and* statement, or some combination of them. Then we can apply our formulas for $P(E)$, $P(E')$, $P(E \text{ or } F)$, $P(E \mid F)$, and $P(E \text{ and } F)$, depending upon the circumstances. We list these basic formulas. E and F represents events, and S represents the sample space.

$$P(E) = \frac{n(E)}{n(S)}$$
$$P(E) = 1 - P(E') \text{ or } 1 - P(\text{not } E) \text{ or } 1 - P(E \text{ does not occur})$$
$$P(E') = 1 - P(E) \text{ or } 1 - P(E \text{ does occur})$$
$$P(E \text{ or } F) = P(E) + P(F) - P(E \text{ and } F) \qquad \text{(any two events)}$$
$$P(E \text{ or } F) = P(E) + P(F) \qquad \text{(any two disjoint events)}$$
$$P(E \mid F) = \frac{P(E \text{ and } F)}{P(F)}$$
$$P(E \text{ and } F) = P(E) \cdot P(F \mid E) \qquad \text{(any two events)}$$
$$P(E \text{ and } F) = P(E) \cdot P(F) \qquad \text{(any two independent events)}$$

Figure 5.5

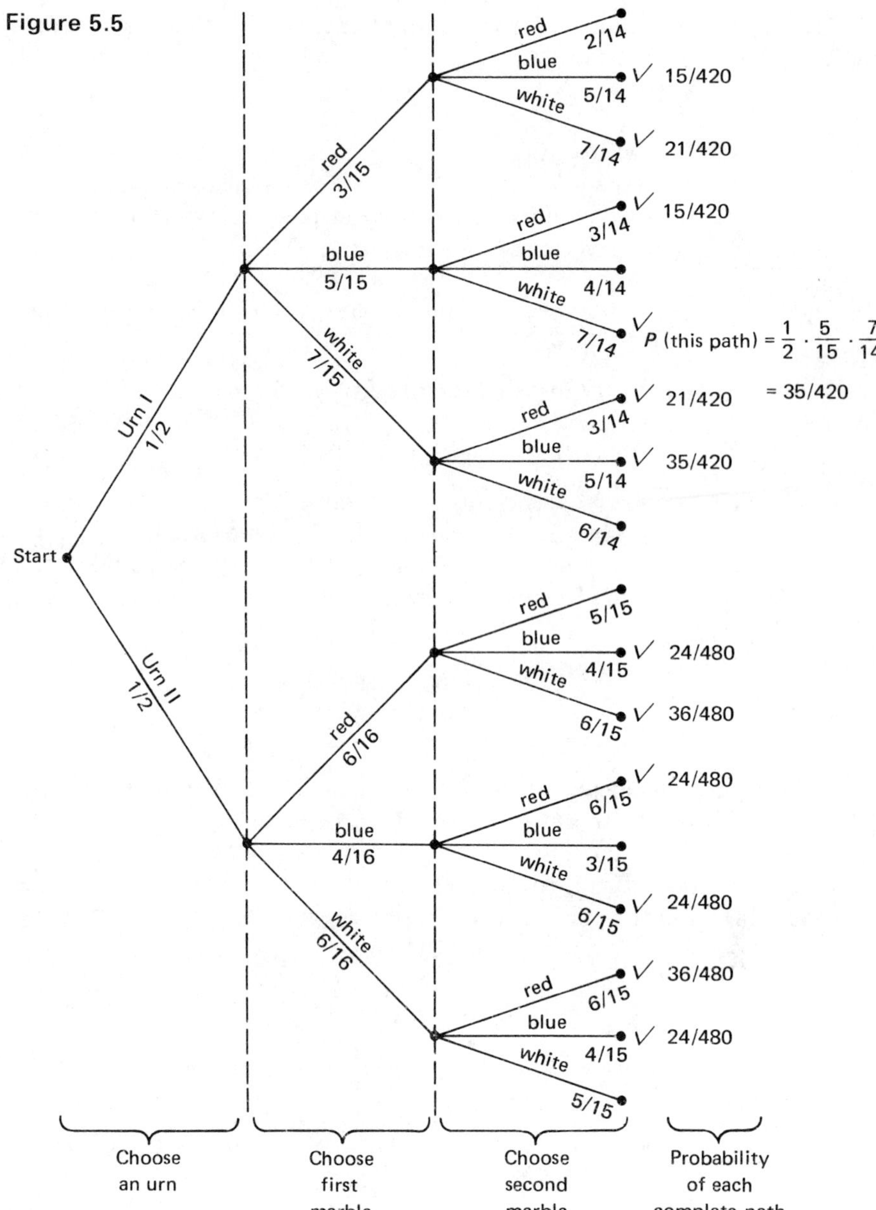

Problem set 5–3

1. A card is drawn at random from a standard deck. Find the probability that the card is
   a. red or a heart.
   b. black or a queen.
   c. a 7 or a diamond.
   d. a club or red.

2. Use the basic formula of probability calculation and the appropriate properties of sets to show that if A and B are any two sets, then

$$P(A \text{ or } B) = P(A) + P(B) - P(A \text{ and } B).$$

3. "John stays in town to work on 70% of the week nights and has a date in town on 60% of the week nights. Therefore, the probability that he stays in town on Wednesday night is 1.30." What is wrong with this answer? What additional information is needed to calculate the probability mentioned?
4. Pacifica Airlines has a very old biplane with two engines, and the plane will not fly unless both engines run.

$$P(\text{port engine will quit}) = .15$$

and

$$P(\text{starboard engine will quit}) = .20.$$

The engines operate independently of each other.

a. Find the probability that the plane will safely complete its next flight.
b. Assuming that the plane will fly on one engine, find the probability that the plane will safely complete its next flight.
c. Suppose that

$$P(\text{port quits} \mid \text{starboard quits}) = .24$$

and

$$P(\text{starboard quits} \mid \text{port quits}) = .30$$

because of the increased strain on the remaining engine. Suppose that both engines still must run to keep the plane in the air. What is the probability of a complete flight with these conditions?
d. Let the conditional probabilities in this section be the same as those in part (c). If the plane can fly on one engine, what is the probability of a complete flight now? Hint: $P(E \mid F') = 1 - P(E' \mid F')$.

5. We may generalize

$$P(E \text{ or } F) = P(E) + P(F) - P(E \text{ and } F)$$

to any three events E, F, and G of an associated sample space by the equation

$$P(E \text{ or } F \text{ or } G) = P(E) + P(F) + P(G) - P(E \text{ and } F) - P(E \text{ and } G) - P(F \text{ and } G) + P(E \text{ and } F \text{ and } G).$$

This equation is true because of the results of Problem 14 in Problem set 2–3 and because of the basic formula of probability calculation.

Suppose that Pacifica Airlines' biplane in the previous problem is destroyed in a crash landing after both engines have quit, and a Ford Trimotor takes its place. The probability that the port engine quits is .12, the probability that the middle engine quits is .10, and the probability that the starboard engine runs is .85. Assume that the failure of any engine is independent of the failure of any other engine.

a. If the plane must have all three engines running, then find the probability of completing a flight.
b. If the plane will run with one engine out, then find the probability of completing a flight.
c. If the plane can land in the desert safely after two engines quit, then find the probability of a safe ending for the trip.
d. If the middle engine is replaced with a new one whose reliability is 99%, then how do the answers to questions (a), (b), and (c) change?

6. Two cards are drawn without replacement from a standard deck of fifty-two cards. Find the probability that
   a. both cards are 6's
   b. the first card is a red 3 and the second is a club
   c. the first card is a club and the second is a red 3
   d. both cards are the same
   e. both cards are different
   f. both cards are red
   g. both cards are diamonds
   h. the first card is a red 3 and the second is a heart
7. Rework Problem 6, but this time assume that the two cards are drawn with replacement of the first and reshuffling of the deck before the second card is drawn.
8. If A and B are any two events with $P(A) = .6$, $P(B) = .3$, and $P(B \mid A) = .4$, then find
   a. $P(A \text{ and } B)$
   b. $P(A \mid B)$
   c. $P(A \text{ or } B)$
9. John Daniels awakens one morning and successively pulls two socks at random out of a drawer with ten black, six brown, and two blue socks. What is the probability that he has a matched pair?
10. Return to the problem of the fifty students and their breakfast habits (Example 7). Verify that these general properties hold true in the problem by substitution of appropriate values into these formulas.
    a. $P(C \mid C) = 1$
    b. $P(\varnothing \mid D) = 0$
    c. $P(D' \mid C) = 1 - P(D \mid C)$
    d. $P(C \mid D') = \dfrac{P(C) - P(C \text{ and } D)}{1 - P(D)}$
11. Rework the problem of the urn and the two draws from it (Example 8), but this time find the probability that the marbles are the same color, assuming that there is replacement.
12. Rework Example 8, but here find the probability that the marbles are of different colors with replacement.
13. Rework Example 8 by constructing a tree diagram with the correct probabilities written in the diagram.

14. If the batting average of a baseball player is .400, then find the probability of at least one hit in his next
    a. time at bat
    b. two times at bat (Hint: $P(\text{at least 1 hit}) = 1 - P(\text{no hits})$)
    c. three times at bat
    d. four times at bat
    e. five times at bat
    f. n times at bat
15. A committee of five is to be chosen from eight Democrats and six Republicans by drawing five names out of a hat where each of the pool of fourteen has his name written on an identical piece of paper. Find the probability that
    a. there are three Republicans on the committe of five
    b. there are four Democrats on the committee of five
    c. there are more Democrats than Republicans on the committee of five
    d. there are more Republicans than Democrats on the committee of five
    e. same as (d), only there are now eight Democrats and eight Republicans in the pool
    f. same as (d), only there are now eight Democrats and ten Republicans in the pool
    g. same as (d), only there are now eight Democrats and twelve Republicans in the pool
    h. same as (d), only there are now eight Democrats and fourteen Republicans in the pool
    i. same as (d), only there are now eight Democrats and sixteen Republicans in the pool
    j. same as (d), only there are now eight Democrats and r Republicans in the pool where $r \geqslant 3$.
    k. same as (d), only there are now d Democrats and r Republicans in the pool where $d \geqslant 3$ and $r \geqslant 3$.
    l. same as (d), only there are now d Democrats and r Republicans in the pool and seven people on the committee with $d \geqslant 4$ and $r \geqslant 4$.
    m. same as (d), only there are now d Democrats and r Republicans in the pool and nine people on the committee with $d \geqslant 5$ and $r \geqslant 5$.
    n. same as (d), only there are now d Democrats and r Republicans in the pool and c people on the committee, c being an odd number greater than 1 with $d \geqslant (c+1)/2$ and $r \geqslant (c+1)/2$.
16. A high school senior applies for admission to college A and to college B. She estimates the probability of her acceptance at A to be .7, the probability of her acceptance at B to be .4, and the probability of the rejection of at least one of her applications to be .6. Find the probability that she will be admitted to at least one of the colleges.
17. Two defective radio tubes are mixed up with three good ones. You test them one by one until you locate both defective tubes.
    a. Construct a tree diagram for the outcomes of this experiment. What is the probability that the second bad tube will be the

b. second one tested?
c. third one tested?
d. fourth one tested?
e. fifth one tested?

18. A fair coin is tossed three times. What is the probability that
    a. the coin lands heads on all three tosses?
    b. the coin lands heads on the third toss if it has landed heads on the first two tosses?

19. Every student at Empire State College is classified according to his grade-point average and class. Here is a summary of the results of that cross-classification, where appropriate abbreviations are used.

Class \ Average	D 0.00–0.99	C 1.00–1.99	B 2.00–2.99	A 3.00–3.99
F-freshman	37	98	275	46
S-sophomore	21	39	246	53
J-junior	4	11	196	62
N-senior	0	4	189	60

One person is chosen at random from the student body. Find these probabilities about him:

a. $P(F)$
b. $P(C)$
c. $P(D \mid F)$
d. $P(F \mid D)$
e. $P(N \text{ and } B)$
f. $P(D \mid S \text{ or } J)$
g. $P(N \mid A)$
h. $P(J \mid A)$
i. $P(S \mid A)$
j. $P(F \mid A)$
k. $P(A \mid N)$
l. $P(A \mid J)$
m. $P(A \mid S)$
n. $P(A \mid F)$
o. $P(B \text{ or } A \mid F)$
p. $P(B \text{ or } A \mid N)$
q. $P(N \text{ or } B)$

20. A five-card poker hand is drawn from a standard deck. Find the probability of a hand that has
    a. two pairs (one pair of each of two different face values plus another card of a third face value).
    b. three of a kind (exactly three cards of one face value plus two more different face values).
    c. a full house (three cards of one face value and two cards of one other face value).
    d. four of a kind (four cards of one face value).
    e. straight flush (five cards in sequence and of the same suit).

21. Two boxes each contain six red poker chips and four white poker chips. One chip is selected at random from the first box and placed in the second. Find the probability that
    a. a chip selected at random from the second box will be white.
    b. a chip selected at random from the first box will be white.
    c. two chips selected at random from the second box will be white.
    d. two chips selected at random from the first box will be white.

22. One of the classical problems of probability is the birthday problem. In it we choose a group of n people and ask for the probability of at least one

duplication (or triplication, quadruplication, etc.) of birthdays (month and day only) among these n people. We assume that the people are chosen at random from those whose birthdays are not February 29, and we also assume that the birthdays are equally distributed among the 365 possibilities. A person born on 16 February 1955 is assumed to have the same birthday as a person born on 16 February 1956.

As an example of this situation, we will examine a group of three randomly chosen people whom we name A, B, and C for identification purposes. We seek

$$P(\text{at least one duplication of birthdays among these three people})$$

and will calculate this probability by subtracting P(no duplications of birthdays among these three people) from 1, since we saw earlier that $P(E)$ equals $1 - P(\text{not } E)$. Since

$$P(\text{no duplication of birthdays})$$

is the same as

$$P(\text{all three birthdays are different})$$

these statements can be rewritten as

$$\begin{aligned}P(&A \text{ has a birthday } \textit{and } B \text{ has a birthday different}\\ &\text{from } A\text{'s } \textit{and } C \text{ has a birthday different from } A\text{'s}\\ &\text{and from } B\text{'s}).\end{aligned}$$

Since the three events which form the one event in the above pair of parentheses are all independent of one another, we use the multiplication principle to write

$$\begin{aligned}&P(\text{all three birthdays are different})\\ &\qquad = P(A \text{ has a birthday}) \cdot P(B \text{ has a birthday different from } A\text{'s})\\ &\qquad\qquad \cdot P(C \text{ has a birthday different from } A\text{'s and from } B\text{'s}),\end{aligned}$$

which is

$$1 \cdot \frac{364}{365} \cdot \frac{363}{365}$$

or, for the sake of consistency,

$$\frac{365}{365} \cdot \frac{364}{365} \cdot \frac{363}{365} = \frac{365 \cdot 364 \cdot 363}{365^3}.$$

It is true that the numerator of the previous fraction is the same as ${}_{365}P_3$ (verify this fact by evaluating the permutation symbol). Thus our answer work so far says

$$P(\text{all 3 birthdays are different}) = \frac{{}_{365}P_3}{365^3} = .9918.$$

Then the probability of at least one duplication of birthdays among these randomly chosen people is

$$1 - .9918 = .0082.$$

a. Write out an analysis similar to the previous one for the special case of $n = 4$ randomly chosen people. Your conclusion should be

$$P(\text{dup}) = 1 - \frac{{}_{365}P_4}{365^4}$$

where $P(\text{dup})$ is an abbreviation for

$$P(\text{at least one duplication of birthdays}).$$

Computer calculations say that $P(\text{dup})$ in this case is

$$1.0000 - .9836 = .0164.$$

We could continue through a step-by-step process to construct the following table. However, a computer has done the enormous calculations for us.

If the number of people in the group is $n =$	Then the probability of at least one duplication of birthdays is $P(\text{dup}) =$
5	$1 - {}_{365}P_5/(365^5) = .0271$
6	$1 - {}_{365}P_6/(365^6) = .0405$
7	$1 - {}_{365}P_7/(365^7) = .0562$
8	$1 - {}_{365}P_8/(365^8) = .0743$
9	$1 - {}_{365}P_9/(365^9) = .0946$
10	$1 - {}_{365}P_{10}/(365^{10}) = .1169$
15	$1 - {}_{365}P_{15}/(365^{15}) = .2529$
20	$1 - {}_{365}P_{20}/(365^{20}) = .4114$
22	$1 - {}_{365}P_{22}/(365^{22}) = .4757$
23	$1 - {}_{365}P_{23}/(365^{23}) = .5073$
25	$1 - {}_{365}P_{25}/(365^{25}) = .5687$
30	$1 - {}_{365}P_{30}/(365^{30}) = .7063$
35	$1 - {}_{365}P_{35}/(365^{35}) = .8144$
40	$1 - {}_{365}P_{40}/(365^{40}) = .8912$
50	$1 - {}_{365}P_{50}/(365^{50}) = .9704$
60	$1 - {}_{365}P_{60}/(365^{60}) = .9941$
100	$1 - {}_{365}P_{100}/(365^{100}) = 1.0000$
⋮	⋮
n	$1 - {}_{365}P_n/(365^n)$

The results are not at all what we would expect from common sense. If twenty-three people are chosen at random, we intuitively say that the probability of a duplication of birthdays is $23/365 = .0630$, instead of the true .5073. This example is one of the few instances where common sense does not fit well with the true state of affairs. Incidentally, Presidents James Polk and Warren Harding were born on November 2, and Presidents John Adams, James Monroe, and Thomas Jefferson died on July 4.

b. If the number of people in your class is approximately thirty or more, then have each silently (that is, independently!) write his or her birthday and the birthday of a close friend or relative on opposite sides of the same piece of paper. Collect the pieces and check to see if there is a duplication of birthdays among the students. Then check to see if there is a duplication of birthdays among the students' friends and relatives.

c. Construct the graph of the table from part (a) with n on the horizontal axis and $P(\text{dup})$ on the vertical axis.

23. We can generalize from the last problem. Suppose that we have a large pool of people to choose from and that we define a duplication of birthdays to occur when only the day of the month determines a duplication. For example, two people born on 13 July 1953 and on 13 September 1955 are assumed to have the same birthday. We exclude from our pool those born on the 29th, 30th, or 31st of any month.

a. Write out a detailed analysis to find the probability of at least one duplication of birthdays among three randomly chosen people, and calculate it.

b. Same as (a), only here four people are chosen at random.

c. Same as (a), only here five people are chosen at random.

d. Same as (a), only here seven people are chosen at random. Do not evaluate your expression, but it equals .56.

e. Same as (a), only here ten people are chosen at random. Do not evaluate your expression, but it equals .84.

f. Same as (a), only here fourteen people are chosen at random. Do not evaluate your expression, but it equals .98.

g. A person walks into a series of Howard Johnson's restaurants where twenty-eight ice cream flavors are possible, and he chooses a flavor by randomly choosing one out of a bag containing twenty-eight pieces of paper, one flavor written on each piece of paper. He then replaces the slip after each drawing. What is the probability of his repeating at least one flavor after three visits?

h. Same as (g), but after four visits.

i. Same as (g), but after five visits.

j. Same as (g), but after seven visits. (Do not evaluate your answer.)

k. Same as (g), but after fourteen visits. (Do not evaluate your answer.)

l. Same as (g), but after n visits, where n is a positive whole number less than twenty-nine.

m. Same as (g), but after one visit.

n. Answer part (m) by replacing n by one in the formula you obtained in part (l). Is your answer the same as common sense would predict?

o. On the basis of everything in this problem and the previous one, complete the following statement. Suppose that a sample of n items is chosen with replacement from a set with c possible choices where $2 \leqslant n \leqslant c$. Then the probability of at least one duplication of items among the n in the sample is $P(\text{dup}) =$ ______________

24. Refer to Problem 30 of the last problem set. What is the probability that the entire shipment will be rejected according to the owner's decision rule?

The binomial law and acceptance sampling

We saw in the last section that we could calculate

$$P(E_1 \text{ and } E_2 \text{ and } E_3 \text{ and } \dots \text{ and } E_n)$$

where the n individual events $E_1, E_2, E_3, \dots, E_n$ are independent of one another by multiplying their respective probabilities. In symbols,

$$\begin{aligned} &P(E_1 \text{ and } E_2 \text{ and } E_3 \text{ and } \dots \text{ and } E_n) \\ &= P(E_1) \cdot P(E_2) \cdot P(E_3) \cdot \cdots \cdot P(E_n). \end{aligned}$$

We will investigate some of the more important consequences of this multiplication property in this section.

Example 15 Carl takes a five-question multiple choice test with four possible responses (A, B, C, or D) to each question. Only one response is correct for each test item, and since Carl has not prepared for this test he answers each question by pure guessing. What is the probability that he guesses the first two questions correctly and the last three questions incorrectly?

Solution We label the question Q_1, Q_2, Q_3, Q_4, and Q_5. With this labeling in mind we want to find

$$\begin{aligned} &P(Q_1 \text{ is correct } \textit{and } Q_2 \text{ is correct } \textit{and } Q_3 \text{ is incorrect} \\ &\qquad \textit{and } Q_4 \text{ is incorrect } \textit{and } Q_5 \text{ is incorrect}). \end{aligned}$$

Since answering each question is an event unaffected by and independent of the answer to any other question, the probability of our overall event is

$$\begin{aligned} &P(Q_1 \text{ is correct}) \cdot P(Q_2 \text{ is correct}) \cdot P(Q_3 \text{ is incorrect}) \\ &\qquad \cdot P(Q_4 \text{ is incorrect}) \cdot P(Q_5 \text{ is incorrect}) \\ &= \frac{1}{4} \cdot \frac{1}{4} \cdot \frac{3}{4} \cdot \frac{3}{4} \cdot \frac{3}{4} \\ &= \left(\frac{1}{4}\right)^2 \cdot \left(\frac{3}{4}\right)^3. \end{aligned}$$

Example 16 We have the same conditions as in Example 15. What is the probability that Carl guesses Q_2 and Q_5 correctly and the others incorrectly?

Solution We want to find

$$P(Q_1 \text{ is incorrect } \textit{and } Q_2 \text{ is correct } \textit{and } Q_3 \text{ is incorrect } \textit{and } Q_4 \text{ is incorrect } \textit{and } Q_5 \text{ is correct}).$$

Since answering each question is an event unaffected by and independent of the answer to any other question, the probability of our overall event is

$$\begin{aligned} &P(Q_1 \text{ is incorrect}) \cdot P(Q_2 \text{ is correct}) \cdot P(Q_3 \text{ is incorrect}) \cdot P(Q_4 \text{ is incorrect}) \cdot P(Q_5 \text{ is correct}) \\ &= \frac{3}{4} \cdot \frac{1}{4} \cdot \frac{3}{4} \cdot \frac{3}{4} \cdot \frac{1}{4} \\ &= \left(\frac{1}{4}\right)^2 \cdot \left(\frac{3}{4}\right)^3. \end{aligned}$$

Example 17 We have the same conditions as in Examples 15 and 16. What is the probability that Carl guesses any two *specific* questions correctly and the three others incorrectly?
Solution Notice the analysis used in the previous two solutions. Whenever we have specified which two questions are to be answered correctly this means the remaining three must be answered incorrectly. The probability of answering each of the two specified questions correctly is 1/4, and thus the probability of answering each of the three remaining questions incorrectly is 3/4. We multiplied these five probabilities of independent events (two 1/4's and three 3/4's) to obtain, in each case, $(1/4)^2 \cdot (3/4)^3$ as the final probability. Surely, we will always obtain

$$\left(\frac{1}{4}\right)^2 \cdot \left(\frac{3}{4}\right)^3$$

(which is .0264) in any such case with two specified questions to be answered correctly.

Example 18 We have the same conditions as in Examples 15, 16, and 17. What is the probability that Carl guesses exactly two questions correctly?
Solution This problem is a generalization of the questions presented in the previous three examples. Notice that we do not care *which* two questions he answers correctly. We write, as the key to this problem,

$$\begin{aligned} &P(\text{guesses exactly two questions correctly}) \\ &\quad = P(\text{guesses only } Q_1 \text{ and } Q_2 \text{ correctly } \textit{or} \\ &\qquad \text{guesses only } Q_1 \text{ and } Q_3 \text{ correctly } \textit{or} \\ &\qquad \text{guesses only } Q_1 \text{ and } Q_4 \text{ correctly } \textit{or} \\ &\qquad \cdots\cdots\cdots\cdots\cdots\cdots \textit{or} \\ &\qquad \text{guesses only } Q_4 \text{ and } Q_5 \text{ correctly}). \end{aligned}$$

Since each of these individual events is disjoint from any other event, we next write

$$\begin{aligned}&P(\text{guesses exactly two questions correctly})\\&\quad= P(\text{guesses only } Q_1 \text{ and } Q_2 \text{ correctly})\\&\qquad+ P(\text{guesses only } Q_1 \text{ and } Q_3 \text{ correctly})\\&\qquad+ P(\text{guesses only } Q_1 \text{ and } Q_4 \text{ correctly})\\&\qquad+ \cdots\cdots\cdots\\&\qquad+ P(\text{guesses only } Q_4 \text{ and } Q_5 \text{ correctly}).\end{aligned}$$

We know that each of these probabilities is equal to $(1/4)^2 \cdot (3/4)^3$, so the only additional information we have to know is the number of such probabilities in the right-hand side of the previous equation. There are as many probabilities as there are ways of choosing which two specific questions out of five are to be answered correctly. Pictorially,

$$\overline{}\ \overline{}\ \overline{}\ \overline{}\ \overline{}$$
$$Q_1\quad Q_2\quad Q_3\quad Q_4\quad Q_5$$

we must choose two ___'s out of five to fill with the word "correct," and this automatically determines which three ___'s we fill with the word "incorrect." The number of ways of choosing two correct answers (and hence three incorrect answers) is ${}_5C_2$ or $\binom{5}{2}$, which equals 10. Finally, we conclude that we have the sum of $\binom{5}{2}$ terms, each of which is $(1/4)^2 \cdot (3/4)^3$, so our answer is

$$\begin{aligned}&P(\text{guesses exactly two questions correctly})\\&\quad= \binom{5}{2} \cdot \left(\frac{1}{4}\right)^2 \cdot \left(\frac{3}{4}\right)^3\\&\quad= 10 \cdot .0264\\&\quad= .264.\end{aligned}$$

With these examples in mind we can begin to vary the three independent variables of the examples. They are n, the number of questions on each test; p, the probability of Carl (or any unprepared person) answering each of the questions correctly by pure guessing; and r, the number of correct answers on the entire test which we are interested in. First, we rewrite slightly Example 18's answer of

$$\binom{5}{2} \cdot \left(\frac{1}{4}\right)^2 \cdot \left(\frac{3}{4}\right)^3$$

as

$$\binom{5}{2} \cdot \left(\frac{1}{4}\right)^2 \cdot \left(1 - \frac{1}{4}\right)^{5-2}$$

to aid in reaching a future general formula involving letters instead of specific numbers. Since 1/4 is the probability of guessing correctly on any question, we can say that the probability of guessing incorrectly on any question must be 3/4 or $1 - 1/4$. If there are two correct answers, then we know that the other three, or $5 - 2$, are incorrect.

Example 19 If there are seven questions on the test instead of five, then similar reasoning says that the probability of exactly two correct answers, symbolized by

$$P(7, 2, 1/4),$$

is

$$\binom{7}{2} \cdot \left(\frac{1}{4}\right)^2 \cdot \left(1 - \frac{1}{4}\right)^{7-2}$$

because we must choose two ___'s out of seven ___'s to label with "correct," the probability of guessing correctly on any question is 1/4, and the probability of guessing incorrectly on any question is $1 - 1/4$. There are two 1/4's to multiply together with $7 - 2$, or 5, 3/4's and with $\binom{7}{2}$ to obtain the final answer, approximately .312.

Example 20 If we are interested in the probability of Carl's guessing four questions correctly on a five-question test, then we write

$$\begin{aligned} P(5, 4, 1/4) &= \binom{5}{4} \cdot (1/4)^4 \cdot (1 - 1/4)^{5-4} \\ &\doteq .015. \end{aligned}$$

Example 21 Carl knows enough about the material to eliminate one wrong response on each question but must still guess at which one of the remaining three is correct, so that his probability of guessing any one question correctly is 1/3. If we want to know the probability of exactly two correct answers on the test, we have

$$\begin{aligned} P(5, 2, 1/3) &= \binom{5}{2} \cdot (1/3)^2 \cdot (1 - 1/3)^{5-2} \\ &\doteq .329. \end{aligned}$$

As expected, $P(5, 2, 1/3)$ is greater than $P(5, 2, 1/4)$ because the probability of guessing a question correctly on each trial is greater here than in Example 18.

Predicting test outcomes

It is interesting to see how well these four calculated probabilities actually let us predict the outcomes of taking tests by guessing. Three summaries follow in which a computer has "taken" certain multiple choice tests 1000 times by purely guessing on each question. The computer "guesses" on each question (in Example 18, for instance) by performing mathematical calculations which are equivalent to choosing a letter at random from the set $\{A, B, C, D\}$. If the choice is the correct one (with probability 1/4), the computer records this fact and continues to the next question. When five successive questions have been answered by this process the number of correct answers is tabulated and the next test of five questions is taken. The table entries corresponding to each of the last four examples (18 through 21) are indicated. The teletypewriter, which sends the numbers to and receives the numbers from the computer used, distinguishes between the letter "0" (O) and the numeral "zero" (Ø).

```
DO YOU WANT INSTRUCTIONS ? YES
WHEN THE QUESTION MARK APPEARS TYPE IN T, THE NUMBER OF
  TIMES YOU WISH TO HAVE THE COMPUTER TAKE THIS TEST BY
  GUESSING ON EACH MULTIPLE CHOICE QUESTION; Q, THE NUMBER
  OF QUESTIONS ON THE TEST (Q <= 5Ø); AND R, THE NUMBER
  OF POSSIBLE RESPONSES TO EACH QUESTION.

?1ØØØ,5,4
  T  Q R

NUMBER OF TESTS WITH                PERCENT OF TESTS WITH
THESE NUMBERS OF                    THESE NUMBERS OF
CORRECT ANSWERS                     CORRECT ANSWERS

    Ø     --    249                         25
    1     --    389                         39
    2     --    259  Example 18             26
    3     --    37                          9
    4     --    14   Example 20             1
    5     --    2                           Ø

END OF TEST TAKING

DO YOU WANT INSTRUCTIONS ? NO

?1ØØØ,7,4
  T  Q R

NUMBER OF TESTS WITH                PERCENT OF TESTS WITH
THESE NUMBERS OF                    THESE NUMBERS OF
CORRECT ANSWERS                     CORRECT ANSWERS

    Ø     --    115                         12
    1     --    3Ø4                         3Ø
    2     --    317  Example 19             32
    3     --    189                         19
    4     --    7Ø                          7
    5     --    4                           Ø
    6     --    1                           Ø
    7     --    Ø                           Ø

END OF TEST TAKING
```

```
DO YOU WANT INSTRUCTIONS ? NO

?1000, 5, 3
   T  Q R

NUMBER OF TESTS WITH                PERCENT OF TESTS WITH
THESE NUMBERS OF                    THESE NUMBERS OF
CORRECT ANSWERS                     CORRECT ANSWERS

    0    --    140                        14
    1    --    340                        34
    2    --    310 Example 21             31
    3    --    172                        17
    4    --    36                         4
    5    --    2                          0

END OF TEST TAKING
```

In Example 18 $P(5, 2, 1/4)$ is .264, which we may interpret to mean that 26.4% of the time Carl takes this test he should theoretically get exactly two correct answers. Since 26.4% of 1000 is 264, we expect 264 out of the 1000 tests to have exactly two correct answers. Instead, 259 is the observed frequency, and it is different from 264 because of variations in the experimental process. We will examine this variation of five tests more closely in the next chapter.

The binomial law

We are now ready for a very important generalization of our work which is called the Binomial Law or the Repeated Trials Formula or the Bernoulli Trials Formula. It says that if

1. an experiment is independently repeated n times, and
2. an outcome of interest has probability p at each trial or repetition of the experiment, and
3. we are interested in the occurrence of this specific outcome exactly r times in the n trials $(0 \leqslant r \leqslant n)$,

then

$$P(n, r, p) = \binom{n}{r} \cdot p^r \cdot (1 - p)^{n-r}.$$

We emphasize that before we can use this formula we must know that the trials are *independent* of one another and that p does *not* change from trial to trial. In example 18 $n = 5$, $r = 2$, and $p = 1/4$; in example 19 $n = 7$, $r = 2$, and $p = 1/4$; in example 20 $n = 5$, $r = 4$, and $p = 1/4$; and in example 21 $n = 5$, $r = 2$, and $p = 1/3$. Table XI in the Appendix gives the values of $P(n, r, p)$ corresponding to selected triplets of values of n, r, and p.

Example 22 A person takes a ten-question multiple choice test with five possible answers for each question. Only one answer is correct, and the person answers each question by guessing. Find the probability that he passes the test if six or more correct answers result in a passing mark.
Solution We use a direct application of the Binomial Law with $n = 10$, $r \in \{6, 7, 8, 9, 10\}$, and $p = 1/5 = .20$.

$$\begin{aligned}
&P(6 \text{ or more correct})\\
&\quad = P(\text{exactly 6 correct } \textit{or} \text{ exactly 7 correct } \textit{or} \text{ exactly}\\
&\qquad\quad \text{8 correct } \textit{or} \text{ exactly 9 correct } \textit{or} \text{ exactly 10 correct})\\
&\quad = P(\text{exactly 6 correct}) + P(\text{exactly 7 correct})\\
&\qquad\quad + P(\text{exactly 8 correct}) + P(\text{exactly 9 correct})\\
&\qquad\quad + P(\text{exactly 10 correct})\\
&\quad = P(10, 6, .20) + P(10, 7, .20) + P(10, 8, .20)\\
&\qquad\quad + P(10, 9, .20) + P(10, 10, .20)\\
&\quad = .0055 + .0008 + .0001 + .0000^* + .0000^*\\
&\quad = .0064
\end{aligned}$$

or less than one chance out of a hundred.

Example 23 We have the same conditions as in Example 22, but assume here that there are four possible answers for each question.
Solution The only variation in our input to the Binomial Law is that $p = 1/4 = .25$.

$$\begin{aligned}
&P(6 \text{ or more correct})\\
&\qquad = P(10, 6, .25) + P(10, 7, .25) + P(10, 8, .25)\\
&\qquad\quad + P(10, 9, .25) + P(10, 10, .25)\\
&\qquad = .0162 + .0031 + .0004 + .0000 + .0000\\
&\qquad = .0197,
\end{aligned}$$

which is an increase from .0064.

Example 24 We have the same conditions as in Example 22, but assume here that there are three possible answers for each question.
Solution The only variation in our input to the Binomial Law is that

$p = 1/3 \doteq .33$.

$$\begin{aligned}
&P(6 \text{ or more correct})\\
&\quad = P(10, 6, .33) + P(10, 7, .33) + P(10, 8, .33)\\
&\qquad + P(10, 9, .33) + P(10, 10, .33)\\
&\quad = .0569 + .0163 + .0031 + .0001 + .0000 \quad (\text{right?})\\
&\quad = .0764.
\end{aligned}$$

* P(exactly 9 correct) and P(exactly 10 correct) are not exactly .0000000 ..., but they are less than .00005, so that when rounded to the nearest .0001 they become .0000 in Table XI.

Example 25 We have the same conditions as in Example 22, but assume here that there are two possible answers for each question. This would be the case on a true-false test.

Solution The only variation in our input to the Binomial Law is that $p = 1/2 = .50$.

$$\begin{aligned} &P(6 \text{ or more correct}) \\ &\quad = P(10, 6, .50) + P(10, 7, .50) + P(10, 8, .50) \\ &\qquad + P(10, 9, .50) + P(10, 10, .50) \\ &\quad = .2051 + .1172 + .0439 + .0098 + .0010 \\ &\quad = .3770. \end{aligned}$$

As we might expect, Examples 22 through 25 illustrate that as a person's chances of guessing correctly on each question increase, so do the corresponding probabilities of his guessing six or more questions correctly on the entire test.

One of the more important applications of the Binomial Law is the *acceptance-sampling* formula. We investigate by means of examples.

Example 26 A quality-control inspector selects a random sample of twenty bolts from a large lot and accepts the entire lot if the sample contains at most two defectives. Otherwise, he rejects the lot. If 5% of the entire lot is known to be defective, find the probability that he accepts the entire lot.

Solution If 5% of the lot is defective, then the probability that any one item in the lot or in the sample is defective is .05. We now have a direct application of the Binomial Law with $n = 20$, $r \in \{0, 1, 2\}$, and $p = .05$.

$$\begin{aligned} &P(\text{accepts entire lot}) \\ &\quad = P(\text{at most 2 defectives in the sample of 20}) \\ &\quad = P(\text{exactly 0 defectives } \textit{or} \text{ exactly 1 defective } \textit{or} \\ &\qquad \text{exactly 2 defectives}) \\ &\quad = P(\text{exactly 0 defectives}) + P(\text{exactly 1 defective}) \\ &\qquad + P(\text{exactly 2 defectives}) \\ &\quad = P(20, 0, .05) + P(20, 1, .05) + P(20, 2, .05) \\ &\quad = .3585 + .3774 + .1887 \qquad (\text{from Table XI}) \\ &\quad = .9246. \end{aligned}$$

We now vary the input variable of p, the fraction of the entire lot that is defective, to see how this variation affects the corresponding values of $P(20, \max 2, p)$. The term "max 2" means two or fewer defectives. Be sure to check these calculations by using Table XI in the appendix.

Table 5.3

p	$P(20, \max 2, p)$
.05	.9246
.10	.6770
.20	.2060
.30	.0354
.40	.0036
.50	.0002
.60	.0000

Of course, if $p = .00$ then he would always have a defective-free sample, so that $P(20, \max 2, .00)$ would be 1.00.

Acceptance sampling formula

It seems logical that the probability of accepting an entire lot of goods based on the characteristics of a random sample drawn from it depends on three variables. They are n, the size of the sample, which in Example 26 is 20; x, the *maximum* number of defectives tolerated in the sample, which in Example 26 is 2, meaning that we tolerate exactly 0 or exactly 1 or exactly 2 defectives; and p, the probability that any one item in the entire lot (and hence the sample itself as well) is defective, which in Example 26 is .05. We now make the generalization to the *Acceptance Sampling Formula*:

$$\begin{aligned} P(n, \max x, p) &= P(n, 0, p) + P)n, 1, p) \\ &\quad + P(n, 2, p) + \cdots + P(n, x, p) \\ &= \binom{n}{0} \cdot p^0 \cdot (1-p)^{n-0} + \binom{n}{1} \cdot p^1 \cdot (1-p)^{n-1} \\ &\quad + \binom{n}{2} \cdot p^2 \cdot (1-p)^{n-2} + \cdots + \binom{n}{x} \cdot p^x \cdot (1-p)^{n-x} \end{aligned}$$

We emphasize that the $P(n, \max x, p)$ on the left-hand side of this equation represents the probability of at most x defectives, meaning x *or fewer*, in the sample; the $P(n, x, p)$ on the right-hand side represents the probability of *exactly* x defectives in the sample.

The next table gives us the results of applying the Acceptance Sampling Formula to these values of n, x, and p as defined in the previous paragraph:

$$\begin{aligned} &n \in \{5, 10, 20, 30\}; \\ &x \in \{0, 1, 2, 3, 4, 5\}; \\ &p \in \{.01, .02, .05, .10, .20, .30, .40, .50\}. \end{aligned}$$

Table 5.4

N	X	P	P(N, MAX X, P)
5	0	.01	.951
5	0	.02	.904
5	0	.05	.774
5	0	.1	.59
5	0	.2	.328
5	0	.3	.168
5	0	.4	.078
5	0	.5	.031
5	1	.01	.999
5	1	.02	.996
5	1	.05	.977
5	1	.1	.919
5	1	.2	.737
5	1	.3	.528
5	1	.4	.337
5	1	.5	.188
5	2	.01	1
5	2	.02	1
5	2	.05	.999
5	2	.1	.991
5	2	.2	.942
5	2	.3	.837
5	2	.4	.683
5	2	.5	.5
5	3	.01	1
5	3	.02	1
5	3	.05	1
5	3	.1	1
5	3	.2	.993
5	3	.3	.969
5	3	.4	.913
5	3	.5	.813
5	4	.01	1
5	4	.02	1
5	4	.05	1
5	4	.1	1
5	4	.2	1
5	4	.3	.998
5	4	.4	.99
5	4	.5	.969
5	5	.01	1
5	5	.02	1
5	5	.05	1
5	5	.1	1
5	5	.2	1
5	5	.3	1
5	5	.4	1
5	5	.5	1

Table 5.4 Continued

N	X	P	P(N, MAX X, P)
10	0	.01	.904
10	0	.02	.817
10	0	.05	.599
10	0	.1	.349
10	0	.2	.107
10	0	.3	.028
10	0	.4	.006
10	0	.5	.001
10	1	.01	.996
10	1	.02	.984
10	1	.05	.914
10	1	.1	.736
10	1	.2	.376
10	1	.3	.149
10	1	.4	.046
10	1	.5	.011
10	2	.01	1
10	2	.02	.999
10	2	.05	.988
10	2	.1	.93
10	2	.2	.678
10	2	.3	.383
10	2	.4	.167
10	2	.5	.055
10	3	.01	1
10	3	.02	1
10	3	.05	.999
10	3	.1	.987
10	3	.2	.879
10	3	.3	.65
10	3	.4	.382
10	3	.5	.172
10	4	.01	1
10	4	.02	1
10	4	.05	1
10	4	.1	.998
10	4	.2	.967
10	4	.3	.85
10	4	.4	.633
10	4	.5	.377
10	5	.01	1
10	5	.02	1
10	5	.05	1
10	5	.1	1
10	5	.2	.994
10	5	.3	.953
10	5	.4	.834
10	5	.5	.623

Table 5.4 Continued

N	X	P	P(N, MAX X, P)
20	0	.01	.818
20	0	.02	.668
20	0	.05	.358
20	0	.1	.122
20	0	.2	.012
20	0	.3	.001
20	0	.4	0
20	0	.5	0
20	1	.01	.983
20	1	.02	.94
20	1	.05	.736
20	1	.1	.392
20	1	.2	.069
20	1	.3	.008
20	1	.4	.001
20	1	.5	0
20	2	.01	.999
20	2	.02	.993
20	2	.05	.925
20	2	.1	.677
20	2	.2	.206
20	2	.3	.035
20	2	.4	.004
20	2	.5	0
20	3	.01	1
20	3	.02	.999
20	3	.05	.984
20	3	.1	.867
20	3	.2	.411
20	3	.3	.107
20	3	.4	.016
20	3	.5	.001
20	4	.01	1
20	4	.02	1
20	4	.05	.997
20	4	.1	.957
20	4	.2	.63
20	4	.3	.238
20	4	.4	.051
20	4	.5	.006
20	5	.01	1
20	5	.02	1
20	5	.05	1
20	5	.1	.989
20	5	.2	.804
20	5	.3	.416
20	5	.4	.126
20	5	.5	.021

Table 5.4 Continued

N	X	P	P(N, MAX X, P)
30	0	.01	.74
30	0	.02	.545
30	0	.05	.215
30	0	.1	.042
30	0	.2	.001
30	0	.3	0
30	0	.4	0
30	0	.5	0
30	1	.01	.964
30	1	.02	.879
30	1	.05	.554
30	1	.1	.184
30	1	.2	.011
30	1	.3	0
30	1	.4	0
30	1	.5	0
30	2	.01	.997
30	2	.02	.978
30	2	.05	.812
30	2	.1	.411
30	2	.2	.044
30	2	.3	.002
30	2	.4	0
30	2	.5	0
30	3	.01	1
30	3	.02	.997
30	3	.05	.939
30	3	.1	.647
30	3	.2	.123
30	3	.3	.009
30	3	.4	0
30	3	.5	0
30	4	.01	1
30	4	.02	1
30	4	.05	.984
30	4	.1	.825
30	4	.2	.255
30	4	.3	.03
30	4	.4	.002
30	4	.5	0
30	5	.01	1
30	5	.02	1
30	5	.05	.997
30	5	.1	.927
30	5	.2	.428
30	5	.3	.077
30	5	.4	.006
30	5	.5	0

For example, if there are $n = 10$ items in a sample drawn at random from a population where fraction $p = .20$ is defective and we allow at most $x = 3$ defectives in the sample, then the probability that the sample has at most 3 defectives is .879 (right?).

As illustrated next, it is possible to program a computer to work with any triplet $(n, \max x, p)$ to calculate $P(n, \max x, p)$.

```
DO YOU WANT INSTRUCTIONS? (1 = YES, 2 = NO) ? 1
WHEN THE QUESTION MARK APPEARS, PRINT THE VALUES OF
  THESE VARIABLES IN THE ORDER SPECIFIED: N, THE
  NUMBER OF ITEMS IN THE SAMPLE; X, THE MAXIMUM NUMBER
  OF DEFECTIVES IN THE SAMPLE THAT YOU WILL TOLERATE; AND
  P, THE PROBABILITY THAT ANY ONE ITEM IN THE SAMPLE
  IS DEFECTIVE.  THE OUTPUT CONSISTS OF THE PROBABILITIES
  OF EXACTLY Ø DEFECTIVES, EXACTLY 1 DEFECTIVE,
  EXACTLY 2 DEFECTIVES, ... , EXACTLY X DEFECTIVES, AND
  ON THE FINAL LINE APPEARS THE SUM OF THESE INDIVIDUAL
  PROBABILITIES WHICH IS THE TOTAL ANSWER.




INPUT N, MAX X, P
?65,7,.Ø8


NUMBER OF
DEFECTIVES, R          P(N, R, P)

     Ø                    .ØØ44
     1                    .Ø25
     2                    .Ø696
     3                    .1272
     4                    .1714
     5                    .1818
     6                    .1531
     7                    .1159

TOTAL PROBABILITY IS .8534     = P( 65    , MAX 7     , .Ø8).
```

We close this section by noting that we assume the random sampling is without replacement from a very large lot. This assumption can lead to difficulties which are discussed in problem 9 of the following problem set. Moreover, we are often more interested in determining the fraction of defectives in the entire lot on the basis of the fraction of defectives in the sample, instead of estimating sample characteristics based on the characteristics of the population. This interest arises because we often do not know, for example, the exact percentage of defectives that a machine will produce in a given lot. We will return to this important area of estimating population properties based on sample properties in the next chapter.

Problem set 5–4

For the first four problems assume that a quality-control inspector will accept an entire lot of flashbulbs if a random sample contains at most one defective flashbulb.

1. Find the probability that a sample of size $n = 5$ contains at most one defective for the following values of p, the probability that any one flashbulb is defective in the sample (which is the same as the fraction of the entire lot that is defective).
   a. .01 b. .02 c. .05 d. .10
   e. .20 f. .30 g. .40 h. .50
2. Find the probability that a random sample of size $n = 10$ contains at most one defective for the following values of p, the probability that any one flashbulb in the sample is defective.
   a. .01 b. .02 c. .05 d. .10
   e. .20 f. .30 g. .40 h. .50
3. Find the probability that a random sample of size $n = 20$ contains at most one defective for the following values of p, the probability that any one flashbulb in the sample is defective.
   a. .01 b. .02 c. .05 d. .10
   e. .20 f. .30 g. .40 h. .50
4. Find the probability that a random sample of size $n = 30$ contains at most one defective for the following values of p, the probability that any one flashbulb in the sample is defective.
   a. .01 b. .02 c. .05 d. .10
   e. .20 f. .30 g. .40 h. .50
5. On a large sheet of paper graph the results of the previous four problems on the same set of axes with p on the horizontal axis and $P(n, \max 1, p)$ on the vertical axis. Label the four curves with $n = 5$, $n = 10$, $n = 20$, $n = 30$. Draw a vertical line through $p = .10$. What generalization do the successive values of the y coordinate of these points of intersection of the line with the curves suggest?
6. Repeat problems 1 through 5, but this time assume that the sample may contain at most two defectives and still have the entire lot accepted.
7. a. Complete this table.

n	$P(n, \max 1, .10)$	$P(n, \max 2, .20)$
5		
10		
20		
30		

   b. What generalization, consistent with common sense, does this table suggest?

8. Even after inspection, the quality-control manager of Brand T motorcycles knows that 10% of his machines are defective because of sloppy workmanship and tremendous vibration problems inherent in the two-cylinder, four-cycle engine. He does not care to remedy this situation, but hopes to sell a large lot of the machines to the Big Twin Favorites motorcycle club. The club president samples seven of the bikes in this shipment to check for missing parts, loose nuts, oil leaks, exposed wiring, etc. He agrees to accept the entire shipment if the sample contains at most two defectives.
   a. What is the probability of his accepting the shipment?
   b. Suppose he will tolerate three defectives in the sample. What is the probability of accepting the shipment now?
   c. Suppose that the president insists that the sample have at most one defective. What is P(acceptance of entire lot) now?
   d. If the president insists on zero defectives, what is P(acceptance of entire lot)?
9. A recently married couple is trying to predict the sex composition of their future children. Complete this table, assuming $P(\text{boy}) = P(\text{girl}) = .50$, where n is the number of children in the family, b is the number of boys, and g is the number of girls.

n	b	g	P(family with b boys and g girls)
3	3	0	
3	2	1	
3	1	2	
3	0	3	
4	4	0	
4	3	1	
4	2	2	
4	1	3	
4	0	4	
5	more boys than girls		
5	equal numbers of boys and girls		
6	more boys than girls		
6	equal numbers of boys and girls		

10. We have insisted all along that the lot from which we select the random sample must be "sufficiently large." This is because of the necessity of sampling without replacement, so that the same item is not tested twice. As an example, suppose that a random sample of size $n = 2$ is drawn without replacement from a lot of 1000 items with 100 defectives. Then

$$P(\text{defective on second draw} \mid \text{first was good}) = 100/999$$

and

$$P(\text{defective on second draw} \mid \text{first was defective}) = 99/999,$$

so that P(defective on second draw) depends on what has happened before ($100/999 \neq 99/999$). Strictly speaking, then, sampling without replacement does not lead to independent events, but as long as the sample is "large enough" we will not be "too far off" by assuming the sample drawings are independent of each other. More advanced work with probability would enable us to quantify the qualitative words enclosed in quotes. The following examples help us to understand this situation a little better.

a. Suppose a large lot contains 10% defective items and a random sample of size n is drawn without replacement. Complete the table where a maximum of x defectives is tolerated in the sample.

n	x	$P(n, \max x, .10)$
1	1	
2	1	
2	2	
3	1	
3	2	
3	3	
4	1	
4	2	
4	3	
4	4	
5	1	
5	2	
5	3	

b. Suppose a lot of forty ball bearings contains four defectives and a random sample of size n is drawn without replacement. Complete the table where a maximum of x defectives is tolerated in the sample.

n	x	$P(n, \max x, .10)$
1	1	
2	1	
2	2	
3	1	
3	2	
3	3	
4	1	
4	2	
4	3	
4	4	
5	1	
5	2	
5	3	

11. Suppose that a person takes a multiple choice test where she estimates the probability of guessing any question correct to be .6 because of her

prior preparation for the test. If the test contains ten questions, find the probability that she will obtain

a. seven correct answers. Hint: $(.6)^7 \cdot (.4)^3 = (.4)^3 \cdot (1-.4)^7$ and ${}_{10}C_7 = {}_{10}C_3$. In general, if $p > .5$, then
$$P(n, r, p) = P(n, n-r, 1-p).$$
b. six correct answers.
c. seven or more correct answers.
d. three or less correct answers.
e. four or five correct answers.
f. Find the sum of the answers to parts (d), (e), (b), and (c).

12. Repeat Problem 11, only this time assume the probability of her guessing each question correct to be .8.
13. A law school professor with a class of thirty students calls on six students chosen at random without duplication every day. What is the probability that one unprepared student, Mark McElroy, will escape detection for
    a. one day?
    b. two consecutive days?
    c. three consecutive days?
    d. four consecutive days?
    e. five consecutive days?
    f. n consecutive days?
14. A law school professor with a class of thirty students calls on six students chosen at random without duplication every day. Suppose there is a group of five students in the class who are always unprepared. What is the probability that everyone in this group will escape detection for
    a. one day?
    b. two consecutive days?
    c. three consecutive days?
    d. four consecutive days?
    e. five consecutive days?
    f. n consecutive days?
15. Suppose that a large group of jurors watched a particular criminal trial and that 20% of these jurors believed the defendant to be guilty before any deliberation. If a jury of twelve were selected from this pool (contrary to what actually happens where the jury is chosen before the trial), then
    a. find the probability that the selection contains a majority of people who believe the defendant to be guilty, and
    b. find the probability that the selection contains a majority of people who believe the defendant to be guilty if the jury size is now six instead of twelve.

Probability and the law

So far in this text we have traveled the main highways in learning mathematical methods and content appropriate for social and management scientists.

We are now leaving topics of general interest for a brief excursion down a path of special interest for future lawyers and political scientists. This excursion also illustrates to all of us how mathematical principles can turn up in rather unanticipated ways. In this investigation of probability and the law we will see how basic probability principles (1) convicted and acquitted a couple of a crime and (2) cast serious doubt on the validity of a recent United States Supreme Court decision.

Probability in conviction and acquittal

The following quotation is from *Time* magazine, 8 January 1965, page 42, quoted with permission of the publishers. It is the complete reproduction of the article entitled "TRIALS—The Laws of Probability."

> Around noon one day last June an elderly woman was mugged in an alley in San Pedro, Calif. Shortly afterward a witness saw a blonde girl, her pony tail flying, run out of the alley, get into a yellow car driven by a bearded Negro, and speed away. Police eventually arrested Janet and Malcolm Collins, a married couple who not only fitted the witness's physical description of the fugitive man and woman but also owned a yellow Lincoln. The evidence, though strong, was circumstantial. Was it enough to prove the Collinses guilty beyond a reasonable doubt?
>
> Confidently answering yes, a jury has convicted the couple of second-degree robbery because Prosecutor Ray Sinetar, 30, cannily invoked a totally new test of circumstantial evidence—the laws of statistical probability.
>
> In presenting his case, Prosecutor Sinetar stressed what he felt sure was already in the jurors' minds: the improbability that at any one time there could be two couples as distinctive as the Collinses in San Pedro in a yellow car. To "refine the jurors' thinking" Sinetar then explained how mathematicians calculate the probability that a whole set of possibilities will occur at once. Take three abstract possibilities (A, B, C) and assign to each a hypothetical probability factor. A, for example, may have a probability of 1 out of 3; B, 1 out of 10; C, 1 out of 100. The odds against A, B, and C occurring together are the product of their total probabilities (1 out of $3 \times 10 \times 100$), or 3,000 to 1.
>
> After an expert witness approved Sinetar's technique, the young prosecutor asked the jury to consider the six known factors in the Collins case: a blonde white woman, a pony-tail hairdo, a bearded man, a Negro man, a yellow car, an interracial couple. Then he suggested probability factors ranging from 1-to-4 odds that a girl in San Pedro would be blonde to 1-to-1,000 odds that the couple would be Negro-white. Multiplied together, the factors produced odds of 1 to 12 million that the Collinses could have been duplicated in San Pedro on the morning of the crime.
>
> Public Defender Donald Ellertson strenuously objected on grounds that the mathematics of probability were irrelevant, and that Sinetar's probability factors were inadmissible as assumptions rather than facts. Sinetar, however, merely estimated the factors before inviting the jurors to substitute their own. And the public defender will not appeal because he found no trial errors strong enough to outweigh the strong circumstantial evidence. Convicted by math, Malcolm Collins received a sentence of one year to life. Janet Collins got "not less than one year."

Of course, we recognize the use of the multiplication principle of probability to obtain the equation

$$P(\text{duplication of this couple}) = 1/12{,}000{,}000.$$

There are three possible drawbacks to accepting this equation; we can recognize them through our previous work with probability. First of all, in order to conclude that

$$P(E_1 \text{ and } E_2 \text{ and } E_3 \text{ and } E_4 \text{ and } E_5 \text{ and } E_6) = P(E_1) \cdot P(E_2) \cdot P(E_3) \cdot P(E_4) \cdot P(E_5) \cdot P(E_6),$$

as Prosecutor Sinetar did, we must assume that the six events of

blonde white woman, ponytail hairdo, bearded man,
Negro man, yellow car, interracial couple

are all independent of one another. For example, if blondes are more likely to wear their hair in a ponytail than nonblondes, then

$$P(\text{ponytail} \mid \text{blonde}) \neq P(\text{ponytail})$$

and

$$P(\text{blonde and ponytail}) = P(\text{blonde}) \cdot P(\text{ponytail} \mid \text{blonde}) \neq P(\text{blonde}) \cdot P(\text{ponytail}).$$

Second, on what evidence from population studies and surveys did the prosecutor conclude that, for example,

$$P(\text{a girl in San Pedro is blonde}) = 1/4?$$

Third, the prosecution states that

$$P(\text{duplication of this couple}) = 1/12{,}000{,}000,$$

and we recall the results of our work in Problem set 5–3 (Problem 22) where the probability of a duplication of birthdays among twenty-three people was much higher (.51) than we intuitively expected. Perhaps, then, 1/12,000,000 is too small a number.

Probability in legal appeals

We now return to *Time*, 26 April 1968, which is reproduced with permission of the publishers. The following quotations begins after the "DECISIONS—Trial by Mathematics" article has summarized the previous *Time* article of 1965.

> The logic of it all seemed overwhelming, and few disciplines pay as much homage to logic as do the law and math. But neither works right with the wrong premises. Hearing an appeal of Malcolm Collins' convictions, the California Supreme Court recently turned up some serious defects, including the fact that not even the odds were all they seemed.
>
> To begin with, the prosecution failed to supply evidence that "any of the individual probability factors listed were even roughly accurate." Moreover, the factors were not shown to be fully independent of one another as they must be to satisfy the

mathematical law; the factor of a Negro with a beard, for instance, overlaps the possibility that the bearded Negro may be part of an interracial couple. The 12 million to 1 figure, therefore, was just "wild conjecture." In addition, there was not complete agreement among the witnesses about the characteristics in question. "No mathematical equation," added the court, "can prove beyond a reasonable doubt (1) that the guilty couple *in fact* possessed the characteristics described by the witnesses, or even (2) that only *one* couple possessing those distinctive characteristics could be found in the entire Los Angeles area."

To explain why, Judge Raymond Sullivan attached a four-page appendix to his opinion that carried the necessary math far beyond the relatively simple formula of probability. Judge Sullivan was willing to assume it was unlikely that such a couple as the one described existed. But since such a couple did exist—and the Collinses demonstrably did exist—there was a perfectly acceptable mathematical formula for determining the probability that another such couple existed. Using the formula and the prosecution's figure of 12 million, the judge demonstrated to his own satisfaction and that of five concurring justices that there was a 41% chance that at least one other couple in the area might satisfy the requirements. (The proof involved is essentially the same as that behind the common parlor trick of betting that in a group of 30 people, at least two will have the same birthday; in that case, the probability is 70%.)[2]

"Undoubtedly," said Sullivan, "the jurors were unduly impressed by the mystique of the mathematical demonstration but were unable to assess its relevancy or value." Neither could the defense attorney have been expected to know of the sophisticated rebuttal available to them. Janet Collins is already out of jail, has broken parole and lit out for parts unknown. But Judge Sullivan concluded that Malcolm Collins, who is still in prison at the California Conservation Center, had been subjected to "trial by mathematics" and was entitled to a reversal of his conviction. He could be tried again, but the odds are against it.

In reading the second article we can see where Judge Sullivan and the California Supreme Court used familiar probability principles to overturn the earlier conviction. We should note that probability, like so much of mathematics, can occur in unusual ways.

Probability and jury decisions

We now look back to problem 15 of the previous section to introduce the next application of probability theory to the law. The idea for and explanation of this application come from David F. Walbert's "The Effect of Jury Size on the Probability of Conviction: an Evaluation of Williams v. Florida," which appeared in the *Case Western Reserve Law Review*, Volume 22, Number 3 (April 1971), quoted by permission. Essentially, Mr. Walbert challenges the United States Supreme Court's decision in *Williams v. Florida*, 399 U.S. 78 (1970), in which the Court said that a defendant in a criminal case can receive as fair a trial with a jury of six as he can with a traditional jury of twelve, all other factors in the decision-making process being equal.

[2] For the complete explanation, see *People v. Collins*, 68 Cal 2d 319, 66 Cal Rptr 497, 438 P2d 33, 36 ALR3d 1176, pp. 1191–1193.

Mr. Walbert's article is necessarily comprehensive and studies the many facets of the functions of a jury. For a full explanation of the background circumstances readers should refer to the original article or to a lawyer, for our analysis and understanding of Mr. Walbert's work is limited to the formal mathematical principles he employs as part of his dissent from the Supreme Court's decision. With our limitations of law and knowledge of the *Case* article clearly in mind, we again read Problem 15 from the last section. Be sure to do this before reading any further. The page numbers following each quotation refer to Mr. Walbert's article.

> . . . Of course, the ultimate function of the jury is to resolve the question of the defendant's guilt. And a jury is preferred over other means because it provides the defendant with "an inestimable safeguard against the corrupt or over-zealous prosecutor and against the compliant, biased, or eccentric judge." (p. 534)
>
> . . . The framework for analysis, then, is simple. A criminal trial takes place, and two questions are asked: What would be the outcome if a jury of six had sat in judgment? And what would be the outcome if a jury of twelve had sat in judgment? (p. 540)
>
> . . . A verdict of guilty or innocent requires unanimity in nearly all criminal cases (p. 542).

Mr. Walbert develops his mathematical analysis around a probability model of a jury's decision making from the time it leaves the courtroom, prior to any deliberation, to the pronouncement of one of the three possible verdicts: not guilty (innocent), guilty, or no decision (a hung jury in which a unanimous decision for acquittal or conviction was not reached). Here is his assumption of the relationship between potential jurors, the trial itself, and the jury's pre-deliberation opinions.

> If each potential juror had actually observed the trial, a certain fraction of them would be inclined to consider the defendant guilty at the conclusion of the courtroom proceedings, immediately prior to deliberation. This fraction will be denoted by f_t. Conversely, $1-f_t$ is the fraction of the entire pool that would be inclined to believe the defendant innocent just before deliberation begins. This characterization of the complete group of potential jurors is also sufficient to depict the petit jury because the petit jury is simply a statistical subset of these potential jurors. Thus, the single parameter, f_t, correlates the trial itself to a description of the jury's leanings before it begins deliberation (within the statistical limits involved in randomly selecting a jury from the pool at large). In other words, both the entire complex of events within the trial and the personalities are distilled into this one variable. (pp. 540–541)

A petit jury is an "ordinary" jury that decides on the ultimate guilt or innocence in a specific case; in contrast, a grand jury can only indict, not convict or acquit, a defendant. Also,

f_t = the fraction of *all potential jurors* in the pool who, *if they were to witness the defendant's trial*, would believe him guilty *just prior to deliberation*

= P(any one juror would believe the defendant to be guilty).

We may also think of f_t as the probability that a single person on the jury believes the defendant to be guilty just before the jury begins deliberation. It is therefore a measure of the effectiveness of the prosecution. From this interpretation we easily see that f_t is entirely independent of the number of jurors who surround any one person on the jury.

> Because a particular jury is merely a randomly drawn subset of all the potential jurors, the probabilities of obtaining petit juries with various fractions of conviction-prone members can be easily calculated. Stated in another way, one fact is known about each juror as he leaves the jury box to begin deliberation: The probability that he believes the defendant to be guilty is exactly equal to the value of f_t. This one parameter determines the likelihood of a majority of jurors being conviction prone . . . . (pp. 540–541)

Before going any further we assume that majority persuasion is an important factor in our analysis of jury size. Majority persuasion means that regardless of the size of the jury the initial majority opinion almost always emerges as the final unanimous opinion because of the influence of the majority on the minority during deliberation. The best available evidence, referenced on page 544 of the article, states that the minority opinion becomes majority opinion in only about 3% of jury trials; assuming it happens 0% of the time will therefore not appreciably change our final results. What about a jury of twelve where initially six believe the defendant guilty and six believe him innocent? Here we assume—and there is historical evidence to support this assumption—that half of those situations result in a final verdict for conviction and the other half of those situations result in a final verdict for acquittal. A similar situation exists for a jury of six, with three members initially for conviction and the other three favoring acquittal.

Deriving formulas

We are now ready to derive the formulas for the probability of conviction dependent on the initial leanings of the posttrial, predeliberation juries. As an example, suppose we have a twelve-person jury chosen from a pool of which 20% of the people would believe the defendant to be guilty. Then our previous analysis allows us to express the related numbers as follows:

P(defendant is convicted)

$= P$(twelve-person jury selected from the pool with $f_t = .20$ contains a majority leaning toward conviction)

$+ \frac{1}{2} \cdot P$(twelve-person jury selected from the pool with $f_t = .20$ contains six jurors leaning toward conviction)

$= P$(twelve-person jury contains twelve people leaning toward conviction)

$+ P$(twelve-person jury contains eleven people leaning toward conviction)

+ P(twelve-person jury contains ten people leaning toward conviction)
+ P(twelve-person jury contains nine people leaning toward conviction)
+ P(twelve-person jury contains eight people leaning toward conviction)
+ P(twelve-person jury contains seven people leaning toward conviction)
+ $\frac{1}{2} \cdot P$(twelve-person jury contains six people leaning toward conviction).

As a typical expression to evaluate let us calculate

P(twelve-person jury contains nine people leaning toward conviction).

This is really an application of the Binomial Law because we are independently choosing $n = 12$ people where

P(any member of the pool believes the defendant guilty) $= f_t = p = .20$

and

$$r = 9.$$

We are interested in the probability of exactly nine occurrences of a single event, whose individual probability of occurrence is .20, in twelve independent trials. Thus

$$\begin{aligned} P(12, 9, .20) &= \binom{12}{9} \cdot (.20)^9 \cdot (1 - .20)^{12-9} \\ &= .0001 \qquad \text{(from Table XI)} \end{aligned}$$

This calculation typifies the fact that we are really saying

$$\begin{aligned} &P(\text{defendant is convicted}) \\ &\quad = P(12, 12, .20) + P(12, 11, .20) + P(12, 10, .20) \\ &\qquad + P(12, 9, .20) + P(12, 8, .20) + P(12, 7, .20) \\ &\qquad + \frac{1}{2} \cdot P(12, 6, .20) \\ &\quad = .0000 + .0000 + .0000 + .0001 + .0005 + .0033 + \frac{1}{2} \cdot .0155 \\ &\quad = .0117. \end{aligned}$$

If we now hold everything constant except for changing the jury size to six, then we have

P(defendant is convicted)
$= P$(six-person jury selected from the pool with $f_t = .20$ contains a majority leaning toward conviction)
$+ \frac{1}{2} \cdot P$(six-person jury selected from the pool with $f_t = .20$ contains three jurors leaning toward conviction)

The difference in probabilities is .0580 − .0117 = .0463 or, more important, .0580/.0117 = 4.970, which means that with all other factors held constant and with $f_t = .20$, a defendant is almost five times as likely to be convicted with a six-person jury as he is with a twelve-person jury!

Table 5.5 shows the computer-calculated results of evaluating appropriate expressions of P(defendant is convicted) for values of $f_t = p$ from 0.00 to 1.00 in increments of .05. Following this table of numerical calculations is a graphical representation of the data (overleaf), and then Table 5.6 on page 209 shows the ratio comparison between P(conviction, jury size = 6) and P(conviction, jury size = 12).

Table 5.5

P = F T	P(CONVICTION), JURY = 6	P(CONVICTION), JURY = 12
Ø	Ø	Ø
.Ø5	.ØØ12	Ø
.1	.ØØ86	.ØØØ3
.15	.Ø266	.ØØ27
.2	.Ø579	.Ø117
.25	.1Ø35	.Ø343
.3	.1631	.Ø782
.35	.2352	.1487
.4	.3174	.2465
.45	.4Ø69	.3669
.5	.5	.5
.55	.5931	.6331
.6	.6826	.7535
.65	.7648	.8513
.7	.8369	.9218
.75	.8965	.9657
.8	.9421	.9883
.85	.9734	.9973
.9	.9914	.9997
.95	.9988	1
1	1	1

For $p = f_t = .02$ and .04 the table entries in columns 2 and 3 are not entirely correct because the numbers are rounded off to six decimal places before they are printed. The numbers in the fourth column are correct.

Notice that the only time

$$P(\text{conviction}, n = 6) = P(\text{conviction}, n = 12)$$

is when $p = f_t$ is 0, .5, or 1. In all other cases, which form the vast majority of actual situations,

$$P(\text{conviction}, n = 6) \neq P(\text{conviction}, n = 12).$$

Figure 5.6

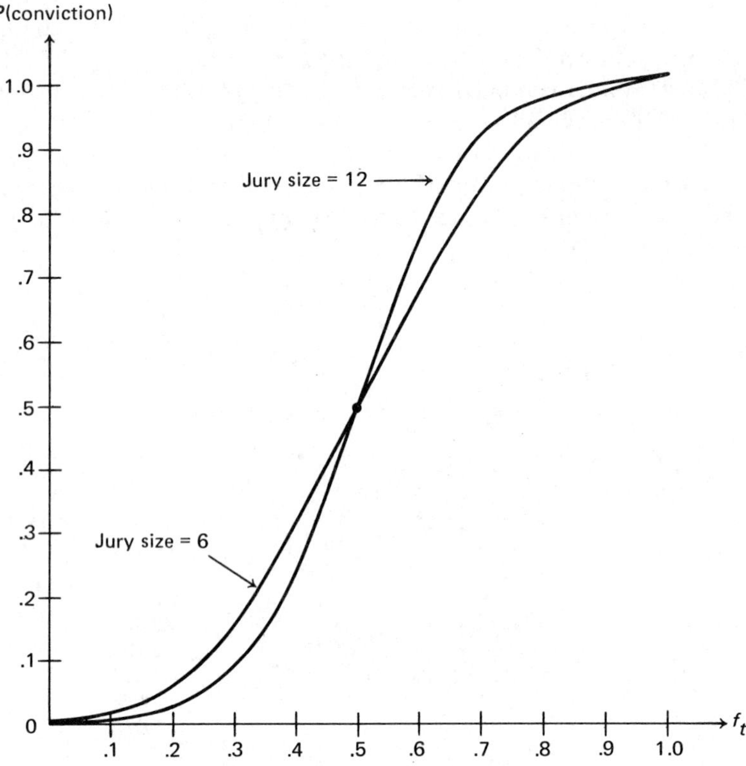

It is precisely from the results of these calculations based on a familiar formula that Mr. Walbert refutes the Supreme Court's decision that jury size can be reduced from twelve to six for a saving of time and money without affecting the justice due the defendant.

> It is evident from these numerical results that serious differences exist between the six- and 12-man juries. For nearly all values of f_t, the size of the jury is substantially related to the probability of conviction. If f_t is larger than .5, the defendant has a greater chance of acquittal with a six-man jury; if it is less than .5, the six-man jury increases the likelihood of conviction. . . .
>
> Thus, the Court's conclusion that both juries would return the same verdict is erroneous. . . .
>
> When the number of jurors is reduced from 12 to six, the decreased representation makes the defendant's fate more a matter of the chance involved in selecting the petit jury. Consequently, the actual verdict is less likely to reflect the opinion of a "representative cross section of the community. . . ." (p. 547)

Problem set 5–5

1. Bill Clarendon is a criminal lawyer defending a "client" who is guilty of armed robbery; the client knows it and so does Bill. However, the

Table 5.6

P	P(CON), J=6	P(CON), J=12	COL2 / COL3
0	0	0	-------
.02	.000078	0	2861.99
.04	.000602	.000002	378.784
.06	.00197	.000017	118.948
.08	.004525	.000085	53.2375
.1	.00856	.000296	28.9477
.12	.014319	.000804	17.8098
.14	.022	.001843	11.9366
.16	.031759	.003727	8.52036
.18	.043707	.006847	6.3834
.2	.05792	.011654	4.96988
.22	.074434	.018643	3.99264
.24	.093251	.028322	3.29251
.26	.114343	.041189	2.77604
.28	.137648	.057696	2.38573
.3	.16308	.078225	2.08476
.32	.190526	.103056	1.84877
.34	.219851	.13235	1.66113
.36	.250897	.16613	1.51024
.38	.283491	.204272	1.38781
.4	.31744	.246502	1.28778
.42	.35254	.292399	1.20568
.44	.388575	.341409	1.13815
.46	.425319	.392865	1.08261
.48	.46254	.446003	1.03708
.5	.5	.5	1
.52	.53746	.553996	.970151
.54	.57468	.607135	.946545
.56	.611425	.65859	.928384
.58	.64746	.707601	.915007
.6	.68256	.753498	.905855
.62	.716509	.795727	.900446
.64	.749102	.833869	.898345
.66	.780149	.867649	.899152
.68	.809473	.896944	.90248
.7	.83692	.921775	.907943
.72	.862352	.942303	.915153
.74	.885657	.95881	.923704
.76	.906748	.971678	.933178
.78	.925566	.981357	.943149
.8	.94208	.988346	.953188
.82	.956293	.993153	.962885
.84	.968241	.996273	.971863
.86	.978	.998157	.979805
.88	.985681	.999196	.986474
.9	.99144	.999705	.991733
.92	.995475	.999915	.995559
.94	.99803	.999984	.998046
.96	.999398	.999999	.999399
.98	.999923	1	.999923
1	1	1	1

criminal has money and offers Bill a large fee if he can convince the jury that his client is innocent by harrassment of witnesses, nitpicking about tiny details, legal stalling, and other tactics. Bill estimates that even after his vigorous defense f_t would be .70 as the jurors begin deliberation.

a. What is P(conviction, $n = 12$)? (Use Table XI)
b. What is P(conviction, $n = 6$)? (Use Table XI)
c. Which size jury would Bill argue for?

2. This problem refers to the case of the couple that was convicted and then acquitted. If in San Pedro, California, extensive samples reveal that 25% of women are blondes, 20% of blondes have their hair in ponytails, and 10% of all women wear ponytails, then let B symbolize the event of a woman being blonde and let P symbolize the event of a woman having a ponytail. If a woman is picked at random from the population of San Pedro, then calculate these probabilities.

a. $P(B)$ b. $P(P)$ c. $P(P \mid B)$
d. $P(B \text{ and } P)$ e. $P(B) \cdot P(P)$ f. $P(B \mid P)$
g. What fact does a comparison of the results from parts (d) and (e) tell you?

3. It is one thing to predict the composition of a jury selected from a specific pool and another to actually select the jurors. We could then compare the predicted results with the actual results. A computer has been programmed to pick 10,000 juries at random from specified pools and tabulate the results of this selection process. Essentially, the computer performed mathematical calculations which are equivalent to picking points at random from the interior of a circle with area of one square unit. If the number calculated corresponded to a point in the shaded region below, a conviction-prone juror was recorded; otherwise, the juror was recorded as acquittal-prone. The cycle of calculation and comparison occurred either six or twelve times in order to pick one jury. This process selected 10,000 juries six times under varying conditions, and the results appear in the following six tables. The symbol "#" represents the word "number."

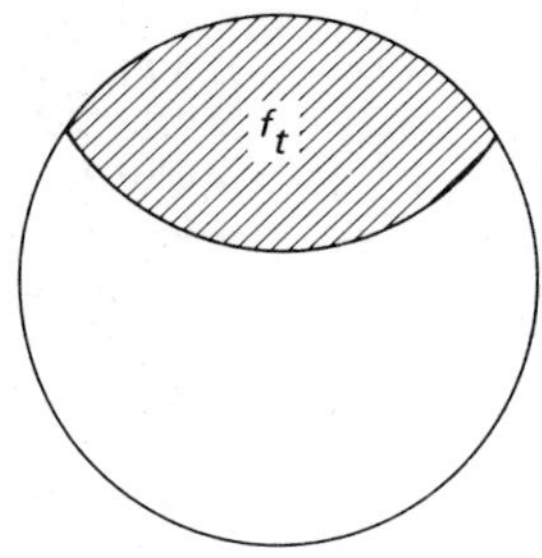

FOR 10000 JURIES OF 6 PEOPLE EACH CHOSEN FROM A POOL
WHERE FRACTION .25 WOULD BELIEVE THE DEFENDANT TO BE GUILTY:

# OF CONVICTION-PRONE JURORS, J	NUMBER OF JURIES WITH J CONVICTION-PRONE JURORS	PERCENT OF JURIES WITH J CONVICTION-PRONE JURORS
0	1716	17.16
1	3593	35.93
2	2953	29.53
3	1359	13.59
4	325	3.25
5	52	.52
6	2	.02

FOR 10000 JURIES OF 12 PEOPLE EACH CHOSEN FROM A POOL
WHERE FRACTION .25 WOULD BELIEVE THE DEFENDANT TO BE GUILTY:

# OF CONVICTION-PRONE JURORS, J	NUMBER OF JURIES WITH J CONVICTION-PRONE JURORS	PERCENT OF JURIES WITH J CONVICTION-PRONE JURORS
0	301	3.01
1	1309	13.09
2	2284	22.84
3	2561	25.61
4	1985	19.85
5	1044	10.44
6	385	3.85
7	106	1.06
8	23	.23
9	2	.02
10	0	0
11	0	0
12	0	0

FOR 10000 JURIES OF 6 PEOPLE EACH CHOSEN FROM A POOL
WHERE FRACTION .5 WOULD BELIEVE THE DEFENDANT TO BE GUILTY:

# OF CONVICTION-PRONE JURORS, J	NUMBER OF JURIES WITH J CONVICTION-PRONE JURORS	PERCENT OF JURIES WITH J CONVICTION-PRONE JURORS
0	166	1.66
1	899	8.99
2	2336	23.36
3	3093	30.93
4	2422	24.22
5	928	9.28
6	156	1.56

```
FOR 10000   JURIES OF 12   PEOPLE EACH CHOSEN FROM A POOL
 WHERE FRACTION .5          WOULD BELIEVE THE DEFENDANT TO BE GUILTY:

# OF CONVICTION-       NUMBER OF JURIES       PERCENT OF JURIES
PRONE JURORS, J        WITH J CONVICTION-     WITH J CONVICTION-
                       PRONE JURORS           PRONE JURORS

       0                      2                      .02
       1                      28                     .28
       2                      188                    1.88
       3                      541                    5.41
       4                      1241                   12.41
       5                      1894                   18.94
       6                      2240                   22.4
       7                      1934                   19.34
       8                      1235                   12.35
       9                      523                    5.23
       10                     134                    1.34
       11                     38                     .38
       12                     2                      .02
```

```
FOR 10000   JURIES OF 6    PEOPLE EACH CHOSEN FROM A POOL
 WHERE FRACTION .75         WOULD BELIEVE THE DEFENDANT TO BE GUILTY:

# OF CONVICTION-       NUMBER OF JURIES       PERCENT OF JURIES
PRONE JURORS, J        WITH J CONVICTION-     WITH J CONVICTION-
                       PRONE JURORS           PRONE JURORS

       0                      3                      .03
       1                      51                     .51
       2                      358                    3.58
       3                      1281                   12.81
       4                      2956                   29.56
       5                      3575                   35.75
       6                      1776                   17.76
```

```
FOR 10000   JURIES OF 12   PEOPLE EACH CHOSEN FROM A POOL
 WHERE FRACTION .75         WOULD BELIEVE THE DEFENDANT TO BE GUILTY:

# OF CONVICTION-       NUMBER OF JURIES       PERCENT OF JURIES
PRONE JURORS, J        WITH J CONVICTION-     WITH J CONVICTION-
                       PRONE JURORS           PRONE JURORS

       0                      0                      0
       1                      0                      0
       2                      1                      .01
       3                      2                      .02
       4                      23                     .23
       5                      116                    1.16
       6                      356                    3.56
       7                      1006                   10.06
       8                      1898                   18.98
       9                      2613                   26.13
       10                     2338                   23.38
       11                     1350                   13.5
       12                     297                    2.97
```

The first table says that

$$\frac{1}{2} \cdot 1{,}359 + 325 + 52 + 2 = 1{,}059$$

of the juries selected were conviction-prone at the very beginning. Now,

$$\frac{1}{2} \cdot P(6,3,.25) + P(6,4,.25) + P(6,5,.25) + P(6,6,.25)$$

$$= .0659 + .0330 + .0044 + .0002 \text{ (from Table XI)}$$
$$= .1035.$$

That is, we theoretically expect 10.35% of all juries selected under these circumstances to be conviction-prone (see also Table 5.5). 10.35% of 10,000 = 1,035 which does not appear to be too far away from the observed frequency of 1,059 ($= 1/2 \cdot 1359 + 325 + 52 + 2$). Incidentally, the computer used here picked and tabulated about eighteen juries per second. Complete this table with reference to the previous tables.

Table	Jury size	f_t	Theoretical number of conviction-prone juries	Actual number of conviction-prone juries
1	6	.25	1,035	1,059
2				
3				
4				
5				
6				

Odds and expectation

The language of odds

Many times in everyday conversation people describe the probability of an event by using the language of "odds." For example, the odds on the horse Margate Charley may be quoted as "4 to 1." What does this mean?

To say that there are 4-to-1 odds on Margate Charley means that, in the opinion of informed people, the horse has four chances of losing and one chance of winning in an upcoming race. Since there are estimated to be five possible chances on the race (four where Margate Charley loses + one where he wins) and only one of these five corresponds to the horse's winning, then

$$P(\text{Margate Charley wins}) = 1 \text{ chance out of } 5 = 1/5.$$

Also,

$$P(\text{Margate Charley loses}) = 1 - P(\text{wins}) = 1 - 1/5 = 4/5.$$

If the odds on a horse are listed as three to five, this means that there are three chances of the horse's losing and five chances of his winning for a total of $3+5=8$ possible outcomes, of which five are favorable to the horse. Then

$$P(\text{horse wins}) = 5/8 = .625.$$

The general definition comes easily from these two examples.

Definition If the *odds on* an event are a to b, then there are a chances of the event *not* occurring and b chances of it occurring; so out of the $(a+b)$ possible outcomes b of these correspond to the event's occurring. Thus

$$P(\text{event occurs whose odds are } a \text{ to } b) = b/(\mathrm{a}+b).$$

Recall that

$$P(\text{Margate Charley wins whose odds were 4 to 1}) = 1/5 = 1/(4+1).$$

Thus "odds on" really means "odds against," since the first number, a, is the number of chances of the event *not* occurring. This interpretation of

$$\text{odds on} = \text{odds against}$$

is the one commonly applied in race tracks and casinos across the country, even though other mathematics textbooks define odds of a to b to mean $a/(a+b)$ as the corresponding probability of the event occurring. On the other hand, if the odds on an event are explicitly stated as odds in favor of, then this reverse method is appropriate.

Example 27 The Cleveland Browns are 3-to-1 favorites in a football game against the New England Patriots Find P(Browns win).

Solution Here the numbers of 3 to 1 are specifically given as odds in favor of, so the 3 represents the number of chances of the Browns' winning and the 1 represents the number of chances of the Browns' losing; since out of the $3+1=4$ possible outcomes 3 correspond to the occurrence of the Browns' winning,

$$P(\text{Browns win}) = 3/4 = .75.$$

Who determines the odds at a race track or on any sporting event involving betting? The bettors themselves do. If, for example, \$20,000 is bet on a particular race and if \$8,000 of that amount is bet on Andy's Folly, then the bettors have determined that

$$P(\text{Andy's Folly wins}) = 8{,}000/20{,}000 = .40$$
$$2/5 = 2/(3+2).$$

The odds on Andy's Folly are 3 (chances of losing) to 2 (chances of winning). If Andy's Folly does win, then \$18,000 (\$20,000 – \$2,000 for the state government and the horse) will be distributed back to the bettors; in this way the \$8,000 bet on the horse has suddenly grown to \$18,000 (assuming that 10% goes to the state and the horse). Since 18,000/8,000 = 4.5/2, a \$2 bet (the minimum) brings the bettor back a profit of \$2.50. A bettor on any other horse, of course, loses all of his money to the government, the winning horse, and those who bet on Andy's Folly. In practice, a bettor can win if his horse comes in first, second, or third. Since the payoffs are always rounded to a fifth of a dollar, in this case the track's management would pay \$4.40 to each \$2 bettor and keep the 10¢ "breakage" for itself. If the ratio of returns from the horse to the amount bet on it were 4.59/2, then the amount paid to each \$2 bettor on the winning horse would still be \$4.40. In this case, the track would receive the 19¢ breakage.

Expectation

Another application of probability theory to financial situations is found in expectation. Two examples best explain this concept.

Example 28 A man pays an entrance fee of 50¢ to enter a lottery in which one card is drawn from a standard deck. He receives \$1.60 if the card drawn is a heart and nothing otherwise. What is his expected gain, or *mathematical expectation*?

Solution His *expectation* is the amount of money he may expect to win. If we temporarily neglect the 50¢ entrance fee, we observe that the probability of his winning \$1.60 is 1/4 because $P(\text{heart}) = 1/4$. Thus we intuitively conclude that he has a 1/4 hold, or ownership, of the \$1.60. Since 1/4 of \$1.60 is 40¢, we say that his *expectation* in one play of this lottery is 40¢. Exactly two outcomes can occur when he enters the lottery for one play.

1. He wins \$1.60, or
2. he wins \$0.00.

We can think of his expected winnings of 40¢ per entry as being a long-run average of many \$.00 gains and fewer \$1.60 gains. Of course, we have to take the entrance fee of 50¢ into account. Since he has an expected gross gain per game of 40¢ and a definite payment of 50¢ to make, then the *expected net gain*, *ENG*, is

$$40¢ - 50¢ = -10¢.$$

Again, he actually nets either \$1.60 − \$.50 = \$1.10 or 0¢ − 50¢ = −50¢ on any one game, so we think of the *ENG* of −10¢ as a long-run average figure. Any commercial lottery always has the expectation, *E*, greater than 0 in order to attract players and always has the *ENG* negative to the player and hence positive to the people running the lottery.

Example 29 A player pays $2.50 to enter a lottery where a pair of dice are rolled. If a sum of 7 occurs he receives $6, and if a sum of 11 occurs he receives $9. What are his expectation and his *ENG*?

Solution Since

$$\text{Expected Net Gain} = \text{Expectation} - \text{Entrance Fee}$$

or

$$ENG = E - EF,$$

we concentrate our efforts on evaluating E.

$$\begin{aligned} P(\text{win}) &= P(\text{roll a sum of 7 or roll a sum of 11}) \\ &= P(\text{roll a sum of 7}) + P(\text{roll a sum of 11}) \\ &= 6/36 + 2/36 \\ &= 1/6 + 1/18. \end{aligned}$$

He has a 1/6 ownership of $6 equal to $1 and a 1/18 ownership of $9 = 50¢. His total expected income per game is therefore $1.50. Then $1.50 – $2.50 = – $1.00 is the final *ENG*.

We are now ready to generalize to the situation of calculating the total expectation of a situation where n disjoint events

$$E_1, E_2, E_3, E_4, \ldots, E_n$$

may occur with corresponding monetary values of

$$v_1, v_2, v_3, v_4, \ldots, v_n$$

and corresponding probabilities of

$$P(E_1), P(E_2), P(E_3), P(E_4), \ldots, P(E_n).$$

Then the total expectation to the beneficiary of having one of the n events occur is

$$E = P(E_1)\cdot v_1 + P(E_2)\cdot v_2 + P(E_3)\cdot v_3 + P(E_4)\cdot v_4 + \cdots + P(E_n)\cdot v_n$$

and the final expected net gain after taking the entrance fee, *EF*, into account is

$$ENG = E - EF.$$

In the above examples we actually evaluated

$$E = P(E_1)\cdot v_1 = P(\text{heart})\cdot(\text{value of a heart}) = 40¢$$

and

$$\begin{aligned} E &= P(E_1)\cdot v_1 + P(E_2)\cdot v_2 \\ &= P(\text{rolling a 7})\cdot(\text{value of a 7}) + P(\text{rolling an 11})\cdot(\text{value of an 11}) \\ &= \$1.00 + \$.50 = \$1.50. \end{aligned}$$

Problem set 5–6

1. Complete this table.

Odds on an event	P(event occurs)
3 to 1	
7 to 1	
1 to 3	
9 to 2	
2 to 3	
4 to 5	

2. Complete this table.

Odds on E	P(E occurs)	P(E does not occur)
5 to 1		
2 to 7		
	1/4	
	1/6	
	3/4	
	5/7	
	2/5	
	1/9	

3. Complete this table.

Odds in favor of E	Odds on E	P(E occurs)
5 to 1		
8 to 1		
5 to 2		
7 to 3		
1 to 5		
1 to 3		

4. What is your mathematical expectation if you receive \$8 for drawing a diamond from a standard deck?
5. What is your mathematical expectation if you receive \$1.30 for drawing a black from a standard deck and receive \$2 for drawing a red face card? What is the maximum entrance fee you should pay in order to expect a long-run profit?
6. A marble is drawn at random from an urn containing five red marbles, three white marbles, and eight blue marbles. What are the odds in favor of the marble's being red? What are the odds on a white marble being drawn?

7. A man chooses a marble at random from a box that contains two red marbles, three white marbles, and five blue marbles. If he is paid 10¢ for a red marble, 20¢ for a white marble, and 30¢ for a blue marble, what is his mathematical expectation?
8. If Mr. Smith, who is twenty-eight years old, buys a $10,000 life insurance policy for one year, what is the minimum amount the insurance company should collect in premiums, disregarding administrative costs, to break even? Use the table in Problem 6 of Problem set 5–1.
9. Assume that you have $10,000 in your life savings account and a wealthy gambler offers you a deal. You are to give him the $10,000 and then draw a card from a standard well-shuffled deck. He will give you $240,000 if you draw a spade.
   a. What is your mathematical expectation?
   b. What is your expected net gain?
   c. Would you accept his offer?
10. A man offers to bet you even money that when rolling a pair of fair dice he will throw a sum of 7 before the sixth roll. Should you take the bet?
11. A deck is made up of cards that are numbered from 1 to 10 inclusive. You draw two cards from the deck at random without replacement. If the sum of the numbers on the cards you draw is even you win $1, and if the sum is odd you lose $1. Is the game fair to you?
12. Same as 11, but this time draw the cards with replacement.
13. A man goes to a horse-racing track and records the amounts of money bet on each horse in a race along with the corresponding odds. A Win bet on a horse returns money to the bettor only if the horse wins. A Place bet on a horse returns money to the bettor only if the horse wins or finishes second. A Show bet on a horse returns money to the bettor only if the horse wins or finishes second or finishes third. Complete this table of actual dollars bet and their associated odds.

Horse number	Win bets	Place bets	Show bets	Odds	P(win)
1	1609	383	118	7 to 2	.222
2	700	243	62	9 to 1	
3	1188	400	92	5 to 1	
4	3058	1338	162	7 to 5	
5	160	97	34	45 to 1	
6	668	306	48	10 to 1	
7	1656	493	108	3 to 1	

Find the sum of the numbers in the sixth column. This sum indicates that probabilities calculated from odds do not always have a sum of 1. Horse 7 won and paid $8.80, $6.20, and $3.60. Horse 5 was second and paid $23.40 and $7.40. Horse 3 was third and paid $4.00. Can you determine how these odds and payoffs were calculated based on how the money was bet?

14. When the weather is rainy and hot, a contractor loses \$600 on a certain type of job, and he makes a \$2,100 profit when the weather is not rainy and not hot. If the weather is rainy and not hot, he loses \$500; if it is hot but not rainy, he makes a profit of \$300. For the days during which the contractor would consider such a job, there is a 40% chance of rain. The probability of hot weather when it is rainy is 40%, but the probability when it is not rainy is 70%. Find his expected value of taking such a job. Hint: Construct a tree diagram with as many branches as there are combinations of weather conditions and place the probabilities and dollar values in appropriate places.

Chapter 6

Statistics

Basic concepts

1. In the Scholastic Aptitude Test (informally known as "College Boards") John Smith received a score of 480 in quantative ability. What percentage of the students who took this test have scores below his?
2. Suppose that one group of children enrolled in the federal government's Head Start program is studying number ideas by conventional teacher-directed means, while another group of children studies the same subject matter by watching the television program *Sesame Street.* Given the results from the two groups on the same test at the end of the period of instruction, can we conclude that watching *Sesame Street* is a better way to learn certain number ideas than the traditional method of instruction? What limitations must we place on our answer?
3. A steel washer, picked at random from a stamping press, is tossed in the air fifty times to see if it is in balance. One side appears thirty-three times instead of the twenty-five one might expect if the washer were perfectly balanced. To what extent can we conclude that the washer is out of balance?

These questions are only a few of the many that frequently arise in the fields of social and management science. The area of study known as statistics grew out of the need to assemble and analyze large volumes of numerical data arising from affairs of the state—hence the name statistics. It has become an almost indispensable tool in such fields as psychology, sociology, economics, education, market research, public opinion polling, and mass communication. The purpose of this chapter is to help us see what some of the basic concepts of statistics are, how it is used, and some of its potential abuses. For example, we will answer questions 1 and 3 before the chapter is finished. There are several definitions we could choose, but statistics essentially is the presentation, description, and analysis of numerical data.

Organizing and presenting data

We begin by plunging into a specific example in order to see how we can organize and present raw data.

Example 1 We have twenty-five examination scores from a class of freshman mathematics students, and the scores are listed in the order that the instructor graded them.

68 52 59 61 64 75 69 54 62 63 47
65 72 66 61 68 51 64 71 48 68 45
58 56 64

It is almost impossible to comprehend the data as they[1] stand, so one way to begin is by placing them in order from smallest to largest. We do this by picking out the smallest number (in this case 45), writing it as the first number in a new list, and crossing it off the original list. We repeat this cycle with the smallest remaining number (in this case 47) and continue in an identical manner until all the numbers are in ascending order in the new list.

45 47 48 51 52 54 56 58 59 61 61
62 63 64 64 64 65 66 68 68 69 71
72 75 78

We pick out the smallest and largest values, which are 45 and 78, and say that the *range* of the data is

$$(\text{largest number} - \text{smallest number}) = 78 - 45 = 33.$$

It is easier to see trends if we next build a table by placing the sorted data into equally spaced subsets, called *intervals* or *classes*, such that each data item belongs in exactly one class. We then indicate how many data items are in each class, thus denoting the *frequency* of the class. The *frequency table* constructed looks like this:

Table 6.1

Interval	Class mark or midpoint, X	Frequency, f
45 up to 51	48	3
51 up to 57	54	4
57 up to 63	60	5
63 up to 69	66	8
69 up to 75	72	3
75 up to 81	78	2

The number in the middle of an interval is its *class mark* or *midpoint*, as shown in the second column of this table. (See Problem 6 for a more detailed explanation of this statement.)

[1] The singular form is *datum* and the plural form is *data*, not *datas*, in accordance with Latin grammar.

Choosing intervals

One question immediately arises: why and how did we choose the six intervals of width six? Unfortunately, there are no universal rules to help us choose the exact number and size of the classes. One source[2] says that statisticians almost always choose between six and fifteen classes, depending on whether the number of data items, N, is sufficiently "small" or "large." Since the range is 33, and we choose to have 6 intervals, a width of 6 for each interval is adequate ($6 \cdot 6 = 36$, and $36 >$ range of 33) to have all the data fall into one of the six intervals. The first interval is 45 up to 51, its midpoint is 48, and by looking at the sorted list we count three numbers in that interval. Notice that the interval "45 up to 51" includes 45, 46, 47, 48, 49, and 50 but excludes 51, which is in the next interval. Thus we advance to the second interval of 51 up to 57, with its midpoint of 54 and its frequency of four data items. At the end of this process we have our frequency table as previously printed.

Are there other ways we could have done this? Certainly. We could have started the first interval at 44 so that the last interval would have ended with 80, even though some of the frequencies would have changed. We also could have chosen seven intervals of width five so that the first interval would have been 45 up to 50—but the midpoint is then a decimal, 47.5. Since the midpoints are used later in calculations, it is desirable, but not absolutely necessary, to have them be whole numbers. The sum of the frequencies should give the number of data items. This fact is expressed by the equation

$$N = \sum f = 25,$$

since each of the N scores is placed into exactly one interval. (Recall Problem 11 of Problem Set 1–2, which discusses the summation notation, $\sum$.) Note that a capital letter X is used to signify the midpoint of an interval, while a small letter x represents an individual datum, or item of data.

Some situations, such as analyzing hundreds of numbers, may not require that the data be sorted before constructing a frequency table. In this case we simply place each unsorted data item into an interval one at a time by making a little slash each time it is placed. Our table would have been constructed with an additional "tally area" column, and we would cross out each data item from the original list as it is placed into the correct interval.

Table 6.2

Interval	Tally area	X	f
45 up to 51	///	48	3
51 up to 57	////	54	4
57 up to 63	~~////~~	60	5
63 up to 69	~~////~~ ///	66	8
69 up to 75	///	72	3
75 up to 81	//	78	2

[2] John E. Freund, *Statistics, A First Course* (Englewood Cliffs, N.J.: Prentice-Hall), 1970, p. 13.

Presenting data geometrically

It is often desirable to present data geometrically as well as algebraically. A convenient way to do this is by a *histogram*, which is a bar graph with midpoints of the data intervals on the horizontal axis and the corresponding frequencies on the vertical axis. For our present example we obtain

Figure 6.1

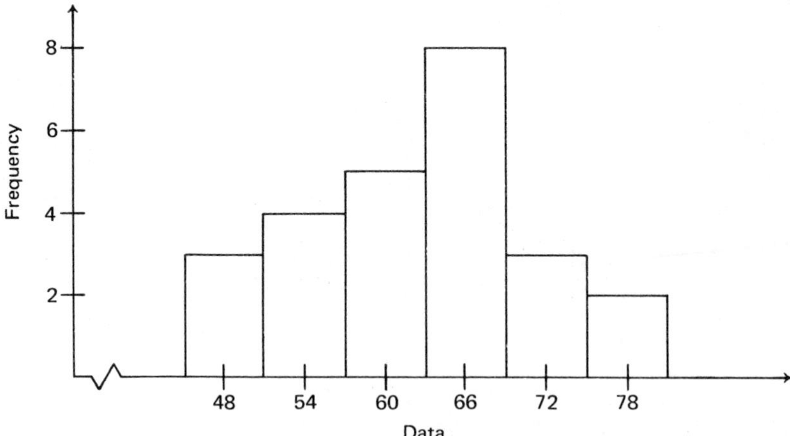

It is immediately evident that the histogram consists of rectangular bars whose bases begin and end at numbers on the horizontal axis corresponding to the intervals in the frequency table. Often used along with, or in place of, a histogram is a *frequency polygon*. To construct a frequency polygon on top of a histogram we pick out the next class marks above and below the existing ones on the horizontal axis, mark the top middle of each bar, and join these points with line segments.

Figure 6.2

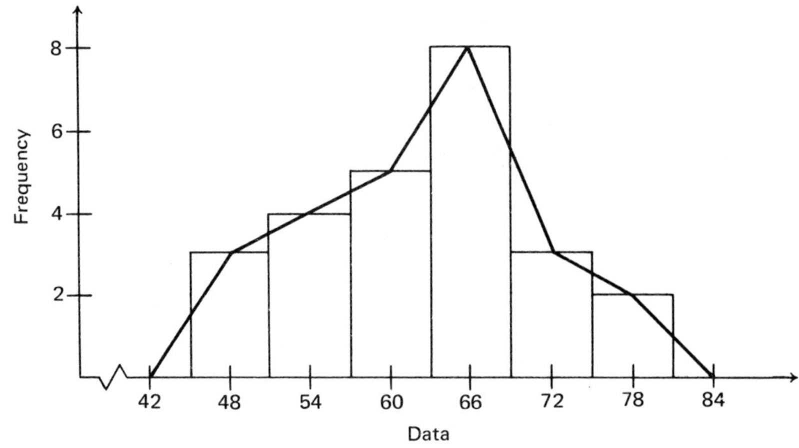

A final addition to our pictorial presentation of the data is a vertical axis to the right of the histogram/frequency polygon which shows the *relative frequencies* of the intervals, that is, the percent of the data items that are in a particular interval. For example, the interval from 51 up to 57 has 4 out of the 25 data items, or $(4/25) \cdot 100 = 16\%$ of the total number. Displaying this *relative frequency polygon* gives us

Figure 6.3

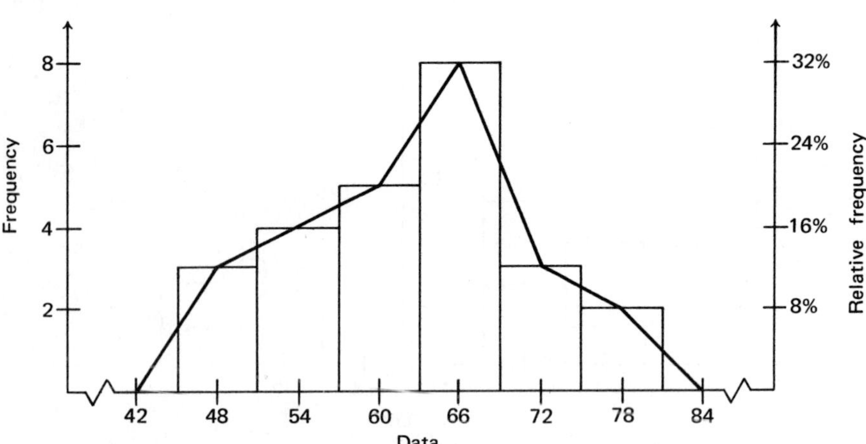

A frequency polygon and a relative frequency polygon are exactly the same geometric entity. They differ only in the scale used on their vertical axes. A glance at the relative frequency polygon shows immediately, for example, that the interval centered at 72 has 12% of all the test scores.

Problem set 6–1

1. Here are some airport temperatures recorded on certain days of March at Logan Airport in Boston.

38	44	22	26	48	51	23	56	19	37	36
45	58	33	30	25	41	35	35	40	22	42
38	37	53	32	18	33	44	47	21	28	37
41	31	46	54	28	34	35				

   a. Sort the data into ascending order.
   b. Construct a frequency table using six intervals with the first one beginning at 18.
   c. Construct a histogram, frequency polygon, and relative frequency polygon based on your table from (b).

2. Refer to Problem 1. Do parts (b) and (c) again, but have the first interval begin at 16.

3. Refer to Problem 1. Do parts (b) and (c) again, but have the frequency table contain 7 intervals beginning at 18.
4. Here are the weights of forty male students in Jones Hall at ABC University.

138	164	150	132	144	125	149	157
146	158	140	147	136	148	152	178
168	126	138	176	163	120	154	165
146	173	142	147	135	153	140	135
161	145	135	142	150	156	145	128

Construct the frequency table, histogram, and frequency polygon directly from the raw data without sorting them. Use 8 classes with a class width of 10 units (pounds).

5. Here is the distribution of hourly wages from the laundry department of a large hospital.

Hourly wages	Number of employees
2.00 up to 2.20	6
2.20 up to 2.40	14
2.40 up to 2.60	19
2.60 up to 2.80	40
2.80 up to 3.00	25
3.00 up to 3.20	9
3.20 up to 3.40	4
3.40 up to 3.60	2
3.60 up to 3.80	1

Construct the histogram, frequency polygon, and relative frequency polygon of the set of wages from this table.

6. Consider the first interval, 45 up to 51, of Table 6.1. If we portray this interval geometrically on a real number line we have, as its graph,

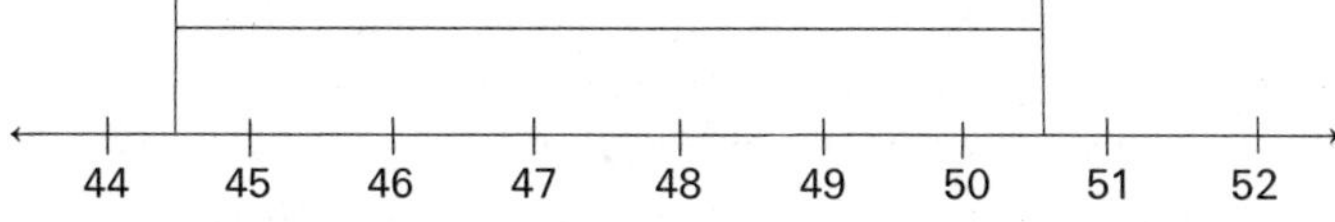

The number 44.8 is included in the graph because, when rounded to the nearest whole number, it becomes 45. Similarly, 50.8 rounded to the nearest whole number is 51. Technically, the midpoint of [44.5, 50.5] is 47.5. We will use 48 as the midpoint of the interval 45 up to 51 because of convenience. What we gain in the way of convenience by this choice is lessened slightly by a loss of accuracy in future calculations with the chosen midpoint. What are the true intervals and their midpoints of the remaining five intervals in Table 6.1?

Describing data—measures of central tendency and dispersion

So far in this chapter we have shown how data are organized and presented in tables and graphs. A logical question to discuss next is "how do we take an entire set of data and represent it algebraically by an 'average' number?"

Whenever we talk about basketball players, baseball pitchers, and college students we often ask what the person's *average* is in terms of points per game, earned runs for each nine innings pitched, and semester or quarter grade-point average. A basketball player may have accumulated twenty different scores over a season of twenty-five games, but we would like to describe, with one number, his performance in all of the twenty-five games. Likewise, earned run averages and grade-point averages vary for the same person and for different people, but we still wish to have one number, called a measure of central tendency, describe the many numbers in a set. The question, then, is what number is at the center of a set?

Mean, median, and mode

The number most commonly used as an average to describe a set of data is the *mean*. It is defined to be the sum of the data items divided by the number of data items. In more formal terms,

$$\bar{x} = \frac{\sum x}{N}$$

where $\bar{x}$ ("x bar") or the mean is the sum ($\sum$) of all the data items (x's) divided by the number of them (N). With reference to our example of the twenty-five test scores from the last section,

$$\bar{x} = \frac{\sum x}{N} = \frac{68 + 52 + 59 + \cdots + 56 + 64}{25} = \frac{1541}{25} = 61.64.$$

One interpretation of $\bar{x} = 61.64$ is that if the total number of points were divided equally among the twenty-five students, then each one would have obtained 61.64 points. Notice also that every value of x is involved in the calculations of $\bar{x}$, and a change in any *one* value of x changes the value of $\bar{x}$.

Another measure of central tendency of a set is the *median*, or middle number in a list of sorted data. It is the number above and below which are half of the data in the list. If $A = \{4, 7, 1, 8, 3\}$, then the sorted set becomes $A = \{1, 3, 4, 7, 8\}$ and the median is 4, since there are just as many data items below it (two) as above it. Note that

$$\bar{x} = 23/5 = 4.6 \neq 4 = \text{median}.$$

In general, the two measures of central tendency are different from each other. Suppose that there is no one middle number, such as in $B = \{7, 4, 4, 5, 9, 8\}$—then what? A diagram answers this question.

$$\{7,4,4,5,9,8\} \longrightarrow \{4,4,\underbrace{5,7},8,9\} \longrightarrow \{5,7\}$$

even number of data — sort — take the two middle numbers — find their mean

$$\longrightarrow 6$$

median

Note that three numbers are below 6 and three are above it. In our example of test scores the median is the thirteenth number in the sorted list $(12+1+12 = 25)$, or 63.

The third measure of central tendency is the *mode*, or most frequently occurring number. In the previous set B, 4 is the mode, and the mode of the twenty-five test scores is 64. If two or more numbers have equally high frequencies, then the set has two or more modes. Set $C = \{2,5,5,6,7,9,9,10,14\}$ has 5 and 9 as its modes. If no number occurs more than once, as in set A, then the set has no mode.

Which measure of central tendency do we use? It depends on the use of the result. The mean is used as an overall indication of the size of the numbers in a set because it is affected by each number. It is used extensively in theoretical statistics and probability along with many applications of the theory. The median is used when the order or location of a number in a set of measures is important. For example, suppose the five inhabitants of Newbury have incomes of \$100, \$150, \$200, \$250, and \$9300. The mean is \$2000, but the median of \$200 is obviously a much better figure descriptive of Newbury's "average" income. The mean can be distorted by extreme values and often is discounted in their presence. The mode would be of interest to the owner of a men's hat shop. He does not particularly care that the mean hat size of his customers is 7.0614 (hat sizes are in increments of 1/8) or that the median hat size is 7 1/8, but he would use the modal size of, say, 7 1/4 as he places his inventory orders. Whenever we use the word "average" from now on, it will always refer to the mean unless we specify otherwise.

Computing the mean of grouped data

What about calculating the mean of a set of data which have been organized into groups? We often meet a set of data for the first time *after* other people

Table 6.3

Interval	X	f	Actual sum of numbers in interval	Estimated sum of numbers in interval, $f \cdot X$
45 up to 51	48	3	140	$3 \cdot 48 = 144$
51 up to 57	54	4	213	$4 \cdot 54 = 216$
57 up to 63	60	5	301	$5 \cdot 60 = 300$
63 up to 69	66	8	522	$8 \cdot 66 = 528$
69 up to 75	72	3	212	$3 \cdot 72 = 216$
75 up to 81	78	2	153	$2 \cdot 78 = 156$

have organized it into classes, and so we return to our example of twenty-five test scores with two columns added to its frequency table.

If we examine the interval 45 up to 51, whose midpoint is 48 and whose frequency is 3, we see that the true sum of the data items 45, 47, 48 is 140. If we made the assumption that each of the data items were equal to the midpoint of 48, then the sum of the three numbers would be

$$48 + 48 + 48 = 3 \cdot 48 = 144.$$

Notice that this estimated sum of 144 was obtained by doing one multiplication instead of two additions. Let us see where this same assumption leads us in the fourth interval, whose true sum is

$$63 + 64 + 64 + 64 + 65 + 66 + 68 + 68 = 522.$$

If we assume that all the numbers are equal to the midpoint of 66, then there are eight 66's, or $8 \cdot 66 = 528$. Again we see that the estimated sum is close to the true sum of 522, and the former is obtained with considerably less work than the latter. The table is already finished for us. Adding the members of the three-right-hand columns, we obtain $\sum f = 25$, $\sum x = 1{,}541$, and $\sum f \cdot X = 1{,}560$, respectively. We see that 1,560 is symbolized by $\sum f \cdot X$ because we added the estimated sums of each interval in order to obtain the estimated total sum of 1,560. Summarizing, we sec that the true sum, $\sum x$, is 1,541, but twenty-four addition operations were necessary; the estimated sum, $\sum f \cdot X$, is 1,560, but only six multiplications and five additions were necessary. Of course, the calculated value of $\bar{x}$ on the basis of the estimated sum gives us $1560/25 = 62.40$, which is different from the true mean of 61.64; this difference, however, is usually insignificant in applications.

We easily generalize this short-cut method of computing the mean of grouped data to give us the formula

$$\bar{x} = \frac{\sum f \cdot X}{N}.$$

We should keep in mind that the capital letter X represents a midpoint of an interval, or a class mark, while f represents the frequency of that same interval and N is the number of data items in the entire set. Not only is this formula generally easier to apply, but also the data quite often have already been grouped when we first obtain them. In this latter case we must work with the grouped data formula.

Example 2 Find the median of the twenty-five test scores, assuming that they have been given solely as grouped data.

Solution The median is the thirteenth number in the sorted set. Since the first twelve numbers $(3+4+5)$ are in the first three classes, we know the median is in the fourth class with its lower limit of 63. We have to go one score beyond the twelve scores to reach the thirteenth score. The width of the class is six points and there are eight scores, so that each score takes up 6/8 of a unit on the horizontal axis, assuming that the eight scores are spread equally over the 6 points. We are one score into the class, so the median is estimated to be

$$63 + 1 \cdot \frac{6}{8} = 63.75.$$

The true median is 63 instead of the estimated 63.75.

Generalizing, we have the formula for calculating the estimated median from a set of grouped data. If

L = lower limit of class in which median is located,
i = number of scores in that class needed to reach median score,
w = width of median class, and
f = frequency of median class,

then

$$\text{median} = L + i \cdot \frac{w}{f}.$$

Mean deviation

So far we have studied measures of central tendency of a set of data. We now study the *measures of dispersion* which are applicable to a set of numbers. An example will illustrate the idea of dispersion.

Example 3 Suppose that $A = \{1, 5, 6, 7, 11\}$ and that $B = \{3, 5, 6, 7, 9\}$ so that their graphs are respectively

Figure 6.4

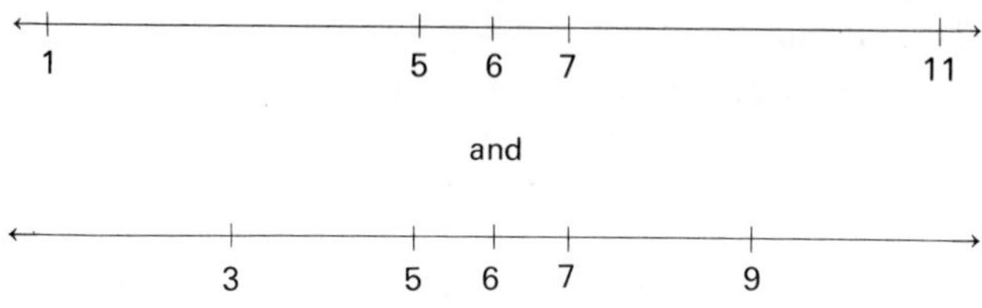

The diagram clearly indicates that set A is more "spread out," or more dispersed, than set B. How do we proceed to formalize this common sense idea by searching for a number which actually measures dispersion?

Solution First of all, the range is $10\,(=11-1)$ for set A and $6\,(=9-3)$ for set B. The range alone is not sufficient to measure the dispersion of a set. For example, $C = \{10, 14, 15, 16, 20\}$ and $D = \{10, 11, 15, 19, 20\}$ have the same range of 10, but the elements of D are obviously more dispersed from the common mean of 15 than the elements of C are. An important idea has been introduced in the last sentence: dispersion means dispersion *from the mean.* Hence a measure of dispersion somehow will measure the dispersion or deviation of the elements of a set from its mean.

We return to our two sets $C = \{10, 14, 15, 16, 20\}$ and $D = \{10, 11, 15, 19, 20\}$ with identical range (10) and mean (15). If we are trying to calculate a measure of dispersion for an entire set, it seems reasonable to begin by calculating the dispersion of each element in the set from its mean.

Table 6.4 Set C

Element x	Mean $\bar{x}$	Dispersion from $\bar{x}$, $\lvert x-\bar{x}\rvert$
10	15	$\lvert 10-15\rvert = 5$
14	15	$\lvert 14-15\rvert = 1$
15	15	$\lvert 15-15\rvert = 0$
16	15	$\lvert 16-15\rvert = 1$
20	15	$\lvert 20-15\rvert = 5$

Table 6.5 Set D

Element x	Mean $\bar{x}$	Dispersion from $\bar{x}$, $\lvert x-\bar{x}\rvert$
10	15	$\lvert 10-15\rvert = 5$
11	15	$\lvert 11-15\rvert = 4$
15	15	$\lvert 15-15\rvert = 0$
19	15	$\lvert 19-15\rvert = 4$
20	15	$\lvert 20-15\rvert = 5$

We take the absolute value of the differences because we are only interested in the distance of the elements from the mean, not the direction. In adding the dispersions we find that for set C

$$\sum |x-\bar{x}| = 5+1+0+1+5 = 12$$

and for set D

$$\sum |x-\bar{x}| = 5+4+0+4+5 = 18.$$

Notice that $\sum (x-\bar{x})$ is 0 in both cases (right?) This fact also explains why we take the absolute values of the individual differences before summing them. Since two sets may not have the same number of elements, we take the total sum and divide it by the number of elements in the set, which tells us the *average* (or mean) dispersion of each element from its mean. In set C the mean deviation is $12/5 = 2.4$, and in set D it is $18/5 = 3.6$. Since common sense told us a while back that set D was more spread out than set C, the measure of "spread-outness" for set D should be greater than the corresponding measure for set C. Our intuition is confirmed by 3.6 being greater than 2.4. More generally, the mean deviation of any set of N data items is defined by the equation

$$MD = \frac{\sum |x-\bar{x}|}{N}.$$

We should convince ourselves that the respective mean deviations of sets A and B are $12/5 = 2.4$ and $8/5 = 1.6$.

Standard deviation

As easily understood as the mean deviation is, though, it is seldom used in applied statistics. There are problems with absolute value in studying the theory of statistics, and another measure is more applicable in analyzing data. This other measure of dispersion is called the *standard deviation*, which we introduce by returning to our set $C = \{10, 14, 15, 16, 20\}$. One way to eliminate the problem of absolute value, and yet to keep the difference of each element from the mean a nonnegative number, is to square each difference.

Table 6.6

Element x	Mean $\bar{x}$	Dispersion from mean = $(x-\bar{x})$	$(x-\bar{x})^2$
10	15	$(10-15) = -5$	25
14	15	$(14-15) = -1$	1
15	15	$(15-15) = 0$	0
16	15	$(16-15) = 1$	1
20	15	$(20-15) = 5$	25

The sum of the squares of the deviations is

$$\sum (x-\bar{x})^2 = 25 + 1 + 0 + 1 + 25 = 52.$$

To find the average of these squares, we divide their sum of 52 by 5 (the number of elements) and obtain 10.4. But 10.4 is much too large a measure of dispersion, since the largest actual dispersion of any element from the mean is 5. The number 10.4 was arrived at through squaring each deviation. In order to partially undo this inflated result caused by squaring we take the square root of 10.4, which is 3.23, to two decimal places. Taking the square root also has the advantage of having the final answer in the same units as the original measurement.

To summarize, we calculated the standard deviation of set C by evaluating the expression

$$\sqrt{\frac{(10-15)^2 + (14-15)^2 + (15-15)^2 + (16-15)^2 + (20-15)^2}{5}}.$$

If we wish to calculate the standard deviation of set D, it seems reasonable to work with the five numbers in it, just as we worked with the elements of set C. First we find the sum of the squares of the individual deviations:

$$\begin{aligned}\sum (x-\bar{x})^2 &= (10-15)^2 + (11-15)^2 + (15-15)^2 + (19-15)^2 + (20-15)^2 \\ &= 82.\end{aligned}$$

We continue by dividing this sum by 5 to obtain 16.4. We finally take the square root of 16.4, which results in the answer of 4.05, to two decimal places. As expected, $4.05 > 3.23$, since the elements of set D are more dispersed from their mean than are the elements of set C.

We make a straightforward generalization from these two examples to obtain the formula for any set of N numbers.

$$\text{standard deviation} = \sqrt{\frac{\sum (x-\bar{x})^2}{N}}$$

Statisticians have agreed to use the symbol σ (the Greek lowercase letter "sigma") for the standard deviation. Some authors use $N-1$ as the denominator instead of N. The *variance* of a set, by definition, is σ^2 or, in a formula,

$$\sigma^2 = \frac{\sum (x-\bar{x})^2}{N}.$$

In set C the variance is 10.4, and in set D $\sigma^2 = 16.4$.

If we want to calculate σ for $E = \{2, 7, 8, 9, 11, 13, 14\}$, we first find the mean, which is $\bar{x} = 64/7$. Then $\sum (x-\bar{x})^2$ would equal

$$(2-64/7)^2 + (7-64/7)^2 + \cdots + (14-64/7)^2,$$

which is rather awkward to calculate. Is there any way to obtain this sum without going through all of the subtractions and squarings? Fortunately, the answer is yes. With a little bit of algebra and work with summation notation we could show that in general

$$\sigma = \sqrt{\frac{\sum (x-\bar{x})^2}{N}} = \cdots \text{ algebra } \cdots = \sqrt{\frac{\sum x^2}{N} - (\bar{x})^2}.$$

Interested readers will find some additional work with this derivation in Problem 17. In the case of set E, $\sum x^2$ is the sum of the squares of the individual x's, or

$$\begin{aligned}\sum x^2 &= 2^2 + 7^2 + 8^2 + 9^2 + 11^2 + 13^2 + 14^2 \\ &= 4 + 49 + 64 + 81 + 121 + 169 + 196 \\ &= 684.\end{aligned}$$

Then

$$\sigma = \sqrt{\frac{684}{7} - \left(\frac{64}{7}\right)^2}$$

$$\sigma = \sqrt{\frac{684}{7} \cdot \frac{7}{7} - \frac{64^2}{7^2}}$$

$$\sigma = \sqrt{\frac{4788}{49} - \frac{4096}{49}}$$

$$\sigma = \sqrt{\frac{692}{49}}$$

$$\sigma = 3.76.$$

In this case we only had to square the numbers themselves rather than to square such numbers as $(11-64/7)$, yet the results obtained by this short-cut formula are identical to the definition formula of

$$\sigma = \sqrt{\frac{\sum (x-\bar{x})^2}{N}}.$$

From now on we will compute σ for ungrouped data by using the shortcut formula

$$\sigma = \sqrt{\frac{\sum x^2}{N} - (\bar{x})^2}.$$

Example 4 What about finding the standard deviation of our twenty-five test scores in ungrouped form? On the surface it looks easy enough to do by just plugging in the numbers in the σ formula so that

$$\sigma = \sqrt{\frac{45^2+47^2+\cdots+78^2}{25} - \left(\frac{45+47+\cdots+78}{25}\right)^2}.$$

There is a lot of arithmetic to do, even though we have used the shortcut formula: $25+1=26$ squarings, $24+24=48$ additions, two divisions, one subtraction, and one square root. Since we were able to cut down on the calculation of $\bar{x}$ by using a grouped data formula, we hopefully ask whether we can do the same for σ. It is also important that we develop a formula for finding σ for grouped data because, as mentioned earlier, we often have *only* grouped data to begin with.

Computing the standard deviation for grouped data

Solution We start by showing the table of grouped data we used before to calculate the mean.

Table 6.7

Interval	X	f	$f \cdot X$
45 up to 51	48	3	144
51 up to 57	54	4	216
57 up to 63	60	5	300
63 up to 69	66	8	528
69 up to 75	72	3	216
75 up to 81	78	2	156

We must find a reliable estimate of the sum of the squares of all the x's. We therefore begin by looking at the three scores in the first interval, which are 45, 47, and 48. Here

$$\begin{aligned}\sum x^2 &= 45^2 + 47^2 + 48^2 \\ &= 2025 + 2209 + 2304 \\ &= 6538.\end{aligned}$$

To obtain $\sum x^2$ for the entire set we would, of course, calculate $\sum x^2$ for each of the other five intervals and then add the six subtotals to end up with $\sum x^2$. If we assume that each of the three numbers in the first interval equals 48, then the sum of their squares would be

$$48^2 + 48^2 + 48^2 = 3 \cdot 48^2 = 3 \cdot 2304 = 6{,}912,$$

which differs from the true value of 6,538 by about 6%. However, we obtained this reasonably accurate estimated sum, symbolized by $f \cdot X^2$, by one squaring and one multiplication, instead of three squaring and two addition operations. (Problem 16 discusses this difference of 374 in more detail.) If we calculate the five remaining estimated subtotals in a similar fashion, our table would then become

Table 6.8

Interval	X	f	$f \cdot X$	Estimated subtotal $= f \cdot X^2$
45 up to 51	48	3	144	$3 \cdot 48^2 = 6{,}912$
51 up to 57	54	4	216	$4 \cdot 54^2 = 11{,}664$
57 up to 63	60	5	300	$5 \cdot 60^2 = 18{,}000$
63 up to 69	66	8	528	$8 \cdot 66^2 = 34{,}848$
69 up to 75	72	3	216	$3 \cdot 72^2 = 15{,}552$
75 up to 81	78	2	156	$2 \cdot 78^2 = 12{,}168$
				Sum = 99,144

Since 99,144 is the estimated sum of the squares of the twenty-five numbers, we plug it into the formula for σ, giving us

$$\sqrt{\frac{99144}{25} - (\bar{x})^2}.$$

Since $\bar{x} = 62.4$ from the grouped data formula, then

$$\bar{x} = \sqrt{\frac{99144}{25} - (62.4)^2}$$

$$\bar{x} = \sqrt{3965.76 - 3893.76} = \sqrt{72} = 8.49.$$

To generalize from this specific example we note that 99,144 is $\sum f \cdot X^2$, 25 is N, and 62.4 is $\bar{x}$; therefore

$$\sigma = \sqrt{\frac{\sum f \cdot X^2}{N} - \left(\frac{\sum f \cdot X}{N}\right)^2}.$$

This is the formula for finding σ for grouped data. Since we made the assumption that each number in an interval was equal to its midpoint, the σ calculated will not exactly equal the true value of σ calculated by the shortcut ungrouped formula. In our case $\sum x^2$ is 96,807, so that σ is truly 8.53 (right?).

Unfortunately, the calculation of means and standard deviations is generally mechanical and full of chances to make arithmetic errors. However, it is necessary that we do the computing because the knowledge of the values of $\bar{x}$ and σ for a set of data is quite useful for making inferences about the set. We will use this knowledge in the next section.

Problem set 6–2

You may wish to use a calculator to help with some of these problems.

1. If $A = \{8, 10, 2, 7, 5, 7\}$, then calculate its mean, median, and mode.
2. Calculate σ for set $D = \{3, 5, 7, 10\}$ by using the definition formula.
3. Calculate σ for set $D = \{3, 5, 7, 10\}$ by using the shortcut formula. Now do you see the advantage of this formula?
4. Let set S be $\{4, 5, 8, 10, 11\}$.
   a. Calculate $\bar{x}$.
   b. If $S' = \{4+3, 5+3, 8+3, 10+3, 11+3\}$, what is its mean?
   c. What is the relationship between the mean of S and the mean of S'?
   d. If T is any set and k is added to each element of T to form a new set T', and $\bar{x}$ is the mean of T, what do you think the mean of T' will be?
5. Let $U = \{4, 5, 8, 10, 11\}$ and let $U' = \{2 \cdot 4, 2 \cdot 5, 2 \cdot 8, 2 \cdot 10, 2 \cdot 11\}$. How are the mean of U and the mean of U' related?
6. Let set V be $\{4, 5, 12\}$.
   a. Calculate σ.
   b. If V' is $\{4+7, 5+7, 12+7\}$, then calculate its standard deviation.
   c. What is the difference between the results?
   d. Can you make a generalization based on the results?
7. Let set W be $\{4, 5, 12\}$. In the previous problem we calculated σ for it.
   a. If W' is $\{9 \cdot 4, 9 \cdot 5, 9 \cdot 12\}$, then calculate σ for it.
   b. How are σ from W and σ from W' related?
8. Refer to problems 6 and 7.
   a. If W'' is $\{-9 \cdot 4, -9 \cdot 5, -9 \cdot 12\}$, then calculate σ for it.
   b. How are σ from W and σ from W'' related?
   c. If S is any set $\{x_1, x_2, x_3, \ldots, x_n\}$ and S' is the set $\{k \cdot x_1, k \cdot x_2, k \cdot x_3, \ldots, k \cdot x_n\}$, then what is the relation between σ from S and σ from S'?
9. Refer to Problem 1 of the previous problem set. Calculate $\bar{x}$ and σ to the nearest .01 for that set.
10. Refer to Problem 2 of the previous problem set. Calculate $\bar{x}$ and σ to the nearest .01 for that set.
11. Refer to Problem 3 of the previous problem set. Calculate $\bar{x}$ and σ to the nearest .01 for that set.
12. Suppose that the Department of Commerce conducts a survey and consequently publishes the following (hypothetical) table about the hourly wages of factory workers for the year 1967.

Wages	Number of workers (in thousands)
under 1.50	30
1.50 up to 1.80	130
1.80 up to 2.10	750
2.10 up to 2.40	1,200
2.40 up to 2.70	1,510
2.70 up to 3.00	1,620
3.00 up to 3.30	1,380
3.30 up to 3.60	1,100
3.60 and over	2,050

The last class is an example of an *open class* because it has no known upper limit and hence no midpoint.

a. Find the mean hourly wage.
b. Find the median hourly wage.

13. Bill's test scores are 75, 79, and 83.
    a. He wants to have an average of 81 after the next test. What score must he get on the next test?
    b. He wants to have an average of 86 after the next test. What score must he get on it? If the maximum number of points on a test is 100, can he obtain the 86 average?
14. Refer to Problem 4 of the previous problem set. Calculate the mean, median, and standard deviation of the weights by using the grouped data formulas.
15. Refer to Problem 5 of the previous problem set. Calculate the mean, median, and standard deviation of the wages by using the grouped data formulas.
16. This problem and Problem 6 of the previous problem set are related. If we find the true midpoint of $[44.5, 50.5]$, the point is 47.5 because the distance from 47.5 to 44.5 is the same as the distance from 47.5 to 50.5. Moreover, the mean of $\{45, 46, 47, 48, 49, 50\}$, the set of allowable integers in the interval 45 up to 51, is 47.5. $(47.5)^2 = 2256.25$, and $3 \cdot 2256.25 = 6768.75$ is the resulting estimate of the sum of the squares of the set of numbers $\{45, 47, 48\}$. $45^2 + 47^2 + 48^2 = 6538$ is the actual sum, and the difference between $\sum x^2$ and its estimated value, $\sum f \cdot X^2$, is 230.75 here instead of 374 which resulted in Example 4 from choosing 48 as the midpoint of the interval 45 up to 51. We actually introduced a systematic but minor error into our calculations by choosing 48 as the midpoint of this interval instead of 47.5. However, it is more convenient to work with 48, 54, 60, etc., in carrying out the arithmetic.
    a. Find the mean of the data in this problem by using 47.5, 53.5, 59.5, 65.5, 71.5, and 77.5 as the midpoints of the intervals.
    b. Use these midpoints and the results of part (a) to calculate the resulting standard deviation of the data.
17. Show that if $x \in \{a, b, c\}$ and the mean of this set is

$$m = \frac{a+b+c}{3},$$

then

$$\frac{\sum (x-m)^2}{3} = \frac{(a-m)^2 + (b-m)^2 + (c-m)^2}{3}$$

$$= \frac{a^2+b^2+c^2}{3} - \left(\frac{a+b+c}{3}\right)^2$$

$$= \frac{\sum x^2}{3} - m^2.$$

The normal curve and its properties

In the last section we not only learned the meanings and the formulas for $\bar{x}$ and σ but also learned that their calculations can be quite time consuming and prone to error. Why, then, do we grind through the arithmetic? It so happens that many different sets of observations have one remarkable property in common—their histograms have similar shapes. Examples of these diverse sets are IQ's, diameters of ball bearings made by an automatic machine, weights, test scores, and sizes of plant stems. The common shape of their histograms is

Figure 6.5

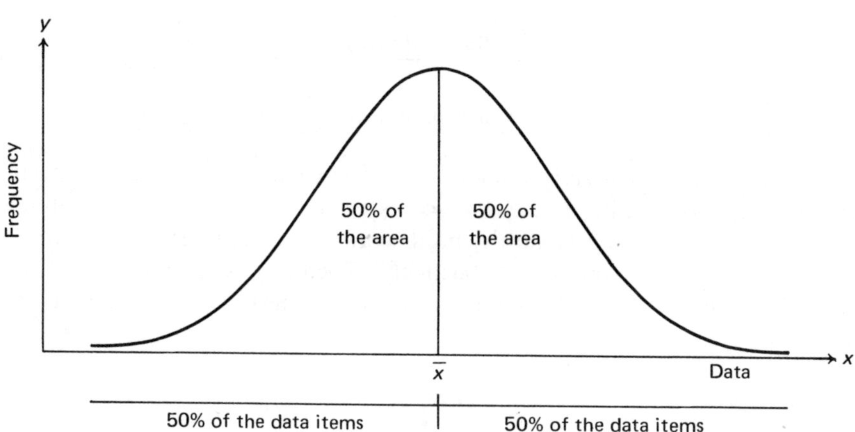

This graph is the so-called *normal* or *bell-shaped* curve. Its equation is

$$y = \frac{1}{\sqrt{2\pi}} \cdot e^{-x^2/2}$$

where e is an irrational number close to 2.718. This number is the same e we encountered in Chapter 3 when we discussed exponential functions. We will not discuss here why so many widely different phenomena from the world around us have similar distributions of their numerical values, but we just accept it as a fact of nature and proceed to inquire into the properties of this normal distribution. Problem 9 gives a brief discussion of the origin of one bell-shaped curve.

Figure 6.5 illustrates that the mean is also the median (since half the data items are above it and half are below it) and the mode (since the mean has the highest frequency). Statisticians and mathematicians have proven, by using the equation of the normal curve, that the curve has the following properties where $\bar{x}$ is the mean of the set and σ is its standard deviation:

Figure 6.6

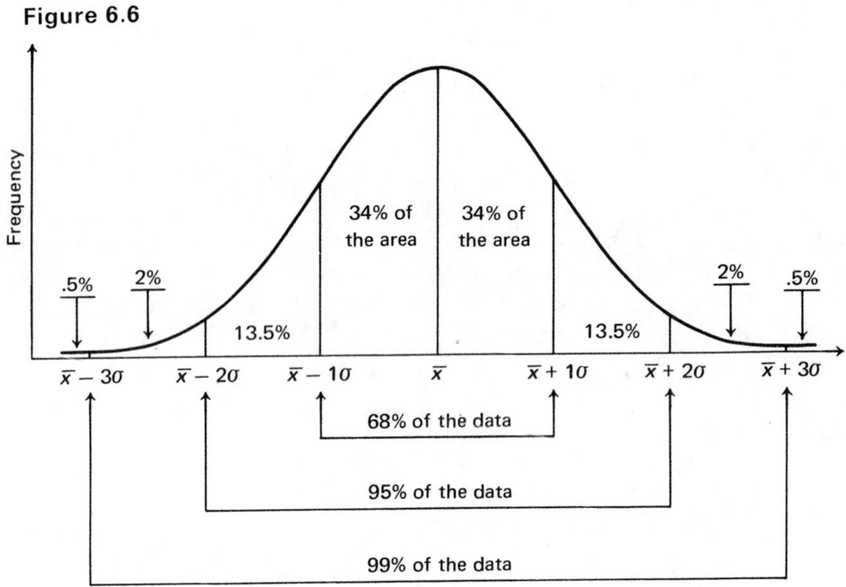

Let us see how this description of the normal curve applies to a specific situation. Suppose we have 1,200 test scores known to be normally distributed and someone has already done the hard work of calculating $\bar{x}$ and σ, whose values are 70 and 6 points, respectively. Then the key scores are even standard deviation units (1 σ-unit = 6) from the mean of 70. Showing this graphically,

Figure 6.7

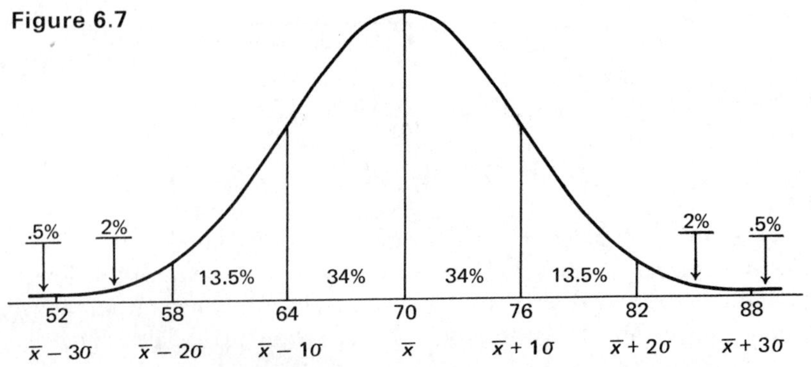

How many scores are between 70 and 76? In order to use our knowledge of the normal curve we think of 70 and 76 in terms of how many σ-units they are from the mean. Since 70 is the mean and 76 is 1σ above it—and between $\bar{x}$ and $\bar{x}+1\sigma$ in any set of normally distributed data there are 34% of all the data items—then 34% of the 1,200 scores, or 408 of them, will be between 70 and 76. Suppose there had been 800 test scores. The analysis would be identical since we must think in terms of how many σ-units an actual score is from the mean; we still find that the *percentage* is 34, but here we multiply it by 800 to get 272 as the *number* of scores between 70 and 76. Some further examples should help us understand this important diagram even better. We find the number of scores between

64 and 76 or between $\bar{x} - 1\sigma$ and $\bar{x} + 1\sigma$ = 68% of 1200 = 816;

64 and 82 or between $\bar{x} - 1\sigma$ and $\bar{x} + 2\sigma$ = $(34 + 34 + 13.5)\%$ of 1200 = 81.5% of 1200 = 978;

58 and 82 or between $\bar{x} - 2\sigma$ and $\bar{x} + 2\sigma$ = 95% of 1200 = 1140;

76 and 88 or between $\bar{x} + 1\sigma$ and $\bar{x} + 3\sigma$ = $(13.5 + 2)\%$ of 1200 = 15.5% of 1200 = 186;

above 82 or above $\bar{x} + 2\sigma$ = $(2 + .5)\%$ of 1200 = 2.5% of 1200 = 30;

below 64 or below $\bar{x} - 1\sigma$ = $(13.5 + 2 + .5)\%$ of 1200 = 16% of 1200 = 192.

The z-score

Suppose we want to know the percent of scores between 70 and 79. 70 is 0 σ-units from the mean and 79 is 1.5 σ-units from the mean of 70 (since $79 = 70 + 1.5 \cdot 6$), but we do not have $\bar{x} + 1.5\sigma$ on our diagram. What do we do now? Fortunately, we can find the value in Table XII in the appendix, which gives the area under the normal curve from the mean up to 3.99 σ-units away from it. The mean is labeled 0 since we can think of the mean of any set of numbers as being 0 σ-units from itself. Since 79 is 1.5 σ-units from 70, we say z_{79} ("z sub 79," the so-called *z-score*, corresponding to 79) is 1.50. The symbol z represents the deviation of a specific score from the mean, as expressed in standard deviation units. Looking in the table under z, we first go down to 1.5 and then go directly into the row to the right of 1.5 until we are in the column headed 0, where we find .4332. This means that 43.32% of the data items in any normally distributed set are between the mean and 1.5 σ-units above it; so our answer is $.4332 \cdot 1200 = 519.84$, which we round off to 520 scores. In order to have some practice in looking up areas corresponding to given z-scores we should convince ourselves that these pairs of z-scores and areas are correct: (0.32, .1255), (1.68, .4535), (2.3, .4893), (3, .4987), (.53, .2019), (3.05, .4989).

Let us do some more examples to see how we should find areas under the normal curve between, above, and below selected values of z. Keep in mind that the equation $z = -2.36$ means, for example, that a certain data item is

2.36 standard deviations below the mean of the set. A diagram is a genuine aid in all of these examples and in future problems involving the interpretation of normally distributed data. We are sure, of course, to look up each value of z in Table XII as we study these examples.

Example 5 Find areas under the normal curve between these z-scores.
Between 0 and 1.43?

Figure 6.8

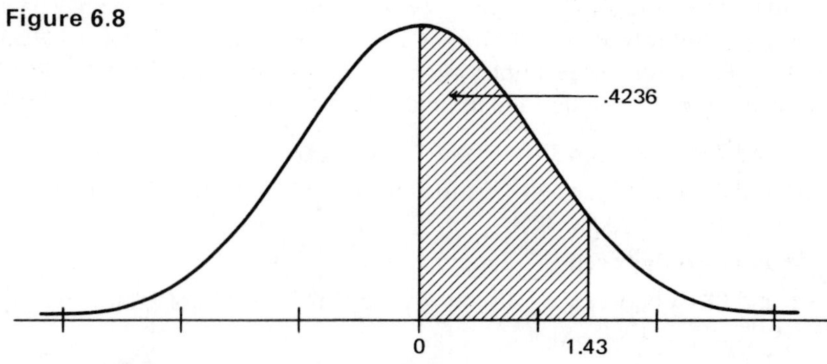

Between 0 and .74?

Figure 6.9

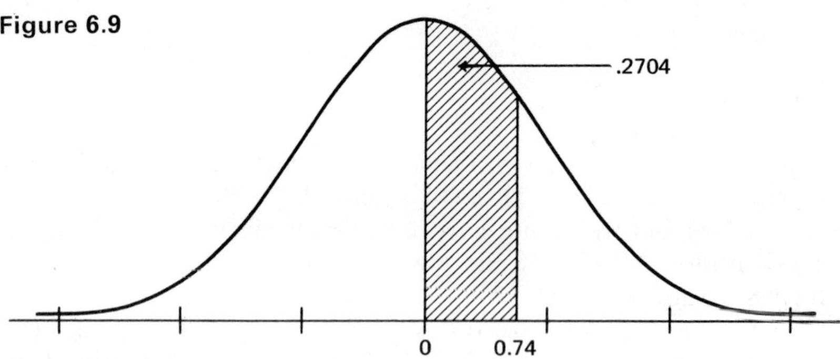

Between 0 and -1.56? This is the same as the area between 0 and 1.56, since the left part of the curve is identical to the right part.

Figure 6.10

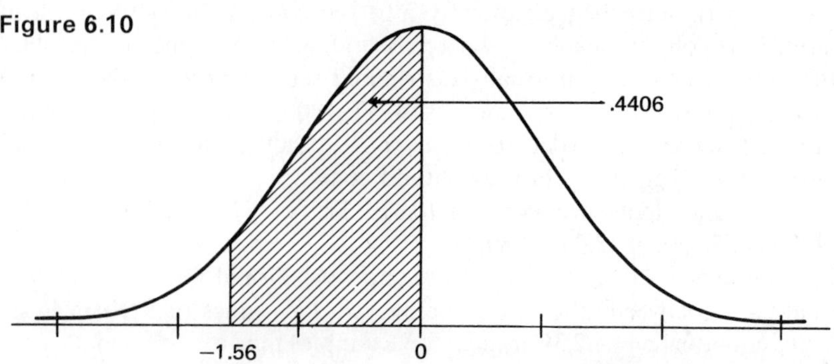

Between $-.58$ and 1.36? The geometry of the situation tells us that the area of the entire shaded region is the sum of the two shaded regions bounded by the vertical line through $z = 0$.

Figure 6.11

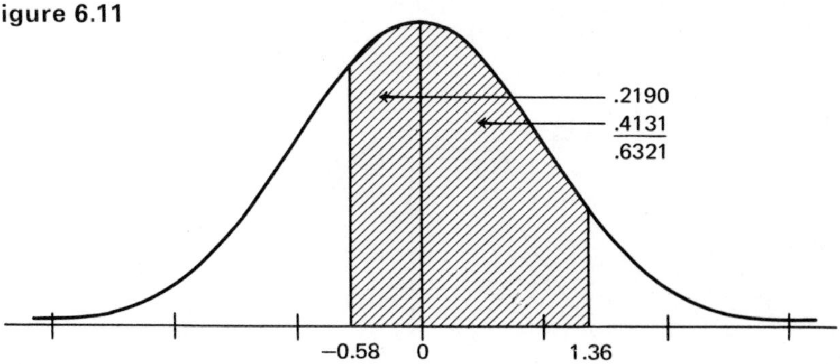

Above -1.78?

Figure 6.12

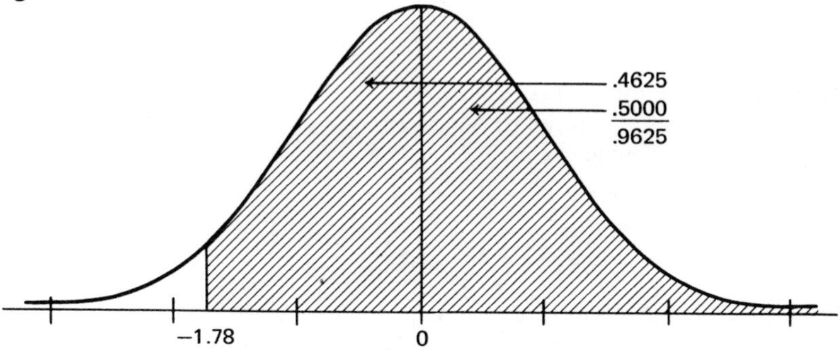

Above .49? The area is not given to us directly, but geometrically.

$$\begin{aligned}\text{shaded area} &= (\text{area of region to the right of } z = 0) \\ &\quad - (\text{area of region between } z = 0 \text{ and } z = .49) \\ &= .5000 - .1879 \\ &= .3121\end{aligned}$$

Figure 6.13

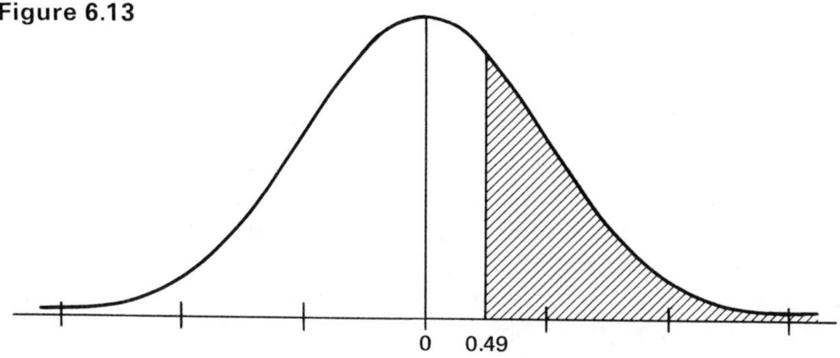

Below -2.35? As in the previous example, the area is not given to us directly, but

$$\begin{aligned}\text{shaded area} &= (\text{area of region to the left of } z = 0)\\ &\quad - (\text{area of region between } z = 0 \text{ and } z = -2.35)\\ &= .5000 - .4906\\ &= .0094.\end{aligned}$$

Figure 6.14

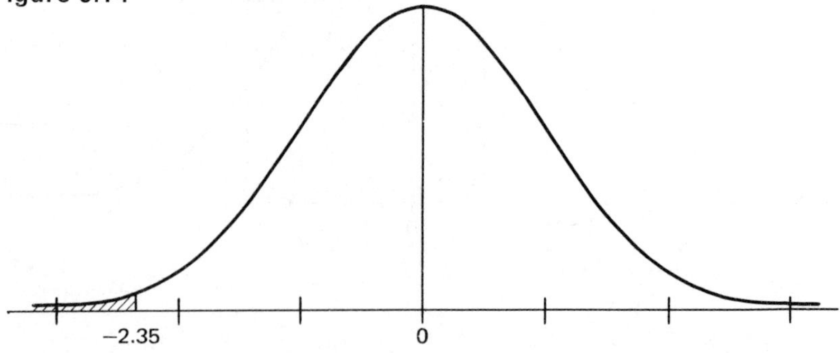

Between .93 and 1.75? Geometrically,

$$\begin{aligned}\text{shaded area} &= (\text{area of region between } z = 0 \text{ and } z = 1.75)\\ &\quad - (\text{area of region between } z = 0 \text{ and } z = .93)\\ &= .4599 - .3238\\ &= .1361.\end{aligned}$$

Figure 6.15

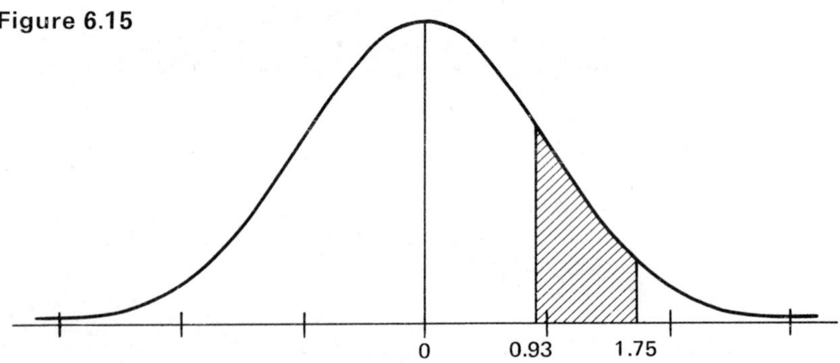

Converting actual numbers to z's

Now that we know how to interpret the normal distribution table as long as the numbers we are interested in are expressed in terms of z-scores, or σ-units from the mean, we look for a way of converting actual numbers in a set (x's) into their corresponding z's. We return to the previous example with $\bar{x} = 70$ and $\sigma = 6$. It is easy to verify that 79 is 1.5 σ-units from 70, but how did we get 1.5 in the first place? Since 79 is 9 units from the mean ($79 - 70 = 9$), we want to know how many σ-units there are in the difference; that is, how many 6's are in 9? We divide 9 by 6 to obtain 1.5. Symbolically,

$$\begin{aligned} z_{79} &= \text{the number of } \sigma\text{-units 79 is from the mean} \\ &= (79-70)/6 \\ &= 1.50. \end{aligned}$$

Generalizing,

$$\begin{aligned} z_x &= \text{the number of } \sigma\text{-units } x \text{ is from the mean of } \bar{x} \\ &= \frac{x-\bar{x}}{\sigma} \end{aligned}$$

where x is a number in a set of normally distributed data with mean $\bar{x}$ and standard deviation σ.

Example 6 We work a few problems under the assumptions that $N = 700$, n.d. (abbreviation for *n*ormally *d*istributed), $\bar{x} = 47.6$, and $\sigma = 13.2$. How many data items are less than 59.8? We begin, as always, by making a diagram with the mean, the numbers we are interested in, and their corresponding z-scores; then we shade in the appropriate region. In this case $z_{47.6}$ is obviously 0, and

$$\begin{aligned} z_{59.8} &= \frac{59.8-47.6}{13.2} \\ &= .92. \end{aligned}$$

Figure 6.16

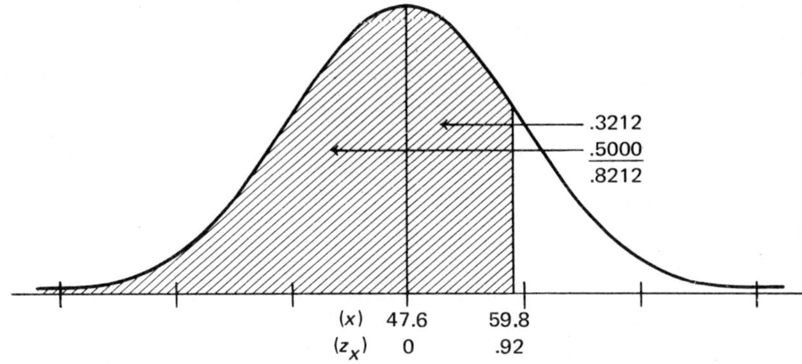

Finally, we multiply .8212 by 700 to get 574.84, which rounds off to 575.

Between 42.5 and 63.7? Here is the diagram with its shaded region and its corresponding z-scores

$$\begin{aligned} z_{42.5} &= \frac{42.5-47.6}{13.2} \\ &= -.39 \end{aligned}$$

and

$$z_{63.7} = \frac{63.7 - 47.6}{13.2}$$
$$= 1.22.$$

Figure 6.17

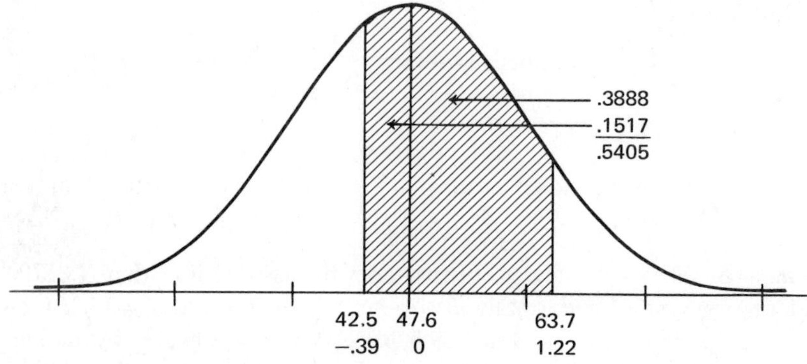

Thus we have

$$.5405 \cdot 700 = 378.35,$$

or 378 data items as the answer.

Between 49 and 73.8? Here is the diagram with its shaded region and its corresponding z-scores

$$z_{49} = \frac{49 - 47.6}{13.2}$$
$$= .11$$

and

$$z_{73.8} = \frac{73.8 - 47.6}{13.2}$$
$$= 1.98.$$

Figure 6.18

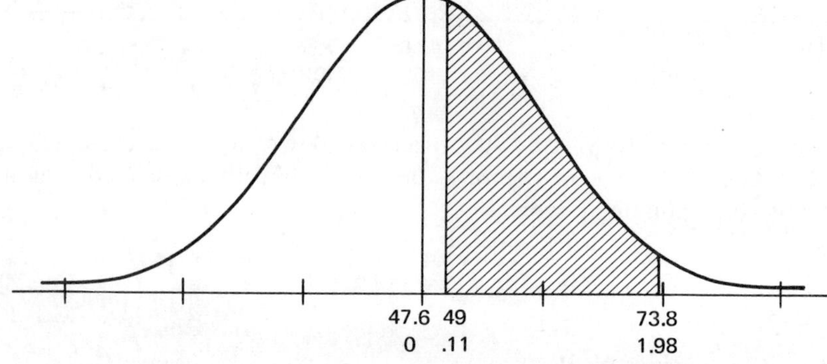

Thus we have

$$(.4761 - .0438) \cdot 700 = 302.61,$$

or 303 data items as the answer.

To summarize this section, we have studied the normal distribution curve because of its representation of a surprisingly wide variety of phenomena in the world around us. As long as we know that data are normally or even approximately normally distributed, and as long as we know the values of $\bar{x}$ and σ for the set, we can make conclusions about the distribution of numbers in the set with only a little calculation and use of the normal distribution table.

Problem set 6–3

1. Find the area under the normal curve between these z-scores.
   a. 0 and .34
   b. 0 and 2.3
   c. 0 and -1.31
   d. 0 and $-.6$
   e. -2.45 and 2.93
   f. -1 and $+1$ (Why does the result differ slightly from what we first learned about the normal curve?)
   g. above 1.14
   h. below 1.5
   i. below $-.63$
   j. between $-.87$ and -2.15
   k. below -1.96 or above .46
2. If 500 heights of basketball players measured to the nearest inch are n.d., $\bar{x} = 70$, $\sigma = 6$, then find the number of basketball players whose heights are
   a. below 75.
   b. between 68 and 78.
   c. between 73 and 81.
   d. less than 65.
   e. greater than 60.
   f. equal to 74. (Hint: we must think of heights "equal to 74" as those heights in the interval [73.5, 74.5).)
3. In the example of twenty-five test scores we saw that $\bar{x} = 61.64$ and $\sigma = 8.53$. *If*—and it is an assumption—the twenty-five test scores are normally distributed, we would expect to find 68.26% of the twenty-five test scores, or seventeen of them, within one standard deviation of the mean. "Within one standard deviation of the mean" means

$$\bar{x} - 1\sigma < x < \bar{x} + 1\sigma,$$

which in this specific case is

$$61.64 - 1 \cdot 8.53 < x < 61.64 + 1 \cdot 8.53$$

or

$$53.11 < x < 70.17.$$

From examining the sorted set we see that there are actually 16 numbers in the interval. How do we account for this discrepancy between the expected value of 17 and the actual number of 16?

4. For each part of this problem refer to Problem 3.
   a. How many test scores do we expect to find between $\bar{x} - 2\sigma$ and $\bar{x} + 2\sigma$, and how many are actually in this interval? Determine the expected number according to the assumption of normality in this part of the problem and the remaining three.
   b. How many test scores do we expect to find between $\bar{x} - 3\sigma$ and $\bar{x} + 3\sigma$, and how many are actually in this interval?
   c. How many test scores do we expect to find greater than 1σ above the mean, and how many are actually there?
   d. How many test scores do we expect to find greater than 1σ below the mean, and how many are actually there?
5. Repeat the instructions for all parts of Problem 4 using the data in Problem 4 of Problem Set 6–1. As part (e) of this problem, determine the number of weights we expect to find within one standard deviation unit of their mean and the number which are actually in that interval.
6. On a math test, with scores which were n.d. with $\bar{x} = 75$ and $\sigma = 4$, Bill's score was 79. On his German test, with scores which were n.d. with $\bar{x} = 63$ and $\sigma = 5$, his score was 73. On which test did he do better in relation to the rest of the class?
7. Answer Question 1 at the very beginning of this chapter, using the assumption that scholastic aptitude test scores are normally distributed with $\bar{x} = 500$ and $\sigma = 100$.
8. If IQ's are normally distributed with $\bar{x} = 100$ and $\sigma = 16$, and a test determines Claudia's IQ to be 120, then determine the percentage of people with IQ's which are lower than hers.
9. Suppose a fair coin is tossed ten times and we want to know the probability of each possible outcome where the sample space $S = \{0, 1, 2, 3, \ldots, 10\}$ is the possible number of heads. The calculations are routine, since we are dealing with a straightforward application of the Binomial Law or Repeated Trials Formula. For example,

$$P(\text{exactly 3 heads}) = {}_{10}C_3 \cdot (1/2)^3 \cdot (1 - 1/2)^7$$
$$= .117 \qquad \text{(from Table XI).}$$

If we use Table XI for the ten remaining values of x, we would obtain the following table.

Number of possible heads, x	$P(10, x, 1/2) = {}_{10}C_x \cdot (1/2)^x \cdot (1/2)^{10-x}$
0	.001
1	.010
2	.044
3	.117
4	.205
5	.246
6	.205
7	.117
8	.044
9	.010
10	.001

As expected, the sum of the probabilities is 1.000. Nothing, of course, is new. What we do next is to make a graph from the table with the values of x, *considered to be midpoints of intervals*, on the horizontal axis and with their corresponding probabilities on the vertical axis.

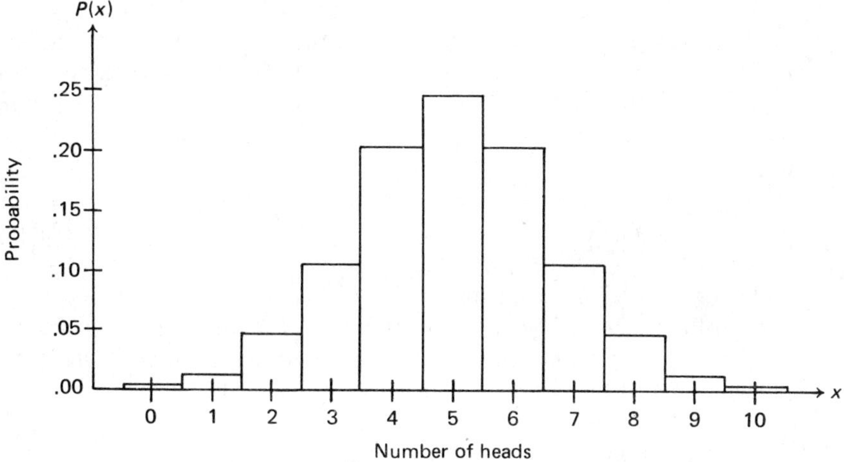

Notice that the histogram is reasonably close to being a bell-shaped curve. This fact becomes even more apparent if we construct the frequency polygon on top of the histogram. If we were to let the number of tosses increase continually and construct the corresponding histograms and frequency polygons, then they would come closer and closer to the normal curve. One standard deviation unit would have to have the same length on any horizontal axis, regardless of the number of tosses. Each histogram would have an area of 1 because the sum of the probabilities is always 1. This geometric approach growing out of one specific example is one way to discuss the origin of the normal curve.

a. Use Table XI to construct a histogram/frequency polygon for the probability distribution of the number of heads when a fair coin is tossed fifteen times.

b. Use Table XI to construct a histogram/frequency polygon for the probability distribution of the number of heads when a fair coin is tossed twenty times.

10. The *interquartile range* of a set of numbers is an interval of the set in which the middle half of the numbers are located after the numbers have been arranged in ascending order. The *quartile deviation*, Q, is the length of an interval beginning at the midpoint of the interquartile range and ending either at the left-hand endpoint or the right-hand endpoint. If the numbers are normally distributed so that the mean and median coincide, then $Q \doteq .67 \cdot \sigma$, as we can see from Table XII.

    a. Calculate Q for the twenty-five test scores, assuming the center of the interquartile range is the mean, $\bar{x} = 61.64$.

    b. How many of the scores are actually in the interval $(\bar{x}-Q, \bar{x}+Q)$? Theoretically, there are half of 25 scores in that interval.

11. It is early November, and the day of the championship football game in the Central Conference between Western College and Eastern College has arrived. Western, trailing 21 to 16, has the ball on Eastern's 2 yard line with time out and 1 second remaining in the game. The coach, Art Nims, has to make a choice between using his fullback, Brutus, or his halfback, Twinkletoes. Brutus' runs over the season are normally distributed, $\bar{x} = 2.5$ yards and $\sigma = .3$ yards. Twinkletoes' runs over the season are normally distributed, $\bar{x} = 3.0$ yards and $\sigma = .8$ yards. Assuming that either player will do as well now as he has done before, which player would you advise Coach Nims to use? Hint: find the percentage of plays for which Brutus has exceeded 2 yards, do the same for Twinkletoes, and make your decision accordingly.

12. Refer to problem 2. If one basketball player is picked at random from the 500 players, and his height is measured to the nearest inch, then find the probability that his height is

    a. below 75.

    b. between 68 and 78.

    c. between 73 and 81.

    d. less than 65.

    e. greater than 60.

    f. equal to 74.

13. A fuse is designed to melt when a current of five amperes passes through it. Just as mass production of this fuse begins the first 400 fuses are tested with increasing current until they melt. The mean of these 400 melting currents is 5.0 amperes and the standard deviation is .0019 amperes. Would these statistics support the advertisement that not more than one out of one hundred of the fuses melts at a current which differs from 5.0 amperes by more than .005 amperes?

Least square lines

In analyzing the results of a scientific experiment or tests of behavioral attitudes we often look for a general pattern to grow out of specific results. For example, if we have the mathematical aptitude scores and corresponding first semester mathematics grades for 300 college freshmen, we look for one general equation that relates the two variables and best fits this set of specific pairs of data. Then we would ask just how well the equation fits the set of pairs, and we would attempt to use the resulting equation to predict future final math grades of all college freshmen from their entrance exam scores.

Example 7 Here are six freshman students at a state college with their entrance exams (x) and final semester grades (y) in the same math course as given in this table. What is the equation which best describes the relationship existing between x and y so that we can use the values of the predict*or* variable (x) to estimate the values of the predict*ed* variable (y) for future freshman math students in this course?

Table 6.9

Entrance exam x	Final grade (letter)	Final grade (numerical equivalent), y
58	D	1.0
60	C−	1.7
67	D	1.0
74	B	3.0
78	B	3.0
86	A−	3.7

Figure 6.19

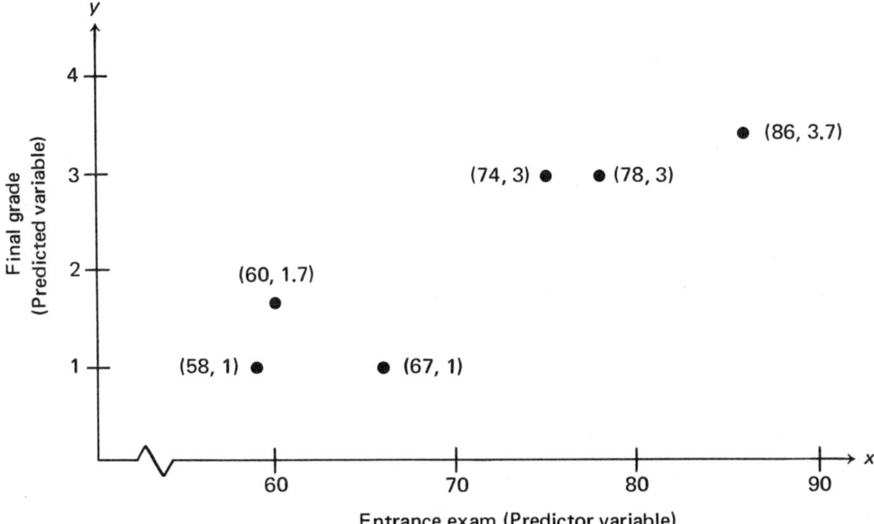

Solution The relationship existing between x and y will become more apparent if we make what is known as a *scatter diagram* by plotting the data pairs (x, y) as shown in Figure 6.19. The algebraic question "what linear equation best describes the relationship between x and y?" has as its geometric equivalent "what line best fits the points (x, y), even though it may not contain any of them?" Two questions face us immediately: exactly what is meant by the "best-fit" line, and how do we find its equation, which in turn will be the equation we accept as the one best relating x and y?

The best-fit line and its equation

The "best-fit" line we choose is the one which minimizes the sum of the squares of the deviations. Experience has shown that the least squares standard is the best one to work with, even though other standards are available. A *deviation* is the length of the vertical line segment between the line and a point (x, y).

Figure 6.20

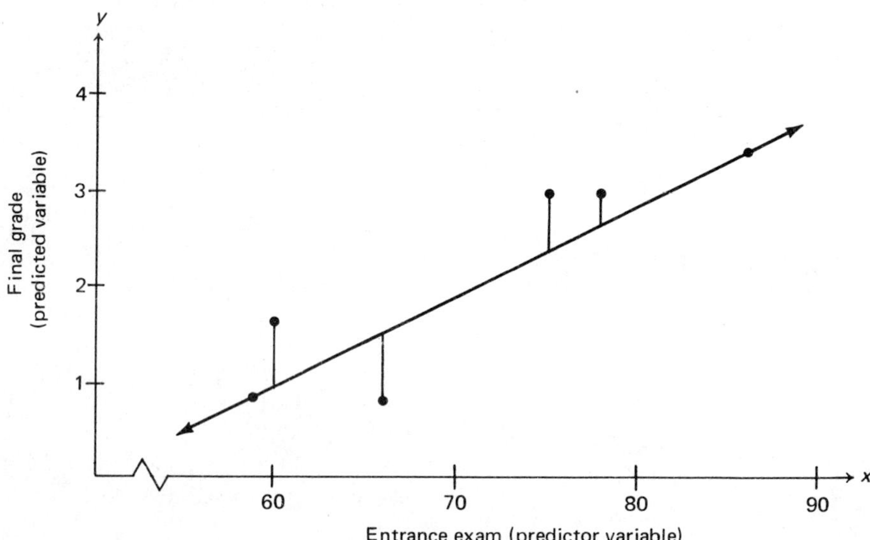

In this diagram we have *guessed* that the best-fit line passes through the left-most and right-most points. Its slope is

$$\frac{y_2 - y_1}{x_2 - x_1} = \frac{3.7 - 1.0}{86 - 58} = \frac{2.7}{28} \doteq .096. \tag{A}$$

It passes through the point with coordinates (58, 1), and, by the results of Problem 15 of Problem set 3–3, its equation is

$$y - 1 = .096 \cdot (x - 58)$$

or

$$y = .096x + -4.568.$$

We next extend the previous table in light of our graph.

Table 6.10

Exam score, x	Final grade, y	y-coordinate of point on guessed best-fit line, $.096 \cdot x + -4.568$	Deviation |column 2| − |column 3|	Deviation squared
58	1.0	1.000	.000	.000000
60	1.7	1.192	.508	.258064
67	1.0	1.864	.864	.746496
74	3.0	2.536	.464	.215296
78	3.0	2.920	.080	.006400
86	3.7	3.688	.012	.000144

We expect that the last deviation of .012 would be .000 because we deliberately chose our guessed best-fit line to pass through the point (86, 3.7). This small discrepancy arises because of rounding off 2.7/28 to .096 in equation (A) when, to 5 decimal places, the quotient is .09643. Also, the sum of the squares of the deviations is 1.226400.

Of course, our temporary best-fit line with its equation

$$y = .096 \cdot x + -4.568$$

is only a guessed line. There may well be another line such that the sum of the squares of its deviations will be less than 1.226400.

Our search for the best-fit line leads us to accept without proof the following theorem.

Theorem The equation of the best-fit line (taking the form $y = a \cdot x + b$) which minimizes the sum of the squares of the deviations is determined by evaluating the expressions

$$a = \frac{n \cdot \sum xy - \sum x \cdot \sum y}{n \cdot \sum x^2 - (\sum x)^2}$$

$$b = \frac{\sum y - a \cdot \sum x}{n}.$$

We continue our example by using this theorem to find those values of a and b which will determine the equation of the best-fit line. A close look at the expressions for a and b tells us what the column headings in the work table must be.

Table 6.11

x	y	xy	x^2
58	1.0	58	3364
60	1.7	102	3600
67	1.0	67	4489
74	3.0	222	5476
78	3.0	234	6084
86	3.7	318.2	7396
$\sum x = 423$	$\sum y = 13.4$	$\sum xy = 1001.2$	$\sum x^2 = 30409$

Since n is the number of *pairs* of data, $n = 6$ in our case. Also,

$$\left(\sum x\right)^2 = (423)^2 = 178{,}929.$$

Hence,

$$a = \frac{6 \cdot 1001.2 - 423 \cdot 13.4}{6 \cdot 30409 - 178929} = \frac{6007.2 - 5668.2}{182454 - 178929} = \frac{339}{3525} = .096.$$

Similarly,

$$b = \frac{13.4 - .096 \cdot 423}{6} = \frac{13.4 - 40.61}{6} = \frac{-27.21}{6} = -4.535.$$

Finally,

$$y = .096 \cdot x + -4.535.$$

Prediction equations

There is a small but significant point to take up next. While

$$y = .096x + -4.535$$

is the equation of the least squares line, the prediction equation itself is

$$y' = .096x + -4.535.$$

The symbol y represents an *observed* final grade corresponding to a given x, while y' represents the *predicted* final grade corresponding to a given x. In tabular form,

Table 6.12

Exam score, x	Final grade, y	Predicted final grade, $y' = .096x + -4.535$	Error of prediction, $\|y - y'\|$
58	1.0	1.033	.033
60	1.7	1.225	.475
67	1.0	1.897	.897
74	3.0	2.569	.431
78	3.0	2.953	.047
86	3.7	3.721	.021

The last column says that while the prediction equation

$$y' = .096x + -4.535$$

is the best linear one available, it is not perfect because there are differences between the observed and predicted final grades. The sum of the squares of these errors of prediction is 1.219734, which is less than the 1.226400 from our guessed best-fit line.

Just how good is this prediction equation which best describes the linear relationship between x and y? Can we find a number which measures the goodness of fit of this best-fit line? We will answer these questions in the next section.

Graphs other than straight lines are sometimes more appropriate as we seek to find the equation which best fits a set of data pairs. For example, if a scatter diagram of raw data gives us

Figure 6.21

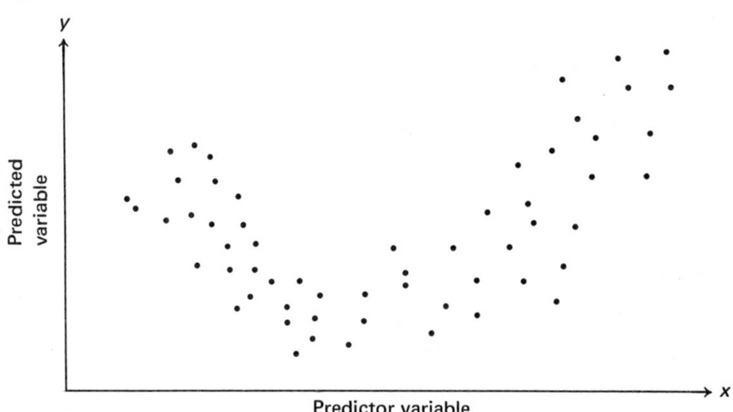

then a parabola appears to fit the points better than a line would. The best prediction equation could thus be of the form

$$y' = ax^2 + bx + c.$$

Our example has only one predictor variable and one predicted variable. In reality researchers study several predictor variables together to determine what combination of them is the best arrangement for estimating the one predicted variable. For example, a college may keep records of many freshmen with their high school grade, high school rank, American College Testing scores, and first semester grade point averages in order to find the best combination of the first three variables to be used in predicting the last one. The techniques are basically extensions of what we have done but with much more involved mathematics. Prediction equations are sometimes called *regression* equations. Much of applied statistics deals with the interrelationships of several predictor variables and the one predicted variable, and computers are used extensively to do the enormous amounts of arithmetic.

Problem set 6–4

1. In the sample problem of this section suppose a student has an entrance exam score of 70. What is his predicted final math grade?
2. Consider set $S = \{(1,3), (2,4), (3,5), (4,6), (5,7)\}$.
   a. Determine the equation of the best-fit line for these number pairs (x, y).
   b. Extend your calculation table to have columns headed $|y - y'|$ and $(y - y')^2$.
   c. Determine the value of $\sum (y - y')^2$.
3. Consider set $S = \{(4,6), (5,7), (7,16), (8,10), (11,13)\}$.
   a. Determine the equation of the best-fit line for these number pairs (x, y).
   b. Extend your calculation table to have columns headed $|y - y'|$ and $(y - y')^2$.
   c. Determine the value of $\sum (y - y')^2$.
4. Consider set $S = \{(8,17), (10,16), (13,14), (14,11)\}$.
   a. Determine the equation of the best-fit line for these number pairs (x, y).
   b. Extend your calculation table to have columns headed $|y - y'|$ and $(y - y')^2$.
   c. Determine the value of $\sum (y - y')^2$.
5. Consider the set $S = \{(-2,4), (-1,1), (0,0), (1,1), (2,4)\}$. Do the same steps with this set as you did in Problem 2.
6. Consider the set $S = \{(3,4), (3,8), (5,4), (5,8), (9,4), (9,8)\}$. Do the same steps with this set as you did in Problem 2.

Correlation

In our example of the last section we saw that the linear equation which best fits the data pairs of the form $(x, y) =$ (entrance exam, final grade) was

$$y' = .096x + -4.535.$$

The choice of $a = .096$ and $b = -4.535$ minimized the value of $\sum(y - y')^2$. We return now to the question, posed to us in the last section, "Can we find a number which measures the goodness of fit of the best-fit line?" The answer is yes, but we delay the specific details of how to find the number until we learn some additional background knowledge.

Total variation and explained variation

Consider the set of possible values of the predicted variable y, which is $\{1, 1.7, 1, 3, 3, 3.7\}$. There is variation of the values of y since they do not all equal one number. We define the *total variation* of this set to be the number $\sum(y - \bar{y})^2$ or, since $\bar{y} = 2.233$,

$$\begin{aligned} \text{total variation} &= \sum (y - \bar{y})^2 \\ &= (1 - 2.233)^2 + (1.7 - 2.233)^2 + \cdots \\ &\quad + (3.7 - 2.233)^2 \\ &= 6.653. \end{aligned}$$

This result of 6.653 is not to be confused with the variance of the set, which is

$$\sigma^2 = \frac{\sum(y - \bar{y})^2}{n} = \frac{6.653}{6} = 1.109,$$

or with the standard deviation of the set, which is

$$\sigma = \sqrt{\frac{\sum(y - \bar{y})^2}{n}} = \sqrt{1.109} = 1.053.$$

Here is the best fit line along with the original data pairs.

Figure 6.22

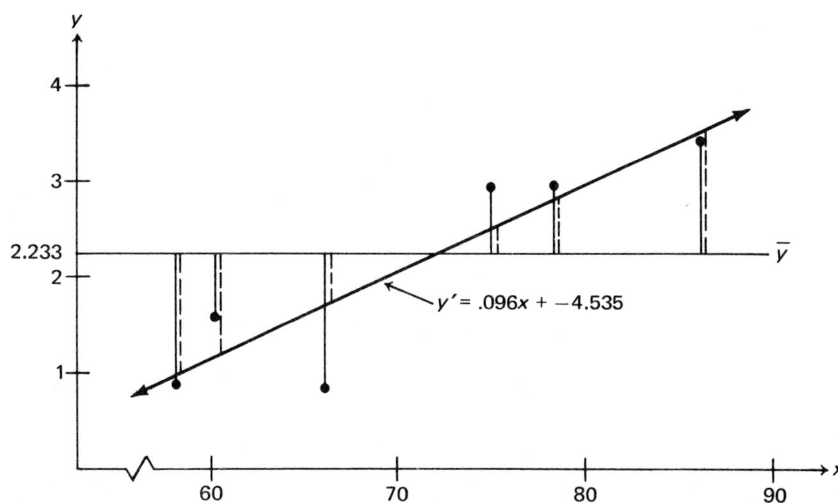

The sum of the squares of the lengths of the solid line segments is the total variation of the final grades from their mean of 2.233. The sum of the squares of the lengths of the dotted line segments (which segments actually lie on top of the solid ones) is defined to be the *explained variation* of the final grades from their mean. Intuitively, some of the total variation of the final grades can be accounted for by the linear relationship existing between x and y; the linear relationship does predict some deviation of a value of y' from the mean of the values of y. Algebraically, the definition of explained variation is given by the equation

$$\text{explained variation} = \sum (y' - \bar{y})^2.$$

In our problem, then,

$$\begin{aligned}\sum (y' - \bar{y})^2 &= (1.033 - 2.233)^2 + (1.225 - 2.233)^2 + \cdots \\ &\quad + (3.721 - 2.233)^2 \\ &= 5.414.\end{aligned}$$

Notice that if all the points were on the best fit line the total variation and the explained variation would be equal because the linear relationship accounts for all the variation among the final grades. Such a situation is very rare in the actual worlds of social and management science.

Coefficient of determination and the coefficient of correlation

What we do next is to find the fraction of the total variation of the final grades which is explained by the linear relation between the values of x and y. We form the ratio

$$\frac{\text{explained variation}}{\text{total variation}} = \frac{\sum (y' - \bar{y})^2}{\sum (y - \bar{y})^2},$$

which is defined to be the *coefficient of determination*. It is symbolized by r^2. In our case, then,

$$\begin{aligned} r^2 &= \frac{\text{explained variation}}{\text{total variation}} \\ &= \frac{5.414}{6.653} \\ &= .814. \end{aligned}$$

Another interpretation of the equation $r^2 = .814$ is that 81.4% of the variation in the final exam scores is accounted for by the variation in the entrance exam scores under the assumption that the relation is best described by an equation of the form $y = a \cdot x + b$.

The *unexplained variation* is the sum of the squares of the deviations of the values of y' from the actual values of y. $\sum (y - y')^2$, which is 1.220 for this example, represents the error of prediction of the best linear prediction equation. In general,

$$\text{total variation} = \text{explained variation} + \text{unexplained variation}.$$

In our example,

$$5.414 + 1.220 = 6.634 \neq 6.653$$

because of accumulated rounding errors.

The *coefficient of correlation,* which is more commonly used than the coefficient of determination as a measure of the relationship between two variables, is defined to be the square root of the coefficient of determination. That is,

$$r = \pm\sqrt{\frac{\text{explained variation}}{\text{total variation}}} = \pm\sqrt{\frac{\sum(y' - \bar{y})^2}{\sum(y - \bar{y})^2}}$$

Since the explained variation is always a nonnegative number less than or equal to the total variation, we conclude that

$$-1 \leqslant r \leqslant 1.$$

It is worthwhile to examine the graphical interpretation of a few values of r under the assumption that we are working with a linear relationship between y and x. Here are some scatter diagrams along with their best-fit lines and corresponding values of r.

Figure 6.23

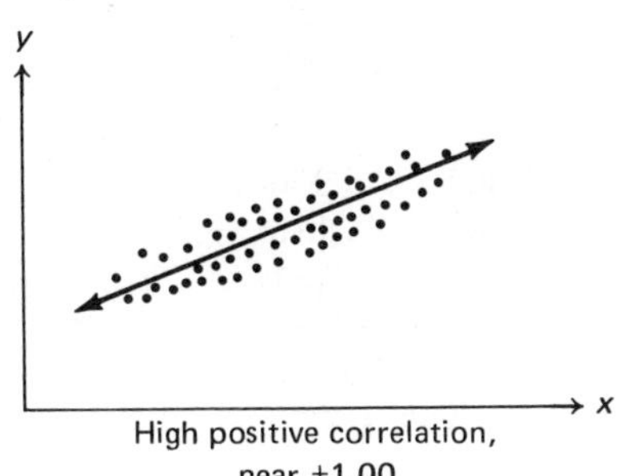

High positive correlation, near +1.00

Figure 6.24

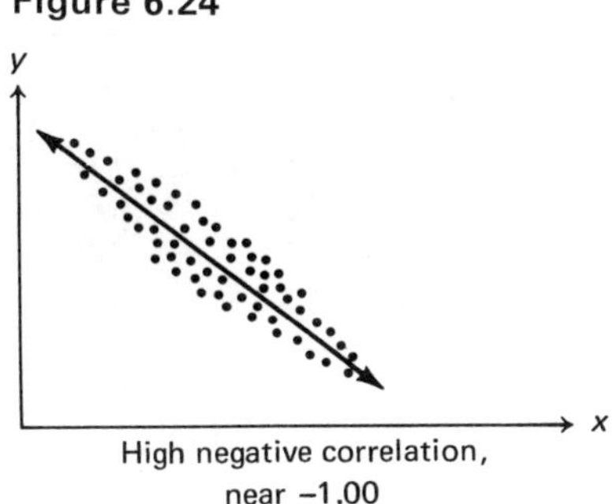

High negative correlation, near −1.00

Figure 6.25

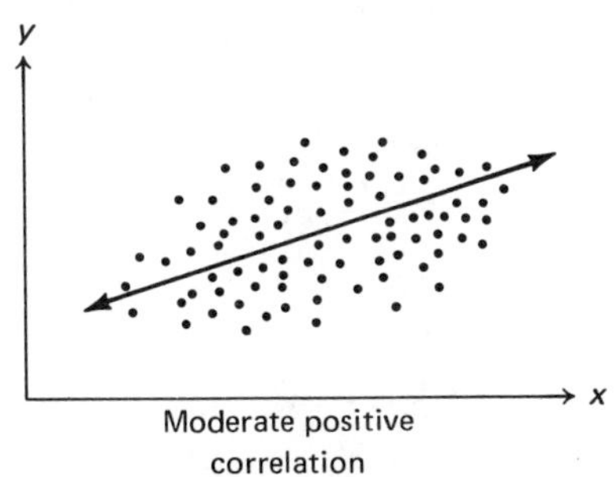

Moderate positive correlation

Figure 6.26

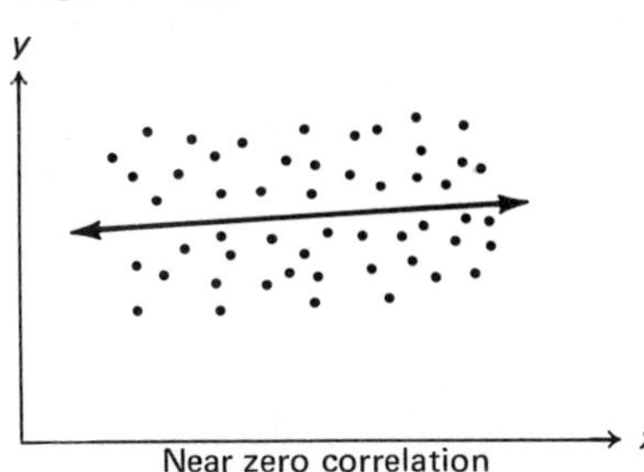

Near zero correlation

We have seen that the calculation of r from a set of pairs of data is very involved. There is a tremendous amount of arithmetic in finding the equation of the best-fit line, $\sum(y'-\bar{y})^2$, and $\sum(y-\bar{y})^2$. Fortunately, there is a formula which considerably shortens the amount of effort needed to calculate a linear correlation coefficient. It is

$$r = \frac{n\cdot\sum xy - \sum x\cdot\sum y}{\sqrt{[n\cdot\sum x^2-(\sum x)^2]\cdot[n\cdot\sum y^2-(\sum y)^2]}}.$$

Example 8 A farmer wishes to know if there is any linear relation between the number of bags of fertilizer he applies and the amount of insect damage per acre. He keeps a record of four fields for a season with the results of this table.

Table 6.13

Fertilizer per acre	8	10	13	14
Insect damage per acre	17	16	14	11

Calculate the linear correlation coefficient. How would we interpret the results of these numbers to him?

Solution We use our previous computational formula for r to determine the column headings of our work table. The value of n is 4 because we have 4 *pairs* of data.

Table 6.14

x	y	xy	x^2	y^2
8	17	136	64	289
10	16	160	100	256
13	14	182	169	196
14	11	154	196	121
$\sum x = 45$	$\sum y = 58$	$\sum xy = 632$	$\sum x^2 = 529$	$\sum y^2 = 862$

$$r = \frac{4\cdot 632 - 45\cdot 58}{\sqrt{[4\cdot 529-(45)^2]\cdot[4\cdot 862-(58)^2]}}$$

$$r = \frac{2528-2610}{\sqrt{[2116-2025]\cdot[3448-3364]}}$$

$$r = \frac{-82}{\sqrt{91\cdot 84}}$$

$$r = \frac{-82}{87.43}$$

$$r = -.94$$

Since $r = -.94$, the coefficient of determination is

$$\begin{aligned} r^2 &= (-.94)^2 \\ &= .88 \end{aligned}$$

which, when explained to the farmer, should let him know that 88% of the variation in insect damage is explained by the variation in the fertilizer applied.

The word *linear* has applied to all the best-fit curves, coefficients of determination, and coefficients of correlation that we have worked with so far. It is entirely possible for a best-fit curve to be different from a line. There are formulas for finding the best-fitting parabola ($y' = ax^2 + bx + c$), the best-fitting exponential curve ($y' = a \cdot e^{bx}$), the best-fitting power curve ($y' = a \cdot x^b$), etc., to a set of pairs of data.More advanced techniques are required to find best-fit curves for these nonlinear relationships. We will not investigate this area.

Misuse of *r*

A word of caution is in order before we conclude that r is a magical number that measures cause-and-effect relations. Such a conclusion often is totally unwarranted. A study several years ago showed that there was a high positive linear correlation between the price of rum in Havana, Cuba, and the salaries of Protestant ministers in New England, but the relation is obviously not causal. On the other hand, scores of studies have shown that there is a strong positive relationship between the number of cigarettes smoked and the incidence of lung cancer among the people studied. In both cases we must beware of the *post hoc* method of reasoning. *Post hoc* is an abbreviation of the Latin phrase *post hoc, ergo propter hoc*, which is translated "after this, therefore because of this." It means that just because event B occurs after (or with) event A, this does not automatically imply that B occurred because of A.

We have conformed with tradition in the use of the ratio

$$r^2 = \frac{\text{explained variation}}{\text{total variation}}$$

and in the use of such phrases as "70% of the variation of y *is explained by* the variation of x." Perhaps it would be better to say "70% of the variation of y *occurs with* the variation of x." We should caution the farmer mentioned in Example 8 that other factors, such as an increase in the insects' natural enemies, may also enter into the picture. Another example of the misuse of r is to say that since people with higher incomes have college educations, education is the primary reason for high incomes. (This situation, incidentally, applies to educational levels through the bachelor's degree but not always proportionately beyond, since people with master's degrees often become teachers.) This situation requires a sharpening of our common sense analysis of the variables involved in a relationship, along with some more involved statistical concepts, before we can conclude that a high value of r means that a change in variable x causes a change in variable y. Nevertheless, knowing the value of r can be useful in describing the relationship between two sets.

Problem set 6–5

1. Indicate whether the linear correlation coefficient for these pairs of sets is positive, negative, near zero, or should not be computed.
   a. height and weight
   b. foot size and intelligence
   c. age (up to age twenty) and ability to learn a foreign language
   d. amount of alcohol consumed and ability to control a car
   e. zodiac sign and personality
2. Calculate the linear correlation coefficient between x and y as given in this table.

x	5	7	8	10
y	10	11	13	16

3. Calculate the linear correlation coefficient between x and y as given in this table.

x	5	7	8	10	13
y	10	11	13	16	12

   Do you see the difference that one exceptional pair of numbers can make in r when n is small?
4. Does an r of $+.60$ describe a relation that is three times the strength of a relation described by an r of $+.20$?
5. Consider the data pairs given in this table.

x	0	0	2	2	4	4
y	4	6	9	7	10	12

   a. Calculate the value of a linear r.
   b. Determine the equation of the best-fit line.
   c. Graph the best-fit line and the data pairs on the same set of coordinate axes.
6. Carry through the same instructions as in Problem 5, but apply them to the data of Problem 2 in the last problem section.
7. Carry through the same instructions as in Problem 5, but apply them to the data of Problem 6 in the last problem section.

8. a. Calculate the value of a linear r for the data of Problem 5 in the last problem section.
   b. Determine the goodness of fit (that is, calculate r^2) of the curve $y = x^2$ to the data of Problem 5 in the last problem section.
9. Verify that $r^2 = .814$ in our example of entrance exam and final grades by using the shortcut formula to calculate r. The square of this obtained r should be very close to .814.

Chi-square hypothesis testing

Up to this point our study of statistics has dealt with describing sets of data. We are now in a position to make inferences about the characteristics of a population, or universal set, based upon the characteristics of a set of data which may or may not be a sample chosen from the population. We know that it is not necessary to drink all the soup to know how it tastes—provided our spoonful is representative of the entire bowl—but there is always the possibility of getting an unrepresentative sample and hence making the wrong conclusion about the bowl. A *population* is defined to be the complete set of individuals, objects, or measurements which have a characteristic in common. A set may or may not be a *sample*, or subset, of a certain population. These questions of relationship between a population and a sample are illustrated in the following example.

Example 9 John Smith has a penny and he wants to see if it is a fair coin. To do so, he flips it fifty times and observes that thirty-two heads and eighteen tails appear. Is this coin fair? Keep in mind that we are trying to infer the unknown characteristics of a population (the infinite string of H's and T's that flipping this one coin could produce) from the known characteristics of a sample (the finite string of H's and T's that flipping this coin fifty times produced).

Solution If this coin were unbiased so that

$$P(\text{head}) = 1/2$$

on any toss, then we would expect half of a series of consecutive tosses to be heads and the other half to be tails. In this case we expect one-half of fifty tosses to produce heads and one-half of fifty tosses to produce tails. We organize our work on this problem into a diagram.

Figure 6.27

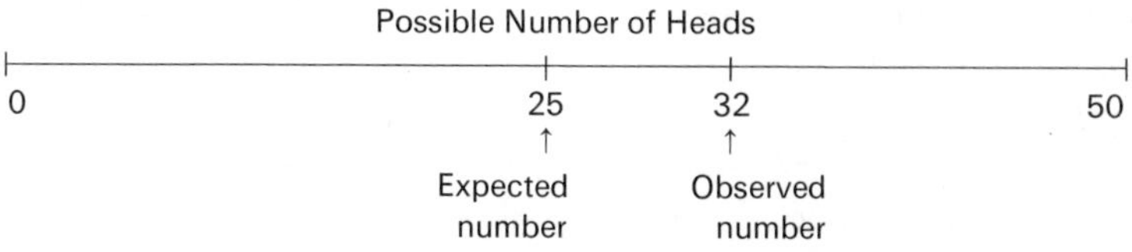

Null hypothesis and alternate hypothesis

There are two mutually exclusive explanations for the observed number of thirty-two heads instead of the expected number of twenty-five heads.

1. The coin is really fair. It is not always going to yield exactly twenty-five heads in each fifty tosses, and thirty-two heads out of fifty tosses is just a chance occurrence, an unusual sample, or a random fluctuation that is still representative of this fair coin.
2. The coin is really unfair. It is biased in favor of heads, and the sample containing thirty-two heads is truly representative of the unfair coin. Chance alone, or sampling variation, is not responsible for the difference of seven heads.

We can also diagram these explanations.

Figure 6.28

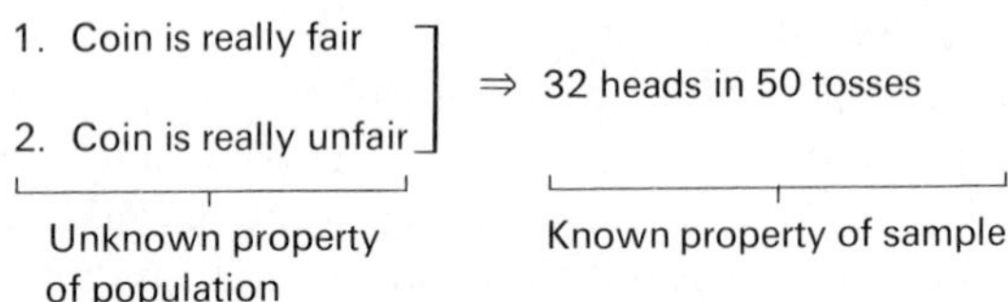

It is our task to determine whether (1) or (2) is the cause of the known effect. To do this we will calculate the probabilities of the two possible causes which produce the sample outcome and then possibly choose the cause with the higher probability. In the language of the statistician, the two possible causes are *hypotheses*, or tentative explanations of observable phenomena. We will *temporarily* accept (1) as the cause and immediately determine the probability of its being the cause. Statement (1) is called the *null* hypothesis since it effectively states that there is *no* (null) true difference between the characteristics of the population and the characteristics of the sample. That is, the coin is unbiased even though a sample of its tosses happens to contain more heads than tails, and the difference between the expected and observed results is attributable to chance alone. Hence, we say

$$H_0\text{: the coin is fair}$$

and we proceed to calculate the probability that a fair coin will yield a sample with thirty-two or more heads in fifty tosses.

How do we calculate this probability? The method is a bit roundabout, but it does work (see Problem 16 for a brief justification of this method). We first calculate a number called χ^2 (chi-square, pronounced like "sky" without the "s"), which is obtained by using 25 and 32 as input numbers, and then using a table to convert the resulting value of χ^2 into a probability.

The formula for χ^2 is

$$\chi^2 = \sum \frac{(f_o - f_h)^2}{f_h}$$

where

$$f_o = \text{observed frequencies in the sample}$$

and

$$f_h = \text{hypothetical frequencies in the sample under the acceptance of the null hypothesis.}$$

A close look at the formula tells us that as the values of $(f_o - f_h)$ increase—that is, as the differences between the observed and hypothetical frequencies increase—the corresponding values of χ^2 increase as well. This means that χ^2 is a measure of the difference between what we observe and what we expect in the sample. We can calculate χ^2 for this experiment by setting up a table with the outcomes of the experiment in the first column.

Table 6.15

Outcome	f_o	f_h	$f_o - f_h$	$(f_o - f_h)^2$	$(f_o - f_h)^2/f_h$
Heads	32	25	7	49	49/25 = 1.96
Tails	18	25	−7	49	49/25 = 1.96
					$\chi^2 = 3.92$

How, then, do we convert $\chi^2 = 3.92$ into our desired probability by using the previously mentioned table? Before taking up this question we must understand the concept of *degrees of freedom.* Suppose we have to fill up four empty boxes with numbers whose sum is 30. We can choose the first three numbers with freedom (e.g., 13, 5, and 2), but then there is no freedom left in choosing the fourth number (which is 10 in this example). Hence we say that there are 3 degrees of freedom, or $df = 3$. Generalizing, we say that if n boxes are to be filled so that the n numbers in them have a predetermined sum, then there are $(n - 1)$ degrees of freedom. As an equation we have

$$df = (\text{number of outcomes of an experiment}) - 1.$$

In our coin example $df = 2 - 1 = 1$, since there are two possible outcomes, heads or tails, and the sum of their occurrences must be 50.

In order to use the table on the next two pages we locate our χ^2 of 3.92 between two table entries in the horizontal row with $df = 1$, which in this case are 3.84 and 5.41. Above 3.84 and 5.41 are 5% and 2%, respectively. The correspondence of $3.84 \leftrightarrow 5\%$ means that the probability of obtaining a χ^2 of 3.84 or more in an experiment such as ours is 5%, while $5.41 \leftrightarrow 2\%$ means

$$P(\chi^2 = 5.41 \text{ or more with } df = 1) = 2\%.$$

Table 6.16 Percentile values (χ^2) for chi-square with df degrees of freedom

df	.5%	1%	2%	5%	10%	25%	50%	75%	90%	95%	98%	99%	99.5%
1	7.88	6.63	5.41	3.84	2.71	1.32	.455	.102	.0158	.0039	.0006	.0002	.0000
2	10.6	9.21	7.82	5.99	4.61	2.77	1.39	.575	.211	.103	.0404	.0201	.0100
3	12.8	11.3	9.84	7.81	6.25	4.11	2.37	1.21	.584	.352	.185	.115	.072
4	14.9	13.3	11.7	9.49	7.78	5.39	3.36	1.92	1.06	.711	.429	.297	.207
5	16.7	15.1	13.4	11.1	9.24	6.63	4.35	2.67	1.61	1.15	.752	.554	.412
6	18.5	16.8	15.0	12.6	10.6	7.84	5.35	3.45	2.20	1.64	1.13	.872	.676
7	20.3	18.5	16.6	14.1	12.0	9.04	6.35	4.25	2.83	2.17	1.56	1.24	.989
8	22.0	20.1	18.2	15.5	13.4	10.2	7.34	5.07	3.49	2.73	2.03	1.65	1.34
9	23.6	21.7	19.7	16.9	14.7	11.4	8.34	5.90	4.17	3.33	2.53	2.09	1.73
10	25.2	23.2	21.2	18.3	16.0	12.5	9.34	6.74	4.87	3.94	3.06	2.56	2.16
11	26.8	24.7	22.6	19.7	17.3	13.7	10.3	7.58	5.58	4.57	3.61	3.05	2.60
12	28.3	26.2	24.1	21.0	18.5	14.8	11.3	8.44	6.30	5.23	4.18	3.57	3.07
13	29.8	27.7	25.5	22.4	19.8	16.0	12.3	9.30	7.04	5.89	4.77	4.11	3.57
14	31.3	29.1	26.9	23.7	21.1	17.1	13.3	10.2	7.79	6.57	5.37	4.66	4.07
15	32.8	30.6	28.3	25.0	22.3	18.2	14.3	11.0	8.55	7.26	5.99	5.23	4.60
16	34.3	32.0	29.6	26.3	23.5	19.4	15.3	11.9	9.31	7.96	6.61	5.81	5.14
17	35.7	33.4	31.0	27.6	24.8	20.5	16.3	12.8	10.1	8.67	7.26	6.41	5.70
18	37.2	34.8	32.2	28.9	26.0	21.6	17.3	13.7	10.9	9.39	7.91	7.01	6.26
19	38.6	36.2	33.7	30.1	27.2	22.7	18.3	14.6	11.7	10.1	8.57	7.63	6.84

20	40.0	37.6	35.0	31.4	28.4	23.8	19.3	15.5	12.4	10.9	9.24	8.26	7.43
21	41.4	38.9	36.3	32.7	29.6	24.9	20.3	16.3	13.2	11.6	9.92	8.90	8.03
22	42.8	40.3	37.7	33.9	30.8	26.0	21.3	17.2	14.0	12.3	10.6	9.54	8.64
23	44.2	41.6	39.0	35.2	32.0	27.1	22.3	18.1	14.8	13.1	11.3	10.2	9.26
24	45.6	43.0	40.3	36.4	33.2	28.2	23.3	19.0	15.7	13.8	12.0	10.9	9.89
25	46.9	44.3	41.6	37.7	34.4	29.3	24.3	19.9	16.5	14.6	12.7	11.5	10.5
26	48.3	45.6	42.9	38.9	35.6	30.4	25.3	20.8	17.3	15.4	13.4	12.2	11.2
27	49.6	47.0	44.1	40.1	36.7	31.5	26.3	21.7	18.1	16.2	14.1	12.9	11.8
28	51.0	48.3	45.4	41.3	37.9	32.6	27.3	22.7	18.9	16.9	14.8	13.6	12.5
29	52.3	49.6	46.7	42.6	39.1	33.7	28.3	23.6	19.8	17.7	15.6	14.3	13.1
30	53.7	50.9	48.0	43.8	40.3	34.8	29.3	24.5	20.6	18.5	16.3	15.0	13.8
40	66.8	63.7	61.1	55.8	51.8	45.6	39.3	33.7	29.1	26.5	23.7	22.2	20.7
50	79.5	76.2	73.5	67.5	63.2	56.3	49.3	42.9	37.7	34.8	31.5	29.7	28.0
60	92.0	88.4	85.7	79.1	74.4	67.0	59.3	52.3	46.5	43.2	39.4	37.5	35.5
70	104.2	100.4	97.7	90.5	85.5	77.6	69.3	61.7	55.3	51.7	47.4	45.4	43.3
80	116.3	112.3	109.6	101.9	96.6	88.1	79.3	71.1	64.3	60.4	55.6	53.5	51.2
90	128.3	124.1	121.5	113.1	107.6	98.6	89.3	80.6	73.3	69.1	63.9	61.8	59.2
100	140.2	135.8	133.2	124.3	118.5	109.1	99.3	90.1	82.4	77.9	72.2	70.1	67.3

Source: Catherine M. Thompson, *Table of percentage points of the χ^2 distribution*, Biometrika, Vol. 32 (1941), by permission of the author and publisher.

Since 3.92 is between 3.84 and 5.41, then $P(\chi^2$ of 3.92 or more) is between the corresponding probabilities of 5% and 2%. As we successively rewrite this table-reference operation in different words, we have

$$2\% < P(\chi^2 \text{ of } 3.92 \text{ or more}) < 5\% \quad \text{or}$$
$$2\% < P(32 \text{ or more heads out of 50 tosses by chance}) < 5\% \quad \text{or}$$
$$2\% < P(\text{coin is really fair}) < 5\%.$$

Since

$$P(\text{coin is really fair}) < 5\%,$$

then the probability of the complementary event that the coin is really unfair is greater than 95%. Symbolically,

$$P(\text{coin is really unfair}) > 95\%.$$

Hence we *reject* the null hypothesis H_o since the probability that a fair coin produces a sample outcome such as ours is so small; we accept the *alternate* hypothesis, symbolized here by H_1, and conclude that the coin is very likely biased.

Is this conclusion absolutely guaranteed? Have we *proven* that the coin is genuinely biased? No. It is entirely physically *possible* for a fair coin to give thirty-two heads in fifty tosses, but it is highly *improbable* that chance alone would cause so great a number of heads. Hence our conclusion must be stated in terms of probability. With a probability of .95, we conclude that the coin is biased, realizing that the probability of our conclusion being wrong still exists, though it is less than .05. That is, if the experiment of flipping a fair coin fifty times were repeated many thousands of times (see Problem 17), then the number of outcomes with thirty-two or more heads would tend to constitute less than 5% of all the sample outcomes. Notice that, because the value of χ^2 is less than 5.41, we cannot be 98% certain of our conclusion. We will agree with most statisticians that, unless otherwise stated, the probability that a specified population yields a known outcome must be less than or equal to the *significance level* of .05 before we reject the associated null hypothesis. We use the term *significance* in the sense of "not attributable to chance." Of course, if we can reject a null hypothesis with a 98%, 99%, or 99.5% degree of confidence, we will do so.

Model for hypothesis testing

We do another example to use as a model for future hypothesis-testing experiments. The five-step method we use here can be used in any similar situation.

Example 10 Over a long period of time the grades given by a group of instructors has averaged 12% A's, 18% B's, 40% C's, 18% D's, and 12% F's. A new instructor gives 22 A's, 34 B's, 66 C's, 16 D's, and 12 F's in teaching the same course. Determine at a 5% level of significance whether the new instructor is following the grading pattern set by the instructors before him.

Solution

1. Clearly state the null hypothesis in a complete sentence so that we can calculate the probability that a known sample is drawn from the specified population. Remember that we often state a null hypothesis solely for the purpose of rejecting it. Here, the null hypothesis is

   H_o: there is no significant difference between the old instructors' grading pattern (population) and the new instructor's grading pattern (sample).

   In other words, the difference between the two grading patterns is due to chance alone.
2. Determine the degrees of freedom. Since there are five possible grades as the outcome of the grading process,

   $$df = 5 - 1 = 4.$$

3. State the decision rule in terms of probabilities and significant values of χ^2. As usual, unless we are told otherwise, we use the 5% rule: If our probability of obtaining such a sample from a specified population is less than 5%, then reject H_o; otherwise, accept it. The significant table value of χ^2, 5%, $df = 4$ is 9.49; so if χ^2 from the calculations is greater than or equal to 9.49, we will reject the null hypothesis.
4. Calculate χ^2 by using the formula

   $$\chi^2 = \sum \frac{(f_o - f_h)^2}{f_h}.$$

Table 6.17

Grade	f_o	f_h	$f_o - f_h$	$(f_o - f_h)^2$	$(f_o - f_h)^2/f_h$
A	22	18	4	16	.89
B	34	27	7	49	1.81
C	66	60	6	36	.60
D	16	27	−11	121	4.48
F	12	18	−6	36	2.00
					$\chi^2 = 9.78$

An immediate question is "where did the values of f_h come from?" Recall that the null hypothesis is one of no difference in the grading patterns; the values of f_h are determined under this assumption. Since the new instructor gives a total of 150 grades, and we assume that he grades the same way as the old instructors, we would expect 12% of the 150 grades, or 18, to be A's. The hypothetical values of the B's,

C's, D's, and F's are determined in the same manner. Carrying out the arithmetic tells us that $\chi^2 = 9.78$.

5. State the conclusion. According to our previously accepted decision rule, we must reject the null hypothesis at the 5% level of significance since 9.78 is greater than 9.49. We conclude, with a 95% level of confidence, that this new instructor does grade differently (in this case, he gives higher grades) than the old instructors.

The test statistic

$$\chi^2 = \sum \frac{(f_o - f_h)^2}{f_h}$$

for an experiment with m possible outcomes may be written as

$$\chi^2 = \frac{(O_1 - E_1)^2}{E_1} + \frac{(O_2 - E_2)^2}{E_2} + \cdots + \frac{(O_m - E_m)^2}{E_m} \tag{A}$$

where

$$O_1, O_2, \ldots, O_m$$

and

$$E_1, E_2, \ldots, E_m$$

represent the values of the observed (f_o) and expected or hypothetical (f_h) frequencies, respectively. According to the summation notation we studied in Problem 11 of Problem set 1–2, we may write equation (A) as

$$\chi^2 = \sum_{i=1}^{m} \frac{(O_i - E_i)^2}{E_i}.$$

We may show that the previous equation is the same as

$$\chi^2 = \sum_{i=1}^{m} \frac{(O_i)^2}{E_i} - n \tag{B}$$

where n, the number of times the sampling occurs, is subtracted from the sum of the m fractions. Equation (B) is usually easier to use to evaluate χ^2 than equation (A), particularly if decimal values of E_i are involved. For example, in the first example of this section

$$\begin{aligned}
\chi^2 &= \sum_{i=1}^{2} \frac{(O_i)^2}{E_i} - n \\
&= \frac{(O_1)^2}{E_1} + \frac{(O_2)^2}{E_2} - 50 \\
&= \frac{(32)^2}{25} + \frac{(18)^2}{25} - 50 \\
&= \frac{1024}{25} + \frac{324}{25} - 50 \\
&= 40.96 + 12.96 - 50 \\
&= 3.92
\end{aligned}$$

Even though equation (A) is easier to apply in this one example, this situation is the exception rather than the rule.

As we close this section we note that we could have considered both examples in terms of analyzing differences. In the first example we analyzed the difference between the expected result of 25 and the actual result of 32, concluding that the difference of 7 heads was very likely caused by bias in the coin itself, rather than by mere chance or sampling fluctuations. It is well worth repeating that we talk only in terms of probabilities, never absolute certainties, as we seek to explain causes of known effects or sample characteristics.

Problem set 6–6

1. In the following cases use the χ^2 table to locate the probability that the sample results in a value of χ^2 which is greater than or equal to the indicated value.
   a. $df = 2,\ \chi^2 = 9.04$
   b. $df = 5,\ \chi^2 = 6.59$
   c. $df = 1,\ \chi^2 = 8.73$
   d. $df = 4,\ \chi^2 = 2.98$
   e. $df = 3,\ \chi^2 = 9.99$
2. Complete these tables, which show the results of hypothesis-testing experiments, and state in your conclusion the level of significance at which we can reject the null hypothesis.

a.

Outcome	f_o	f_h	$(f_o-f_h)^2/f_h$
A	13	10	
B	7	10	
C	11	12	
D	12	10	
E	9	10	

$df =$ ________ $\chi^2 =$ ________

Conclusion:

b.

Outcome	f_o	f_h	$(f_o-f_h)^2/f_h$
W	7	5	
X	8	5	
Y	3	5	
Z	7	10	

$df =$ ________ $\chi^2 =$ ________

Conclusion:

c.

Outcome	f_o	f_h	$(f_o-f_h)^2/f_h$
Q	12	8	
R	14	8	
S	7	8	
T	4	8	
U	3	8	

$df =$ ______ $\chi^2 =$ ______

Conclusion:

3. Answer Question 3 from the very first part of this chapter.
4. Claire Voyant says she has the power of extrasensory perception, or ESP. To test her claim her brother Flam forms a deck of ninety cards with one of three symbols on each card: a triangle, a circle, or a square. He then draws a card, asks her what figure is on the card, which she cannot see, records her answer as right or wrong, and goes on to the next card. At the end of the deck she has forty-six correct answers. What can Flam say about Claire's ESP ability? Use the five-step method outlined in the text.
5. It is late in the afternoon on a hot Sunday in June at Laconia, New Hampshire, where the 100-mile national motorcycle road race has just ended. The riders are discussing the performances of their bikes. Among the top fifty finishers the machines are distributed as follows: thirteen of Brand H, thirteen of Brand K, five of Brand N, four of Brand T, and fifteen of Brand Y. Each brand represents 20% of all the motorcycles entered in the race, and we assume that all other factors, such as rider ability, tires, and gasoline, are equal. The Brand T riders say that the real reason for the unequal distribution of brands was not because of the genuine slowness of their motorcycles, but because it was simply an "off" day. Nonsense, reply the Brand Y riders; Japanese machines are really better than British machines, and today's sample of the top fifty bikes proves this (Brands N and T are made in England). As an impartial mathematician you are to settle the argument between the two groups. Do so, and state your conclusion in a complete sentence.
6. Refer to the first table entry of Problem 3 in Problem set 5–5. What is the probability that the sample of 10,000 juries summarized there is selected from the theoretical population?
7. Repeat the instructions of Problem 6 applied to the second table entry of problem 3 in Problem set 5–5.
8. Repeat the instructions of Problem 6 applied to the third table entry of Problem 3 in Problem set 5–5.
9. Repeat the instructions of Problem 6 applied to the fourth table entry of Problem 3 in Problem set 5–5.

10. Repeat the instructions of Problem 6 applied to the fifth table entry of Problem 3 in Problem set 5–5.
11. Repeat the instructions of Problem 6 applied to the sixth table entry of Problem 3 in Problem set 5–5.
12. The draft lottery of 1 December 1969 assigned a number from 1 to 366, inclusive, to every eligible man by matching the number with the date of birth. For example, a man born on 3 July 1951 had a draft priority number of 115. It was generally agreed that men with numbers between 1 and 183 would be drafted in 1970 and those men with numbers between 184 and 366 would not be drafted. The following table gives the number of men born in the specified month who have draft numbers in the first category.

Jan	Feb	Mar	Apr	May	Jun	Jul	Aug	Sep	Oct	Nov	Dec
12	12	10	11	14	14	14	19	17	13	21	26

What is the probability that the sample of months and numbers of men likely to be drafted are *independent* of each other? What is the probability that a set of 183 equally-distributed birthdays will yield a sample such as this one? Can you obtain from a local Selective Service Board or your Registrar's Office the data needed to construct a table for the current calendar year? If so, set up an appropriate hypothesis and test it.

13. In this problem we examine the months of birth and months of death of 348 famous Americans, such as George Washington, Abraham Lincoln, and Franklin Roosevelt. The quotations and facts are from David P. Phillips' "Deathday and Birthday: An Unexpected Connection," which appears in Judith Tanur's excellent edited volume *Statistics: A Guide to the Unknown* (San Francisco: Holden-Day, Inc., 1972).

> In the movies and in certain kinds of romantic literature, we sometimes come across a deathbed scene in which a dying person holds onto life until some special event has occurred. For example, a mother might stave off death until her long-absent son returns from the wars. Do such feats of will occur in real life as well as in fiction? If some people really do postpone death, how much can the timing of death be influenced by psychological, social, or other identifiable factors? Can deaths from certain diseases be postponed longer than deaths from other diseases?
>
> In this essay we shall see how dying people react to one special event: their birthdays. We want to learn whether some people postpone their deaths until after their birthdays. If we compare the date of death with the date of birth for a large number of people, will we find fewer deaths than expected just before the birthday? If we do find a dip in deaths, we may conclude that some of these people are postponing death until after their birthdays.
>
> We shall use elementary statistical methods in approaching the problem. For example, the comparison of an actual number of events with the number that might be expected is one of these methods; others will be noted later. . . .

Months from birth month	−6	−5	−4	−3	−2	−1	0	+1	+2	+3	+4	+5
Number of deaths	24	31	20	23	34	16	26	36	37	41	26	34

The table shows that, for example, 16 out of the 348 people died in the month before the month in which their birthday would have occurred, whereas 36 people died in the month after their birthday had occurred. If we assume that a person makes no special effort to see one more birthday before his death or to die in a particular month, we would expect $348/12 = 29$ deaths in each month.

a. Suppose we hypothesize that there is no relation between the month of a famous person's death and the month in which he was born. Determine the value of χ^2 under this assumption by using the data in the table.
b. Determine the probability that the hypothesis from Part (a) of this problem could explain the table of months and deaths.

14. If we examine a large collection of cars at a shopping plaza, we might expect the number of dented fenders to be about the same for all four fenders. Suppose we have located 300 cars with a total of 560 dented fenders in the course of one day of investigation with this distribution.

Location of fender	Number
right front	165
right rear	170
left rear	105
left front	120

Does the investigation yield any results that the owner of a warehouse supplying automobile body parts would like to know?

15. Suppose a box of grass seed is labeled as having

30% Kentucky Blue (very good)
30% Perennial Ryegrass (cheaper)
25% Fescue (okay)
15% Annual Ryegrass (cheapest, and only lasts one year).

If a curious gardener planted 1,000 seeds with the following results after sprouting, does he have much reason to doubt the labeling?

280 Kentucky Blue
290 Perennial Ryegrass
270 Fescue
160 Annual Ryegrass

16. In the first example of this section we used $\chi^2 = 3.92$ as an intermediate number which was eventually used to find the probability of obtaining thirty-two or more heads in fifty tosses of a coin that we assumed was

fair. If we realize that tossing a fair coin independently fifty times while wanting to know the probability of thirty-two or more heads fulfills the conditions of the Binomial Law or Repeated Trials Formula, then we have

$$\begin{aligned} &P(32 \text{ or more heads}) \\ &\quad = P(32 \text{ heads}) + P(33 \text{ heads}) + \cdots + P(50 \text{ heads}) \\ &\quad = \binom{50}{32} \cdot (.5)^{32} \cdot (.5)^{18} + \binom{50}{33} \cdot (.5)^{33} \cdot (.5)^{17} + \cdots \\ &\quad\quad + \binom{50}{50} \cdot (.5)^{50} \cdot (.5)^{0} \\ &\quad = .0160 + .0087 + .0044 + .0020 + .0008 \\ &\quad\quad\quad + .0003 + .0001 + 12 \cdot .0000 \\ &\quad = .0323. \end{aligned}$$

The decimal values came from a table which is more extensive than our Table XI. Notice that the result of .0323 is between .0200 and .0500, as expected from the location of $\chi^2 = 3.92$ between $\chi^2 = 5.41$ and $\chi^2 = 3.84$. The probability of obtaining exactly twenty-five heads is

$$P(50, 25, .5) = \binom{50}{25} \cdot (.5)^{25} \cdot (.5)^{25} = .1123,$$

which is a fairly small number. This fact emphasizes that we are interested in the cause of the *deviation* from the expected number, even if the probability of the expected number itself is small.

a. If a dime is tossed twenty times and fourteen tails appear, then use the χ^2 method to calculate the probability that a fair coin would give fourteen or more tails in twenty tosses.
b. If a dime is tossed twenty times and fourteen tails appear, then use the Binomial Law and Table XI to calculate the exact probability that a fair coin would give fourteen or more tails in twenty tosses.
c. If a die is rolled thirty times and nine 4's appear, then use the χ^2 method to calculate the probability that a fair die would give nine or more 4's in thirty tosses.
d. If a die is rolled thirty times and nine 4's appear, then use the Binomial Law and the following table to calculate the exact probability that a fair die would give nine or more 4's in thirty tosses.

x	9	10	11	12	13	14
$P(30, x, .17)$	.0339	.0146	.0054	.0018	.0005	.0001

$P(30, x, .17)$ is .0000 for x greater than 15.

17. In Example 9 we saw that the probability of obtaining thirty-two or more heads in fifty tosses of a fair coin was between .0200 and .0500. Problem 16 showed that the exact probability is .0323; that is, if a fair coin is flipped fifty consecutive times, and if we repeat this experiment a very

large number of times, then the number of strings of fifty consecutive tosses in which thirty-two or more heads are observed is expected to be 3.23% of that very large number.

A computer has been programmed to "flip" a coin by using the random number feature of the BASIC computer language. The outcome of this random number generator is a number in the interval [.000001, .999999]. If the result of generating one random number was less than .500000, a head was recorded as the result of this "flip." This process occurred fifty times to obtain one experimental outcome of $\{0, 1, 2, \ldots, 50\}$ possible heads, and this experiment was repeated 10,000 times. Answer these questions with reference to the following table of results.

a. How many of the experiments resulted in thirty-two or more heads?
b. Theoretically, how many of the experiments should have resulted in thirty-two or more heads?
c. How many random numbers did the computer have to generate?
d. The computer used (a Hewlett-Packard Model 2000C) required 27 minutes of calculating time. How many strings of fifty consecutive tosses did it perform in each second of calculating time?
e. Graph the results shown in the table with the possible number of heads (between ten and forty, inclusive) on the horizontal axis and their corresponding frequencies on the vertical axis.

```
WHEN THE QUESTION MARK APPEARS, TYPE IN THE VALUES
  OF N1, THE NUMBER OF TIMES YOU WANT THE COMPUTER
  TO 'FLIP' A FAIR COIN IN A CONSECUTIVE STRING,
  AND N2, THE NUMBER OF SUCH STRINGS.
? 50,10000
```

POSSIBLE NUMBER OF HEADS	FREQUENCY OF OCCURRENCE
0	0
1	0
2	0
3	0
4	0
5	0
6	0
7	0
8	0
9	0
10	1
11	0
12	0
13	7
14	8
15	16
16	43
17	87
18	162
19	268
20	395

```
21          601
22          773
23          946
24          1034
25          1130
26          1083
27          943
28          799
29          646
30          415
31          292
32          164
33          111
34          27
35          29
36          14
37          5
38          1
39          0
40          0
41          0
42          0
43          0
44          0
45          0
46          0
47          0
48          0
49          0
50          0

END OF COIN TOSSING SIMULATION
```

18. Refer to Example 18 in Chapter 5. Use the results of the computer-simulated guessing and Table XI to calculate the resulting value of χ^2 under the hypothesis that Carl is guessing. Is there any significant difference between the hypothetical and the observed results from the 1000 tests?
19. Refer to Example 19 in Chapter 5. Use the results of the computer-simulated guessing and Table XI to calculate the resulting value of χ^2 under the hypothesis that Carl is guessing. Is there any significant difference between the hypothetical and the observed results from the 1000 tests?
20. Refer to Example 21 in Chapter 5. Use the results of the computer-simulated guessing and Table XI to calculate the resulting value of χ^2 under the hypothesis that Carl is guessing. Is there any significant difference between the hypothetical and the observed results from the 1000 tests?

Contingency tables

In the last section we learned how to test an hypothesis that there is no significant difference between population characteristics and sample characteristics by using the test statistic χ^2. Our next area of concern is testing hypotheses of independence between two classifications of the same universal set. Here we will make use of our earlier work with the cross-tabulation of a group of respondents, as illustrated in Problem 13 of Problem set 2–3. The method of this kind of hypothesis testing is presented by means of one example.

Computing the significant difference

Example 11 Students and faculty at Southern State College are given a questionnaire in order to measure and analyze campus attitudes. One sample statement is, "Students should be allowed to visit in one another's dormitory rooms between 7:00 p.m. and 2:00 a.m. on Friday and Saturday evenings." The possible responses are Agree, Disagree, and No Opinion. The results are cross-tabulated according to response and type of respondent as presented in the following *contingency table.*

Table 6.18

	Agree	Disagree	No opinion	Totals
Faculty	60	22	10	92
Student	160	108	40	308
Totals	220	130	50	400

Is there any significant difference between the responses of the faculty and the responses of the students? In other words, is there any relation between a respondent's academic status and his response?

Solution We must analyze our data in terms of percentages, since there are three times as many students as teachers in the respondent group. There will certainly be differences between the two groups. For example, 60 out of the 92 teachers, or $60/92 = 65.2\%$, agree, while 160 out of the 308 students, or $160/308 = 51.9\%$, agree. If we assume as our null hypothesis, that there is no difference between the responses of the two groups then we would expect 55% of the faculty, or $.55 \cdot 92 = 50.6$, to agree and 55% of the students, or $.55 \cdot 308 = 169.4$, to agree. We obtained 55% because $220/400 = 55\%$ of everyone agrees. Hence we have these two differences (60 and 50.6, 160 and 169.4) between observed and expected frequencies to account for along with four other such differences. We naturally expect χ^2 to appear somewhere in our analysis, since χ^2 is a measure of the differences between observed and expected frequencies.

We back up a little and formally state our null hypothesis.

H_o: There is no significant difference between the attitudes of the teachers and the attitudes of the students toward the statement about parietal hours.

That is, this set of responses comes from a population where the faculty and students have the same opinion. We accept the general statement that in a contingency table with r rows and c columns the number of degrees of freedom is

$$df = (r-1) \cdot (c-1).$$

Thus, in this example,

$$df = (2-1) \cdot (3-1) = 2.$$

As usual we will reject H_o if the probability that the sample comes from the specified population is less than or equal to .05. Our decision rule is to reject H_o if our calculations indicate that χ^2 is greater than the appropriate table value of 5.99 (see Table 6.16). If χ^2 is greater than 5.99, we would conclude that there very likely is a difference in the attitudes of teachers and students. We would also conclude that there very likely is some relationship or dependency of a response on the respondent if $\chi^2 > 5.99$.

In order to calculate χ^2 we need to know the values of f_o, which are given to us in the contingency table itself, and the corresponding values of f_h, which we must calculate on the basis of the null hypothesis. We make a table for the values of f_h, which make up the *expected frequency* table. Notice that the format of the table is identical to that of the observed frequency table, Table 6.18.

Table 6.19

	Agree	Disagree	No opinion	Totals
Faculty	50.6	29.9	11.5	92
Student	169.4	100.1	38.5	308
Totals	220	130	50	400

We calculated the first two entries of 50.6 and 169.4 earlier in this section, and the other four are similarly determined.

Our reasoning proceeds something like this as we seek the hypothetical "Disagree" numbers: since 130 out of all 400 respondents disagree, and since we assume that faculty and students disagree equally in terms of percentages, then 130/400 of the faculty, or 29.9, should disagree. Likewise, accepting H_o leads to the result that 130/400 of the 308 students, or 100.1, disagree. In working with decimal values, which is common in calculating the values of f_h, we round them off to the nearest tenth. Finally, since accepting the null hypothesis temporarily means 50/400 of everyone has no opinion, we would expect 50/400 of 92 or 11.5 faculty to have no opinion and 50/400 of 308 or 38.5 students to have no opinion. Here the row and column totals agree with those of the contingency table, although slight differences may exist in other problems because of rounding errors in working out the values of f_h.

$$\chi^2 = \frac{(60-50.6)^2}{50.6} + \frac{(22-29.9)^2}{29.9} + \frac{(10-11.5)^2}{11.5}$$

$$+ \frac{(160-169.4)^2}{169.4} + \frac{(108-100.1)^2}{100.1} + \frac{(40-38.5)^2}{38.5}$$

$$= 1.75 + 2.09 + .20 + .52 + .62 + .06$$

$$= 5.24.$$

Here our χ^2 of 5.24 is less than the 5.99 needed to reject the null hypothesis at the 5% level of significance; so we accept it at the 5% level. Note also that since $5.24 > 4.61$ (the χ^2 value for a 10% probability, as shown in Table 6.16), we can be 90% certain that there is a true difference between faculty and student attitudes. Our arbitrary 95% figure cannot be claimed.

Computing the hypothetical frequency

There is an easier way to compute the hypothetical frequencies, taking advantage of the fact that the row totals, column totals, and grand totals must be identical in both the observed and hypothetical frequency tables. It is true that the product of the row total and the column total in which a hypothetical frequency is located, when divided by the grand total, will have the quotient equal to the hypothetical frequency.

Example 12 Find the hypothetical number of faculty who have no opinion by using the method of the previous paragraph.

Solution The row total of the cell is 92, the column total of the cell is 50, and the grand total is 400. Thus

$$f_h = \frac{92 \cdot 50}{400} = 11.5,$$

as expected. The fraction

$$\frac{92 \cdot 50}{400}$$

is the same as

$$\frac{50}{400} \text{ of } 92.$$

We add, as a final note on contingency tables, that higher statistical theory tells us we should have all the values of f_h greater than 5 in any table used to calculate χ^2. If necessary, we should combine outcomes so that their hypothetical frequencies are 5 or more, thus maximizing the validity of results from using the test.

This section concludes our work with statistics. There is much more we could go into, especially in the areas of prediction equations and hypothesis testing, but time and space constraints prohibit these explorations. Many of us will have to take a separate course in statistics before we are finished with degree requirements or before we can completely design and analyze experiments in human behavior. In closing we note that this chapter has emphasized the uses of statistics, but it has not explained much about its potential and real misuses. Darrell Huff's delightful book *How to Lie With Statistics* (New York: W. W. Norton and Company, 1954) is valuable reading on this subject. Another book with dozens of interesting and valuable applications is the one mentioned in Problem 13 of the last section.

Many of the following problems require a large amount of arithmetic. It would be a good idea to have a calculator at hand with multiplication and division capabilities.

Problem set 6–7

1. Calculate χ^2 for the problem in this section if the number of faculty and students disagreeing were changed to 20 and 110, respectively. By comparing this answer with χ^2 from the original problem, notice the sensitivity of χ^2 to relatively small changes in the contingency table data.
2. A drug company claims that its product, Morning After, is effective in curing hangovers. A group of TGIF (Thank Goodness It's Friday) celebrants who overcelebrated are interviewed at 7:00 a.m. Saturday morning. Of the fifty-one people who took Morning After, nine were not cured two hours later while forty-two were. In the control group of forty-five who did not take the pill there were only sixteen clear heads two hours later. Organize this information into a contingency table and test the hypothesis that taking the drug and the curing of the hangover are independent, i.e., that there is no significant difference between the cures of the users and the nonusers. Carry out the test at the .05 and .01 significance levels.
3. A political scientist decides to see if a person's political party has any effect on his views toward adopting an amendment to the county charter. He selects a random sample from the two major parties and obtains these results from interviewing people in the sample.

	In favor	Opposed	No opinion
Democrat	84	80	36
Republican	66	50	24

What is your conclusion about the relationship between a person's political views and his views toward this particular issue? Begin the construction of the hypothetical frequency table with the row and column totals filled in. Then, use the formula

$$f_h = \frac{\text{row total} \times \text{column total}}{\text{grand total}}$$

to calculate the expected number of Democrats who are in favor and the expected number of Democrats who are opposed. Notice that the remaining four hypothetical frequencies can be determined by simple addition and subtraction of the row and column totals with the previously-obtained values of f_h. The formula for f_h has to be used only twice. How many degrees of freedom are there for this contingency table?

4. An educator wants to determine whether there is any relation between a recent graduate's grades and the ratings of his supervisors in electrical engineering. By reviewing transcripts and questionnaires returned by employers he obtains these tabulations.

			Ratings		
Grade average	Excellent	Good	Average	Poor	Fired
1.70–2.19	7	11	5	16	11
2.20–2.69	18	16	24	13	9
2.70–3.19	35	52	41	16	6
3.20–4.00	20	11	10	5	4

Carry through this problem as an exercise in χ^2 hypothesis testing.

5. A survey was conducted by a heating contractor to determine preferences in the method of heating a house. Five hundred homeowners were asked to declare how well they liked the method of heating they were using. The following results were obtained.

	Very good	Satisfactory	Adequate	Poor
Oil	56	65	51	28
Gas	79	105	46	20
Electric	15	30	3	2

Do these results indicate that there are different types of preferences for the different heating methods?

6. A sociologist surveyed a group of students graduated from high school. He asked them if their parents had attended college and if they were planning to attend college. The following results were obtained.

	Yes	No	Maybe
Only father attended	36	27	37
Only mother attended	6	8	11
Both attended	18	5	52

Is there a relationship between attendance by the parents and anticipated attendance by the children?

7. A final examination was taken by thirty-five male students and thirty-one female students. The instructor noted whether a student was sitting in the front rows or the back rows with the following results.

	Front rows	Back rows
Male students	12	23
Female students	24	7

Assume that the sixty-six students represent a random sample from the population of students taking the type of course involved. Set up an appropriate hypothesis and test it.

8. Suppose that we are interested in finding out if, at a certain college, the choice of a student's major field of interest is related to his class in college. We ask random samples of 100 freshmen, sophomores, juniors, and seniors at the college for their major field of concentration with the results given in the following table. Is there any relationship between a person's class and his major?

	Natural sciences	Social sciences	Humanities
Freshman	45	25	30
Sophomore	36	33	31
Junior	26	41	33
Senior	28	38	34

9. The data in the previous table can be condensed in the following manner.

	Other than humanities	Humanities
Freshman	70	30
Sophomore	69	31
Junior	67	33
Senior	66	34

10. In a pre-election poll which studied the influence of age on voter preference for two presidential candidates, the following results were obtained.

	Prefer candidate A	Prefer candidate B	Undecided
18–29	67	117	16
30–49	109	74	17
Over 49	118	64	18

Set up an appropriate hypothesis and test it.

11. A total of 751 business administration graduates was studied. The final grade point average in college of each of the 751 graduates was determined from the registrar's records. Each graduate who responded indicated the income bracket he was in and the responses were tallied into the following table.

	Under $7,000	$7,000 to $10,000	$10,000 to $20,000	Over $20,000
3.0 to 4.0	22	31	31	8
2.5 to 3.0	67	80	73	17
2.0 to 2.5	124	161	122	15

Source: Robert D. Mason, *Alumni Study* (Toledo, Ohio: University of Toledo, College of Business Administration, 1964) p.44.

Test the hypothesis of no relation between the graduate's grades and their incomes.

12. A student looking for an easy statistics teacher was told by the departmental office that all three statistics teachers passed the same proportions of students. The student did some research and came up with the following results.

	Number passed	Number failed
Professor A	42	8
Professor B	43	5
Professor C	38	14

Should the student believe what the office told him?

13. A subscription service stated that preferences for different national magazines were independent of geographical location. A survey was taken in which 300 persons randomly chosen from three areas were given a choice among three different magazines. Each person expressed his or her favorite. The following results were obtained.

	Magazine X	Magazine Y	Magazine Z
New England	75	50	175
Mid Atlantic	120	85	95
Southeastern	105	110	85

Do you agree with the subscription service's assertion?

14. The editor of *The Record* did research to determine whether social class has any effect on newspaper buying. He took a poll of 150 people from each of three social classes and ascertained whether they read his paper or its competitor, *The Herald*. Here are the results.

	The Record	*The Herald*
Lower	80	70
Middle	90	60
Upper	50	100

Test the hypothesis that choice of newspaper and social class are independent.

15. At Parker Junior College a sizable number of parents of incoming freshmen are given an attitude survey each summer where they respond to certain statements. One of the fifty statements is, "The college physician should dispense birth control pills to female students who request them." Each parent is classified on six criteria. They are:
    sex (father, mother),
    background (rural, urban, suburban),
    general philosophy (conservative, moderate, liberal),
    number of children in college (1, 2, 3),
    education (less than high school, high school, some college, college degree, master's degree, doctorate), and
    annual income (less than $7,000, $7,000 up to $10,000, $10,000 up to $15,000, $15,000 up to $20,000, $20,000 up to $25,000, more than $25,000).

Each of the six parts of this problem gives the observed frequencies after tabulating the questionnaires. Set up an appropriate hypothesis and test it for each table.

a.

Sex	Agree	Disagree	No opinion
Fathers	53	61	14
Mothers	47	113	20

b.

Background	Agree	Disagree	No opinion
Rural	17	22	8
Urban	6	16	0
Suburban	77	136	26

c.

Philosophy	Agree	Disagree	No opinion
Conservative	20	44	8
Moderate	54	114	23
Liberal	26	16	3

d.

Children in college	Agree	Disagree	No opinion
1 Child	74	136	27
2 Children	23	32	7
3 Children	3	6	0

e.

Education	Agree	Disagree	No opinion
Less than HS	3	6	2
HS graduate	26	57	14
Some college	27	51	12
College grad	30	36	2
Master's	9	20	3
Doctorate	5	4	1

f. The capital letter M is sometimes used as an abbreviation for 1,000. It is convenient to do so here.

Annual income	Agree	Disagree	No opinion
Less than $7M	0	4	0
$7M up to $10M	8	12	3
$10M up to $15M	10	47	9
$15M up to $20M	29	36	11
$20M up to $25M	19	36	6
More than $25M	34	39	5

Chapter 7

Systems of Linear Equations

Solving systems of 2×2 equations

In this section we review some familiar algebraic topics and learn others that are necessary for future applications of solving equations. We know, for example, that it is easy to determine whether the pair of numbers

$$x = 2, \quad y = 5 \text{ or } (2, 5)$$

is the solution to the system of equations

$$\begin{aligned} 3x - 4y &= -14 \\ 7x + 9y &= 59 \end{aligned}$$

merely by replacing x by 2 and y by 5 in both equations:

$$\begin{cases} 3 \cdot 2 - 4 \cdot 5 \stackrel{?}{=} -14 \\ 7 \cdot 2 + 9 \cdot 5 \stackrel{?}{=} 59 \end{cases}$$

$$\begin{cases} 6 - 20 \stackrel{?}{=} -14 \\ 14 + 45 \stackrel{?}{=} 59. \end{cases}$$

Since the answer to both of the last two questions is *yes*, then we know that (2, 5) is the solution of the original equations.

We can also show that (0, 3.5) is not a solution because substitution of its numerical values for x and y does not yield a true statement in *both* equations.

$$\begin{cases} 3 \cdot 0 - 4 \cdot 3.5 \stackrel{?}{=} -14 \\ 7 \cdot 0 + 9 \cdot 3.5 \stackrel{?}{=} 59 \end{cases}$$

$$\begin{cases} 0 - 14 \stackrel{?}{=} -14 \\ 0 + 31.5 \stackrel{?}{=} 59 \end{cases}$$

Even though the first question is answered *yes*, the *no* answer to the second one means that the pair (0, 3.5) is still not a solution for the original system.

Now that we know how to determine if a pair of numbers is a solution to a system of two equations in two unknowns, we will seek ways of finding those pairs of numbers (which pairs, as we will see later, may not even exist). There are many ways we could approach this topic, but we will begin with the traditional process of *elimination* and *substitution*.

Example 1 Solve the following system.

$$\begin{aligned} 5x - 2y &= 32 \\ 3x + 7y &= -30 \end{aligned}$$

Solution We arbitrarily decide to eliminate the y. We multiply the top equation by 7 and the bottom equation by 2.

$$\begin{cases} 7 \cdot (5x - 2y = 32) \\ 2 \cdot (3x + 7y = -30) \end{cases}$$

$$\begin{cases} 35x - 14y = 224 \\ 6x + 14y = -60 \end{cases}$$

This last system of equations has the same solution, or is *equivalent* to, the original system. We justify our work by noting that it is permissible to multiply or divide both sides of an equation by a nonzero number. Next we add the two equations together, an operation whose justification is the rule "equals added to equals give equals."

$$\begin{array}{r} (35x - 14y = 224) \\ +\,(\ 6x + 14y = -60) \\ \hline (41x + \ 0y = 164) \end{array} \qquad \text{(A)}$$

Notice that y has been *eliminated* in the derivation of equation (A). We now replace the first equation of our original pair by (A) after dividing it through by 41.

$$\left.\begin{aligned} 1x + 0y &= 4 \\ 3x + 7y &= -30 \end{aligned}\right\} \qquad \text{(B)}$$

The top equation of system (B) clearly has $x = 4$ as its solution, so we *substitute* 4 for x in the second equation.

$$\begin{aligned} 3x + 7y &= -30 \\ 3 \cdot 4 + 7y &= -30 \\ 12 + 7y &= -30 \\ 7y &= -42 \\ y &= -6 \end{aligned}$$

Hence, $x = 4$ and $y = -6$, or $(4, -6)$, is our solution. If we check our work by replacing x by 4 and y by -6 in the original system, we will obtain two true statements.

Example 2 Solve the system

$$\begin{aligned} 5x - 2y &= 32 \\ 3x + 7y &= -30 \end{aligned}$$

by converting it into *triangular form.*

Solution To convert a system to triangular form means to rewrite it in such a way that when expressed as an equivalent system of simultaneous equations the nonzero coefficients would take the form of a triangle. The zero coefficients would not be included within this triangle. Thus, if we form the triangle with vertices at the coefficients 5, 7, and 3, we wish to rewrite the system so that the resulting equivalent system has all the other coefficients of the variables outside this triangle—in this case only the -2 —equal to 0.

Table 7.1

Now	Goal—triangular form
$5x - 2y = 32$	$\_\_\cdot x + \ 0\cdot y = \_\_$
$3x + 7y = -30$	$\_\_\cdot x + \_\_\cdot y = \_\_$

In working with a system of two equations in two unknowns the composite operation of *replacing any equation with the sum or difference of any nonzero multiple of itself and a multiple of the other equation* results in an equivalent system. What we actually did in Example 2 was to replace

$$5x - 2y = 32$$

by the sum of

$$7 \cdot (5x - 2y = 32)$$

and

$$2 \cdot (3x + 7y = -30),$$

which resulted in

$$\left.\begin{aligned} 41x + 0y &= 164 \\ 3x + 7y &= -30 \end{aligned}\right\} \text{ Triangular Form}$$

or

$$\begin{aligned} 1x + 0y &= 4 \\ 3x + 7y &= -30. \end{aligned}$$

Thus we return to our system (B) in Example 1. Notice that now we can directly solve the first equation for x and immediately substitute this result for x in the second equation to obtain y. The second equation remained unchanged, but this is not always the case in using this triangular form approach. We may easily verify that $(4, -6)$ is the solution of the original system by direct substitution into both equations. This system is known as a *determinative* system since it *determines* exactly one pair of equation-satisfying numbers. Again, we justify our manipulations by recalling that it is permissible to add equal numbers or expressions to both sides of an equation and to multiply both sides of an equation by the same nonzero number.

There are certain advantages to using the triangular form approach of solving two equations in two unknowns, as we will see in the next section of this chapter.

Example 3 Use the triangular form approach to solve the following system.

$$\begin{aligned} x - 7 &= -2y \\ 16 - 6y &= 3x \end{aligned}$$

Solution We first rewrite the equations in *standard form*, i.e., $ax + by = c$.

$$\begin{aligned} x + 2y &= 7 \\ 3x + 6y &= 16 \end{aligned}$$

If we replace the top equation by the sum of -3 times itself and 1 times the bottom equation, we will have a 0 where the 2 is presently located.

$$\left.\begin{array}{r} -3(x+2y=7) \\ +\ \underline{1(3x+6y=16)} \end{array}\right\} \to 0x + 0y = -5$$

Thus

$$\begin{aligned} 0x + 0y &= -5 \\ 3x + 6y &= 16. \end{aligned}$$

The top equation says

$$0x = -5,$$

which has no solutions since $0 \cdot (\text{every real number})$ is always 0, never -5. This means that the original equivalent system has no solutions either. It is an example of an *inconsistent* system.

Example 4 Solve the system

$$\begin{aligned} 2x + 5y &= -13 \\ -6x - 15y &= 39. \end{aligned}$$

Solution We can replace 5, the coefficient of y in the first equation, by 0 if we add the bottom equation to 3 times the top equation. This replacement is the initial step in solving the system of equations.

$$\left.\begin{array}{r} 3 \cdot (2x+5y = -13) \\ +\ \underline{1 \cdot (-6x-15y = 39)} \end{array}\right\} \to 0x + 0y = 0$$

Thus

$$\begin{aligned} 0x + 0y &= 0 \\ -6x - 15y &= 39. \end{aligned}$$

The top equation says

$$0 \cdot x = 0,$$

which is true for every real number x. Since this equation has an infinite number of solutions, the original equivalent system has an infinite number of solutions as well. This system is known as a *dependent* system because the second equation in the original system is really just the first equation multiplied through by -3; thus the two equations are really equivalent. One equation in two unknowns with nonzero coefficients of x and y will always have an infinite solution set. Note that while the solution set is an infinite set it does not include *all* ordered pairs of real numbers. The pairs $(0, -13/5)$ and $(-24, 7)$ are in the solution set, but $(4, 9)$ and $(1.7, -6.8)$ are not. Once x has been picked, y is uniquely determined by the equation

$$y = \frac{-6x - 39}{15}.$$

That is, the solution set is the infinite set whose members are pairs of numbers of the form

$$\left(x, \frac{-6x - 39}{15}\right) \quad \text{or} \quad \left(x, \frac{-2x - 13}{5}\right).$$

The previous three examples illustrate the three possible situations that may exist in solving systems of linear equations. A linear equation is so called because any equation of the form

$$ax + by = c, \qquad b \neq 0$$

can be rewritten as

$$y = (-a/b) \cdot x + c/b,$$

and from Chapter 3 we know that its graph is a straight line. In general, a linear equation in n unknowns $x_1, x_2, x_3, \ldots, x_n$ has the form

$$a_1 \cdot x_1 + a_2 \cdot x_2 + a_3 \cdot x_3 + \cdots + a_n \cdot x_n = k,$$

where $a_1, a_2, a_3, \ldots, a_n$, and k are constants and the exponent of the unknowns is always 1.

It is worthwhile to examine the geometric counterparts of the previous three examples of linear equations. The first system, Example 2, had exactly one solution and the graph of the two equations is two lines intersecting in just one point. The second system, Example 3, had no solutions, and the two lines in its graph do not intersect (parallel lines). The third system, Example 4, had an infinite number of solutions and the two lines in its graph lie on top of each other, so their intersection is the line itself containing an infinite number of points. We summarize this paragraph in a table.

Table 7.2

Number of solutions	Name of system	Geometric counterpart
0	inconsistent	2 parallel lines
1	determinative	2 intersecting lines
infinite	dependent	2 concurrent lines

Problems 16 and 17 involve additional work with Examples 3 and 4 and their graphs.

Augmented matrix approach

In working with the previous systems of linear equations we could have shortened the amount of time and writing required if we had worked only with the coefficients and constants after dropping the x's, y's, and operation signs as being "excess baggage." An example shows us how.

Example 5 Solve the system

$$\begin{aligned} 1x + 2y &= 9 \\ 4x - 7y &= -39. \end{aligned}$$

Solution We first rewrite the system in augmented matrix format to obtain

$$\left[\begin{array}{cc|c} 1 & 2 & 9 \\ 4 & -7 & -39 \end{array}\right].$$

A *matrix* is simply a rectangular array of real numbers. Here we have two (horizontal) rows and three (vertical) columns in this matrix. It is *augmented* because the constants 9 and -39 *augment* (supplement) the coefficients 1, 2, 4, and -7 in the matrix. Here we have dropped the x's, y's, operation signs, and equal signs because the locations of the numerals let us replace the missing symbols whenever required. To solve the system by converting it to an equivalent and simpler one we rewrite the coefficients to the left of the vertical line into triangular form using the same methods of equation multiplication and addition as before.

$$\left[\begin{array}{cc|c} 1 & 2 & 9 \\ 4 & -7 & -39 \end{array}\right] \rightarrow \left[\begin{array}{cc|c} 7 & 14 & 63 \\ 8 & -14 & -78 \end{array}\right]$$

$$\rightarrow \left[\begin{array}{cc|c} 15 & 0 & -15 \\ 4 & -7 & -39 \end{array}\right] \rightarrow \begin{aligned} 15x &= -15 \\ 4x - 7y &= -39 \end{aligned}$$

The solution is easily found to be $(-1, 5)$ (right?).

If we solve the two now familiar systems

$$\begin{aligned} x + 2y &= 7 \\ 3x + 6y &= 16 \end{aligned} \quad \text{and} \quad \begin{aligned} 2x + 5y &= -13 \\ -6x - 15y &= 39 \end{aligned}$$

according to the augmented matrix format we obtain, respectively,

$$\left[\begin{array}{cc|c} 0 & 0 & -5 \\ 3 & 6 & 16 \end{array}\right] \quad \text{and} \quad \left[\begin{array}{cc|c} 0 & 0 & 0 \\ -6 & -15 & 39 \end{array}\right],$$

which, when written out with the variables and equal signs, give identical results of inconsistency and dependency as before.

This augmented matrix approach may seem a little different and awkward but it is very helpful in the next section for solving systems of more than two unknowns.

Problem set 7–1

Solve each of the first nine problems by using the augmented matrix approach. If you think the system is determinative, then check your answer by substitution.

1. $x + y = 4$
   $2x - y = 2$
2. $3x + 2y = 10$
   $x + y = 0$
3. $-10 - 4x = 3y$
   $4x + 5y = 1$
4. $4x + 5y = 1$
   $x = -y$
5. $x + 2y = 7$
   $3x + 6y = 21$
6. $2x = 6 - 3y$
   $3y - 6 = -2x$
7. $(1/3)x = (1/2)y$
   $2x - 3y = 6$
8. $x + y = 0$
   $2x - 3y = 0$
9. $.2x - .56y = 1$
   $.01x - .25y = .5$
10. The sum of two numbers is 10 and their difference is 40. Find the two numbers.
11. The combined ages of John and Mary are thirty-six years. Three times Mary's age exceeds twice John's age by eight years. Find the age of each.
12. Use a system of two equations in two unknowns to find what quantities of coffee worth 75¢ a pound and $1.15 a pound are needed to produce a blend worth 85¢ a pound.
13. The sum of the digits of a two-digit integer is 13. The number formed by reversing the digits is 9 greater than the original number. Find the number.
14. Solve these four systems graphically by first carefully constructing the graphs of each equation and then estimating the coordinates of the points of intersection of the graphs. Then solve the systems algebraically.

a. $4x - y = 10$
$-2x + 5y = 4$

b. $x - 3.5y = 4.5$
$1.5x - 5.25y = 3.75$

c. $1.4x - y = .6$
$5y - 7x = -2$

d. $3x - 6y = 24$
$7x + y = -19$

15. Find a value of c for which the system

$$\begin{aligned} 2x - 7y &= 13 \\ 4x - 14y &= c \end{aligned}$$

is
a. dependent.
b. inconsistent.

16. Construct the graphs of the two linear equations in Example 3.
17. Construct the graphs of the two linear equations in Example 4.

Solving systems of three and four linear equations

In this section we generalize our previous work with the augmented matrix approach to solving systems of linear equations. We will discuss the cases of solving three linear equations in three unknowns and of solving four linear equations in four unknowns. We proceed with examples.

Solving 3×3 equations

Example 6 Solve the system of equations

$$\begin{aligned} x - 6y + 2z &= 5 \\ 2x - 3y + z &= 4 \\ 3x + 4y - z &= -2. \end{aligned}$$

Solution We begin by writing the system in augmented matrix format.

$$\left[\begin{array}{ccc|c} 1 & -6 & 2 & 5 \\ 2 & -3 & 1 & 4 \\ 3 & 4 & -1 & -2 \end{array}\right]$$

The three vertices of our triangle in the coefficient section of the matrix are 1, -1, and 3. As before, we wish to replace all the coefficients which are excluded from the triangle's vertices, its sides, or its interior—here these excluded coefficients are -6, 2, and 1—by 0 in an equivalent system. We can then easily obtain the desired triplet of numbers (x, y, z) which satisfies all three equations. Which numbers shall we work on first? It is often easiest to work first on those entries in the right-most column before working on the column of constants. Here those entries are 2 and 1. Replacing row 1 with the sum of itself and 2 times row 3, and replacing row 2 with the sum of itself and row 3 yields

$$\left[\begin{array}{rrr|r} 7 & 2 & 0 & 1 \\ 5 & 1 & 0 & 2 \\ 3 & 4 & -1 & -2 \end{array}\right],$$

which contains coefficients of 0 in two desired locations. There are several other approaches we could have used such as replacing row 1 by the sum of itself and -2 times row 2 which would introduce a coefficient of 0 into row 1. It is well worth our while to be fastidious about the arithmetic since one small error can invalidate all the work. We proceed next by replacing the new first row by the sum of itself and -2 times row 2 in order to replace the 2 in the first row with a 0. Notice that the upper right-hand 0 remains 0; this is why we worked with obtaining 0's in the right-most column in the previous step.

$$\left[\begin{array}{rrr|r} -3 & 0 & 0 & -3 \\ 5 & 1 & 0 & 2 \\ 3 & 4 & -1 & -2 \end{array}\right]$$

Now we are almost finished. Converting from augmented matrix format back to the original format gives

$$\begin{aligned} -3x &= -3 \\ 5x + y &= 2 \\ 3x + 4y - z &= -2. \end{aligned}$$

Then

$$-3x = -3,$$

or

$$x = 1.$$

Substitution of $x = 1$ into the second equation says

$$5 \cdot 1 + y = 2$$

or

$$y = -3.$$

Substitution of $x = 1$ and $y = -3$ into the third equation says

$$3 \cdot 1 + 4 \cdot -3 - z = -2$$

or

$$z = -7 \text{ (right?)}.$$

We may verify that $x = 1$, $y = -3$, $z = -7$ or $(1, -3, -7)$ is the solution of the system by substituting these values into all three original equations to obtain a true statement in each case.

If this *row-reduction* method (so called because we *reduce* certain elements in certain *rows* to 0) had given us, for example,

$$0 \cdot x = 7 \quad \text{or} \quad 0 \cdot x = 0,$$

then, just as before, we would conclude that the systems were inconsistent and dependent, respectively.

Example 7 Solve the system of equations

$$\begin{aligned} 2x - 3y + z &= 4 \\ x - 4y - z &= 3 \\ x - 9y - 4z &= 5. \end{aligned}$$

Solution We write the system in augmented matrix format.

$$\left[\begin{array}{ccc|c} 2 & -3 & 1 & 4 \\ 1 & -4 & -1 & 3 \\ 1 & -9 & -4 & 5 \end{array}\right]$$

Now, we replace the first row by the sum of itself and the second row, and we also replace the second row by the sum of itself and the first row.

$$\left[\begin{array}{ccc|c} 3 & -7 & 0 & 7 \\ 3 & -7 & 0 & 7 \\ 1 & -9 & -4 & 5 \end{array}\right]$$

We replace the first row by the difference between itself and the second row,

$$\left[\begin{array}{ccc|c} 0 & 0 & 0 & 0 \\ 3 & -7 & 0 & 7 \\ 1 & -9 & -4 & 5 \end{array}\right],$$

and rewrite in the traditional format.

$$\begin{aligned} 0x &= 0 \\ 3x - 7y &= 7 \qquad \text{(A)} \\ 1x - 9y - 4z &= 5 \qquad \text{(B)} \end{aligned}$$

The first equation, $0x = 0$, clearly has an infinite number of solutions. Thus this system and the original equivalent also have an infinite number of solutions. However, not all triplets of numbers (x, y, z) are solutions of this dependent system. If, for example, we arbitrarily let $x = 14$, then the first equation becomes $0 \cdot 14 = 0$, which is a true statement. Now we substitute $x = 14$ into equation (A).

$$\begin{aligned} 3 \cdot 14 - 7y &= 7 \\ y &= 3 \cdot 14/7 - 1 \; (=5) \end{aligned}$$

When we substitute $x = 14$ and $y = (3 \cdot 14/7 - 1)$ into equation (B) we obtain

$$\begin{aligned} 1 \cdot 14 - 9 \cdot (3 \cdot 14/7 - 1) - 4z &= 5 \\ z &= 1 - 5 \cdot 14/7 \; (= -9). \end{aligned}$$

Thus the triplet $(14, 5, -9)$ is a solution. Generalizing, we may let x equal any real number, so that when equation (A) is solved for y in terms of x we obtain

$$y = 3x/7 - 1 \quad \text{(right?)}.$$

If we replace y by the above expression in equation (B) and solve the resulting equation for z in terms of x, we obtain

$$z = 1 - 5x/7 \quad \text{(right?)}.$$

We conclude that the solution set is

$$\{(x, y, z) \mid x \in R,\ y = 3x/7 - 1,\ z = 1 - 5x/7\}.$$

Solving 4×4 equations

A 4×4 system of linear equations is handled in exactly the same fashion.

Example 8 Solve the system of equations

$$\begin{aligned} 2x - y &= w - z - 1 \\ x - 2z + 5 &= -3y \\ 3x - 2y + 4w &= 1 \\ y - x + 6 &= 3z + w. \end{aligned}$$

Solution We first rewrite each equation in the system according to the format

$$ax + by + cz + dw = k$$

where the coefficients of the variables x, y, z, and w are a, b, c, and d, respectively, and where k is the constant term of the equation. Then

$$\begin{aligned} 2x + -1y + 1z - 1w &= -1 \\ 1x + 3y + -2z + 0w &= -5 \\ 3x + -2y + 0z + 4w &= 1 \\ -1x + 1y + -3z + -1w &= -6. \end{aligned}$$

In augmented matrix format,

$$\left[\begin{array}{rrrr|r} 2 & -1 & 1 & -1 & -1 \\ 1 & 3 & -2 & 0 & -5 \\ 3 & -2 & 0 & 4 & 1 \\ -1 & 1 & -3 & -1 & -6 \end{array}\right].$$

The six coefficients outside the dotted triangle are the ones we want to replace by zeros, and we concentrate on the fourth column first. Here we will add $-1 \cdot (\text{row } 4)$ to the first row and $4 \cdot (\text{row } 4)$ to the third row. We do nothing to the second row, as it already has a 0 in column 4.

$$\left[\begin{array}{rrrr|r} 3 & -2 & 4 & 0 & 5 \\ 1 & 3 & -2 & 0 & -5 \\ -1 & 2 & -12 & 0 & -23 \\ -1 & 1 & -3 & -1 & -6 \end{array}\right]$$

In order to have the 4 and the -2 in the third column replaced by 0 we now replace the first row by the sum of itself and $2 \cdot (\text{row } 2)$ and replace the present second row by the sum of -6 times itself and the third row.

$$\left[\begin{array}{rrrr|r} 5 & 4 & 0 & 0 & -5 \\ -7 & -16 & 0 & 0 & 7 \\ -1 & 2 & -12 & 0 & -23 \\ -1 & 1 & -3 & -1 & -6 \end{array}\right]$$

Finally, we replace the first row by the sum of 4 times itself and the present second row.

$$\left[\begin{array}{rrrr|r} 13 & 0 & 0 & 0 & -13 \\ -7 & -16 & 0 & 0 & 7 \\ -1 & 2 & -12 & 0 & -23 \\ -1 & 1 & -3 & -1 & -6 \end{array}\right]$$

In conventional format we now have the following system, which is equivalent to the original one.

$$\begin{aligned} 13x &= -13 \\ -7x - 16y &= 7 \\ -x + 2y - 12z &= -23 \\ -x + y - 3z - w &= -6. \end{aligned}$$

Then

$$13x = -13 \quad \text{or} \quad x = -1.$$

Substitution of $x = -1$ into the second equation yields

$$-7 \cdot -1 - 16y = 7 \quad \text{or} \quad y = 0.$$

Substitution of $x = -1$ and $y = 0$ into the third equation yields

$$-(-1) + 2 \cdot 0 - 12z = -23 \quad \text{or} \quad z = 2.$$

We finally substitute $x = -1$, $y = 0$, and $z = 2$ into the fourth equation so that

$$-(-1) + 0 - 3 \cdot 2 - w = -6 \quad \text{or} \quad w = 1.$$

We finally conclude that

$$x = -1, \quad y = 0, \quad z = 2, \quad \text{and} \quad w = 1.$$

Be sure to carry through the substitutions mentioned above.

In more formal terms we say that $\{(-1, 0, 2, 1)\}$ is the solution set of the original system. Of course, it is a good idea to check this solution set by substituting the determined values of all four variables into the original equations to be sure the resulting statements are true ones.

Keep in mind that neatness and accuracy are important; do not try to work with more than one column at a time; and remember that there are always several possible choices for ways to reduce appropriate row entries to 0. Remember that there is nothing wrong with having fractions as answers, and the letters $x_1, x_2, x_3, \ldots$ are sometimes used for variables instead of $x, y, z, \ldots$.

Problem set 7–2

1. Solve the following systems by using the augmented matrix, row-reduction method. If a system is not determinative, then indicate whether it is inconsistent or dependent.

a. $x + y + z = 7$
$2x + 3y - z = 4$
$x + y - z = 3$

b. $x_1 + x_2 + x_3 = 7$
$2x_1 + 3x_2 - x_3 = 4$
$x_1 + x_2 - x_3 = 3$

c. $2x - 3y - 3z = 4$
$3x - 2y - z = -4$
$5x - y + z = -5$

d. $2x - y + z = 1$
$x + 2y - z = 3$
$x + 7y - 4z = 2$

e. $0.1x + 1.7y + 0.1z = 3.2$
$0.7x + 0.3y + 0.1z = 2.1$
$0.5x - 0.36 + 0.2z = 2.4$

f. $2x - y + z = 1$
$x + 2y - z = 3$
$x + 7y - 4z = 8$

g. $3x - 8y + 5z = 78$
$2x + 7y - z = -32$
$5x - 3y - 6z = -26$

h. $7y - 47 = z + 4x$
$45 - 5y + z = -6x$
$x - 3y - 13 = 4z$

2. Solve the following systems by using the augmented matrix, row-reduction method. If a system is not determinative, then indicate whether it is inconsistent or dependent.

a. $2x - 6y + 2z + 10w = -7$
$2x + 4y - z - 5w = 6$
$x - 3y + z + 5w = -3$
$2x + 2y - z + 3w = 4$

b. $x + w = 3$
$y + z = -1$
$x + z = 1$
$y + w = 1$

c. $x + y = 3$
$y + z = 5$
$z + w = 7$
$x - y + z - w = -2$

d. $x + y = 2$
$y + z = -1$
$z + w = 0$
$x - y + z - w = 0$

e. $3x_1 + x_2 - x_3 + x_4 = 1$
$x_1 - 2x_2 + x_3 = 0$
$x_1 - 5x_2 = 0$
$x_2 - 5x_3 + x_4 = 1$

f. $x + 2z = 1$
$3x - y = 3$
$x + 2y + z = 4$

3. Write the following systems in augmented matrix format, but do not solve them.

a. $x_1 - x_5 = 1$
$x_2 + x_3 + x_4 = 6$
$3x_3 - x_4 + x_5 = 7$
$x_2 + x_5 = 3$
$x_1 + x_3 + x_4 - x_5 = -6$

b. $x - z = 1$
$y - w = 1$
$w - x = 1$
$z - y = 1$

4. Find the value of k such that the following system is not determinative.

$$kx + y - z = 1$$
$$x - 2y + z = 4$$
$$3x - 3y + z = 1$$

5. Three brands of dog food, A, B, and C, are offered by a wholesaler. Brand A is advertised as containing 20% protein; brand B 28% protein; and brand C 30% protein. Find out how many cases of each kind can be purchased under these restrictions: 224 cases must be purchased; the buyer wants his 224-case lot to average 25% protein; and the amount of A must be twice that of C.
6. A factory supplies three wholesalers who together demand all the output. The three wholesalers, X, Y, and Z, have asked for the following for a given month: X wants as much as Y and Z together, and Y's order is for 10 percent more than Z's. If the production of the factory is 126 units, then how should the manufacturer divide the month's production to supply the proportionate amounts to each wholesaler?
7. You are told that a bag of thirty coins contains nickels, dimes, and quarters amounting to three dollars, and that there are twice as many nickels as there are dimes. Should you believe it?
8. The sum of the digits of a three-digit number is 14 and the middle digit is the sum of the other two digits. If the last two digits are interchanged, the number obtained is 27 less than the original number.
   a. Find the number.
   b. Can you solve the problem if all the information is the same, except that the number obtained when the last two digits are interchanged is 72 less than the original number?
9. Four high schools, South, East, North, and West, have a total enrollment of 1,000 students. South reports that 10% of the students receive A's, 25% B's, and 50% C's; East reports 15% A's, 25% B's, and 55% C's; North reports 25% A's, 15% B's, and 35% C's; West reports 15% A's, 20% B's, and 40% C's. Is this information consistent with the fact that of the total student enrollment in all four schools 15% receive A's, 25% B's, and 50% C's?
10. An automobile factory makes both six- and eight-cylinder cars that are respectively priced at $2,800 and $3,100. Each cylinder of an automobile engine has one piston. In a one-week period the factory used 2,560 pistons and produced cars whose total value is $1,049,000. How many cars of each type did the plant produce?
11. Solve Problem 1, part (g) by *complete* row reduction. This means to reduce the appropriate row elements to 0 until the following equivalent format is reached:

$$\left[\begin{array}{ccc|c} 1 & 0 & 0 & \text{constant 1} \\ 0 & 1 & 0 & \text{constant 2} \\ 0 & 0 & 1 & \text{constant 3} \end{array}\right]$$

From this format we can immediately conclude, without any substitution and evaluation, that

$$x = \text{constant 1}, \quad y = \text{constant 2}, \quad z = \text{constant 3}.$$

You may achieve complete row reduction by continuing to work with the triangular form solution of Problem 1, part (g) until an equivalent system contains three more 0's as coefficients. Complete row reduction is also known as the Gaussian elimination method. Carl F. Gauss (1777–1855) was a famous German mathematician who made many contributions to mathematics.

12. Solve part (h) of Problem 1 by complete row reduction.
13. Solve part (d) of Problem 1 by complete row reduction.
14. Solve part (f) of Problem 1 by complete row reduction.

Determinants

We saw in the last section that the augmented matrix approach to solving systems of linear equations was a reasonably straightforward and easy one. However, there are certain combinations of numbers where the arithmetic can get rather difficult.

Example 9 Solve for x and y.

$$\begin{aligned} 2.67x + 13.4y &= -16.2 \\ 3.05x - 117y &= 7 \end{aligned}$$

Solution We look for a way to replace the $13.4y$ by $0y$ through allowable operations involving 13.4 and -117. The only way to proceed is to multiply the top equation by 117 and the bottom one by 13.4; but a lot of difficult multiplication is required, so we look for another way. Fortunately, there is a way to solve equations such as these by merely substituting numbers into a rather convenient formula. We will return to this example after we learn a little about *determinants.*

Usefulness of determinants

Definition The *determinant* of a 2×2 matrix of real numbers

$$M = \begin{bmatrix} & b \\ c & d \end{bmatrix}$$

is the *number* $a \cdot d - c \cdot b$, and this number is symbolized by

$$\begin{vmatrix} a & b \\ c & d \end{vmatrix}, \ \det M,\ D_M, \text{ or } \Delta_M.$$

Example 10 If

$$M = \begin{bmatrix} 7 & 3 \\ 5 & 8 \end{bmatrix},$$

then

$$\begin{vmatrix} 7 & 3 \\ 5 & 8 \end{vmatrix} = 7 \cdot 8 - 5 \cdot 3 = 56 - 15 = 41 = \det M = D_M = \Delta_M.$$

Example 11 If

$$M = \begin{bmatrix} 7 & -4 \\ 4 & -3 \end{bmatrix},$$

then

$$\begin{vmatrix} 7 & -4 \\ 4 & -3 \end{vmatrix} = 7 \cdot -3 - 4 \cdot -4 = -21 - -16 = -21 + 16 = -5.$$

We must be *very* careful as we perform the arithmetic. Again we emphasize that the determinant of a 2×2 matrix is a *real number*.

Example 12 Find a matrix that has 34 as its determinant.

Solution There are many ways we can rewrite 34 as the difference of two products: $4 \cdot 7 - 3 \cdot -2$, $4 \cdot 7 - -3 \cdot 2$, $7 \cdot 4 - 3 \cdot -2$, $-4 \cdot -7 - 2 \cdot -3$, $15 \cdot 2 - -2 \cdot 2$, and $8 \cdot 5 - 2 \cdot 3$ are some of these many ways. If we rearrange each quadruplet of numbers in upper-left-to-lower-right, lower-left-to-upper-right format we have these matrices, respectively, as correct results.

$$\begin{bmatrix} 4 & -2 \\ 3 & 7 \end{bmatrix} \quad \begin{bmatrix} 4 & 2 \\ -3 & 7 \end{bmatrix} \quad \begin{bmatrix} 7 & -2 \\ 3 & 4 \end{bmatrix}$$

$$\begin{bmatrix} -4 & -3 \\ 2 & -7 \end{bmatrix} \quad \begin{bmatrix} 15 & 2 \\ -2 & 2 \end{bmatrix} \quad \begin{bmatrix} 8 & 3 \\ 2 & 5 \end{bmatrix}$$

We now proceed to solve a system of two linear equations in two unknowns in a way which reveals the usefulness of determinants.

Example 13 Solve the system

$$\begin{aligned} 7x + 4y &= 34 \\ 3x + 8y &= 46. \end{aligned}$$

Solution We could use the familiar augmented matrix, row-reduction method in a straightforward fashion to obtain (2, 5) as the answer (right?). The method of elimination and substitution would also produce $x = 2$, $y = 5$. Instead, we proceed with a rather awkward, although mathematically correct, method for reasons that are not at all immediately apparent.

$8 \cdot (7x + 4y = 34)$ $4 \cdot (3x + 8y = 46)$	Multiply both equations by a constant.
$8 \cdot 7x + 8 \cdot 4y = 8 \cdot 34$ $4 \cdot 3x + 4 \cdot 8y = 4 \cdot 46$	Remove parentheses and multiply through.

$8 \cdot 7x - 4 \cdot 3x = 8 \cdot 34 - 4 \cdot 46$	Subtract the second equation from the first in order to cancel the $32y$'s.
$x \cdot (8 \cdot 7 - 4 \cdot 3) = 8 \cdot 34 - 4 \cdot 46$	Factor out an x.
$x \cdot (7 \cdot 8 - 3 \cdot 4) = 34 \cdot 8 - 46 \cdot 4$	Reverse the order of the factors.
$x = \dfrac{34 \cdot 8 - 46 \cdot 4}{7 \cdot 7 - 3 \cdot 4}$	Divide both sides by $7 \cdot 8 - 3 \cdot 4$.

If we evaluate this fraction it gives as expected $x = 2$. Now, if we rewrite the numerator and denominator of

$$x = \frac{34 \cdot 8 - 46 \cdot 4}{7 \cdot 8 - 3 \cdot 4}$$

as

$$x = \frac{\begin{vmatrix} 34 & 4 \\ 46 & 8 \end{vmatrix}}{\begin{vmatrix} 7 & 4 \\ 3 & 8 \end{vmatrix}},$$

we see that the *denominator* of the fraction that equals x *is the determinant of the matrix of coefficients*, and the *numerator* of the fraction that equals x *is the determinant of the matrix of coefficients after the column of x-coefficients has been replaced by the column of constants.* The numerator of x is called the *x-determinant* and is symbolized by D_x or Δ_x. The denominator of x is called the *coefficient determinant* and is symbolized by D or Δ. If we proceeded in a similar fashion to solve the original system for y, we would have

$$y = \frac{\begin{vmatrix} 7 & 34 \\ 3 & 46 \end{vmatrix}}{\begin{vmatrix} 7 & 4 \\ 3 & 8 \end{vmatrix}} = \frac{322 - 102}{56 - 12} = \frac{220}{44} = 5$$

where the denominator is as before and the numerator is the determinant of the matrix of coefficients after the column of y-coefficients has been replaced by the column of constants.

Cramer's rule

It is a relatively easy step to make and prove this general statement, which is known as *Cramer's Rule*. The system

$$\begin{aligned} ax + by &= k_1 \\ cx + dy &= k_2 \end{aligned}$$

has as its solutions

$$x = \begin{vmatrix} k_1 & b \\ k_2 & d \end{vmatrix} \Big/ \begin{vmatrix} a & b \\ c & d \end{vmatrix} = D_x/D = \Delta_x/\Delta,$$

and

$$y = \begin{vmatrix} a & k_1 \\ c & k_2 \end{vmatrix} \Big/ \begin{vmatrix} a & b \\ c & d \end{vmatrix} = D_y/D = \Delta_y/\Delta.$$

Example 14 Solve for x and y by applying Cramer's Rule.

$$\begin{aligned} 2.67x + 13.4y &= -16.2 \\ 3.05x - 117y &= 7 \end{aligned}$$

Solution

$$x = D_x/D = \begin{vmatrix} -16.2 & 13.4 \\ 7 & -117 \end{vmatrix} \Big/ \begin{vmatrix} 2.67 & 13.4 \\ 3.05 & -117 \end{vmatrix}$$

$$= \frac{-16.2 \cdot -117 - 7 \cdot 13.4}{2.67 \cdot -117 - 3.05 \cdot 13.4}$$

$$= \frac{1895.4 - 93.8}{-312.39 - 40.87}$$

$$= \frac{1801.6}{-353.26}$$

$$= -5.10$$

$$y = D_y/D = \begin{vmatrix} 2.67 & -16.2 \\ 3.05 & 7 \end{vmatrix} \Big/ \begin{vmatrix} 2.67 & 13.4 \\ 3.05 & -117 \end{vmatrix}$$

$$= \frac{2.67 \cdot 7 - 3.05 \cdot -16.2}{-353.26}$$

$$= \frac{18.69 + 49.41}{-353.26}$$

$$= \frac{68.10}{-353.26}$$

$$= -.19$$

Granted, a few awkward multiplications and divisions are involved, but the desirable aspect of this approach is that we can easily program a computer or use a calculator to do the work. Even if a computer or calculator is not available, there is still less work than doing it by the augmented matrix approach. The complete BASIC computer program shows the simplicity and power of the language.

```
10 DATA 2.67, 13.4, -16.2, 3.05, -117, 7
20 READ A, B, K1, C, D, K2
30 LET X = (K1*D-K2*B)/(A*D-C*B)
40 LET Y = (A*K2-C*K1)/(A*D-C*B)
50 PRINT X, Y
60 END
```

When this program is run the DATA item 2.67 is READ into the variable A, the DATA item 13.4 is READ into the variable B, the DATA item -16.2 is READ into the variable K1, the DATA item 3.05 is READ into the variable C, the DATA item -117 is READ into the variable D, and the DATA item 7 is READ into the variable K2. X and Y are then calculated for this specific set of six DATA items, after which their values are PRINTed.

If we wish to solve the system

$$\begin{aligned} 7.983x + .00687y &= -15.93 \\ 4216.7x - 35.006y &= 163, \end{aligned}$$

we then would merely change the DATA statement to

```
10 DATA 7.983, .00687, -15.93, 4216.7, -35.006, 163
```

and the result of $x = -1.804$, $y = -222.012$ would immediately follow.

If we solve an inconsistent or dependent system with Cramer's Rule we would obtain

$$(\text{nonzero constant})/0 \quad \text{and} \quad 0/0,$$

respectively.

What about a system of three equations in three unknowns, x, y, and z? An example illustrates the use of determinants in this situation.

Example 15 Solve the system

$$\begin{aligned} 3x + 4y - 5z &= -4 \\ 2x - 3y + 4z &= 8 \\ 4x - 6y + 3z &= 1. \end{aligned}$$

Solution With two equations in two variables x and y, $x = D_x/D$ and $y = D_y/D$. Can we define D, D_x, D_y, and D_z so that by analogy x will equal D_x/D, y will equal D_y/D, and z will equal D_z/D where

$$D = \begin{vmatrix} 3 & 4 & -5 \\ 2 & -3 & 4 \\ 4 & -6 & 3 \end{vmatrix},$$

$$D_x = \begin{vmatrix} -4 & 4 & -5 \\ 8 & -3 & 4 \\ 1 & -6 & 3 \end{vmatrix},$$

$$D_y = \begin{vmatrix} 3 & -4 & -5 \\ 2 & 8 & 4 \\ 4 & 1 & 3 \end{vmatrix},$$

and

$$D_z = \begin{vmatrix} 3 & 4 & -4 \\ 2 & -3 & 8 \\ 4 & -6 & 1 \end{vmatrix}?$$

Symbolically, notice that

$$\begin{matrix} -4 \\ 8, \\ 1 \end{matrix}$$

which is the column of constants, might replace the coefficients of x from D so that D becomes D_x. Similarly,

$$\begin{matrix} -4 \\ 8 \\ 1 \end{matrix}$$

might replace the coefficients of y from D so that D becomes D_y, and

$$\begin{matrix} -4 \\ 8 \\ 1 \end{matrix}$$

might replace the coefficients of z from D so that D becomes D_z.

The answer is *yes* to all parts of the previous question. We immediately need to know how to evaluate, for example,

$$D_x = \begin{vmatrix} -4 & 4 & -5 \\ 8 & -3 & 4 \\ 1 & -6 & 3 \end{vmatrix}.$$

We do this evaluation by repeating the first two columns and performing the six multiplications indicated by the arrows. We then sum the first three products, sum the second three products, and find the difference of the two sums, which is D_x.

Figure 7.1

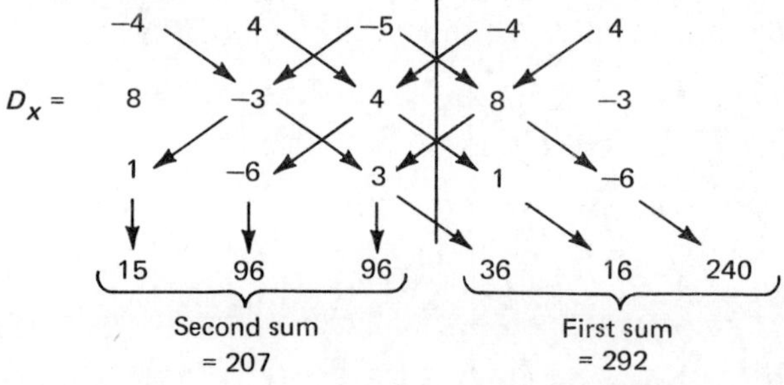

$$\begin{aligned} D_x &= (\text{first sum}) - (\text{second sum}) \\ &= 292 - 207 \\ &= 85 \end{aligned}$$

Similarly,

Figure 7.2

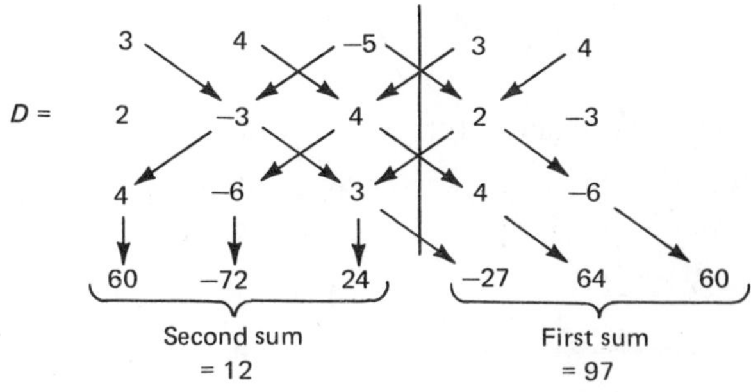

$$\begin{aligned} D &= (\text{first sum}) - (\text{second sum}) \\ &= 97 - 12 \\ &= 85. \end{aligned}$$

Based on a general solution of three equations in three unknowns and a more formal definition of the determinant of a matrix with three rows and three columns, we could prove that $x = D_x/D$. Applying the result of this general statement, we then have

$$x = D_x/D = 85/85 = 1.$$

Continuing, we could show that

$$y = D_y/D = 170/85 = 2$$

and that

$$z = D_z/D = 255/85 = 3 \quad \text{(right on both of these?).}$$

It would be easiest to find y and z by first substituting $x = 1$ into any two of the three original equations and then solving the resulting system of two equations for y and z.

We close by noting again that this scheme of direct multiplication, addition, subtraction, and division lends itself very nicely to a computer or calculator method of solution, and that an inconsistent or dependent system yields (nonzero constant)/0 and 0/0, respectively, in the Cramer's Rule approach. One weakness of using Cramer's Rule is that it tells us nothing about a dependent system with its infinite solution set. If we review Example 4 in the last section, we observe that the row-reduction method lets us specify the infinite solution set. Using Cramer's Rule there would say only that $x = D_x/D = 0/0$. Also, the method of repeating columns and multiplying along certain diagonals can *not* be generalized to permit the evaluation of determinants of matrices with four rows and four columns. Only matrices with as many rows as columns (*square* matrices) have determinants. This fact excludes the use of Cramer's Rule to solve completely and directly, for example, three linear equations in four unknowns.

Problem set 7–3

1. Compute the determinants of these matrices.

a. $\begin{bmatrix} 5 & 2 \\ -3 & 4 \end{bmatrix}$ b. $\begin{bmatrix} -1 & 3 \\ 2 & 4 \end{bmatrix}$

c. $\begin{bmatrix} 3 & 5 \\ 2 & 8 \end{bmatrix}$ d. $\begin{bmatrix} 3 & 7 \\ -4 & -2 \end{bmatrix}$

e. $\begin{bmatrix} 2 & -3 & -1 \\ 1 & 6 & 1 \\ 1 & 6 & 3 \end{bmatrix}$ f. $\begin{bmatrix} 5 & 2 & 1 \\ -3 & 2 & -2 \\ 4 & -1 & 1 \end{bmatrix}$

g. $\begin{bmatrix} 1 & 1 & 1 \\ 4 & -2 & 3 \\ -3 & -3 & -3 \end{bmatrix}$ h. $\begin{bmatrix} 4 & 1 & -3 \\ 4 & 0 & -3 \\ 4 & 0 & -3 \end{bmatrix}$

i. What generalization do part (g) and part (h) suggest?

j. $\begin{bmatrix} 1 & 0 & 0 \\ 0 & 1 & 0 \\ 0 & 0 & 1 \end{bmatrix}$ k. $\begin{bmatrix} 3 & 4 & 2 \\ 1 & -7 & -3 \end{bmatrix}$

2. Use Cramer's Rule to solve the following systems of equations. It is a good idea to evaluate the coefficient determinant first to check for a zero divisor.

a. $2x - 3y = 4$
$x + y = 1$

b. $x - 2y = -1$
$3x + y = 3$

c. $7x - 3y = 5$
$2x + 6y = -6$

d. $3x - 5y = 9$
$-12x + 20y = -36$

e. $4x + 7y = -11$
$14x + 24.5y = -66$

f. $4y = 3x + 22$
$60 - 2x = 8y$

3. Refer back to Example 13, which involves solving the system

$$7x + 4y = 34$$
$$3x + 8y = 46.$$

Use a similar method of reasoning and manipulation to show that

$$y = \frac{7 \cdot 46 - 3 \cdot 34}{7 \cdot 8 - 3 \cdot 4} = \frac{D_y}{D}.$$

4. Use Cramer's Rule to solve the following systems of equations.

a. $2x - y - z = -3$
$x + y + z = 6$
$x - 2y + 3z = 6$

b. $x + 2z = 1$
$3x - y = 3$
$x + 2y + z = 4$

c. $5x + 6y = z$
$7x + 6y - 2z + 8 = 0$
$3x + 2y - z + 5 = 0$

d. $z - 4y = 4$
$3x - 2y - z = 5$
$2x - z = 2$

e. $2a + c + 3 = b$
$a + b = 1 + c$
$a + 2c + 4 = b$

f. $5x + 2y - z = 13$
$4x - 6y - 2z = 5$
$-x + y + 7z = -5$

5. Solve for x.

a. $\begin{vmatrix} 7 & 2 \\ 14 & x \end{vmatrix} = 0$

b. $\begin{vmatrix} x & 3 \\ 2 & x-1 \end{vmatrix} = 6$

6. Consider the system

$$\begin{aligned} 2x + y - 4z &= 21 \\ x + 5y + 2z &= 8 \\ -3x + y + 3z &= -12. \end{aligned}$$

a. Solve the system by evaluating all four determinants required by a complete application of Cramer's Rule.
b. First rewrite the system in triangular form. Then solve the resulting equivalent system by a complete application of Cramer's Rule.
c. First rewrite the system in triangular form. Then use Cramer's Rule to solve for x. Finally, substitute for x in the second and first equations to solve for y and z respectively.
d. Which method took the least amount of time?

7. By generalizing the development you used in Problem 3, prove that the solution to

$$\begin{aligned} ax + by &= k_1 \\ cx + dy &= k_2 \end{aligned}$$

is

$$x = \frac{k_1 d - k_2 b}{ad - cb}, \qquad y = \frac{ak_2 - ck_1}{ad - cb},$$

provided, of course, that $(ad - cb) \neq 0$.

8. Using the text's example as a guide, write the complete BASIC computer program to solve this system.

$$\begin{aligned} 3946.1x + 74.298y &= -19.774 \\ 6003.8x - 123.45y &= 98.007 \end{aligned}$$

9. Show that the three lines whose equations are

$$\begin{aligned} 2x + y + 1 &= 0 \\ 4x + 2y - 3 &= 0 \\ 3x + 4y + 5 &= 0 \end{aligned}$$

do not intersect in one point.

10. A collection of nickels, dimes, and quarters, thirteen coins in all, amounts to \$2.40. If the dimes were nickels, the quarters dimes, and the nickels quarters, the collection would amount to \$1.45. How many of each kind are there?
11. There was a total of forty-six passengers—men, women, and children— on a bus. At the next stop two men got off; then there were as many adults as children. At the next stop twelve children got off; then the number of children was equal to the difference between the numbers of women and men. How many men, women, and children were there at first?
12. The sum of the ages of man, wife, and son is sixty-four years. In six years the father will be three times as old as the son. Four years ago the mother was twelve times as old as the son. How old is each now?
13. A restaurant owner plans to use x tables seating four each, y tables seating six each, and z tables seating eight each for a total of twenty tables. When fully occupied the tables seat 108 customers. If only half of the x tables, half of the y tables and one-fourth of the z tables are used, each fully occupied, then forty-six customers will be seated. What are the values of x, y, and z?
14. A contractor employed twelve men, some of whom he paid \$60 per day, some of whom he paid \$72 per day, and the rest of whom he paid \$80 per day, thus expending a planned payroll of \$876 per day. One day his bookkeeper erroneously interchanged the number of men earning the least with the number earning the most and prepared a payroll of \$816. How many men were hired at each rate?

Solving systems of n×n equations

In the previous three sections of this chapter we have learned various ways to solve linear systems of two equations in two unknowns and three equations in three unknowns. In this brief section we look at the problem of solving linear systems with more than three unknowns.

To begin, we learn some convenient notation about matrices. If a matrix A has m (horizontal) rows and n (vertical) columns, where m and n are positive integers, then a_{ij} signifies the element at the intersection of the ith row and jth column of matrix A.

Example 16 If

$$A = \begin{bmatrix} 2 & 9 & -7 & 4 \\ 6 & 1.4 & 8 & -10 \\ -5 & 0 & 7 & 1 \end{bmatrix}$$

which is a 3×4 matrix (3 rows and 4 columns) then find $a_{21}, a_{12}, a_{23}, a_{33}, a_{31}, a_{22}, a_{34}, a_{24}, a_{42}$.

Solution We know that a_{21} is the element of A that is both in the second row (since $i = 2$ here) and in the first column (since $j = 1$ here), or

$a_{21} = 6$. Similar reasoning tells us that $a_{12} = 9$, $a_{23} = 8$, $a_{33} = 7$, $a_{31} = -5$, $a_{22} = 1.4$, $a_{34} = 1$, and $a_{24} = -10$; a_{42} does not exist since matrix A does not have a fourth row.

Definition The *minor* of an element a_{ij} of a square matrix A is the determinant of the submatrix obtained after removing from A the row and the column that contains a_{ij}.

Example 17 If

$$A = \begin{bmatrix} 2 & 3 & -5 \\ 6 & 0 & 1 \\ 4 & -1 & 7 \end{bmatrix},$$

then the minor of $a_{12} = 3$ is found by removing the first row and second column from A,

$$\begin{bmatrix} 2 & 3 & -5 \\ 6 & 0 & 1 \\ 4 & -1 & 7 \end{bmatrix} \to \begin{bmatrix} 6 & 1 \\ 4 & 7 \end{bmatrix},$$

so that

$$\begin{vmatrix} 6 & 1 \\ 4 & 7 \end{vmatrix} = 38.$$

The minor of $a_{13} = -5$ is

$$\begin{vmatrix} 6 & 0 \\ 4 & -1 \end{vmatrix} = -6,$$

the minor of $a_{23} = 1$ is

$$\begin{vmatrix} 2 & 3 \\ 4 & -1 \end{vmatrix} = -14,$$

and the minor of $a_{31} = 4$ is

$$\begin{vmatrix} 3 & -5 \\ 0 & 1 \end{vmatrix} = 3.$$

Example 18 If

$$B = \begin{bmatrix} 7 & 4 & 6 & 1 \\ -3 & 0 & -2 & 3 \\ 1 & 0 & -1 & 4 \\ -7 & 6 & 5 & 2 \end{bmatrix},$$

then the minor of $b_{32} = 0$ is

$$\begin{bmatrix} 7 & 6 & 1 \\ -3 & -2 & 3 \\ -7 & 5 & 2 \end{bmatrix} = -252.$$

Expansion by minors

We are now ready to compute the value of the determinant of any 3×3 matrix by the process known as *expansion by minors.* To do this we choose any single row (or column), multiply each of its elements a_{ij} by the minor of that element and also by $(-1)^{i+j}$, and add the resulting products. We accept this last sentence as the distillation of much more advanced mathematics than we presently care to deal with.

Example 19 Compute the determinant of matrix A which is

$$\begin{bmatrix} 7 & 6 & 1 \\ -3 & -2 & 3 \\ -7 & 5 & 2 \end{bmatrix}$$

by expansion by minors.

Solution We expand by the first row where $a_{11} = 7$, $a_{12} = 6$, and $a_{13} = 1$.

$$\begin{aligned} \det A &= 7 \cdot (-1)^{1+1} \cdot \begin{vmatrix} -2 & 3 \\ 5 & 2 \end{vmatrix} \\ &\quad + 6 \cdot (-1)^{1+2} \cdot \begin{vmatrix} -3 & 3 \\ -7 & 2 \end{vmatrix} \\ &\quad + 1 \cdot (-1)^{1+3} \cdot \begin{vmatrix} -3 & -2 \\ -7 & 5 \end{vmatrix} \\ &= 7 \cdot 1 \cdot -19 + 6 \cdot -1 \cdot 15 + 1 \cdot 1 \cdot -29 \qquad \text{(A)} \\ &= -133 + -90 + -29 \\ &= -252. \end{aligned}$$

The purpose of multiplying by $(-1)^{i+j}$ is to alternate the signs of the products between + and −. In the above example we could have written

$$\det A = 7 \cdot -19 - 6 \cdot 15 + 1 \cdot -29$$

for equation (A). Expansion by any other row or column will produce an identical result (see Problems 14 and 15). We now proceed to find the determinants of a 2×2 and a 4×4 matrix.

Example 20 Evaluate

$$|B| = \begin{vmatrix} 4 & 5 \\ 6 & 7 \end{vmatrix}$$

by expansion by minors. Use the fact that if, for example,

$$M = [7],$$

then

$$|7| = \det M = 7.$$

Solution We expand by the first row where $b_{11} = 4$ and $b_{12} = 5$.

$$\begin{aligned}\det B &= (4 \cdot (-1)^{1+1} \cdot 7) + (5 \cdot (-1)^{1+2} \cdot 6) \\ &= 28 + -30 \\ &= -2.\end{aligned}$$

Example 21 Evaluate

$$|C| = \begin{vmatrix} 1 & -1 & 3 & 2 \\ 2 & -2 & 2 & -1 \\ 3 & 0 & 0 & -2 \\ 4 & 3 & 1 & 1 \end{vmatrix}$$

by expansion by minors.

Solution We expand by the third row where $c_{31} = 3$, $c_{32} = 0$, $c_{33} = 0$, and $c_{34} = -2$, since the row contains two zeros.

$$\det C = 3 \cdot (-1)^{3+1} \cdot \begin{vmatrix} -1 & 3 & 2 \\ -2 & 2 & -1 \\ 3 & 1 & 1 \end{vmatrix}$$

$$+\, 0 \cdot (-1)^{3+2} \cdot \text{(a determinant we need not evaluate)}^1$$

$$+\, 0 \cdot (-1)^{3+3} \cdot \text{(a determinant we need not evaluate)}^1$$

$$+ -2 \cdot (-1)^{3+4} \cdot \begin{vmatrix} 1 & -1 & 3 \\ 2 & -2 & 2 \\ 4 & 3 & 1 \end{vmatrix}$$

$$\begin{aligned} &= 3 \cdot 1 \cdot -22 + 0 + 0 + -2 \cdot -1 \cdot 28 \quad \text{(Right?)} \\ &= -66 + 0 + 0 + 56 \\ &= -10 \end{aligned}$$

Expanding by the first row would have produced identical results but with much more work and hence more chance for arithmetic errors. This example shows why we should rewrite the original system to obtain as many 0's as possible in a row or column before using determinants. Of course, computers can be programmed to do all the laborious but routine calculations.

In working with a system of four equations in four unknowns it is true that

$$x = D_x/D, \quad y = D_y/D, \quad z = D_z/D, \quad \text{and} \quad w = D_w/w,$$

[1] Why do we not have to list and evaluate these determinants?

as might well be expected. The best way to approach such a system manually is often to solve for one of the variables by Cramer's Rule, so that after substitution into any three equations there are then just three equations in three unknowns. We can then apply Cramer's Rule, so that instead of evaluating five 4×4 determinants we now have just four 3×3 determinants. Problem 6 of the previous section shows possible ways of cutting down even more on the arithmetic.

If we solve the general system of three equations in three unknowns

$$\begin{aligned} ax + by + cz &= k_1 \\ dx + ey + fz &= k_2 \\ gx + hy + iz &= k_3 \end{aligned}$$

by Cramer's Rule (the solution is more monotonous that it is difficult—see Problem 11), then

$$\begin{aligned} D &= aei + bfg + cdh - afh - bdi - ceg \\ D_x &= k_1 ei + bfk_3 + ck_2 h - k_1 fh - bk_2 i - cek_3 \\ D_y &= ak_2 i + k_1 fg + cdk_3 - afj_3 - k_1 di - ck_2 g \\ D_z &= aek_3 + bk_2 g + k_1 dh - ak_2 h - bdk_3 - k_1 eg. \end{aligned}$$

The manual evaluations are very lengthy but relatively easy for a computer.

Using a computer to solve systems of $n \times n$ equations

Solving a 5×5 or larger system should be attempted only with the aid of a computer although Cramer's Rule of successively replacing the column of coefficients of the variable x_i ($i \in \{1, 2, 3, \ldots, n\}$), evaluating the n resulting $n \times n$ determinants, and dividing each of these n numbers by D (still the determinant of the coefficient matrix) remains valid. Surprisingly, solutions of large systems of $n \times n$ equations ($n \geqslant 4$) on a computer are often done by the complete row-reduction method mentioned in Problems 11, 12, 13, and 14 of the first section of this chapter, rather than by Cramer's Rule.

Why? Consider the number of multiplications and divisions involved in solving an $n \times n$ system by each method. Cramer's Rule involves the evaluation of $(n+1)$ $n \times n$ determinants, each requiring $n!$ multiplications, and then n divisions, for a total of $(n+1)! + n$ multiplications and divisions. But complete row reduction requires $n \cdot (n-1+2)$ multiplications to introduce appropriate 0's into the first column and $n \cdot (n-i+2)$ multiplications to introduce appropriate 0's into the ith column for a total of

$$\sum_{i=1}^{n} n \cdot (n-i+2) \doteq n^3$$

multiplications and divisions. Now for 2×2 and 3×3 systems Cramer's Rule is slightly better, and if one chooses a good row or column to compute the determinant, it is quite a bit better. But for general application, especially by a computer, Cramer's Rule is much too long.

Table 7.3

Number of equations and unknowns, n	Number of $\cdot$'s and $\div$'s for solution by Cramer's rule, $(n+1)!$	Number of $\cdot$'s and $\div$'s for solution by complete row reduction, n^3
2	6	8
3	24	27
4	120	64
5	720	125
6	5,040	216
7	40,320	343
8	362,880	512
⋮	⋮	⋮
20	$21! \doteq 10^{20}$	$20^3 = 8{,}000$

Some computers presently in existence can perform 100,000 mixed multiplications and divisions in one second. There are about 32,000,000 seconds in a year, so the computer can perform about, 3,200,000,000,000 mixed multiplications and divisions in a year. Since $10^{20} \div 3{,}200{,}000{,}000{,}000$ is about 30,000,000, then it would take about 30 *million* years for the solution of twenty linear equations in twenty unknowns by Cramer's Rule. In contrast, complete row reduction would require well under one *second* for the solution. Moral: mathematical methods that work in simple cases are not always practical in more general similar cases, even with the aid of a computer.

Problem set 7–4

1. If

$$M = \begin{bmatrix} 7 & 6 & 0 \\ 3 & -5 & 2 \\ 1 & 0 & 6 \\ -3 & -4 & 5 \end{bmatrix},$$

then find the following.

a. m_{23} b. m_{32} c. m_{22} d. m_{11}
e. m_{21} f. m_{43} g. m_{34} h. m_{33}
i. m_{41} j. $(-1)^{2+3}$ k. $(-1)^{4+2}$ l. the second row

2. If

$$E = \begin{bmatrix} 7 & 6 & 2 \\ -4 & 3 & 0 \\ 2 & 5 & -3 \end{bmatrix},$$

then identify these elements of E by their subscripts; for example, since 6 is in the first row and second column we say $6 = e_{12}$.

a. -4 b. -3 c. 5
d. 0 e. 8 f. 2

3. Evaluate the determinants of these matrices by expansion by minors.

a. $A = \begin{bmatrix} 4 & 5 \\ 1 & 3 \end{bmatrix}$ b. $B = \begin{bmatrix} 0 & -3 \\ 2 & 5 \end{bmatrix}$

c. $C = \begin{bmatrix} 17 & -43 \\ 14 & -38 \end{bmatrix}$ d. $D = \begin{bmatrix} 1 & 2 & -1 \\ 2 & 4 & 3 \\ 3 & 6 & -2 \end{bmatrix}$

e. $E = \begin{bmatrix} 2 & 1 & -3 \\ -5 & 2 & 1 \\ 1 & 3 & 4 \end{bmatrix}$ f. $F = \begin{bmatrix} 2 & 6 & 5 \\ 3 & 9 & 4 \\ -4 & -12 & 3 \end{bmatrix}$

g. $G = \begin{bmatrix} 2 & -1 & 3 \\ 1 & 2 & 3 \\ 3 & -2 & 1 \end{bmatrix}$ h. $H = \begin{bmatrix} 1 & 2 & -1 \\ 3 & 1 & 0 \\ -1 & 2 & 3 \end{bmatrix}$

4. Solve for x.

a. $\begin{vmatrix} 1 & x & -1 \\ 2 & 4 & 2 \\ 3 & 6 & -3 \end{vmatrix} = 0$ b. $\begin{vmatrix} 4 & 2x & -1 \\ 2 & x & 3 \\ -2 & -1 & 2 \end{vmatrix} = 0$

5. Solve these systems by complete use of Cramer's Rule and minors in evaluating the determinants.

a. $x + y + z = 4$
$2x - y - 2z = -1$
$x - 2y - z = 1$

b. $2x - y + 3z = 1$
$x + y + z = 1$
$4x - 2y + 6z = 3$

c. $u + v + w = 0$
$2u - v + 5w = 3$
$-u - 2v + w = 0$

d. $x + z + 4 = 0$
$x + y = 1$
$3x + 2y - z = 4$

6. Evaluate by expansion by minors.

a. $\begin{vmatrix} 1 & 1 & 1 & 1 \\ 1 & 0 & -1 & 0 \\ 0 & 1 & 1 & -1 \\ 2 & 0 & -1 & -3 \end{vmatrix}$ b. $\begin{vmatrix} 2 & -2 & 1 & 3 \\ 0 & 2 & -1 & -1 \\ 2 & -3 & 2 & 4 \\ 0 & -1 & 1 & 1 \end{vmatrix}$

c. $\begin{vmatrix} 3 & -2 & 6 & 4 \\ 1 & 0 & 2 & -1 \\ 5 & 4 & 3 & 0 \\ 2 & 2 & -5 & 6 \end{vmatrix}$ d. $\begin{vmatrix} 1 & 0 & 0 & 0 \\ 0 & 1 & 0 & 0 \\ 0 & 0 & 1 & 0 \\ 0 & 0 & 0 & 1 \end{vmatrix}$

7. Solve the following systems.

a. $2x + 3y + z + t = 1$
$x - y - z + t = 1$
$3x + y + z + 2t = 0$
$-x + 0y + z - t = -2.$

Hint: we will introduce three zeros into the third column by replacing the first, third, and fourth rows by their sum with the second row. Then

$$\begin{aligned} 3x + 2y + 0z + 2t &= 2 \\ x - y - z + t &= 1 \\ 4x + 0y + 0z + 3t &= 1 \\ 0x - y + 0z + 0t &= -1 \end{aligned}$$

and

$$D = \begin{vmatrix} 3 & 2 & 0 & 2 \\ 1 & -1 & -1 & 1 \\ 4 & 0 & 0 & 3 \\ 0 & -1 & 0 & 0 \end{vmatrix},$$

which we evaluate by expansion down the third column or across the fourth row. Solve for z by evaluating D_z/D—you should get -1—so that, using the first three equations of the original system,

$$\begin{aligned} 2x + 3y + 1 + t &= 1 \\ x - y + 1 + t &= 1 \\ 3x + y - 1 + 2t &= 0 \end{aligned}$$

$$\begin{aligned} 2x + 3y + t &= 2 \\ x - y + t &= 0 \\ 3x + y + 2t &= 1 \end{aligned}$$

The system is now ready to be solved by Cramer's Rule.

b. $x + y + z + 3t = 3$
$3x + y - z = 0$
$2x - 2y - z + 6t = 4$
$4x - y - 2z - 3t = 0$

c. $x + y + z + t = 3$
$2x - 2y - z + 2t = 0$
$3x - y + 2z + 3t = 3$
$x - y - 2z + t = 0$

8. Find the values of a, b, and c such that the graph of

$$y = ax^2 + bx + c$$

passes through these triplets of points:

a. $(2, -12)$, $(3, -10)$, $(5, 0)$. Hint: since the graph passes through $(2, -12)$ then these coordinates must satisfy the equation and

$$-12 = a \cdot 2^2 + b \cdot 2 + c,$$

or

$$4a + 2b + c = -12.$$

The coordinates of the other two points also satisfy this same equation, and thus we have three equations involving our desired three unknowns, a, b, and c.

b. $(1, 3), (3, 3), (4, 0)$.

c. $(2, -6), (4, 2), (-1, -3)$.

d. $(1, 0), (0, 5), (2, -7)$.

e. $(1, 0), (0, 5)\ (2, -5)$.

9. The height h in feet above sea level and the boiling point of a certain liquid at t degrees Fahrenheit are related by the formula

$$t = ah^2 + bh + c$$

If $t = 212$ when $h = 0$, $t = 210$ when $h = 400$, and $t = 208$ when $h = 500$, then find a, b, and c. If $t = 200$, what is h?

10. Evaluating a 3×3 determinant by expansion by minors requires the evaluation of three 2×2 determinants.
    a. How many 2×2 determinants must be evaluated if we use Cramer's Rule to solve a system of three equations in three unknowns?
    b. How many 3×3 determinants must we evaluate if we use Cramer's Rule on a system of four equations in four unknowns if the 4×4 determinants are evaluated by expansion by minors?
    c. Same as (b), but change 3×3 to 2×2 so that each 3×3 determinant in the answer to (b) is now evaluated by expansion by minors.
11. Solve the general system of three linear equations in three unknowns that is mentioned in the next-to-last paragraph of this section. Use expansion by minors on an appropriate row or column to obtain the results mentioned.
12. If a system of two equations in two unknowns is solved by Cramer's Rule, then
    a. how many different determinants must we evaluate?
    b. how many different determinants must we evaluate for a system of three linear equations in three unknowns?
    c. four equations in four unknowns?
    d. five equations in five unknowns?
    e. n equations in n unknowns?
13. If we solve the system

$$\begin{aligned} 2x - 7y + 4z &= 49 \\ 3x + 2y - 3z &= -7 \\ 5x + 4y - 8z &= -21 \end{aligned}$$

by complete row reduction, then the answer in augmented matrix form is

$$\left[\begin{array}{ccc|c} 1 & 0 & 0 & 3 \\ 0 & 1 & 0 & -5 \\ 0 & 0 & 1 & 2 \end{array}\right]$$

a. Verify this fact by actually carrying out the row reduction; count the number of multiplications and divisions required.

b. Evaluate

$$\sum_{i=1}^{3} 3 \cdot (3 - i + 2),$$

which should equal the answer from (a).

c. Solve the system by a complete application of Cramer's Rule, and count the number of multiplications and divisions required (not including the multiplications by $(-1)^{i+j}$).

d. Compare your answer from part (c) with the appropriate table entries at the very end of this section of text.

14. Rework Example 19 by expanding by the third column.
15. Rework Example 19 by expanding by the second row.

Chapter 8

Linear Programming

Graphic solutions of inequalities

Example 1 Ajax Manufacturing Company produces transmissions and rear axles for the Orion Automobile Company. Each transmission requires one man-hour on the welder, each rear axle assembly requires two man-hours on the same welder, and the welder can not be used more than twenty man-hours per eight-hour shift. The manual installation of the major parts requires five man-hours on a transmission and three man-hours on a rear axle with a maximum of forty-four man-hours available per eight-hour shift. Management also insists that the numerical difference between three times the number of transmissions produced and two times the number of rear axles produced must be at most fifteen because of certain tax advantages. If there is a profit of six dollars on each transmission and seven dollars on each rear axle produced, then how many of each assembly should the production planning department schedule for every eight-hour shift?

This problem is an example of the type which we will study in this chapter. It is an example of a *linear programming* problem and is solved by using techniques that are extensions of what we already know about algebra and geometry. The problem looks bewildering on first inspection because of the mass of data, but it is not really difficult after we have learned some additional mathematics. Thus, before we can return to this specific problem, we must move to the upper half of our rectangle of mathematics and learn some pure mathematics in this section and the next.

In Problem 5 of Problem set 1–2 we reviewed several problems about solving inequalities in one variable. As a prerequisite to solving linear programming problems, we now study linear inequalities in two variables using a graphic approach. In a linear inequality the exponents of all the variables are 1. For example,

$$2x^1 + 5y^1 < 13 \qquad \text{or} \qquad 2x + 5y < 13.$$

The variables may not be multiplied or divided by each other, nor may a variable be in a denominator. Three examples help us to understand linear inequalities and their graphs.

Example 2 Solve the inequality $3y - 5x \leqslant 15$ by constructing its graph.

Solution We break the problem down into separately finding the graphic solution of $3y - 5x = 15$ and of $3y - 5x < 15$, and then uniting the two solutions for the final answer. Technically,

$$\{(x, y) \mid 3y - 5x \leqslant 15\} \qquad \{(x, y) \mid 3y - 5x = 15\} \cup \{3y - 5x < 15\}.$$

If we solve

$$3y - 5x = 15$$

for y, we obtain

$$y = \frac{5}{3}x + 5 \quad \text{(right?)}.$$

From thc discussion immediately following Figure 3.6 in Chapter 3, we know that the graph of this equation is a line whose slope is 5/3. If x is replaced by 0 and then by 2 in the equation

$$y = \frac{5}{3} \cdot x + 5,$$

then the respective corresponding values of y are 5 and 8 1/3 (right?). This means that (0, 5) and (2, 8 1/3) are two points on the lines, so that all the

Figure 8.1

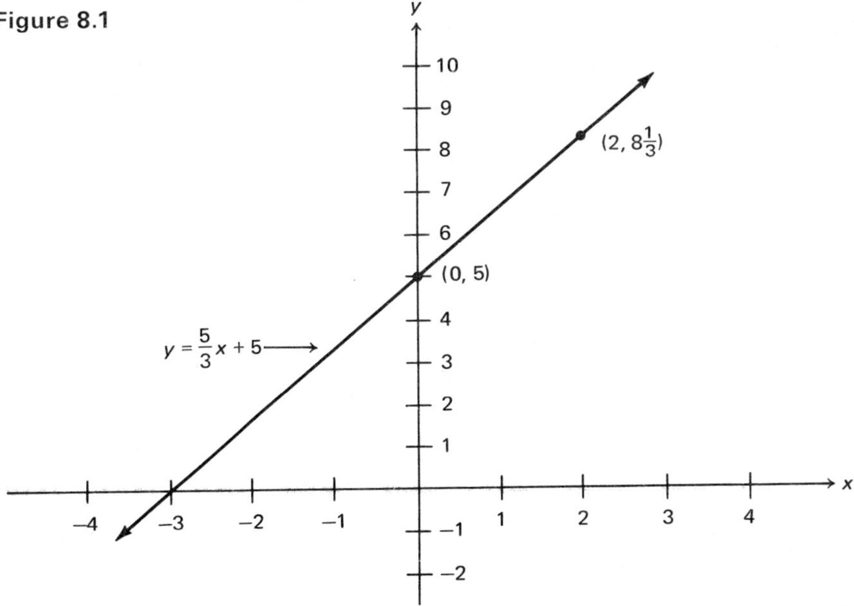

points on the line are now determined (exactly one line passes through two distinct points).

If we now solve the inequality

$$3y - 5x < 15$$

for y, we obtain

$$y < \frac{5}{3}x + 5.$$

If x is replaced by 1, then

$$y < \frac{5}{3} \cdot 1 + 5$$

or

$$y < 6\,2/3,$$

whose graph—a half-line extending down from $(1, 6\,2/3)$—is now added to our previous picture.

Figure 8.2

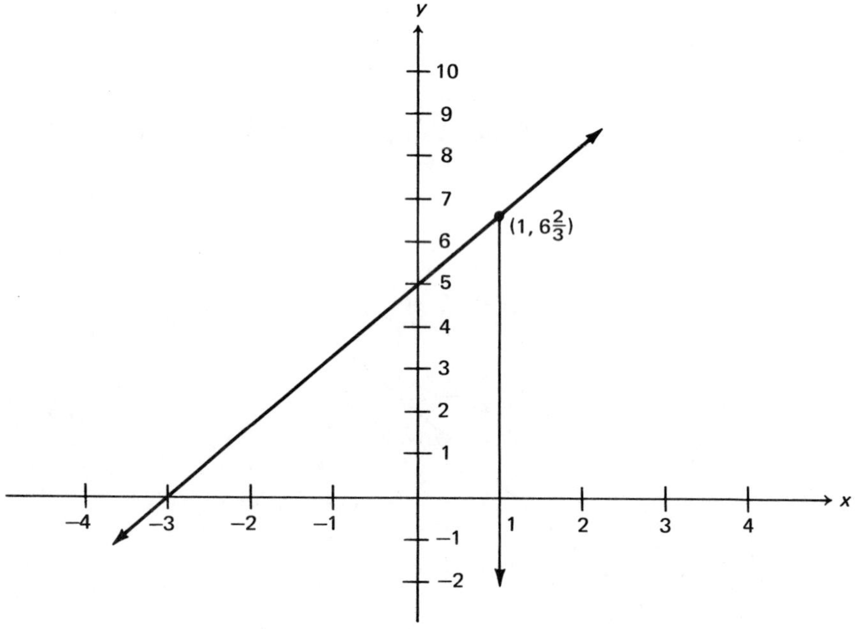

If we continue by letting x take on the specific values, -4, -2, 2 and then solve the resulting inequalities for y, the corresponding, successive graphs would be a series of half-lines extending downward from the line.

Figure 8.3

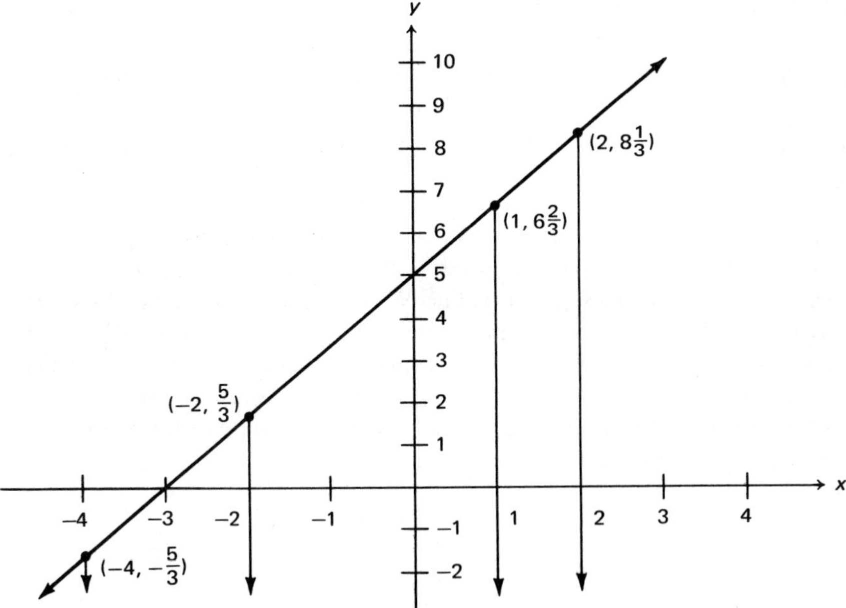

If we continue this process indefinitely, the union of all the infinite half-lines is the half-plane consisting of all points below the line.

Figure 8.4

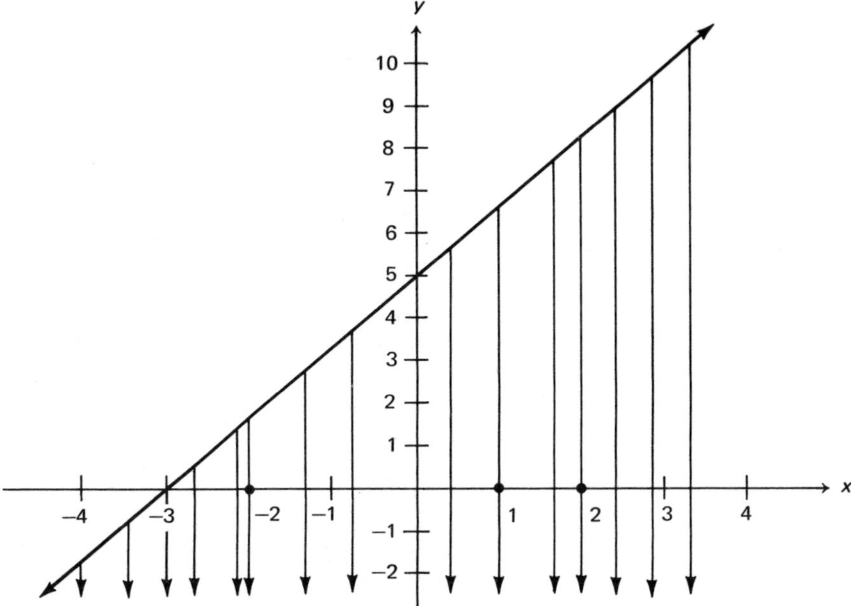

The generalization from this problem is a valid one; that is, the graph of any inequality of the form

$$ax + by \leqslant c \qquad \text{or} \qquad ax + by \geqslant c$$

is the line

$$ax + by = c,$$

along with all points on one side of this line and none of the points on the other side of it.

Example 3 Solve the inequality $7x - 3y \geqslant 2$ by finding its graph.

Solution We know that the graph is the union of the line $7x - 3y = 2$ with all points on one side of that line. We only need two points on the line to locate all of its points, and so we pick the points where the graph crosses the x-axis and where it crosses the y-axis. These points are known respectively as the x-intercept and the y-intercept. Since any point on the x-axis has a y-coordinate of 0, to find the x-coordinate of the x-intercept we replace y by 0 in the equation of the line.

$$\begin{aligned} &7x - 3y = 2 \\ &7x - 3 \cdot 0 = 2 \\ &7x = 2 \\ &x = 2/7 \end{aligned}$$

Then the x-intercept $(2/7, 0)$ is one of the points on the line. To find the y-intercept we similarly realize that its x-coordinate must be 0 (the x-coordinate of any point on the y-axis is 0) and

$$\begin{aligned} &7x - 3y = 2 \\ &7 \cdot 0 - 3y = 2 \\ &-3y = 2 \\ &y = -2/3. \end{aligned}$$

Thus $(0, -2/3)$ is the second point on the line, and we now have all the points on the line. We could have arbitrarily picked other points and numbers to work with, but the choice of the intercepts simplifies the arithmetic after substitution.

What about the graph of $7x - 3y > 2$? Since $(0, 0)$ is not on the line $7x - 3y = 2$, and we know that the graph of the inequality is either all the points on that side of the line containing $(0, 0)$ or else all the points on the opposite side of that line, we substitute 0 for x and 0 for y to see if we have a true statement. If we do, we shade the half-plane containing $(0, 0)$; if we do not, we shade the other half-plane.

$$\begin{aligned} &7x - 3y \overset{?}{>} 2 \\ &7 \cdot 0 - 3 \cdot 0 \overset{?}{>} 2 \\ &0 \overset{?}{>} 2 \end{aligned}$$

Since the answer is *no*, we shade the other side to complete the entire solution.

Figure 8.5

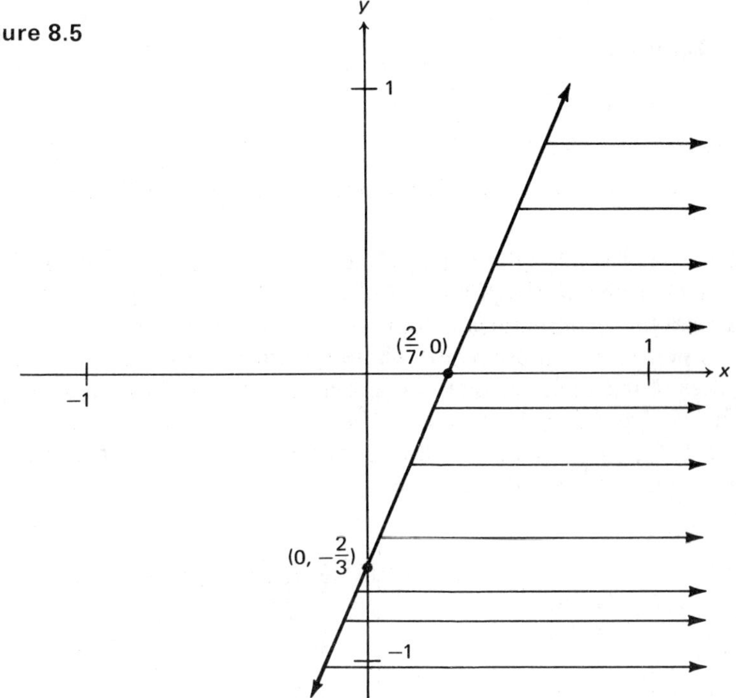

We could have picked any other other point not on the line—(1/2, 2) is clearly above the line—but substitution of (0, 0) makes for easy arithmetic. Notice also that we did not solve the original inequality for y in terms of x, as we did in the previous example. We will do all of our graphing problems of inequalities in two variables by this find-the-intercepts, test-(0, 0) method of Example 3.

Example 4 Solve graphically the simultaneous system of inequalities

$$x - 2y > 4, \qquad 2x + 3y < 6, \qquad x \geqslant 1.$$

Solution We work with the inequalities one at a time. The graph of $x - 2y > 4$ is all the points on one side of the line whose equation is $x - 2y = 4$. Since the line itself is not included in the graph here—$x - 2y > 4$ is a strict inequality—we use a dotted line instead of a solid line in the graph.

Table 8.1

x	y
0	−2
4	0

$$0 - 2 \cdot 0 \overset{?}{>} 4$$

$$0 \overset{?}{>} 4$$

No, so the side opposite (0, 0) is shaded.

The line whose equation is $2x + 3y = 6$ likewise is not included in the graph of $2x + 3y < 6$, so again we will represent its boundary role by a dotted line.

Table 8.2

x	y
0	2
3	0

$$2 \cdot 0 + 3 \cdot 0 \overset{?}{<} 6$$

$$0 \overset{?}{<} 6$$

Yes, so the side including (0, 0) is shaded.

The graph of $x = 1$ is the line through (1, 0) parallel to the y-axis, since we are looking for all the points whose x-coordinates are 1, regardless of the value of the y-coordinates. The graph of $x > 1$ is found by realizing that we are looking for all those points whose x-coordinates are greater than 1, regardless of the values of their y-coordinates, and we thus have the half plane to the right of the line $x = 1$. The point $(3, -4)$, for example, is in the graph of $\{(x, y) \mid x \geqslant 1\}$, since its x-coordinate of 3 is greater than 1 and no restriction is placed on the value of y.

Figure 8.6

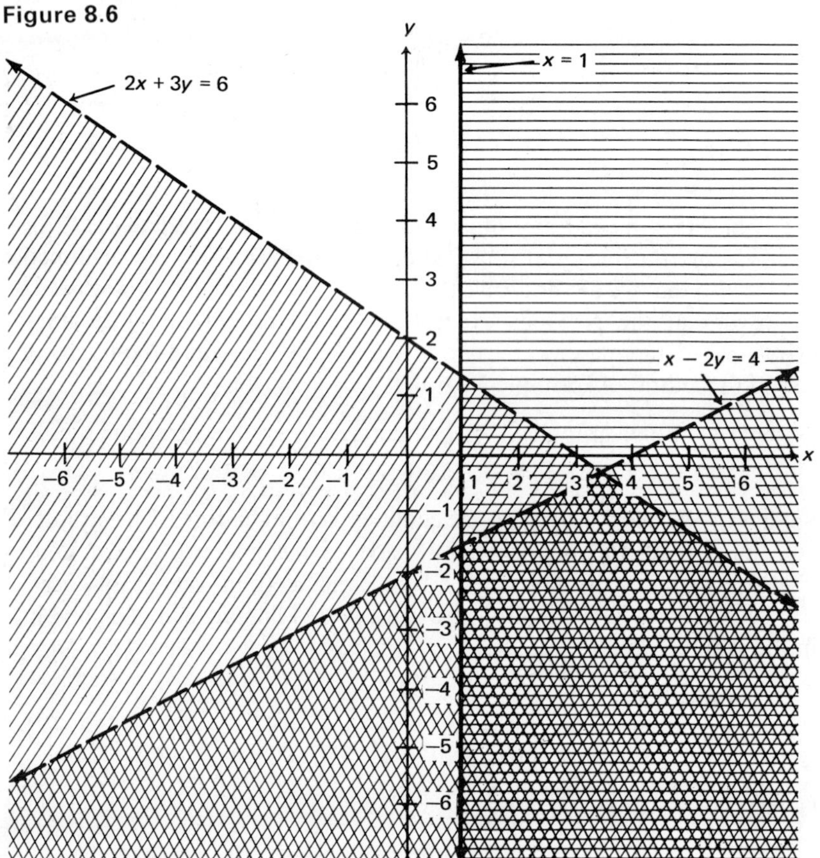

If we place these three graphs on the same set of axes and shade the first one \\\\, the second one ////, and the third one ≡, then the region where *all* three shadings overlap is the graphical solution. The best algebraic description we can give of this region is

$$\{(x, y) \mid (x - 2y > 4) \quad \textit{and} \quad (2x + 3y < 6) \quad \textit{and} \quad (x \geqslant 1)\}.$$

Problem set 8–1

1. Find the intercepts of the graph of each equation.
   a. $x - 3y = 1$
   b. $3x - y = 6$
   c. $x + y = 2$
   d. $2x + 5y = 8$
   e. $x = 4$
   f. $y = -6$
   g. $2x - 3y = 12$
   h. $4.1x - 2.6y = 10$
   i. $14 - 3x = 8y$
   j. $13y + 7 = -5x$
2. There are three points where the boundary lines of the half-planes in Figure 8.6 intersect. Express the coordinates of these three points algebraically.
3. Determine whether these statements are true or false with reference to Example 4.
   a. $(3, -5)$ is in the graph of all three inequalities.
   b. $(7, -2)$ is in the graph of exactly two inequalities.
   c. $(-4, 10)$ satisfies the first inequality but not the last two.
   d. $(3, 4)$ satisfies none of the three inequalities.

   Sketch the graphs of the following seven inequalities.
4. $3x + y \leqslant 1$
5. $-2x + 3y \geqslant -6$
6. $4x - 7y < 5$
7. $x \geqslant 3.6$
8. $y < -2$
9. $y > 2x + 1$
10. $-3x \leqslant 7 - 5y$

    Solve the following eight systems of inequalities graphically.
11. $x - 2y \geqslant -4$
    $x + y \geqslant 2$
12. $x + y \leqslant 8$
    $x - 3y > 6$
13. $3x + y \geqslant 5$
    $2x - 3y \leqslant 7$
14. $x - 3y \geqslant 6$
    $2x - 6y \geqslant 8$
15. $x - 3y \geqslant 6$
    $2x - 6y \leqslant 8$
16. $3x - y \leqslant 3$
    $x + y > 2$
    $x \geqslant -4$

17. $x - 3y \leqslant 6$
    $1.5x + y > 6$
    $x \geqslant 0$
    $y \geqslant 0$
18. $x \geqslant -3$
    $y \leqslant 2$
    $x < 3$
    $y > -4$
19. Find the four inequalities whose graphs intersect to form the interior of the rectangle with vertices at $(1, 3)$, $(5, 3)$, $(1, 7)$, $(5, 7)$.
20. Find the four inequalities whose graphs intersect to form the interior of the parallelogram with vertices at $A = (2, 4)$; $B = (3, 8)$; $C = (5, 1)$; and $D = (4, -3)$.

Solve this system of inequalities graphically.

21. $x + 2y \leqslant 20$
    $5x + 3y \leqslant 44$
    $3x - 2y \leqslant 15$
    $4x + 5y \geqslant 40$
    $4x + 5y \leqslant 50$
    $x \geqslant 0$
    $y \geqslant 0$

Objective functions and their optimal values

We return briefly to the problem of linear programming at the beginning of the previous section. The conditions there were that "each transmission requires one man-hour on the welder, each rear axle assembly requires two man-hours on the same welder, and the welder can not be used more than twenty man-hours per eight-hour shift." If we let x be the number of transmissions and let y be the number of rear axles to produce per shift, then $1 \cdot x$ is the number of man-hours spent by the welder on the transmissions and $2 \cdot y$ is the number of man-hours spent by the welder on the rear axles. The sum of these amounts of time, $1x + 2y$, must be at most 20: $x + 2y \leqslant 20$. We learned how to solve inequalities such as $x + 2y \leqslant 20$ in the previous section precisely so that we can use this learning to solve linear programming problems.

Finding maximum and minimum values

In this same problem the total profit from the transmissions and rear axles produced is $\$6 \cdot x + \$7 \cdot y$ or, in equation form,

$$P = 6x + 7y.$$

There are thus two classes of requirements placed on the production planning department. It must schedule x transmissions and y rear axles per shift to stay within the physical requirements of available machinery and manpower while simultaneously maximizing P. In the previous section we learned how to work with several inequalities in two variables x and y; in this section we learn how to find the maximum values of *objective* functions (the *objective* here is to maximize the profit P) such as

$$P = f(x, y) = 6x + 7y.$$

We proceed with an example.

Example 5 Consider the function

$$C = f(x, y) = 3x + 8y$$

whose domain is the intersection of these four sets:

$$\{(x, y) \mid 5x - 6y \geqslant -30\}$$
$$\{(x, y) \mid 5x + 3y \leqslant 60\}$$
$$\{(x, y) \mid x \geqslant 0\}$$
$$\{(x, y) \mid y \geqslant 0\}.$$

Find the largest value of C (a range element) and the pairs of numbers (x, y) (domain elements) where this largest value of C occurs.

Solution We begin by finding the intersection of the first two sets within the first quadrant only, since x and y must both be nonnegative from the last two sets. We use the intercepts of the two lines given by

$$5x - 6y = -30 \qquad \text{and} \qquad 5x + 3y = 60$$

to locate these lines. We use the algebraic method of elimination and substitution or Cramer's Rule to see that their point of intersection is $(6, 10)$ (right?). Notice the arrows we use to indicate the shadings of the regions.

Figure 8.7

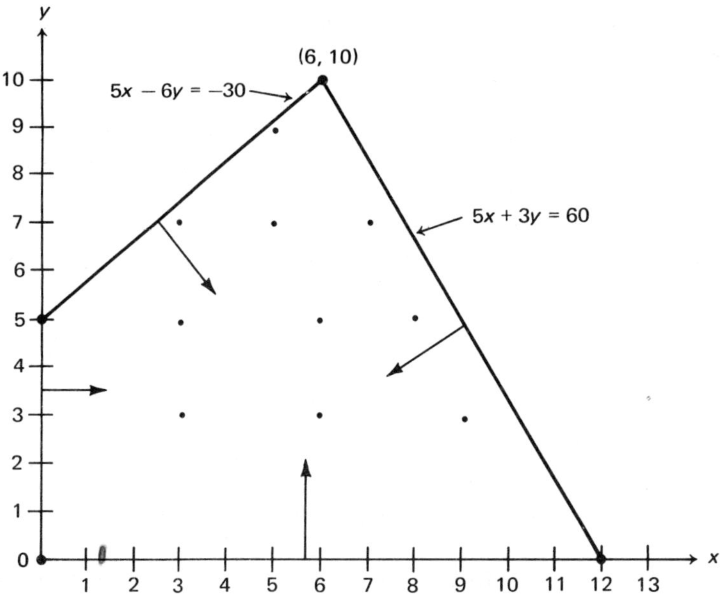

Since we are trying to maximize $C = f(x, y) = 3x + 8y$ for the points (x, y) inside or on the quadrilateral, we list several points (x, y), including all four vertices, for which we evaluate the expression $C = 3x + 8y$. These points are indicated in Table 8.3 below. It is impossible to list all the infinite real number pairs inside or on the quadrilateral, but we hope that the chosen pairs of integers may help us to find the maximum value of C. Those values of C with an * indicate that they occur at a vertex of the quadrilateral.

Table 8.3

x	y	$C = f(x, y) = 3x + 8y$
0	0	0*
3	0	9
6	0	18
9	0	27
12	0	36*
0	3	24
3	3	33
6	3	42
9	3	51
0	5	40*
3	5	49
6	5	58
8	5	64
3	7	65
5	7	71
7	7	77
5	9	87
6	10	98*

Inspection of the values in the right-hand column shows that *the minimum value of C* (0) *and the maximum value of C* (98) *occur at vertices of the quadrilateral,* which is the boundary of the domain of the function $C = f(x, y) = 3x + 8y$. This is a very important property exhibited in this example, and we are ready to state without proof the generalization of this specific case (though Problem 12 should strengthen our conviction of the general property).

Basic linear programming theorem

Let S be a set of points in the coordinate plane bounded by a polygon with convex corners, and let $C = f(x, y) = ax + by + c$ be a function with S as its domain. Then the maximum and minimum values of C occur among the corner points of set S.

By a convex corner we mean one that is "sticking out" (think of "*exit,*" which means "out of") in contrast to a con*cave* ("*cave*d in") polygon. The distinction is clear in Figure 8.8.

Figure 8.8.

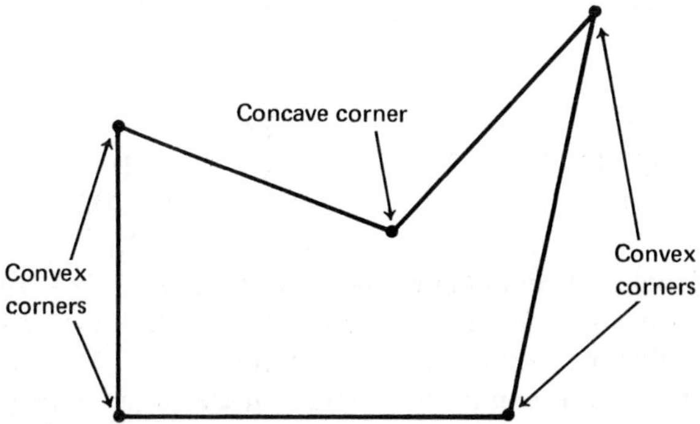

In Example 5 $a = 3$, $b = 8$, and $c = 0$. The proof of this theorem is omitted because of its complexity. There may be valid points in the domain which yield values of C greater than or less than the values of C at certain vertices (corner points) of the valid region, but the maximum or minimum value occurs at *one* of the vertices. For examples, at $(6, 3)$ C has the value 42, which exceeds the value of C at the vertex $(0, 5)$, but the maximum value of C (98) and the minimum value of C (0) occur at the respective vertices $(6, 10)$ and $(0, 0)$.

Now we can summarize, the procedure for finding the maximum and minimum value of a function

$$C = f(x, y) = ax + by + c$$

with a convex polygon as its domain: since the polygon is the intersection of several linear inequalities, locate this polygon by graphing the linear inequalities on the same set of coordinate axes. Use Cramer's Rule if necessary to locate the corner points of the polygon. Then substitute the number pairs (x, y) from these corner points into the expression

$$C = f(x, y) = ax + by + c$$

to find the pair (x, y) that produces the smallest and largest corresponding values of C.

Problem set 8–2

1. Find the maximum and minimum values of the function

$$C = f(x, y) = 5x + 7y$$

with the intersection of these inequalities as the domain. In all cases assume that $x \geqslant 0$ and $y \geqslant 0$, as will always be the case in future linear programming problems. We make this assumption because, for example, it does not makes any sense to produce -3 rear axles, -17 transmissions, or a negative amount of any product.

a. $2x + 3y \leqslant 15$
$3x + y \leqslant 12$

b. $2x + y \leqslant 7$
$3x + y \leqslant 8$

c. $x + y \leqslant 7$
$2x + y \leqslant 9$
$3x + y \leqslant 12$

d. $x + y \geqslant 2$
$2x + 3y \leqslant 15$
$3x + y \leqslant 12$

e. $x + 3y \leqslant 21$
$2x + 3y \leqslant 24$
$2x + y \leqslant 16$

f. $2x + 3y \geqslant 6$
$x + 3y \leqslant 21$
$2x + 3y \leqslant 24$
$2x + y \leqslant 16$

2. Rework Problem 1, but change the function to $D = 7x + 5y$.
3. Rework Problem 1, but change the function to $E = 5x - 7y$.
4. Rework Problem 1, but change the function to $F = 7x - 5y$.

In all the maximum-minimum problems done so far both maximum and minimum values have existed for the objective function. Some problems have an unbounded region for the domain, so that only a minimum value of the objective function C exists:

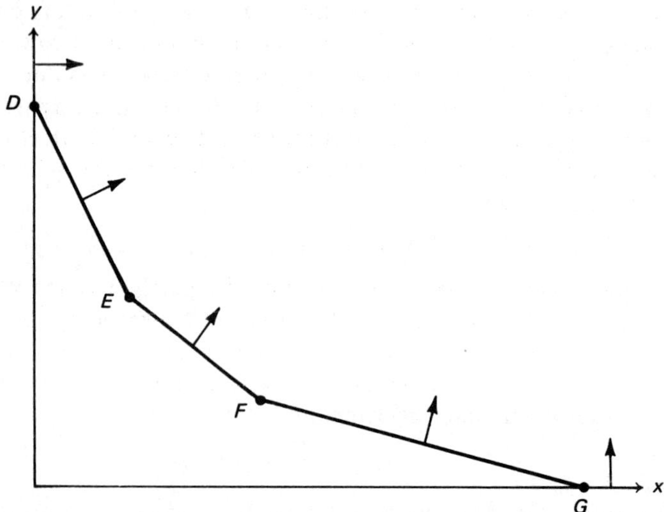

In this case the minimum value of C still exists at one of the corner points D, E, F, and G. There is no maximum value of C because as the valid points are chosen further and further from the origin, the corresponding values of C increase without bound. In rare circumstances the maximum or minimum value of an objective function may be the same at two adjacent corner points. In this case all the points on the line segment joining the two corner points will produce the maximum or minimum value of C; it therefore makes no difference which point on that segment we choose.

5. Find the maximum and minimum values of the function $P = f(x, y) = 4x + 5y + 2$ with the intersection of the inequalities $x \geqslant 0$, $y \geqslant 0$, $x + y \geqslant 5$, and $3x + 5y \geqslant 21$ as its domain.

6. Rework Problem 5, but change the objective function to $Q = 4x + 5y$.
7. Rework Problem 5, but change the objective function to $R = 5x + 4y$.
8. Rework Problem 5, but change the objective function to $S = 4x + 4y + 3$.
9. Optimize (find both max and min values where possible) $C = f(x, y) = 25x + 35y$ subject to the *constraints* (the inequalities involving x and y are often called constraints because they *constrain*, or set limits upon, the permissible values of x and y) $x \geqslant 0$, $y \geqslant 0$, $x + y \geqslant 2$, $2x + 3y \leqslant 15$, and $3x + y \leqslant 12$.
10. Minimize $C = .07x + .05y$ subject to the constraints $x \geqslant 0$, $y \geqslant 0$, $x - 3y \leqslant 0$, $2x - 6y \geqslant 0$.
11. Optimize $C = 2x - 9y$ subject to the constraints $0 \leqslant x \leqslant 7$, $y \geqslant 0$, $x - 3y \leqslant 0$, $2x - 6y + 9 \geqslant 0$.
12. While the general proof of the Basic Linear Programming Theorem is justifiably omitted because of its complexity, certain intuitive considerations might make this theorem seem more plausible than the inspection of only one table, Table 8.3. We can increase the credibility by the following extension of Example 5.

    Suppose we choose an arbitrary value for C, the number we are trying to maximize. Then the graph of $3x + 8y = C$ is a straight line. If this line passes through the region illustrated in Figure 8.7, then there are values of x and y acceptable to the inequalities that will allow C to have this value, and C is thus below (or possibly at) its maximum. If, on the other hand, the line does not pass through the region at all, no admissible x and y will give this value of C, and thus C is above its permissible maximum under the inequalities involving x and y. For example, a value of $C = 74$ occurs at $(6, 7)$, or $x = 6$ and $y = 7$. A value of $C = 104$ is not possible under the circumstances because the graph of $3x + 8y = 104$ is a line beyond the domain of the function.

    As we take various arbitrary values of C, the lines thus formed lie parallel to one another. Their slopes are all equal to $-3/8$, which is seen after rewriting

$$3x + 8y = C$$

as

$$y = -\frac{3}{8} \cdot x + \frac{C}{8}.$$

Since the intercepts of each line are $(0, C/8)$ and $(C/3, 0)$, then larger values of C will have as their graphs lines which lie farther from the origin. For example, the line through $(0, 48/8)$ and $(48/3, 0)$ is farther from the origin than the line through $(0, 24/8)$ and $(24/3, 0)$. The corresponding lines will thus move farther and farther from the origin as C grows, and they will give the maximum permissible C when a line just passes out of the region. The following diagram, which is an expansion of Figure 8.7, shows in geometric fashion that this maximum value of C will occur at a vertex. In this case the vertex has coordinates (6, 10).

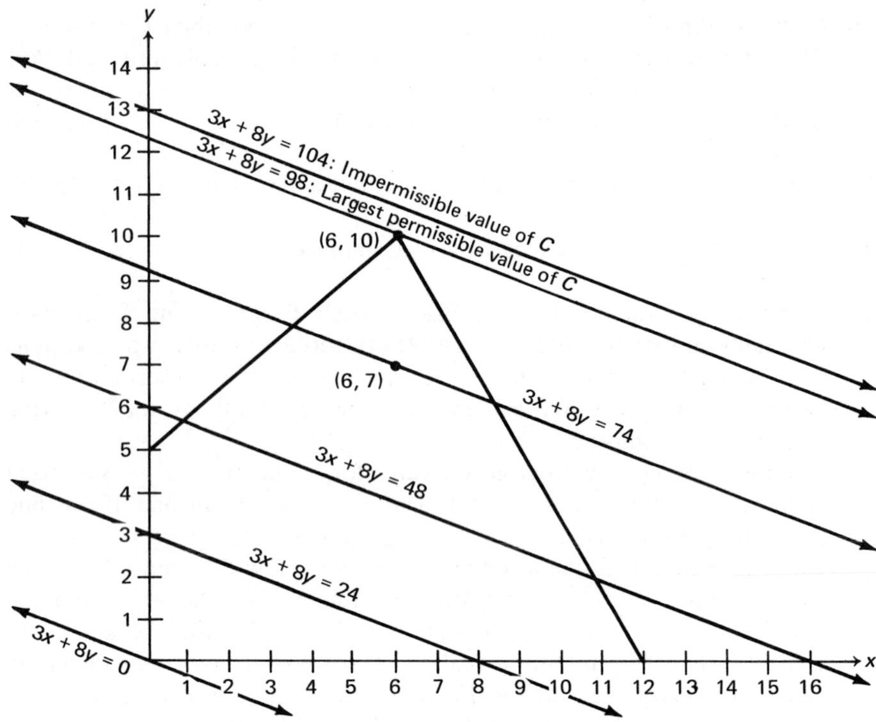

a. Copy Figure 8.7 on a separate sheet of paper with units of 1/2″ for both the horizontal and the vertical axis. Use the length of the paper for the x-axis if you are using an 8 1/2″ by 11″ sheet of paper.
b. Construct the graph of $3x+8y=72$ on top of your copied Figure 8.7.
c. Construct the graph of $3x+8y=90$ on top of your copied Figure 8.7.
d. Construct the graph of $3x+8y=108$ on top of your copied Figure 8.7.
e. Find the point where the objective function $C=f(x,y)=5x+3y$ is maximized in Figure 8.7, and construct the graph of $5x+3y=C$ for this maximum value of C.

Graphic solutions of linear programming problems

We are now, finally, able to solve the Ajax Manufacturing Company's problem about its production schedule of transmissions and rear axles (Example 1). Keep in mind the two major simultaneous requirements placed upon us. We must

1. work within the physical limitations of machines, manpower, and tax requirements, and
2. maximize the profit to the company.

There are five major steps we use to solve this problem and all similar ones.

Five-step method

1. Clearly identify in words the *numbers* asked for by the problem and then represent them by letters. Here, we are trying to find

   the number of transmissions to produce per eight-hour shift

   and

   the number of rear axles to produce per eight-hour shift.

   Then we represent these number by letters.

   Let x = the number of transmissions to produce per eight-hour shift.

   Let y = the number of rear axles to produce per eight-hour shift.

2. Write the constraints (inequalities) involving the numbers from step 1 according to the information in the problem description. This abstraction step is the heart of the process and is also the most difficult. Recall the method we used with the word problems in Chapter 1 when we first wrote *in words* the common sense physical idea exhibited in the problem. There are three basic restrictions placed on the transmissions and rear axles by the welder, the manual installation, and the tax law. Here is how we translate the restriction from the welder.

$$(\text{total number of hours spent on the welder}) \leqslant 20$$

$$\begin{pmatrix}\text{number of hours}\\ \text{welding transmissions}\end{pmatrix} + \begin{pmatrix}\text{number of hours}\\ \text{welding rear axles}\end{pmatrix} \leqslant 20$$

$$\begin{pmatrix}\text{number of hours}\\ \text{welding one transmission}\end{pmatrix} \cdot \begin{pmatrix}\text{number of}\\ \text{transmissions}\end{pmatrix}$$

$$+ \begin{pmatrix}\text{number of hours}\\ \text{welding one rear axle}\end{pmatrix} \cdot \begin{pmatrix}\text{number of}\\ \text{rear axles}\end{pmatrix} \leqslant 20$$

$$1x + 2y \leqslant 20$$

We proceed in a similar manner for the restriction from the manual installation of parts.

$$(\text{total number of hours spent installing parts}) \leqslant 44$$

$$\begin{pmatrix}\text{number of hours installing}\\ \text{transmission parts}\end{pmatrix} + \begin{pmatrix}\text{number of hours installing}\\ \text{rear axle parts}\end{pmatrix} \leqslant 44$$

$$\begin{pmatrix}\text{number of hours installing}\\ \text{parts in one transmission}\end{pmatrix} \cdot \begin{pmatrix}\text{number of}\\ \text{transmissions}\end{pmatrix}$$

$$+ \begin{pmatrix}\text{number of hours installing}\\ \text{parts in one rear axle}\end{pmatrix} \cdot \begin{pmatrix}\text{number of}\\ \text{rear axles}\end{pmatrix} \leqslant 44$$

$$5x + 3y \leqslant 44$$

Now we consider the tax law restriction placed on x and y.

(three times the number of transmissions)
– (two times the number of rear axles) $\leqslant 15$

$3x - 2y \leqslant 15.$

3. Write the objective function—usually to maximize profit or to minimize cost—in an expression of the form $F = ax + by + c$ where x and y are the letters from step 1. Here

$$P = f(x, y) = 6x + 7y,$$

since the total profit is the sum of the total profit from the transmissions $(6 \cdot x)$ and the total profit from the rear axles $(7 \cdot y)$.
4. Graph the constraints from step 3 in the first quadrant, since x and y are always nonnegative, using the intercepts to locate two points on each line and Cramer's Rule, if necessary, to locate the points of intersection of the lines.

$x + 2y \leqslant 20$

Table 8.4

x	y
0	10
20	0

(0, 0) side is shaded

$5x + 3y \leqslant 44$

Table 8.5

x	y
0	44/3
44/5	0

(0, 0) side is shaded

$3x - 2y \leqslant 15$

Table 8.6

x	y
0	−15/2
5	0
10	15/2

(0, 0) side is shaded

If any intercept is outside the first quadrant, then merely pick a positive value for x or y so that its corresponding y or x value is positive, as we did in Table 8.6. Then we solve pairs of equations simultaneously to locate the corner points of the convex set.

$$x + 2y = 20$$
$$5x + 3y = 44$$
$$x = \frac{D_x}{D}$$
$$x = \begin{vmatrix} 20 & 2 \\ 44 & 3 \end{vmatrix} \div \begin{vmatrix} 1 & 2 \\ 5 & 3 \end{vmatrix}$$
$$x = (60 - 88)/(3 - 10)$$
$$x = -28/-7$$
$$x = 4$$

$$5x + 3y = 44$$
$$3x - 2y = 15$$
$$x = \frac{D_x}{D}$$
$$x = \begin{vmatrix} 44 & 3 \\ 15 & -2 \end{vmatrix} \div \begin{vmatrix} 5 & 3 \\ 3 & -2 \end{vmatrix}$$
$$x = (-88 - 45)/(-10 - 9)$$
$$x = -133/-19$$
$$x = 7$$

By Substitution	By Substitution
$4 + 2y = 20$	$5 \cdot 7 + 3y = 44$
$2y = 16$	$35 + 3y = 44$
$y = 8$	$3y = 9$
$(4, 8)$	$y = 3$
	$(7, 3)$

Figure 8.9

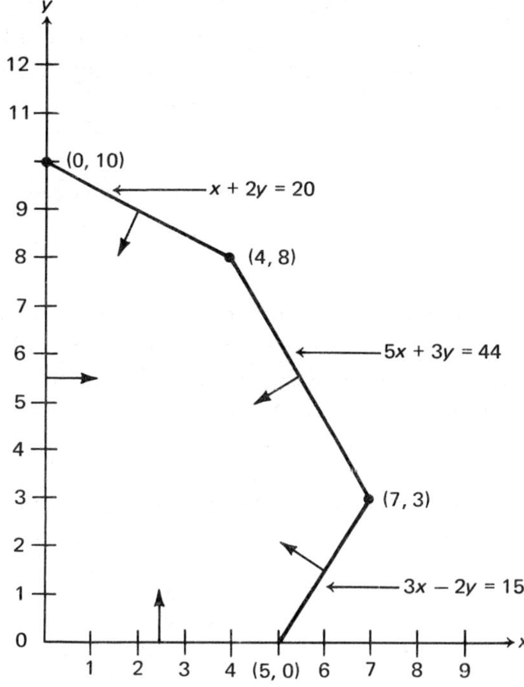

If the region in the diagram was not a convex polygonal one, we could go no further, since the theorem that lets us find the optimal value of the objective function holds true only for convex sets.

5. Evaluate the objective function at the corner points of the convex set from step 4, and determine the optimal value and its point (x, y) of occurrence.

Table 8.7.

Corner point, (x, y)	Profit $= 6x + 7y$
(0, 0)	$0 + 0 = 0$
(0, 10)	$0 + 70 = 70$
(4, 8)	$24 + 56 = 80$
(7, 3)	$42 + 21 = 63$
(5, 0)	$30 + 0 = 30$

The maximum profit is \$80, occurring at $(x = 4,\ y = 8)$, so the interpretation for Ajax is that the company should produce four transmissions and eight rear axles per shift to stay within its physical resources while maximizing profit.

In linear programming problems it is often convenient to call the points with coordinates (x, y) in the domain *feasible* points; that is, the *feasible* points are exactly those which satisfy all the *constraints* (inequalities) of the problem. The function to be optimized is considered defined only on feasible points. The *optimal* point (x, y), which produces the maximum or minimum range value of the function, must be an *extreme* (corner) point of the set of feasible points. (The one possible and necessary modification to this statement is discussed in Problem 4 of Problem set 8–2.) This terminology is useful, general, and quite common.

Finding optimal values

Example 6 A chemical product is a mixture of two ingredients, A and B. The total quantity produced per day must equal or exceed 1,000 gallons but not be more than 1,500 gallons. At least 200 gallons of the mixture must be ingredient A. The mixture must contain at least 300 gallons of pure alcohol. Ingredient A has 40% alcohol and ingredient B has 24% alcohol. If ingredient A costs \$1 per gallon and ingredient B \$2 per gallon, then how much of each should be used in the mixture to minimize costs?

Solution We proceed by the five-step method of the previous example.

1. Let x = number of gallons of ingredient A to be used in the mixture. Let y = number of gallons of ingredient B to be used in the mixture.
2. Since (total cost) = (total cost of A) + (total cost of B), we have

$$C = f(x, y) = 1 \cdot x + 2 \cdot y$$

as the objective function to be minimized.

3. Total quantity produced of the mixture must equal or exceed 1,000 gallons:

$$x + y \geqslant 1{,}000.$$

Total quantity produced of the mixture must not be more than 1,500 gallons:

$$x + y \leqslant 1{,}500.$$

At least 200 gallons of the mixture must be A:

$$x \geqslant 200.$$

A is 40% alcohol, B is 24% alcohol, and 300 gallons of *pure* alcohol must be produced as a minimum:

$$\begin{pmatrix}\text{number of gallons of}\\ \textit{pure}\text{ alcohol from } A\end{pmatrix} + \begin{pmatrix}\text{number of gallons of}\\ \textit{pure}\text{ alcohol from } B\end{pmatrix} \geqslant 300$$

40% of $A + 24\%$ of $B \geqslant 300$

$.40x + .24y \geqslant 300$

4. We construct the graphs of the previous inequalities after doing the preliminary algebra.

$x + y \geqslant 1000$

Table 8.8

x	y
0	1000
1000	0

(0, 0) side is not shaded

$x + y \leqslant 1500$

Table 8.9

x	y
0	1500
1500	0

(0, 0) side is shaded

$x \geqslant 200$

The graph is thc vertical line through (200,0) and all points to the right of it.

$.40x + .24y \geqslant 300$

Table 8.10

x	y
0	1250
750	0

(0, 0) side is not shaded

$x + y = 1{,}000$
$x + y = 1{,}500$
Since the sum of two numbers can not be 1,000 and 1,500 simultaneously there is no point of intersection; the graph is two parallel lines.

$x + y = 1{,}000$
$x = 200$
By substitution
$200 + y = 1{,}000$
$y = 800$
(200, 800)

$x + y = 1{,}000$
$.40x + .24y = 300$
$x = D_x/D$

$$x = \begin{vmatrix} 1{,}000 & 1 \\ 300 & .24 \end{vmatrix} \div \begin{vmatrix} 1 & 1 \\ .4 & .24 \end{vmatrix}$$

$x = (240 - 300)/(.24 - .4)$
$x = -60/-.16$
$x = 375$

$x + y = 1{,}500$
$x = 200$
By substitution
$200 + y = 1{,}500$
$y = 1{,}300$
(200, 1,300)

By substitution
$375 + y = 1{,}00$
$y = 625$
$(375, 625)$

$x + y = 1{,}500$
$.40x + .24y = 300$
By Cramer's Rule and substitution $x = -375$ and $y = 1{,}875$ (Right?). $(-375, 1{,}875)$ is not in the first quadrant so it is not a corner point.

$x = 200$
$.40x + .24y = 300$
By substitution
$.40 \cdot 200 + .24y = 300$
$80 + .24y = 300$
$.24y = 220$
$y = 220/.24$
$y \doteq 917$
$(200, 917)$

Figure 8.10

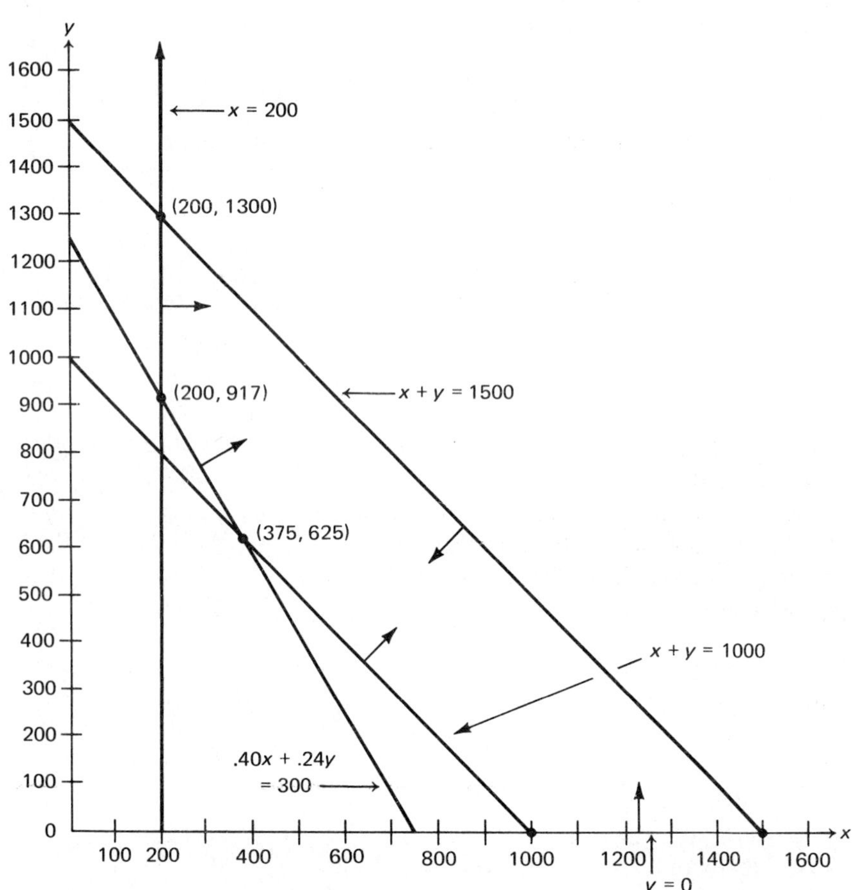

5. We test the corner points for optimal values.

Table 8.11

x	y	$C = x + 2y$
1000	0	1000
375	625	1625
200	1300	2800
200	917	2034
1500	0	1500

Therefore, the chemical company should use 1,000 gallons of ingredient A and 0 gallons of ingredient B to meet the requirements at minimal cost.

Example 7 A manufacturer of decorative glass bottles produces 200 bottles each week. His equipment turns out bottles of three quality grades, I, II, and III. He has a contract to sell twenty-five bottles of grade I and fifty bottles of grade II each week. He can sell all that he produces of grade II quality but can sell at most one hundred bottles of grade I and one hundred bottles of grade III. His profit per bottle on grade I is \$40, on grade II it is \$25, and on grade III it is \$30. How many of each kind of bottle should he produce weekly to maximize his profit?

Solution We are asked for three numbers here—the number of bottles per week of grade I, grade II, and grade III. If we let x, y, and z represent these respective numbers, then

$$x + y + z = 200;$$

since exactly 200 bottles are produced weekly,

$$z = 200 - x - y.$$

For example, if $x = 30$ and $y = 50$ then z would equal $200 - 30 - 50$, or 120. As long as we can express one of the unknowns in terms of x and y, we can solve linear programming problems in three variables. Otherwise, the graphic approach with two axes for the two variables can not be used. We continue with a condensed outline of the five-step method.

1. Let x = number of bottles produced weekly of grade I quality.
   Let y = number of bottles produced weekly of grade II quality.
   Let $(200 - x - y)$ = number of bottles produced weekly of grade III quality.
2. $P = f(x, y) = 40x + 25y + 30(200 - x - y)$
   $= 10x - 5y + 6{,}000$ (Right?)
3. He sells a minimum of 25 and a maximum of 100 bottles of grade I:

   $$25 \leqslant x \leqslant 100.$$

   He sells a minimum of 50 bottles of grade II:

   $$y \geqslant 50.$$

He sells a maximum of 100 bottles of grade III:

$$200 - x - y \leqslant 100 \quad \text{or}$$

$$x + y \geqslant 100.$$

It is implied that the number of grade III bottles is nonnegative:

$$200 - x - y \geqslant 0 \quad \text{or}$$

$$x + y \leqslant 200.$$

4. This graph shows the result of algebraically determining the intercepts of the lines and their points of intersection. Be sure to check the algebra.

Figure 8.11

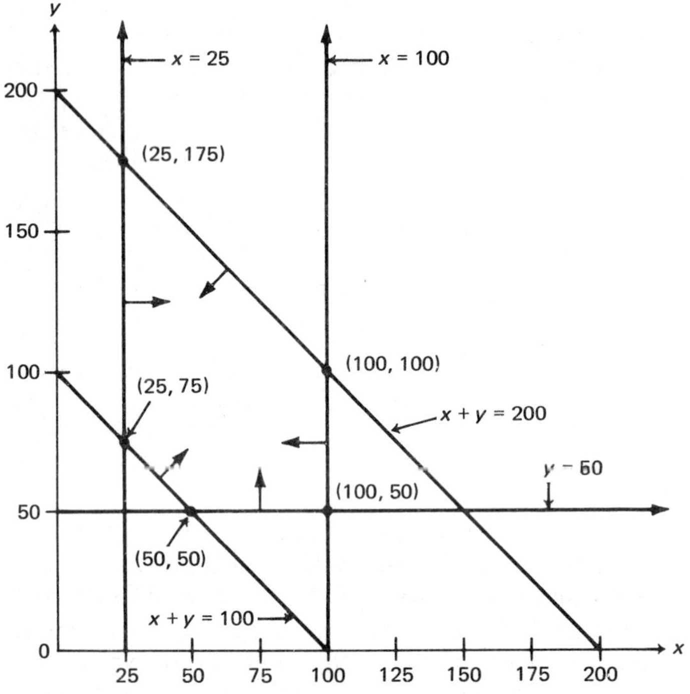

Table 8.12

x	y	$P = 10x - 5y + 6000$
25	75	5875
25	175	5375
100	100	6500
100	50	6750
50	50	6250

The maximum profit of \$6,750 occurs where $x = 100$ and $y = 50$, so

$$z = 200 - x - y = 200 - 100 - 50 = 50.$$

The final interpretation of this problem is that the manufacturer should produce one hundred grade I bottles, fifty grade II bottles, and fifty grade III bottles for maximum profit.

Problem set 8–3

1. A manufacturer makes two kinds of chairs, high-back and rockers. A high-back requires three hours on the lathe and two hours on the sander, while each rocker requires two hours on the lathe and four hours on the sander. The profit on high-back chairs is \$14 each, and the profit on each rocker is \$12. If the lathe operates twelve hours daily and the sander operates sixteen hours daily, then how many of each kind of chair should be produced to maximize the profit?
2. A machine accounting firm employs day and night shifts. The two shifts work only as required to comply with three contracts of the firm, which we designate as Contracts A, B, and C. The day shift can perform six assignments for Contract A, ten assignments for Contract B, and twelve assignments for Contract C per day. The night shift can perform eight assignments for Contract A, nine assignments for Contract B, and four assignments for Contract C per day. During a one-month period Contract A calls for 200 assignments, Contract B 250 assignments, and Contract C 300 assignments. Assuming that the costs of operating the day and night shifts are identical at \$400 per shift, how many days should each shift be required to work during the month for the most economical operation, assuming a month of thirty days?
3. The maximum daily production of a chemical refinery is 1,000 barrels. The refinery produces two grades of chemicals, A and B. A minimum of 500 barrels of Chemical A is required to meet scheduled demands. A minimum of 200 barrels per day of Chemical B is required to meet scheduled demands. The profit on Chemical A is \$2 per barrel, and on Chemical B it is \$1 per barrel. What is the maximum possible profit per day which can be realized from the refinery?
4. The PCD Company manufactures two types of products, gidgets and widgets. Three machines, M_1, M_2, and M_3 are used for the manufacture of each product. One gidget requires one hour on M_1, one hour on M_2, and two hours on M_3; one widget requires three hours on M_1, one hour on M_2, and one hour on M_3. On any working day M_1 is available for twenty-four hours, M_2 is available for ten hours, and M_3 is available for sixteen hours. Each gidget brings in a profit of \$20, and each widget brings in a profit of \$30. Find the number of gidgets and widgets to produce daily that will maximize the profit. How many hours a day is each of the machines M_1, M_2, and M_3 being utilized when (x, y) produces a maximum value of P?

5. The King Concrete Company manufactures bags of concrete from two ingredients, A and B. Each pound of ingredient A costs 6¢ and contains four units of fine sand, three units of coarse sand, and five units of gravel. Each pound of B costs 10¢ and contains three units of fine sand, six units of coarse sand, and two units of gravel. Each bag of concrete must contain at least twelve units of fine sand, twelve units of coarse sand, and ten units of gravel. Graphically find the best combination of ingredients A and B which will meet the minimum requirements of fine sand, coarse sand, and gravel at the least cost; indicate the cost per bag and per pound of concrete.
6. A wholesale distributor of art reproductions manufactures her own wooden frames, two kinds, which she gives away free with the paintings. It takes twenty, ten, and ten minutes to cut, assemble, and finish the first kind of frame, which cost her $1.20 each; and ten, twenty, and sixty minutes to cut, assemble, and finish the other kind of frame, which costs her $3.60 each. If a certain production run calls for at least eight hours of cutting, at least ten hours of assembling, and at least eighteen hours of finishing, how many frames of each kind should she schedule so as to minimize her cost? What is this minimum cost?
7. The Conlon Corporation makes chairs and tables. The company can realize a profit of $5 on each chair and $10 on each table. Each chair requires one hour on the machine and two hours of skilled labor. Each table requires three hours on the machine and one hour of skilled labor. The company has a maximum of nine hours on the machine available and a maximum of eight hours of skilled labor available. Determine the numbers of chairs and tables that should be produced in order to maximize Mr. Conlon's profit.
8. Work Problem 7, given the revised information that the company can realize a profit of $5 on each chair and $15 on each table.
9. Work Problem 7, given the revised information that the company can realize a profit of $5 on each chair and $20 on each table.
10. The quarterback on a football team is faced with the following situation: there are 145 seconds of time remaining in the game; there are 40 yards to go for a touchdown; and his team is behind by 5 points. In a previous serious of plays, off-tackle slants averaged 6 yards per try, while end runs averaged 12 yards per try. However, he knows that it takes 20 seconds to run off-tackle and 30 seconds for an end run. What possible combinations of end runs and off-tackle slants would yield a touchdown in the allotted time if the averages were maintained? If the opposition has an excellent pass completion record and all three of its time-outs left, which of these combinations should the quarterback use so that scoring exactly one 6-point touchdown while using as much of the 145 seconds as possible is the objective?
11. A small furniture manufacturer makes only two products, desks and chairs. The manufacturer has four departments: a fabricating department in which both chairs and desks are worked on; an assembly department in which

both chairs and desks are assembled; an upholstery department through which only the chairs are processed; and a linoleum department in which the linoleum desk tops are made. During the next production period there are 27,000 minutes available in each of the departments, and this available time cannot be changed within the scope of the problem. Firm time standards have been established in each of the departments: in fabricating 15 minutes is required per chair and 40 minutes per desk; in assembly 12 minutes is required per chair and 50 minutes per desk; in upholstery 18.75 minutes is required per chair; and in linoleum 56.25 minutes is required per desk. If the profit is $25 per chair and $75 per desk, then how many of each should be produced to maximize the profit? What is this maximum total profit?

12. Complete this table with reference to Problem 11, where your conclusion should have been that the maximum profit occurs with the production of $x = 1{,}000$ chairs and $y = 300$ desks.

Department	Unused time, in minutes, for $x = 1{,}000$ and $y = 300$
Fabricating	$27{,}000 - (15 \cdot 1{,}000 + 40 \cdot 300) = 0$
Assembly	
Upholstery	
Linoleum	

13. Work Problems 11 and 12 again with the one change that each chair contributes a profit of $25 and each desk contributes a profit of $40.
14. A cosmetic company is making a new hair rinse called HyPer. To be competitive they must be able to advertise their product as having a minimum of the "miracle" ingredients krypton (10%), thallium (30%), and zenof (not more than 25%). The company has two sources for these ingredients. One supplier will sell them a basic material, Major, which is 20% krypton, 20% thallium, and 20% zenof. Another supplier has a basic material, Minor, which is 5% krypton, 30% thallium, and 50% zenof. Major costs $1.50 per gallon, and Minor costs $2.25 per gallon. In order to minimize costs, what mixture of these two (along with a water filler, if needed) should be used to manufacture 100 gallons of HyPer?
15. A television producer is trying to arrange a half-hour show. He plans to use a comedian, an orchestra, and commercials to fill the time. The comic requires a guarantee of $2,000 and works at the rate of $200 a minute, so he will have to be used for at least ten minutes. The orchestra is paid at the rate of $500 per minute and requires a $2,500 guarantee. The commercials cost $100 per minute to produce, and the sponsor demands at least three minutes of commercials. The producer refuses to produce a show having more than ten minutes of commercials. Let x symbolize the number of minutes the comic is used and y the number of minutes the orchestra is used. Then $30 - x - y$ is the number of minutes of commercials. Find the

time arrangement of the show which allows the producer to meet all requirements at a minimum cost to the sponsor.

16. When the leader from the previous problem learns that the producer plans to use his band for only five minutes, he tells the producer that if he will use the band for at least ten minutes, then the band will work for $300 per minute. Does this fact affect the producer's earlier decision?
17. Suppose that in Problem 15 there are 100,000 people who will watch the television show, no matter what is on, but that 1,000 of these turn off their sets for each minute of commercials. Suppose, also, that for every minute the band is on, 30,000 more viewers will watch the show and that for every minute the comedian is on, 25,000 more viewers will watch the show. Choose x and y so that the number of viewers of the show is a maximum.
18. A farmer decides to raise a total of 1,000 capons, geese, and turkeys, with not more than 200 geese included. His facilities require him to raise at least as many turkeys as capons, and at most 600 turkeys. He anticipates profits of $2 per capon, $1.50 per turkey, and $2.25 per goose. How many of each type of fowl should he raise in order to obtain the largest possible profit? What is this profit?
19. The President of the Ruff Dog Food Company wishes to prepare a food mixture, and considers using three compounds, R, S, and T, which contain, by weight, the various percentages of the components carbohydrates, fats, and protein shown here.

Compound	Carbohydrates	Fats	Protein
R	45%	20%	35%
S	45%	45%	10%
T	5%	25%	35%

The mixture is to contain at least 25% of each component. The costs per pound of R, S, and T are 86¢, 58¢, and 94¢, respectively. Determine how many pounds of R, S, and T should be used in 100 pounds of the mixture to minimize the cost while still meeting the dietary requirements. What amounts of R, S, and T should be used in 200 pounds of the mixture? 300 pounds? 500 pounds? P pounds?

20. A drug company plans to produce a cough syrup containing in diluted liquid form an antihistamine, A, a barbituate compound, B, and an aspirin compound, C. In the mixture, by weight, at least 20% but not more than 50% should be A; at least 30% should be B, and the amount of B should be greater than the amount of A; and the amount of C plus the amount of A should exceed 30% of the mixture. The costs of A, B, and C per ounce are $5, $3, and $4, respectively. What percentages of the syrup by weight should be A, B, and C in order to minimize the cost in a batch of 100 ounces? 200 ounces? 300 ounces? 1000 ounces? P ounces?

Simplex algorithm

The linear programming problems we have done so far have been somewhat limited in their practical value since few problems from the real world are as simple and direct as the ones we can do. There are often hundreds of variables and hundreds of constraints in a problem involving, say, a metal stamping plant with thousands of possible products and dozens of machines to form the metal parts. At the present time we can only work with problems having two variables. We could learn enough about solid geometry to let us solve linear programming problems with three products. However, a problem involving the manufacture of four or more products simply can not be done using a purely graphic approach.

Fortunately, there is available a strictly algebraic process to solve linear programming problems in two variables or two hundred variables and with two constraints or two hundred constraints. It is the *Simplex Algorithm* and proceeds solely along algebraic lines with no graphing needed. This algorithm, or series of repeated steps, can require thousands of relatively easy but time-consuming manual calculations that are made to order for a computer. Working through the Simplex Algorithm not only gives us the product mix which optimizes the objective function, but it also gives us additional information, such as the unused capabilities of a particular machine in the manufacturing process; see Problem 12 of the previous section for an example. It also shows us which particular machine among the many manufacturing a product limits the overall manufacturing process.

We do not have time to learn the Simplex Algorithm here. It is important, though, to know that a purely algebraic approach to solving linear programming problems does exist, even if we are not learning it. Dr. J. Ronald Frazer's book, *Applied Linear Programming* (Englewood Cliffs: Prentice-Hall, 1968), is an excellent text which gives a very complete and understandable explanation of the Algorithm. Dr. Frazer has included a computer program, based on the Simplex Algorithm, which will solve linear programming problems with a relatively large number of products and constraints.

Chapter 9

Differential Calculus

Introduction to maximum-minimum problems

Example 1 A salesman believes that the annual sales rate of one of his company's products, metal washers, has been exactly determined as a function of time according to the formula

$$S = f(t) = t^3 - t^2 - 10t + 17.$$

Here $-3 \leqslant t \leqslant 5$, t is the number of years in the past or future (the present is $t = 0$), and S is the annual sales rate in thousands of washers. If we completely accept his assumption, then what are the maximum and minimum values of S corresponding to the specified domain elements, and for what values of t do these values of S occur?

Solution It seems reasonable to calculate S corresponding to those whole numbers t between -3 and 5 inclusive in the domain, plot the resulting points, join them in a natural fashion, and then answer the question presented to us in the problem. For example, if $t = 4$, then its corresponding value of S is

$$S = f(4) = 4^3 - 4^2 - 10 \cdot 4 + 17 = 64 - 16 - 40 + 17 = 25.$$

Carrying out the calculations for the remaining values of t gives us the following table.

Table 9.1

t	−3	−2	−1	0	1	2	3	4	5
$S = f(t)$	11	25	25	17	7	1	5	25	67

Next, we construct the graph.

Figure 9.1

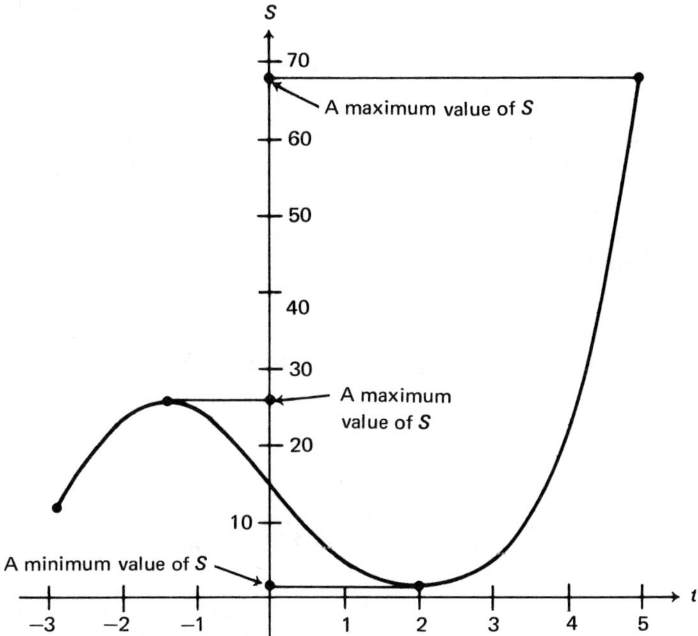

Notice that we also calculated and plotted the point $(-1.5, f(-1.5)) = (-1.5, 26.375)$ (right?) because we did not know whether the graph dipped or rose as a point moved along the curve from $(-2, 25)$ to $(-1, 25)$.

It appears that the minimum value of S in this interval is about 1, occurring at about $t = 2$, and that the maximum value of S in this interval is clearly 67, occurring at $t = 5$. There is a "turning" point or plateau at about $(-1.5, 26.375)$ where the graph is at a maximum distance from the t-axis in the interval of the domain $\{t \mid -2 \leqslant t \leqslant -1\}$. Likewise, at about $(2, 1)$ is a turning point or plateau where the graph is at a minimum distance from the t-axis in the interval of the domain

$$\{t \mid 1 \leqslant t \leqslant 3\}.$$

The last paragraph is very intuitive. *Minimum value*, *maximum value*, *turning point*, and *plateau* are all undefined but nevertheless qualitatively understood terms. The purpose of this chapter is to quantify precisely the qualitative terms mentioned above. We will then be able to find quickly the exact coordinates of the plateau points mentioned above by a process known as *differentiation*.

Locating the turning points

Until we learn this process of differentiation we still must try to locate the turning points exactly. We see from the graph that one of the turning points is located above some value of t between -2 and -1, so we use a computer to evaluate the expression

$$S = t^3 - t^2 - 10t + 17$$

for those values of t from -2 to -1 in steps, or increments, of .1. We can think of this evaluation process as examining the interval

$$\{t \mid -2 \leqslant t \leqslant -1\}$$

under a magnifying glass.

Table 9.2

t	−2	−1.9	−1.8	−1.7	−1.6	−1.5
S	25	25.531	25.928	26.297	26.344	26.375

t	−1.4	−1.3	−1.2	−1.1	−1.0
S	26.296	26.223	25.832	25.459	25.000

As t increases from -1.6 to -1.5, S increases from 26.344 to 26.375. As t increases from -1.5 to -1.4, S decreases from 26.375 to 26.296, so the graph turns from increasing to decreasing in the interval

$$\{t \mid -1.6 \leqslant t \leqslant -1.4\}.$$

Now we use the computer again as a stronger magnifying glass to examine S as t is stepped by .02 in this interval.

Table 9.3

t	−1.60	−1.58	−1.56	−1.54	−1.52	−1.50
S	26.344	26.3593	26.37	26.3761	26.3778	26.375

t	−1.48	−1.46	−1.44	−1.42	−1.40
S	26.3678	26.3563	26.3404	26.3203	26.296

From examining this table we conclude that the turning point has coordinates close to $(-1.52, 26.3778)$. We could strengthen the magnifying glass by calculating $t^3 - t^2 - 10t + 17$ for those values of t between -1.54 and -1.50 in increments of .001. We will not do so but will settle instead for two-decimal-place accuracy for our estimate of the value of t which produces a maximum value of S.

What about the turning point or plateau point close to (2, 1)? We examine the values of S corresponding to values of t between 1 and 3 in steps of .2.

Table 9.4

t	1	1.2	1.4	1.6	1.8	2
S	7	5.288	3.784	2.536	1.592	1

t	2.2	2.4	2.6	2.8	3
S	.808	1.064	1.816	3.112	5

The turning point must have a t-coordinate between 2.0 and 2.4 because the values of S decrease between $t = 2.0$ and $t = 2.2$ and then increase between $t = 2.2$ and $t = 2.4$. Thus, our magnifying glass, the computer, is focused on these values of t in steps of .02.

Table 9.5

t	2	2.02	2.04	2.06	2.08
S	1	.962006	.928062	.898216	.872513

t	2.1	2.12	2.14	2.16
S	.850998	.833725	.820744	.812096

t	2.18	2.2	2.22	2.24
S	.807831	.807999	.812649	.821823

We do not go any higher in our calculations since at $t = 2.2$ the S values are now beginning to increase. We conclude that (2.20, .807999) is very close to the true turning point of the graph.

The calculations have been considerable in our investigation, and this trial-and-error method, which gives only approximate results, is a very awkward one to use. A knowledge of differential calculus will eliminate the trial-and-error approach and give exact results. In our example the true turning points have t-coordinates of $(1-\sqrt{31})/3$ and $(1+\sqrt{31})/3$, with respective S-coordinate approximations of 26.37782928 and .80735591. The maximum value of S is 67.

Problem set 9–1

1. Verify that these points are on the graph of

$$S = f(t) = t^3 - t^2 - 10t + 17.$$

a. (3, 5) b. (−3, 11) c. (−1.4, 26.296)
d. (−1.52, 26.3778) e. (2.1, .851) f. (0, 17)

Find the maximum and minimum values of the ranges of the following functions within the specified domains. Test the smallest and largest value of the domain to see if it produces a maximum or minimum range value. Use the value of x to the nearest .1, obtained by the trial-and-error process, while checking for turning or plateau points. In problems 2 through 8 there is at most one plateau point in the interval. In problems 9 through 12 there are at most two plateau points in the interval. We will justify these two statements in a later section of this chapter. Also, sketch each graph.

2. $y = f(x) = x^2 - 4x + 7, \quad -3 \leqslant x \leqslant 5$
3. $y = f(x) = x^2 + 6x + 13, \quad -5 \leqslant x \leqslant 1$
4. $y = f(x) = x^2 - 8x + 14, \quad -1 \leqslant x \leqslant 7$
5. $y = f(x) = x^2 - 5.2x + 8.26, \quad 1 \leqslant x \leqslant 4$
6. $y = f(x) = x^2 + 6.8x + 13.56, \quad -5 \leqslant x \leqslant -1$
7. $y = f(x) = x^2 + 6.8x + 6.56, \quad -5 \leqslant x \leqslant -1$
8. $y = f(x) = x^2 - 6.8x + 11.56, \quad 2 \leqslant x \leqslant 6$
9. $y = f(x) = x^3 - 12x, \quad -4.1 \leqslant x \leqslant 4.1$
10. $y = f(x) = x^3 - 6x^2 + 12x - 7, \quad 0 \leqslant x \leqslant 6$
11. $y = f(x) = x^3 - 9x^2 + 24x - 14, \quad 0 \leqslant x \leqslant 6$
12. $y = f(x) = 2x^3 + 3x^2 - 36x + 6, \quad -4 \leqslant x \leqslant 4$
13. The complete BASIC computer program needed to do the calculations for Table 9.1 is listed below.

```
10 FOR T = −3 TO 5 STEP 1
20 LET S = T↑3 − T↑2 − 10*T + 17
30 PRINT T, S
40 NEXT T
50 END
```

Statement 10 instructs the machine to assign values to T starting at -3 and stopping at 5, inclusive, in STEPs of 1. Statement 20 says to evaluate $t^3 - t^2 - 10t + 17$ for the current value assigned to t (the operation of exponentiation, such as t raised to the exponent 3, is represented by an up-arrow: ↑) and place the numerical result in S. Statement 30 PRINTs the values of T and S in that order on the same line. Statement 40 causes the computer to get the NEXT value of T and repeat the above two-step process of calculating and printing. The cycle continues until T has equalled 5. Statement 50 ENDs the program and tells the computer to carry out all previous instructions listed before it.

Write the complete BASIC program to evaluate

$$S = t^3 - t^2 - 10t + 17$$

for these values of t: $-2, -1.9, -1.8, -1.7, \ldots, -1.1, -1$. The program should print these values of t along with their corresponding values of S.

14. Write the complete BASIC program to evaluate and print

$$S = t^3 - t^2 - 10t + 17$$

for these values of t: $-1.6, -1.58, -1.56, \ldots, -1.42, -1.4$.

15. Write the complete BASIC program to evaluate and print

$$y = x^3 - x^2 - 10x + 17$$

for these values of x: -1.6, -1.58, -1.56, ..., -1.42, -1.4. When the computer actually executes these instructions, do you think there will be any difference between the output from Problem 14 and the output from Problem 15?

16. Write the complete BASIC program to evaluate

$$y = x^2 - 6.8x + 11.56$$

for these values of x: 2, 2.2, 2.4, 2.6, ..., 5.6, 5.8, 6.

Limits

We are learning some mathematical ideas and concepts that are necessary for the solution of certain real-world problems. The connections among limits, rates of change (the next section), and maximum or minimum values of a function are not easily apparent, but by the end of this chapter we will see their relationships.

Example 2 Consider the function defined by the equation

$$y = f(x) = x^2 - 4x + 9.$$

If $x = 3$, then $y = f(3) = 3^2 - 4 \cdot 3 + 9 = 6$, or $f(3) = 6$. Suppose we observe the independent variable as it begins at $x = -4$ and increases in value toward $x = 3$. What happens to the corresponding values of $y = f(x)$ as x changes in value?

Solution We use a computer to investigate since the calculations are somewhat time consuming.

Table 9.6

X	X↑2 - 4*X + 9
-4	41
-3	30
-2	21
-1	14
0	9
1	6
1.5	5.25
2	5
2.2	5.04
2.4	5.16
2.6	5.36
2.8	5.64
2.9	5.81
2.92	5.8464
2.94	5.8836
2.96	5.9216
2.98	5.9604
2.99	5.9801
2.992	5.98406
2.994	5.98804
2.996	5.99202
2.998	5.996
2.999	5.998

The results are consistent with common sense. After all, if $f(x)$ is 6 when $x=3$ we expect that the values of $f(x)$ for x near 3 should be near 6. In Table 9.6 x approached 3 from below; let us see what happens when x approaches 3 from above, as indicated in Table 9.7.

Table 9.7

X	X↑2 - 4*X + 9
10	69
9	54
8	41
7	30
6	21
5	14
4.5	11.25
4	9
3.8	8.24
3.6	7.56
3.4	6.96
3.2	6.44
3.1	6.21
3.08	6.1664
3.06	6.1236
3.04	6.0816
3.02	6.0404
3.01	6.0201
3.008	6.01606
3.006	6.01204
3.004	6.00802
3.002	6.004
3.001	6.002

Again, the results are consistent with common sense.

If we summarize these last two tables, one way of expressing the pattern of specific results is to say that as the values of x approach 3, either from below or from above, then the corresponding values of $y=f(x)=x^2-4x+9$ approach 6. Symbolically,

$$\operatorname*{limit}_{x\to 3}(x^2-4x+9) = 6.$$

This equation is read "the limit, as x approaches 3, of x^2-4x+9 is 6."

We may generalize to the case of any function f, any specific domain element a to be approached from below *and* from above by x, and the specific number L approached by the y's or $f(x)$'s corresponding to and depending upon the values of x in the domain.

Definition of a limit $\text{limit}_{x\to a} f(x) = L$ if, as x comes closer and closer to a from below and from above, the corresponding values of $y=f(x)$ come closer and closer to L.

In the previous example we were able to find $\text{limit}_{x\to 3} f(x) = L = 6$ by merely plugging the domain element value of $x=3$ into the expression

x^2-4x+9 and arithmetically evaluating the resulting $3^2-4\cdot 3+9$ to obtain $\text{limit}_{x\to 3} f(x) = L = 6$. In many situations involving relatively simple and "well-behaved" functions this plug-in method gives the correct value for L. We must be cautious about applying it indiscriminately, as the following example shows.

Example 3 Find

$$\underset{x\to 3}{\text{limit}} \frac{x^2-2x-3}{x^2-x-6}.$$

Solution By plugging $x=3$ into the expression we have

$$\frac{3^2-2\cdot 3-3}{3^2-3-6} = \frac{0}{0},$$

which is not defined. However, we subtly creep up on $x=3$, as shown in Table 9.8.

Table 9.8

x	$\frac{x^2-2x-3}{x^2-x-6}$
6	.875000
5	.857143
4.5	.846154
4	.833333
3.8	.827586
3.6	.821429
3.4	.814815
3.2	.807692
3.1	.803922
3.06	.802372
3.02	.800797
3.01	.800399
3.006	.800231
3.003	.800120
3.001	.800040
----	------
1	.666667
2	.750000
2.5	.777778
2.9	.795918
2.99	.799599
2.999	.799960

The conclusion is that

$$\underset{x\to 3}{\text{limit}} \frac{x^2-2x-3}{x^2-x-6} = .8.$$

This example shows that the ultimate concern is what value x forces the dependent variable y to come toward as x *approaches* a fixed number, rather than what happens to y when x actually *arrives* at the fixed number.

Five basic limit theorems

There are five basic theorems that we can deduce from the previous definition. However, we state these theorems without proof but with examples.

Limit Theorem 1 If $f(x) = x^n$, where n is a nonnegative integer, then

$$\lim_{x \to a} f(x) = a^n.$$

Example 4 If $f(x) = x^2$, then $\text{limit}_{x\to 4} f(x) = 4^2 = 16$.

Example 5 If $f(x) = x^3$, then $\text{limit}_{x\to 6} f(x) = 6^3 = 216$.

Example 6 If $f(x) = x^5$, then $\text{limit}_{x\to 2} f(x) = 2^5 = 32$.

Limit Theorem 2 If $H(x)$ is the sum of functions $F(x)$ and $G(x)$: $H(x) = [F(x)+G(x)]$, then

$$\begin{aligned}\lim_{x\to a} H(x) &= \lim_{x\to a}[F(x)+G(x)]\\ &= \lim_{x\to a} F(x) + \lim_{x\to a} G(x),\end{aligned}$$

if all the limits exist.

Example 7 Let $F(x) = 3x+5$ and let $G(x) = 4x^2 - 6x + 1$, so that

$$\begin{aligned}H(x) &= [F(x)+G(x)] = [(3x+5)+(4x^2-6x+1)]\\ &= 4x^2 - 3x + 6.\end{aligned}$$

Then $\text{limit}_{x\to 7} H(x) = 181$, $\text{limit}_{x\to 7} F(x) = 26$, and $\text{limit}_{x\to 7} G(x) = 155$, as illustrated by both the plug-in method (right?) and the numerical calculations in Table 9.9.

Table 9.9

X	F(X) = 3*X + 5	G(X) = 4*X↑2 - 6*X + 1	H(X) = [F(X) + G(X)] = 4*X↑2 - 3*X + 6
2	11	5	16
3	14	19	33
4	17	41	58
5	20	71	91
5.5	21.5	89	110.5
6	23	109	132
6.2	23.6	117.56	141.16
6.4	24.2	126.44	150.64
6.6	24.8	135.64	160.44
6.8	25.4	145.16	170.56

6.9	25.7	150.04	175.74
6.92	25.76	151.026	176.786
6.94	25.82	152.014	177.834
6.96	25.88	153.006	178.886
6.98	25.94	154.002	179.942
6.99	25.97	154.5	180.47
6.992	25.976	154.6	180.576
6.994	25.982	154.7	180.682
6.996	25.988	154.8	180.788
6.998	25.994	154.9	180.894
6.999	25.997	154.95	180.947

Since $181 = 26 + 155$, we see that

$$\underset{x\to 7}{\text{limit}}\, H(x) = \underset{x\to 7}{\text{limit}}\, F(x) + \underset{x\to 7}{\text{limit}}\, G(x)$$

is true in this case. In words we say, "The limit of the sum is the sum of the limits." This theorem also applies where $H(x)$ is the sum of any number of functions.

Limit Theorem 3 If $H(x)$ is the difference between functions $F(x)$ and $G(x)$: $H(x) = [F(x) - G(x)]$, then

$$\underset{x\to a}{\text{limit}}\, H(x) = \underset{x\to a}{\text{limit}}\, [F(x) - G(x)]$$

$$= \underset{x\to a}{\text{limit}}\, F(x) - \underset{x\to a}{\text{limit}}\, G(x),$$

if all the limits exist.

Example 8 If $F(x) = 3x^2 - 2$ and $G(x) = 5x + 7$, then

$$H(x) = [F(x) - G(x)] = [(3x^2 - 2) - (5x + 7)]$$
$$= 3x^2 - 5x - 9.$$

Thus $\text{limit}_{x\to 4} F(x) = 46$, $\text{limit}_{x\to 4}\, G(x) = 27$, and $\text{limit}_{x\to 4}\, H(x) = 19$ by the plug-in method. Since

$$19 = 46 - 27,$$

we see that

$$\underset{x\to 4}{\text{limit}}\, H(x) = \underset{x\to 4}{\text{limit}}\, F(x) - \underset{x\to 4}{\text{limit}}\, G(x).$$

In words we say, "The limit of the difference is the difference of the limits."

Limit Theorem 4 If $G(x) = c \cdot F(x)$, then

$$\underset{x\to a}{\text{limit}}\, G(x) = \underset{x\to a}{\text{limit}}\, c \cdot F(x) = c \cdot \underset{x\to a}{\text{limit}}\, F(x),$$

again assuming that all the limits exist.

Example 9 Let $F(x) = -4x^2 + 2x - 7$, let $c = 3$, and let $a = 5$, so that we want to compare

$$\lim_{x \to 5} 3 \cdot (-4x^2 + 2x - 7)$$

with

$$3 \cdot \lim_{x \to 5} (-4x^2 + 2x - 7).$$

The second column of Table 9.10, as well as the plug-in method, illustrates that

$$\lim_{x \to 5} 3 \cdot (-4x^2 + 2x - 7) = -291.$$

Table 9.10

X	3*(-4*X↑2 + 2*X - 7)	(-4*X↑2 + 2*X - 7)
Ø	-21	-7
1	-27	-9
2	-57	-19
3	-111	-37
3.5	-147	-49
4	-189	-63
4.2	-207.48	-69.16
4.4	-226.92	-75.64
4.6	-247.32	-82.44
4.8	-268.68	-89.56
4.9	-279.72	-93.24
4.92	-281.957	-93.9856
4.94	-284.203	-94.7344
4.96	-286.459	-95.4864
4.98	-288.725	-96.2416
4.99	-289.861	-96.6204
4.992	-290.089	-96.6962
4.994	-290.316	-96.7722
4.996	-290.544	-96.8481
4.998	-290.772	-96.924
4.999	-290.886	-96.962

The third column of Table 9.10, as well as the plug-in method, illustrates that

$$3 \cdot \lim_{x \to 5} (-4x^2 + 2x - 7) = 3 \cdot -97 = -291.$$

The two limit expressions are equal because they both equal -291. In words this example says, "The limit of a constant times a function is that constant times the limit of that function."

Limit Theorem 5 If $F(x) = c$, then

$$\lim_{x \to a} F(x) = c.$$

Example 10 Let $F(x) = c = 7$ and let $a = 4$. Recall that $7 = 7 \cdot 1 = 7 \cdot x^0$, and then examine Table 9.11.

Table 9.11

x	$F(x) = 7 = 7 \cdot x^0$
1	$7 \cdot (1)^0 = 7$
1.5	$7 \cdot (1.5)^0 = 7$
2	$7 \cdot (2)^0 = 7$
2.5	$7 \cdot (2.5)^0 = 7$
3	$7 \cdot (3)^0 = 7$
3.5	$7 \cdot (3.5)^0 = 7$
3.7	$7 \cdot (3.7)^0 = 7$
3.9	$7 \cdot (3.9)^0 = 7$
3.95	$7 \cdot (3.95)^0 = 7$
3.99	$7 \cdot (3.99)^0 = 7$

Clearly, $\text{limit}_{x \to 4} 7 = 7$.

We have used the phrase "if these limits exist" without giving examples of where $\text{limit}_{x \to a} f(x)$ did not exist. Pay special attention to Problems 21 through 25 in this problem set.

Problem set 9–2

Evaluate these limits. Use the plug-in method in conjunction with the limit theorems, but also be prepared to work with a pattern of specific results.

1. $\underset{x \to 1}{\text{limit}} (4x + 7)$
2. $\underset{x \to 3}{\text{limit}} \dfrac{3 - x}{6}$
3. $\underset{x \to -4}{\text{limit}} (x/2 + 5)$
4. $\underset{x \to 7}{\text{limit}} (3x^2 - 2x)$
5. $\underset{x \to 4}{\text{limit}} (8)$
6. $\underset{x \to 5}{\text{limit}} (8)$
7. $\underset{x \to 6}{\text{limit}} (8)$
8. $\underset{x \to 2}{\text{limit}} \dfrac{x^2 + x - 6}{x - 2}$
9. $\underset{x \to 2}{\text{limit}} (x + 3)$
10. Factor the numerator of 8, and then compare the original answer with Problem 9.
11. $\underset{x \to 1/2}{\text{limit}} \dfrac{2x^2 + 5x - 3}{4x - 2}$
12. $\underset{x \to 3}{\text{limit}} \dfrac{2x^2 - x - 15}{7x^2 + 5}$
13. $\underset{x \to 3}{\text{limit}} (x^2)$
14. $\underset{x \to 3}{\text{limit}} (5x^2)$
15. $\underset{x \to 3}{\text{limit}} (5x^2 + 7x)$
16. $\underset{x \to 3}{\text{limit}} (5x^2 + 7x - 2)$
17. $\underset{x \to 2}{\text{limit}} (x^2 + 2x^3)$
18. $\underset{x \to 0}{\text{limit}} (3x^2 - 7x + 5)$
19. $\underset{x \to -5/3}{\text{limit}} \dfrac{3x^2 - x - 10}{3x + 5}$
20. $\underset{x \to -3}{\text{limit}} (7x + 10)$

21. $\lim_{x\to 3}(10/(x-3))$ Does the limit exist? Be sure to try $x = 2.7, 2.8, 2.9, 2.95, 2.99, 3.01, 3.05, 3.1, 3.2, 3.3$

22. Graph the function defined by the equation

$$f(x) = \frac{x^2+x-6}{x-2}, \qquad x \neq 2$$

for the domain $\{x \mid -4 \leqslant x \leqslant 4\}$, using the integers in that interval to obtain several $y = f(x)$ values. Also pay particular attention to values near $x = 2$ by using decimal values of x in that region.

23. Often a shipper offers lower rates per pound on large shipments. Ajax Fast Freight sets its freight rates by the schedule

$$C(x) = \begin{cases} .40x, & 0 < x < 200 \\ .35x, & 200 \leqslant x < 500 \\ .32x, & 500 \leqslant x < 1000, \end{cases}$$

where x represents the weight of one item and $C(x)$ is the corresponding cost of shipping that item.

a. Graph this function for $0 < x < 1000$.
b. Evaluate $\text{limit}_{x\to 200+} C(x)$. (This means that x approaches 200 through decreasing values of x which are greater than 200: 203, 202, 201, …: an example of a *right-hand limit.*)
c. Evaluate $\text{limit}_{x\to 200-} C(x)$. (This means that x approaches 200 through increasing values of x which are less than 200: 197, 198, 199, …: an example of a *left-hand limit.*)

Limit$_{x\to 200+} C(x) \neq \text{limit}_{x\to 200-} C(x)$, so $\text{limit}_{x\to 200} C(x)$ does not exist because the definition of a limit, which requires that the separate left-hand and right-hand limits be equal, is not satisfied.

24. Consider the function defined by

$$y = f(x) = |x|/x, \quad x \neq 0.$$

a. Evaluate $\text{limit}_{x\to 0-} f(x)$.
b. Evaluate $\text{limit}_{x\to 0+} f(x)$.
c. What is $\text{limit}_{x\to 0} f(x)$?

25. Consider the function defined by

$$y = f(x) = \frac{x-3}{x^2-3x}, \quad x \neq 3.$$

a. Evaluate $\text{limit}_{x\to 3-} f(x)$.
b. Evaluate $\text{limit}_{x\to 3+} f(x)$.
c. What is $\text{limit}_{x\to 3} f(x)$?

26. If $f(x) = 3x^2 - 5x$, then

$$\begin{aligned} f(x+h) &= 3(x+h)^2 - 5(x+h) \\ &= 3x^2 + 6xh + 3h^2 - 5x - 5h \quad \text{(right?)}, \end{aligned}$$

and thus

$$\frac{f(x+h)-f(x)}{h} = 6x + 3h - 5 \quad \text{(right?).}$$

Find

$$\lim_{h \to 0} \frac{f(x+h)-f(x)}{h} \quad \text{for} \quad f(x) = 3x^2 - 5x.$$

27. Evaluate

$$\lim_{x \to 3} \frac{x+3}{x^2-9}.$$

28. Evaluate

$$\lim_{x \to 3} \frac{x-3}{x^2-9}.$$

29. Refer to Example 3. Factor the numerator and denominator of the fractional expression, and evaluate the resulting expression at $x = 2.9$, 3, and 3.1. Does this work indicate more clearly why the limiting value of the original expression is $L = 6$ as x comes closer and closer to 3?

Rate of change of a function

In the introduction to Chapter Five we noted that two certainties of life were death and taxes. To this meager list we may add change. The world is constantly changing in such areas as population growth, communication devices, attitudes, values, and pollution. Calculus was originally developed to analyze changes in physical locations of objects along with applications to maximum-minimum problems. We will conduct our study of calculus by analyzing changes in functions descriptive of the world around us before proceeding to maximum-minimum problems.

Average rate of change

Definition Let $(a, f(a))$ and $(b, f(b))$ be two points on the graph of a continuous function given by $y = f(x)$. The number

$$\frac{f(b)-f(a)}{b-a},$$

which is symbolized by $\Delta y/\Delta x$ (pronounced "delta y over delta x"), is the *average rate of change of f in the interval* $[a, b]$.

A continuous function, as far as we are concerned, is one whose graph can be drawn without lifting the pencil from the paper. A good example of a discontinuous function is found in Problem 23 of the last section.

Example 11 Consider the function defined by

$$y = f(x) = -x^2 + 6x + 4$$

in the interval [1, 4]. What is its average rate of change there?
Solution According to the definition,

$$\frac{\Delta y}{\Delta x} = \frac{f(b) - f(a)}{b - a}.$$

Since $a = 1$ and $b = 4$, then $f(a) = f(1) = 9$ and $f(b) = f(4) = 12$. Hence

$$\frac{\Delta y}{\Delta x} = \frac{12 - 9}{4 - 1} = \frac{3}{3} = 1.$$

In words, as x has changed from 1 to 4 (Δx symbolizes the changes in the variable x, so $\Delta x = 3$), it has forced a change in the corresponding y values from 9 to 12 ($\Delta y = 3$). The ratio of Δy to Δx is 1 in this interval. The *pair* of symbols Δx symbolizes *one* number, which is the change in x. We can no more separate this pair into Δ and x than we can separate, for example, the letter "d" into "c" and "l".

Example 12 Consider the function defined by the equation

$$y = f(x) = -x^2 + 6x + 4$$

in the interval [1, 3.5]. What is its average rate of change there?
Solution Again we evaluate the expression

$$\begin{aligned}\frac{\Delta y}{\Delta x} &= \frac{f(b) - f(a)}{b - a}\\ &= \frac{f(3.5) - f(1)}{3.5 - 1}\\ &= \frac{12.75 - 9}{3.5 - 1}\\ &= \frac{3.75}{2.5}\\ &= 1.5.\end{aligned}$$

In words, as x changes from 1 to 3.5 ($\Delta x = 2.5$), it forces a change in the corresponding value of y from 9 to 12.75 ($\Delta y = 3.75$). This *average* figure of 1.5 units of change in y per 1 unit of change in x is not a *constant* figure throughout the interval. If x changes by 1 unit from 1.4 to 2.4, then y changes from 10.44 to 12.64 (Right?), which is a change of 2.2 units.

It is worthwhile to examine Example 12 from a geometrical viewpoint.

Figure 9.2

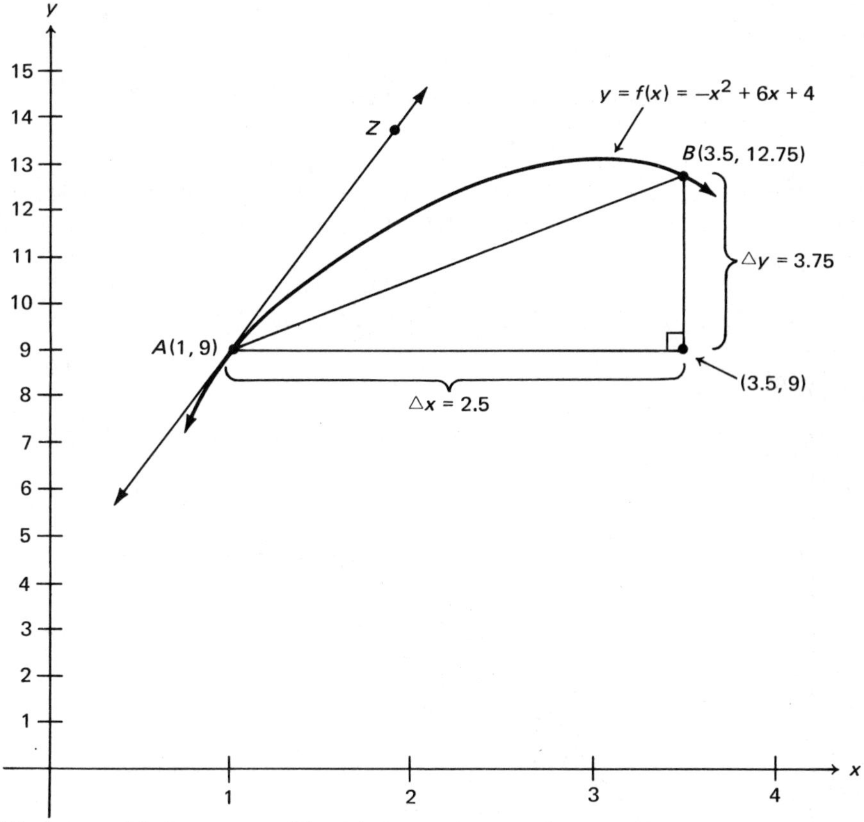

We have added some auxiliary information to the graph in addition to the curve joining $(1, 9) = (1, f(1))$ and $(3.5, 12.75) = (3.5, f(3.5))$. Line AB is an example of a *secant* line—a line joining two points on the curve. Line AZ is the *tangent* line to the graph at the point $(1, 9)$. Line AZ is the only tangent line which touches or "brushes" the curve at $(1, 9)$, while AB is one of many possible secant lines to the curve through $(1, 9)$. We will have a more formal definition of a tangent line later in this chapter.

If a point moves along the curve from $(1, 9)$ to $(3.5, 12.75)$ its x-coordinate changes from 1 to 3.5 or by 2.5 units—$\Delta x = 2.5$; at the same time its y-coordinate changes from 9 to 12.75 or by 3.75 units—$\Delta y = 3.75$. It is also true that the slope of the line AB is 1.5 since

$$(y_2 - y_1)/(x_2 - x_1) = 1.5.$$

Thus we see that

$$\begin{aligned}\frac{\Delta y}{\Delta x} &= \text{average rate of change of } f \text{ in } [1, 3.5] = 3.75/2.5\\ &= 1.5\\ &= \text{slope of the secant line through } (1, 9) \text{ and } (3.5, 12.75)\\ &= \text{slope of the secant line through } (1, f(1)) \text{ and } (3.5, f(3.5)).\end{aligned}$$

Instantaneous rate of change

This fact that

$$\frac{\Delta y}{\Delta x} = \text{slope of the secant line through } (1, f(1)) \text{ and } (3.5, f(3.5))$$

is a very important one, and it holds true for *any* continuous function in *any* interval. That is, if $(a, f(a))$ and $(b, f(b))$ are any two points on the graph of any continuous function, then the slope of the secant line joining them is given by

$$\frac{\Delta y}{\Delta x} = \frac{f(b) - f(a)}{b - a}.$$

Figure 9.3

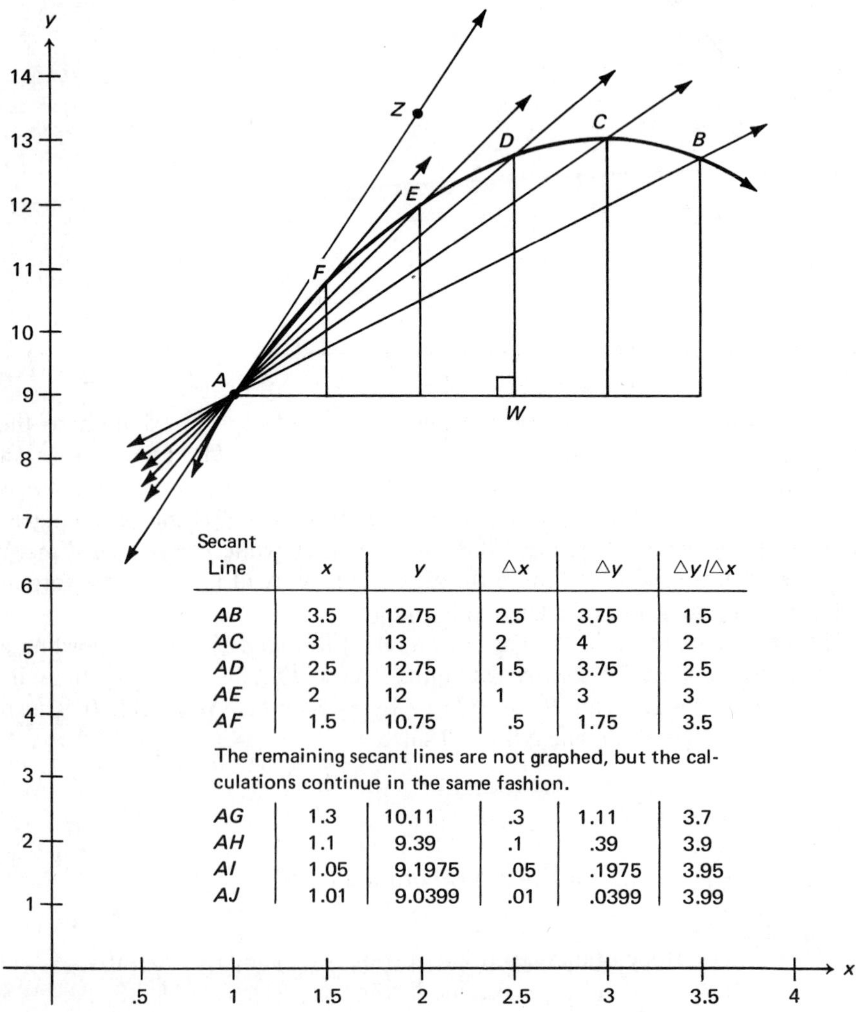

<table>
<tr><th>Secant Line</th><th>x</th><th>y</th><th>Δx</th><th>Δy</th><th>$\Delta y/\Delta x$</th></tr>
<tr><td>AB</td><td>3.5</td><td>12.75</td><td>2.5</td><td>3.75</td><td>1.5</td></tr>
<tr><td>AC</td><td>3</td><td>13</td><td>2</td><td>4</td><td>2</td></tr>
<tr><td>AD</td><td>2.5</td><td>12.75</td><td>1.5</td><td>3.75</td><td>2.5</td></tr>
<tr><td>AE</td><td>2</td><td>12</td><td>1</td><td>3</td><td>3</td></tr>
<tr><td>AF</td><td>1.5</td><td>10.75</td><td>.5</td><td>1.75</td><td>3.5</td></tr>
<tr><td colspan="6">The remaining secant lines are not graphed, but the calculations continue in the same fashion.</td></tr>
<tr><td>AG</td><td>1.3</td><td>10.11</td><td>.3</td><td>1.11</td><td>3.7</td></tr>
<tr><td>AH</td><td>1.1</td><td>9.39</td><td>.1</td><td>.39</td><td>3.9</td></tr>
<tr><td>AI</td><td>1.05</td><td>9.1975</td><td>.05</td><td>.1975</td><td>3.95</td></tr>
<tr><td>AJ</td><td>1.01</td><td>9.0399</td><td>.01</td><td>.0399</td><td>3.99</td></tr>
</table>

Suppose that we keep point A fixed in our previous diagram and move point B along the curve toward A, stopping occasionally to calculate $\Delta y/\Delta x$ to obtain the slopes of these successive secant lines. Since the secant lines through A are approaching coincidence with the tangent line at A, it appears, intuitively at least, that the slopes of these secant lines will also approach the slope of the tangent line at A as their limit.

By looking at Figure 9.3 (fourth, first, and sixth columns, in that order), we see that as Δx approaches 0 the secant lines approach the tangent line and their respective slopes, $\Delta y/\Delta x$, approach 4 as a limit. In symbols,

$$\operatorname*{limit}_{\Delta x \to 0} \frac{\Delta y}{\Delta x} = 4.$$

The plug-in method for evaluating

$$\operatorname*{limit}_{\Delta x \to 0} \frac{\Delta y}{\Delta x} = 4$$

does not apply since if $\Delta x = 0$ then Δy also equals 0 (the two points on the curve have become one and have equal coordinates), and 0/0 is not defined. Negative values of Δx are entirely possible; we work with them in Problem 2 of the problem set.

This method of calculating the slope of a tangent line by creeping (some people prefer to say sneaking) up on it with secant lines is the only one available to us at present. The slope of the tangent line at a specific point (x, y),

$$\operatorname*{limit}_{\Delta x \to 0} \frac{\Delta y}{\Delta x},$$

is also the *instantaneous rate of change of a function at a point.* This is reasonable, for saying that the slope of the curve is 4 at (1, 9) means that if a point moves along the curve and leaves it to travel along the tangent line at (1, 9), then the rate of change of the moving point at (1, 9) will be 4 units of vertical change for 1 unit of horizontal change. Choosing other points of tangency will lead to different instantaneous rates of change. In contrast, the *average* rate of change of the function in the *interval* [1, 3.5] is

$$\frac{\Delta y}{\Delta x} = 1.5.$$

Finding the slope of the tangent line

Suppose now that we want to find the slope of the line tangent to the graph of this same function at the point (.5, 6.75). We could do it by letting Δx take on values of 3, 2.5, 2, 1.5, 1, .5, .3, .1, .05, and .01, which would have the respective secant lines approach the tangent line through (.5, 6.75) while the ten numerical values of $\Delta y/\Delta x$ would also approach the slope of this tangent line as a limit (see Problem 11).

There must be a better and easier way, and there is (!). To understand this more efficient way we have to think of and rewrite our specific results in more general ways. We look more closely at secant line AD in Figure 9.4.

Figure 9.4

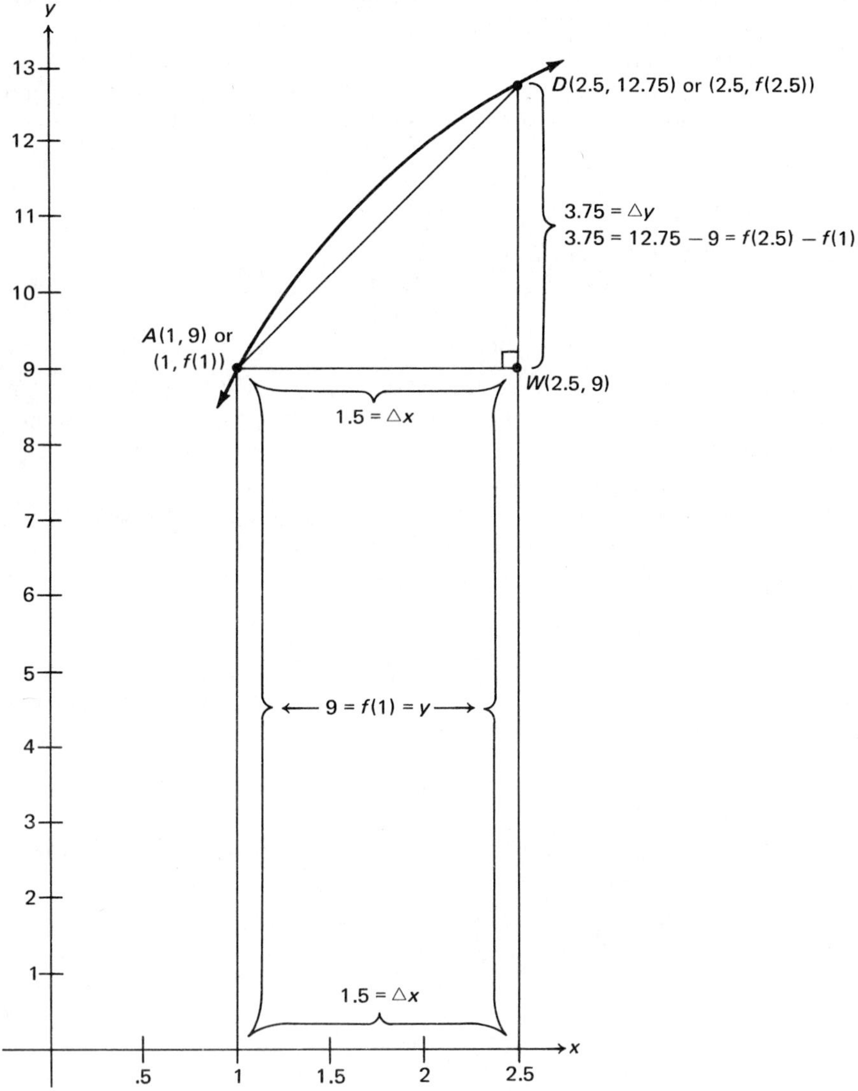

It is true here that

$$\frac{\Delta y}{\Delta x} = \frac{f(2.5) - f(1)}{2.5 - 1}.$$

The key to our generalization is to think of 2.5 as $1 + 1.5$. Then

$$\frac{\Delta y}{\Delta x} = \frac{f(1 + 1.5) - f(1)}{(1 + 1.5) - 1},$$

or

$$\frac{\Delta y}{\Delta x} = \frac{f(1+1.5) - f(1)}{1.5}. \tag{A}$$

Equation (A) can be generalized slightly if we realize that 1.5 plays the role of Δx, or the difference between the x-coordinate of the base point $(1, 9)$ and the x-coordinate of any other point on the curve. Thus the slope of the secant line joining $(1, 9)$ and $(1.4, 10.44)$—where $\Delta x = .4$—is

$$\frac{\Delta y}{\Delta x} = \frac{f(1+.4) - f(1)}{.4} \quad (=3.6, \text{ right?});$$

the slope of the secant line joining $(1, 9)$ and $(1.05, 9.1975)$—where $\Delta x = .05$—is

$$\frac{\Delta y}{\Delta x} = \frac{f(1+.05) - f(1)}{.05};$$

and the slope of the secant line joining $(1, 9)$ and $(1+\Delta x, f(1+\Delta x))$—where $\Delta x = \Delta x$—is

$$\frac{\Delta y}{\Delta x} = \frac{f(1+\Delta x) - f(1)}{\Delta x}. \tag{B}$$

Equation (B) is even more important and general than equation (A). Since equation (B) gives the slope of *any* secant line through $(1, 9)$, and since as Δx approaches 0 the resulting secant lines tend to coincidence with the tangent line through $(1, 9)$, then if we evaluate

$$\operatorname*{limit}_{\Delta x \to 0} \frac{\Delta y}{\Delta x} = \operatorname*{limit}_{\Delta x \to 0} \frac{f(1+\Delta x) - f(1)}{\Delta x}$$

this limit will be the slope of the tangent line at $(1, 9)$. For our function

$$f(x) = -x^2 + 6x + 4$$

considerable algebra—be *sure* to do it—yields

$$\begin{aligned} \frac{f(1+\Delta x) - f(1)}{\Delta x} &= \frac{[-(1+\Delta x)^2 + 6\cdot(1+\Delta x) + 4] - [-1^2 + 6\cdot 1 + 4]}{\Delta x} \\ &= 4 - \Delta x. \end{aligned}$$

Then, as expected,

$$\begin{aligned} \operatorname*{limit}_{\Delta x \to 0} \frac{\Delta y}{\Delta x} &= \operatorname*{limit}_{\Delta x \to 0} \frac{f(1+\Delta x) - f(1)}{\Delta x} \\ &= \operatorname*{limit}_{\Delta x \to 0} (4 - \Delta x) \\ &= \operatorname*{limit}_{\Delta x \to 0} (4) - \operatorname*{limit}_{\Delta x \to 0} (\Delta x) \\ &= \operatorname*{limit}_{\Delta x \to 0} 4 \cdot (\Delta x)^0 - \operatorname*{limit}_{\Delta x \to 0} (\Delta x) \\ &= 4 - 0 \\ &= 4. \end{aligned}$$

Notice how the limit theorems of the previous section have been used here.

Since the slope of the tangent line to the graph at $(1, 9) = (1, f(1))$ is

$$\lim_{\Delta x \to 0} \frac{f(1+\Delta x) - f(1)}{\Delta x},$$

we would expect the slope of the tangent line to the graph at $(2, f(2))$ to be

$$\lim_{\Delta x \to 0} \frac{f(2+\Delta x) - f(2)}{\Delta x};$$

we would expect the slope of the tangent line to the graph at $(7, f(7))$ to be

$$\lim_{\Delta x \to 0} \frac{f(7+\Delta x) - f(7)}{\Delta x};$$

and, in general, we would expect the slope of the tangent line to the graph at $(x, f(x))$ to be

$$\lim_{\Delta x \to 0} \frac{f(x+\Delta x) - f(x)}{\Delta x}. \tag{C}$$

Expression (C) is the culmination of all our work in this section. The beauty of it is that it not only represents the slope of the tangent line to the graph of the function

$$y = f(x) = -x^2 + 6x + 4$$

at any point (x, y) on the graph, but it is also valid for *any* continuous function $y = f(x)$ at *any* point (x, y) where we want to calculate the slope of the tangent line. Of course, the appropriate limit must exist.

Delta process

We examine a geometrical interpretation of expression (C). See figure opposite.

If we start out at a point P with coordinates (x, y) or $(x, f(x))$ and move horizontally away from it Δx units to point A, then we must also move vertically Δy units from A to get back on the curve at point P_1 with coordinates $(x+\Delta x, y+\Delta y)$ or $(x+\Delta x, f(x+\Delta x))$. Then, since

$$CA + AP_1 = CP_1$$

or

$$AP_1 = CP_1 - CA,$$

according to the length of these line segments,

$$\Delta y = f(x+\Delta x) - f(x).$$

Division of both sides by Δx ($\Delta x \neq 0$) gives

$$\frac{\Delta y}{\Delta x} = \frac{f(x+\Delta x) - f(x)}{\Delta x},$$

Figure 9.5

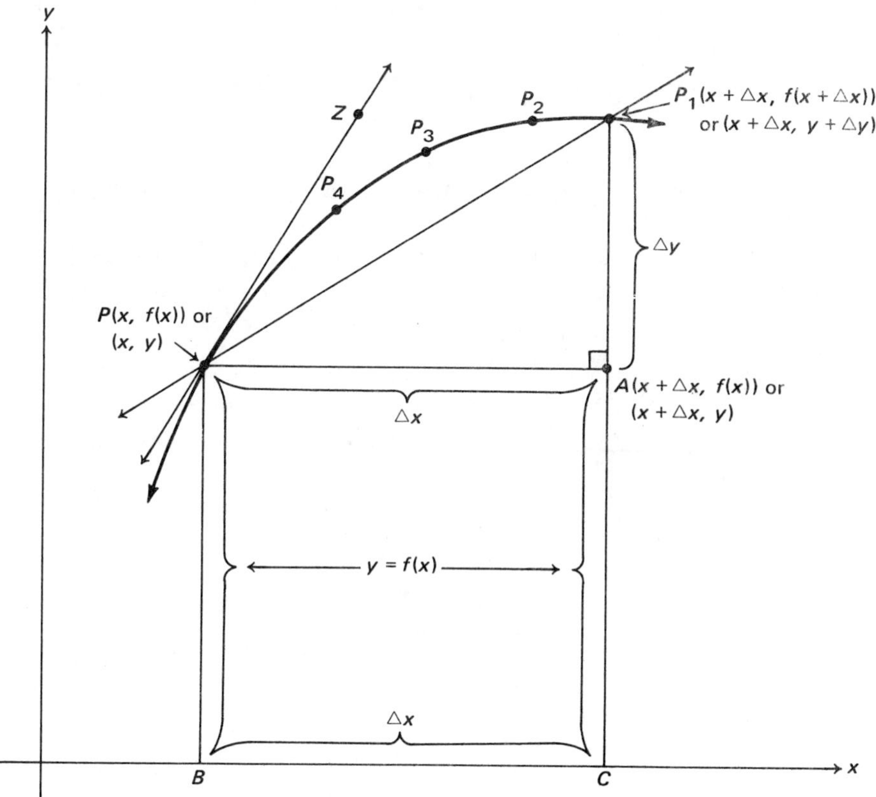

which is the slope of the secant line through P and P_1. Now, if we force Δx to approach 0 so that P_1 coincides with and passes through $P_2, P_3, P_4 \ldots$, then

$$\lim_{\Delta x \to 0} \frac{\Delta y}{\Delta x} = \lim_{\Delta x \to 0} \frac{f(x + \Delta x) - f(x)}{\Delta x},$$

which is the slope of the tangent line to the graph of $y = f(x)$ at the point (x, y).

Example 13 Find the slope of the tangent line to the graph of

$$y = f(x) = -x^2 + 6x + 4$$

at any point (x, y) on the graph.

Solution We use a four-step method, called the *delta process*, which is a summary of our geometry and algebra used in the previous diagram and paragraph.

Step 1. Pick a point $(x, y) = (x, f(x))$ on the curve so that

$$y = f(x) = (-x^2 + 6x + 4).$$

Step. 2. Give x a change of Δx units, which forces a change of Δy units in y (geometrically this means we move from (x, y) on the curve to a point with coordinates $(x+\Delta x, y+\Delta y)$).

$$\begin{aligned} y + \Delta y &= f(x+\Delta x) = (-(x+\Delta x)^2 + 6(x+\Delta x) + 4) \\ &= (-x^2 - 2x\Delta x - (\Delta x)^2 + 6x + 6\Delta x + 4) \quad \text{(right?)} \end{aligned}$$

Notice that $(\Delta x)^2$ is *not* equal to $\Delta^2 \cdot x^2$.

Step 3. Subtract $y = f(x)$ from both sides of the equation in step 2 and then divide both sides by Δx, which is not 0, to form the ratio $\Delta y/\Delta x$. This ratio represents the slope of the secant line through (x, y) and $(x+\Delta x, y+\Delta y)$.

$$\begin{aligned} y &= f(x+\Delta x) - f(x) \\ &= (-x^2 - 2x\Delta x - (\Delta x)^2 + 6x + 6\Delta x + 4) - (-x^2 + 6x + 4) \\ &= -2x\Delta x - (\Delta x)^2 + 6\Delta x \quad \text{(after considerable cancellation)} \end{aligned}$$

$$\begin{aligned} \frac{\Delta y}{\Delta x} = \frac{f(x+\Delta x) - f(x)}{\Delta x} &= \frac{-2x\Delta x - (\Delta x)^2 + 6\Delta x}{\Delta x} \\ &= \frac{\Delta x(-2x - \Delta x + 6)}{\Delta x} \\ &= -2x - \Delta x + 6 \quad (\Delta x \neq 0) \\ &= 6 - 2x - \Delta x \end{aligned}$$

Step 4. Find the limit, if it exists, of $\Delta y/\Delta x$ as Δx approaches 0. This limit is the slope of the tangent line to the curve at $(x, y) = (x, f(x))$.

$$\begin{aligned} \operatorname*{limit}_{\Delta x \to 0} \frac{\Delta y}{\Delta x} = \operatorname*{limit}_{\Delta x \to 0} \frac{f(x+\Delta x) - f(x)}{\Delta x} &= \operatorname*{limit}_{\Delta x \to 0} (6 - 2x - \Delta x) \\ &= \operatorname*{limit}_{\Delta x \to 0} (6) - 2 \cdot \operatorname*{limit}_{\Delta x \to 0} (x) - \operatorname*{limit}_{\Delta x \to 0} (\Delta x) \quad \text{(D)} \\ &= 6 - 2 \cdot x - 0 \quad \text{(E)} \\ &= 6 - 2x \end{aligned}$$

We went from equation (D) to equation (E) by using the limit theorems from the last section. Note that since

$$(6) = (6 \cdot 1) = (6 \cdot (\Delta x)^0)$$

and

$$(x) = (x \cdot 1) = (x \cdot (\Delta x)^0),$$

their respective limits as Δx approaches 0 are 6 and x. Then the slope of the tangent line at $(.5, f(.5))$ is

$$6 - 2 \cdot .5 = 6 - 1 = 5;$$

the slope of the tangent line at $(2.1, f(2.1))$ is

$$6 - 2 \cdot 2.1 = 6 - 4.2 = 1.8;$$

the slope of the tangent line at $(1, f(1))$ is

$$6 - 2 \cdot 1 = 4 \quad \text{(as expected from previous work); and}$$

the slope of the tangent line at $(4, f(4))$ is

$$6 - 2 \cdot 4 = 6 - 8 = -2.$$

This means (recall the section *Graphs of Functions* from Chapter 3) that the tangent line is decreasing from left to right. The slope of the tangent line at $(3, f(3))$ is

$$6 - 2 \cdot 3 = 6 - 6 = 0.$$

In the section after next we will learn the importance of this four-step method for calculating the slope of the graph of a continuous function at any point on that graph.

Problem set 9–3

1. We will work further with our now-familiar function

$$y = f(x) = -x^2 + 6x + 4.$$

Calculate the average rate of change of the function, $\Delta y/\Delta x$, which is also the slope of the secant line through the two corresponding points on the graph, for these intervals of x.

a. $[1, 2]$ b. $[2, 3]$ c. $[3, 4]$
d. $[1, 2.5]$ e. $[2.5, 4]$ f. $[1, 4]$

Comparing the results of (a), (b), and (c) shows that the average rate of change of this function decreases as the intervals of x shift from left to right. Comparing the results of (d) and (e) also illustrates this same fact. The result of (f), which we did in Example 11, says that the overall rate of change of the function in the large interval $[1, 4]$ is an *average* of the rates of change of the function in its partition into subintervals.

2. Complete this table for the same function as in Problem 1. It indicates the slope of the tangent line at $(1, 9)$ by having the secant lines creep up on the tangent line from left to right, instead of from right to left as in the text. If we consider the secant line passing through $(-2, -12)$ and the fixed base point $(1, 9)$, then $\Delta x = -3$ because we have to *add* -3 to 1 to get -2: $1 + -3 = -2$. We know that Δy is -21 because as we move along the graph *from* the base point $(1, 9)$ *to* the second point $(-2, -12)$, the change in y is 21 units downward, or -21. We also note that the change in x occurring in such a move is from 1 to -2, which is 3 units to the left, or -3. Thus the first row in the table is correct.

x	Δx	y	Δy	$\Delta y/\Delta x$
-2	-3	-12	-21	7
-1				
0				
	$-.5$			
	$-.3$			
	$-.1$			
	$-.05$			

3. Consider the graph of the function

$$y = f(x) = x^2 - 10x + 27.$$

Draw its graph for $[2, 9]$. Find the slope of the tangent line to the curve at $(6, 3)$ by calculating the slopes of the successive secant lines through $(6, 3)$ and these points.

a. $(9, f(9))$
b. $(8, f(8))$
c. $(7.5, f(7.5))$
d. $(7, f(7))$
e. $(6.5, f(6.5))$
f. $(6.3, f(6.3))$
g. $(6.1, f(6.1))$
h. $(6.05, f(6.05))$
i. $(6.01, f(6.01))$
j. $(6.001, f(6.001))$
k. Calculate Δx for parts (a) through (j) of this problem.

4. Rework Problem 3, choosing these successive points on the curve.

a. $(3, f(3))$
b. $(4, f(4))$
c. $(4.5, f(4.5))$
d. $(5, f(5))$
e. $(5.5, f(5.5))$
f. $(5.7, f(5.7))$
g. $(5.9, f(5.9))$
h. $(5.95, f(5.95))$
i. $(5.99, f(5.99))$
j. $(5.999, f(5.999))$
k. Calculate Δx for parts (a) through (j) of this problem.

5. By using the delta process of the text work with the function

$$y = f(x) = x^2 - 10x + 27$$

in a general fashion by finding the expression in x (call it y') that represents the slope of the tangent line to the graph at any point (x, y) on the graph. After carrying out the delta process your answer should be

$$y' = 2x - 10.$$

Then find the slopes of the tangent line to the graph at these points.

a. $(7, f(7))$
b. $(6, f(6))$
c. $(5, f(5))$
d. $(4, f(4))$
e. $(3, f(3))$
f. $(-8, f(-8))$
g. $(6.468, f(6.468))$ (Can you imagine doing this one by the method of successive secant lines?)

6. Same as Problem 5, but work with the function

$$y = f(x) = 2 + 8x - x^2.$$

Here the general expression should be

$$y' = 8 - 2x.$$

In the next four problems find the slope of the tangent line to the graph of these functions at any point (x, y) or $(x, f(x))$ on them. Use the delta process to do this. Then find the slope of the tangent line at the specific point $(x, f(x))$ mentioned.

7. $y = f(x) = 5x^2 - 3$

a. $(2, f(2))$
b. $(-4, f(-4))$
c. $(0, f(0))$
d. $(.5, f(.5))$
e. $(.3, f(.3))$
f. $(.1, f(.1))$

8. $y = f(x) = 4x^2 + 6x - 7$
   a. $(-3, f(-3))$
   b. $(-2, f(-2))$
   c. $(1, f(1))$
   d. $(4, f(4))$
9. $y = f(x) = 3x - 5$
   a. $(1, f(1))$
   b. $(6, f(6))$
   c. $(-2, f(-2))$
   d. $(1.78, f(1.78))$
   e. What can you say about the graph of the function, the graphs of successive secant lines through $(1, f(1))$, and the tangent line to the graph at $(1, f(1))$?
10. $y = f(x) = 3\ (= 3 \cdot x^0)$
    a. $(2, f(2))$
    b. $(-4, f(-4))$
    c. $(5, f(5))$
    d. $(1.5, f(1.5))$
    e. Explain why your results of (a) through (d) are consistent with the graph.
11. Do the calculations for $f(x) = -x^2 + 6x + 4$ to find the slope of its tangent line at $(.5, 6.75)$, as mentioned in the text.

Velocity: An alternate interpretation of rate of change

The basic idea we are trying to understand in this optional section is the instantaneous rate of change of a function. Our vehicle to the understanding of

$$\operatorname*{limit}_{\Delta x \to 0} \frac{\Delta y}{\Delta x} = \operatorname*{limit}_{\Delta x \to 0} \frac{f(x + \Delta x) - f(x)}{\Delta x}$$

has been a geometric description of moving secant lines with their corresponding slopes $\Delta y / \Delta x$. It is also revealing to discuss the instantaneous rate of change of a function from the viewpoint of a falling body. However, this section may be omitted since only a few exercises in future problem sets refer back to it. Those exercises will be identified.

One of the basic laws of motion says that if an object is dropped in a vacuum near sea level so that it can freely fall in a vertical direction, and if $t = 0$ on a stopwatch the instant the body is released, then the distance in feet (s) it has fallen t seconds later is given by

$$s = f(t) = 16 \cdot t^2.$$

If $t = 3$, then $s = f(3) = 16 \cdot 3^2 = 16 \cdot 9 = 144$, which says that the object—let us assume it is a steel ball—has fallen 144 feet after 3 seconds.

Average velocity

The *average* velocity of the ball in this interval of time from 0 to 3 seconds is (144 feet)/(3 seconds) = 48 feet per second since average velocity = (distance)/(time). Common sense says that the ball is *not* falling at a *constant* rate of 48 ft/sec, since it is clearly moving faster at the end of the 3-second time interval than at the beginning of it.

Suppose we want to find the velocity exactly 1.5 seconds after the ball is dropped. The formula

$$\text{average velocity} = (\text{distance})/(\text{time})$$

cannot be used here since in the interval from $t = 1.5$ to $t = 1.5$ the distance traveled is 0 feet, the elapsed time is 0 seconds, and 0/0 is not defined. Here the plug-in method for calculating velocity *at an instant* does not work. We retreat slightly and draw a picture to see how we may possibly calculate the instantaneous rate of change of the position of the ball (s) when $t = 1.5$.

Figure 9.6

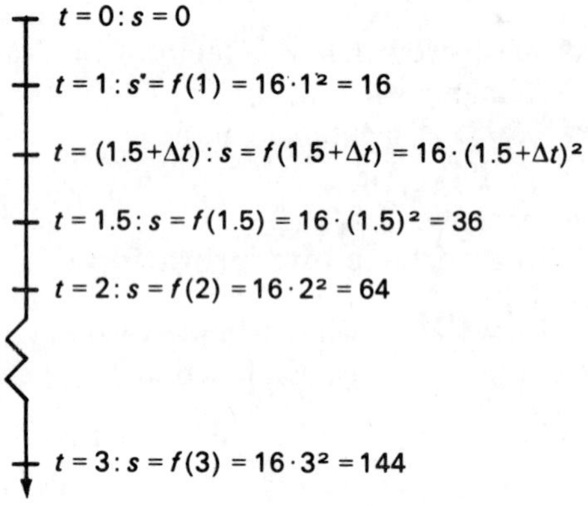

$t = 0 : s = 0$

$t = 1 : s = f(1) = 16 \cdot 1^2 = 16$

$t = (1.5 + \Delta t) : s = f(1.5 + \Delta t) = 16 \cdot (1.5 + \Delta t)^2$

$t = 1.5 : s = f(1.5) = 16 \cdot (1.5)^2 = 36$

$t = 2 : s = f(2) = 16 \cdot 2^2 = 64$

$t = 3 : s = f(3) = 16 \cdot 3^2 = 144$

We are going to place a pair of photoelectric cells along the side of a cliff which are connected to an electronic timer. One of the cells will be permanently located at 36 feet below the top of the cliff, which is the location of the ball after exactly 1.5 seconds of flight. The other cell is placed above the fixed cell, so that at a known time in flight the ball passes it. For example, we could place this cell $s = 16 \cdot (1.4)^2 = 31.36$ feet below the cliff. Here $t = -.1$ second since the ball passes by the first cell .1 second *before* it passes the second (fixed) cell. The cell at 31.36 feet is 4.64 feet away from 36.00 feet, and the average velocity of the ball in this interval is

$$\frac{31.36 - 36}{1.4 - 1.5} = \frac{-4.64}{-.1} = \frac{4.64}{.1} = 46.4 \text{ feet/sec.}$$

It may seem more natural to write the first fraction as

$$\frac{36 - 31.36}{1.5 - 1.4},$$

which also equals 46.4 feet/sec, but an important generalization is more easily made if we have the quotient of these two negative numbers. Since 31.36 is $f(1.5 + -.1)$, 36 is $f(1.5)$, 1.4 is $(1.5 + -.1)$, and 1.5 is 1.5, then

$$\frac{31.36 - 36}{1.4 - 1.5}$$

may be rewritten as

$$\frac{f(1.5+-.1)-f(1.5)}{(1.5+-.1)-1.5}=\frac{f(1.5)+-.1)-f(1.5)}{-.1}=\frac{\Delta s}{\Delta t}.$$

If we want to find the average velocity in an interval beginning .05 seconds before $t = 1.5$, then the answer is

$$\frac{\Delta s}{\Delta t}=\frac{f(1.5+-.05)-f(1.5)}{-.05}\quad(=47.2\text{ feet/sec}).$$

If we want to find the average velocity in an interval beginning .01 seconds before $t = 1.5$, then the answer is

$$\frac{\Delta s}{\Delta t}=\frac{f(1.5+-.01)-f(1.5)}{-.01}\quad(=47.84\text{ feet/sec}).$$

In general the average velocity in an interval beginning $|\Delta t|$ seconds before $t = 1.5$ is

$$\frac{\Delta s}{\Delta t}=\frac{f(1.5+\Delta t)-f(1.5)}{\Delta t}.\qquad(\Delta t<0)\qquad\text{(A)}$$

In our present case

$$s = f(t) = 16t^2,$$

so the last equation, (A), may be written as

$$\begin{aligned}\frac{\Delta s}{\Delta t}&=\frac{f(1.5+\Delta t)-f(1.5)}{\Delta t}\\&=\frac{16\cdot(1.5+\Delta t)^2-16\cdot(1.5)^2}{\Delta t}\\&=\frac{16\cdot(2.25+3\Delta t+(\Delta t)^2)-36}{\Delta t}\\&=\frac{36+48\Delta t+16\cdot(\Delta t)^2-36}{\Delta t}\\&=\frac{\Delta t\cdot(48+16\Delta t)}{t}\\&=48+16\Delta t.\quad\text{(The cancellation is allowed since }\Delta t\neq 0.)\end{aligned}$$

Instantaneous velocity

As we move the top photoelectric cell toward the bottom one according to the desired values of Δt, the ratios of the two corresponding distance and time intervals, $\Delta s/\Delta t$, represent the corresponding average velocities of the falling body in those time intervals Δt seconds long. It is entirely natural to investigate the values of $\Delta s/\Delta t$ as Δt approaches 0 since these ratios, representing average velocities of the ball in the successive time intervals, will approach the instantaneous velocity of the ball at the point (1.5 seconds, 36 feet) = (1.5, 36) as a limiting value.

$$\begin{aligned}\operatorname*{limit}_{\Delta t\to 0}\frac{\Delta s}{\Delta t} &= \operatorname*{limit}_{\Delta t\to 0}\frac{f(1.5+\Delta t)-f(1.5)}{\Delta t}\\ &= \operatorname*{limit}_{\Delta t\to 0}(48+16\Delta t)\\ &= 48 + 16\cdot 0\\ &= 48.\end{aligned}$$

Our conclusion is that the instantaneous velocity of the ball at the point (1.5, 36) is 48 feet per second. By a similar line of reasoning we could show that the instantaneous velocity of the ball at (3 seconds, 144 feet) is

$$\begin{aligned}\operatorname*{limit}_{\Delta t\to 0}\frac{\Delta s}{\Delta t} &= \operatorname*{limit}_{\Delta t\to 0}\frac{f(3+\Delta t)-f(3)}{\Delta t}\\ &= \operatorname*{limit}_{\Delta t\to 0}(96+16\Delta t) \quad \text{(Right?)}\\ &= 96 \text{ feet/sec.}\end{aligned}$$

Equation (A) is the average velocity of a moving body given by a specific law of motion (that of a freely falling body in a vacuum at sea level), $s = f(t) = 16t^2$, in a specific time interval based on $t = 1.5$. We may hastily but safely generalize to the *average* velocity of a moving body given by *any* law of motion $s = f(t)$ in any *interval* based at $(t, f(t))$. Thus

$$\frac{\Delta s}{\Delta t} = \frac{f(t+\Delta t)-f(t)}{\Delta t}$$

is the average velocity in an interval, and

$$\operatorname*{limit}_{\Delta t\to 0}\frac{\Delta s}{\Delta t} = \operatorname*{limit}_{\Delta t\to 0}\frac{f(t+\Delta t)-f(t)}{\Delta t}$$

is the *instantaneous* velocity at a *point* $(t, f(t))$ in the path of the moving body. We think of this limit in terms of instantaneous rate of change because

$$\operatorname*{limit}_{\Delta t\to 0}\frac{\Delta s}{\Delta t}$$

gives the rate of change of distance per unit of change of time at a specific instant in time.

Problem set 9–4

1. A point moves along a number line so that its distance (s feet) from the point t seconds after a stopwatch begins is given by the equation

$$s = f(t) = 4 + 15t - 3t^2, \quad 0 \leqslant t \leqslant 10.$$

a. Evaluate

$$\underset{\Delta t \to 0}{\text{limit}} \frac{\Delta s}{\Delta t} = \underset{\Delta t \to 0}{\text{limit}} \frac{f(t + \Delta t) - f(t)}{\Delta t}$$

by the delta process, which gives the expression for the instantaneous velocity and for the direction of the point for any value of t in the indicated domain. If the resulting expression, which we symbolize by s', is positive for a specific value of t, then the point is moving from left to right along the line. If s' is negative for a specific value of t, then the point is moving from right to left along the line. Find the location (feet from the origin), instantaneous velocity, and direction of the point for these values of t.

b. $t = 0$ c. $t = 1$ d. $t = 2$
e. $t = 2.5$ f. $t = 3$ g. $t = 5$
h. $t = 7$ i. $t = 9$ j. $t = 10$

2. Complete this table for an example in which one of the photoelectric cells is 144 feet below the cliff and we want to find the average velocity of the ball in the resulting intervals as the second photoelectric cell is moved up from below. The initial position of the second cell is 400 feet below the cliff.

	t	Δt	s	Δs	$\Delta s/\Delta t$
a.	5	2	400	256	128
b.	4.5				
c.		1			
d.		0.5			
e.		0.3			
f.		0.1			
g.		0.05			
h.		0.01			
i.		0.001			

j. Calculate $\text{limit}_{\Delta t \to 0} \Delta s/\Delta t$ for $s = f(t) = 16t^2$ by using the delta process. The resulting expression, which is the instantaneous velocity at *any* positive point of time, should be

$$v = 32t.$$

Find the velocity of the ball at these points in time.

k. $t = 1.5$ l. $t = 1.7$ m. $t = 2$
n. $t = 3$ o. $t = 3.784$ p. $t = 6$

The derivative

This chapter began with the problem of finding maximum and minimum values of a function. Recognizing that additional pure mathematics was necessary, we have been working entirely inside the upper half of our general diagram for the last three sections on limits and the instantaneous rate of change of a function at a point (x, y). The next section will show us how to solve max/min problems relatively easily *and* show us the necessity for this preliminary work.

At the beginning of the last section we used a rather lengthy and cumbersome but pattern-revealing method to find the slope of the tangent line to the graph of the specific function

$$y = f(x) = -x^2 + 6x + 4$$

at the specific point $(1, f(1)) = (1, 9)$. Then we saw that the slope of this specific graph at $(1, f(1))$ could be found by evaluating the expression

$$\underset{\Delta x \to 0}{\text{limit}} \frac{\Delta y}{\Delta x} = \underset{\Delta x \to 0}{\text{limit}} \frac{f(1+\Delta x) - f(1)}{\Delta x},$$

so that as a generalization the slope of the tangent line to this specific graph at *any* point $(x, f(x))$ is given by

$$\underset{\Delta x \to 0}{\text{limit}} \frac{\Delta y}{\Delta x} = \underset{\Delta x \to 0}{\text{limit}} \frac{f(x+\Delta x) - f(x)}{\Delta x}.$$

We then generalized to conclude that the slope of the tangent line to the graph of *any* function $y = f(x)$ at *any* point $(x, f(x))$ is given by

$$\underset{\Delta x \to 0}{\text{limit}} \frac{\Delta y}{\Delta x} = \underset{\Delta x \to 0}{\text{limit}} \frac{f(x+\Delta x) - f(x)}{\Delta x},$$

provided that the limit exists. The fraction $\Delta y/\Delta x$ and its limit as Δx approaches 0 are found by the four-step method known as the *delta process*, since it begins by giving x a change of Δx units.

Differentiating a function

What does the delta process accomplish for us? It is a procedure by which we derive an expression for a new function from the expression for an old one. For example, we saw that if

$$y = -x^2 + 6x + 4,$$

then the derived equation

$$y' = -2x + 6$$

gives the slope of the tangent line at any point (x, y) on the graph of the function defined by $y = -x^2 + 6x + 4$. However, the equation $y' = -2x + 6$ defines a *new* function which has been *derived* from the old one, and we call this new function, resulting from the application of the delta process to an original function, the *derivative* of that original function. If the limit (as mentioned previously in Step 4) does not exist at some point x in the domain, then we say that the derivative function is *not defined* at that point, or that the derivative does not exist at that point. This new function is symbolized by y' or $f'(x)$ or dy/dx ("y prime" or "f prime of x" or "dee y dee x").

$$\underbrace{y = f(x)}_{\text{Original function}} \quad \xrightarrow[\text{process}]{\text{delta}} \quad \underbrace{y' = f'(x) = \frac{dy}{dx}}_{\text{Derivative}}$$

The following table of examples (9.12) shows the distinction between the geometric interpretations of $f(x)$ and $f'(x)$ for the function defined by

$$y = f(x) = -x^2 + 6x + 4.$$

Convince yourself that

$$f(x + \Delta x) = -(x + \Delta x)^2 + 6 \cdot (x + \Delta x) + 4$$

and, consequently,

$$\frac{\Delta y}{\Delta x} = \frac{f(x + \Delta x) - f(x)}{\Delta x} = -2x + 6 - \Delta x.$$

In general, the expression

$$\frac{\Delta y}{\Delta x} = \frac{f(x + \Delta x) - f(x)}{\Delta x}$$

for any function f is known as the *difference quotient*. This terminology is appropriate, since $\Delta y/\Delta x$ is the *quotient* of the *difference* between values of y and the *difference* between values of x.

We now give the formal definition of a derivative.

Definition of the derivative of a function Suppose a function f is defined by the equation $y = f(x)$. The *derivative* of this function is a new function, denoted f' or $y' = f'(x)$, which is obtained by using the delta process as follows.

1. To find the value of f' at a point x, we must check to make sure that x is in the domain of f. If it is, then we will find $f'(x)$.
2. Form the difference quotient, $\Delta y/\Delta x$, using the delta process from before, at x.

$$\begin{aligned}
&y = f(x) && \\
&y + \Delta y = f(x + \Delta x) && \text{Change } x \text{ by } \Delta x \text{ units} \\
&y + \Delta y - y = f(x + \Delta x) - f(x) && \text{Subtract } y = f(x) \\
&\Delta y = f(x + \Delta x) - f(x) && y - y = 0 \\
&\frac{\Delta y}{\Delta x} = \frac{f(x + \Delta x) - f(x)}{\Delta x} && \text{Divide by } \Delta x \neq 0
\end{aligned}$$

3. Calculate the limit, if it exists, of $\Delta y/\Delta x$ as $\Delta x \to 0$. This limit is $f'(x)$, and thus the value of the new function f' at x ($f'(x)$ is a range element) has been calculated. We say that the function f is *differentiable* at x.

Table 9.12

	General function	Application of the delta process $\lim_{\Delta x \to 0} \frac{\Delta y}{\Delta x} = \lim_{\Delta x \to 0} \frac{f(x+\Delta x) - f(x)}{\Delta x}$ $= \lim_{\Delta x \to 0} (-2x+6-\Delta x)$ $= -2x+6$	Vertical distance of $(x, f(x))$ from x-axis	Slope of tangent line at $(x, f(x))$
	$y = f(x) = -x^2+6x+4$	$f'(x) = -2x+6$	$y = f(x) = -x^2+6x+4$	$y' = f'(x) = -2x+6$
Example A	$x = 1$: $f(1) = 9$	$f'(1) = 4$	$f(1) = 9$	$f'(1) = 4$
Example B	$x = 3$: $f(3) = 13$	$f'(3) = 0$	$f(3) = 13$	$f'(3) = 0$
Example C	$x = 8$: $f(8) = -12$	$f'(8) = -10$	$f(8) = -12$	$f'(8) = -10$
Example D	$x = -4$: $f(-4) = -36$	$f'(-4) = 14$	$f(-4) = -36$	$f'(-4) = 14$
Example E	$x = -2.7$: $f(-2.7) = -19.49$	$f'(-2.7) = 11.4$	$f(-2.7) = -13.49$	$f'(-2.7) = 11.4$

4. If the limit does not exist, then f' is not defined at x, and we say that the function f is *not differentiable* at x.
5. Summarizing,

$$f'(x) = \lim_{\Delta x \to 0} \frac{\Delta y}{\Delta x} = \lim_{\Delta x \to 0} \frac{f(x+\Delta x) - f(x)}{\Delta x}$$

whenever this limit exists.

Notice once again that, geometrically, $f(x)$ is the vertical distance of the graph of the function from the x-axis at $(x, f(x))$, while $f'(x)$ is the slope of the tangent line to the graph of the function at $(x, f(x))$. (For those of us who studied the last section, $s(t)$ is the vertical distance of the freely falling ball from the top of the cliff at t seconds of time and $s(t)$ feet of distance, while $s'(t)$ is the velocity of the ball at this point of time and distance.) Problem 31 introduces another geometrical interpretation of the derivative of a function.

The mechanics of the delta process are not too difficult for relatively simple functions such as those defined by

$$y = f(x) = 3x^2 - 4x + 5.$$

If we wanted to *differentiate*, or find the derivative of, the function defined by

$$y = f(x) = \frac{x^2 - 3x + 17}{5x^3},$$

the algebraic manipulations required by the delta process would get downright awkward. If we examine this fraction we see that it is composed of powers of x, constants, and the four arithmetic operations. It is reasonable to inquire if we can somehow differentiate the entire expression in terms of the derivatives of these component parts; the answer is yes, and we will return to this question in Problem 32 of Problem set 9–5. We first accept without proof eight theorems on differentiation, which are short cuts to avoid the delta process while quickly giving us the correct derivatives. We assume in all eight cases that the theorems, which are derived by general applications of the delta process, are valid as long as the appropriate limits exist as Δx approaches 0.

Eight theorems on differentiating a function

Theorem 1 If $y = f(x) = x$, then $f'(x) = 1$.
Explanation The graph of

$$y = f(x) = x = 1 \cdot x + 0$$

is a line with slope 1. Problem 9 of Problem set 9–3 indicated that the graph of a linear function, any secant lines through any point on the graph, and the tangent line to the graph at that point all coincide. Since the tangent line and the graph itself coincide, and the slope of the graph is 1, then the slope of the tangent line at $(x, f(x))$ is also 1. The fact that the slope is 1 at any point $(x, f(x))$ implies that the derivative, $f'(x)$, is 1 for all values of x in its domain.

Theorem 2 If $y = f(x) = c$, where c is a constant, then $f'(x) = 0$.

Explanation Recall Problem 10 of the section before last. The graph of, for example, $f(x) = 3$ is a horizontal line 3 units above the x-axis and parallel to it. If we move from one point on the line to another, then $\Delta x \neq 0$ while $\Delta y = 0$; so $\Delta y/\Delta x$, the slope of the line, is always 0 regardless of the value of x. The value of $\Delta y/\Delta x$ will remain 0 as Δx approaches 0.

Theorem 3 If $y = f(x) = x^n$, where n is any integer, then $f'(x) = n \cdot x^{n-1}$.

Example 14 If $y = f(x) = x^2$, then

$$\frac{\Delta y}{\Delta x} = \frac{\Delta x \cdot (2x + \Delta x)}{\Delta x} = 2x + \Delta x \quad \text{(Right?)}$$

and

$$\operatorname*{limit}_{\Delta x \to 0} \frac{\Delta y}{\Delta x} = f'(x) = \operatorname*{limit}_{\Delta x \to 0} (2x + \Delta x) = 2x = 2 \cdot x^{2-1}$$

or, as predicted by the theorem,

$$f'(x) = 2 \cdot x^{2-1}.$$

Problems 28 and 29 of this section generalize this theorem to other values of n.

Theorem 4 If $y = g(x) = c \cdot f(x)$, then $g'(x) = c \cdot f'(x)$. In words, the derivative of a constant times a function is the constant times the derivative of the function.

Example 15 If $c = 6$ and $f(x) = x^3$, so that $y = g(x) = 6 \cdot x^3$ and $f'(x) = 3x^2$, then

$$\frac{\Delta y}{\Delta x} = 18x^2 + 18x\Delta x + 6(\Delta x)^2 \quad \text{(Right?)}$$

and

$$\operatorname*{limit}_{\Delta x \to 0} \frac{\Delta y}{\Delta x} = g'(x) = 18x^2 = 6 \cdot (3x^2) = 6 \cdot f'(x)$$

or, as predicted by the theorem,

$$g'(x) = 6 \cdot f'(x).$$

Example 16 If $c = 4$ and $f(x) = x$, so that $y = g(x) = 4 \cdot x$ and $f'(x) = 1$, then

$$\frac{\Delta y}{\Delta x} = 4 \quad \text{(Right?)}$$

and

$$\operatorname*{limit}_{\Delta x \to 0} \frac{\Delta y}{\Delta x} = g'(x) = 4 = 4 \cdot 1 = 4 \cdot f'(x)$$

or, as predicted by the theorem,

$$g'(x) = 4 \cdot f'(x).$$

Theorem 5 If $h(x) = f(x) + g(x)$ then

$$h'(x) = f'(x) + g'(x).$$

In words, the derivative of the sum of two functions is the sum of their derivatives.

Example 17 If $f(x) = x^3$ and $g(x) = 7$, so that $h(x) = x^3 + 7$, then

$$\begin{aligned} h'(x) &= f'(x) + g'(x) \\ &= 3x^2 + 0 \\ &= 3x^2. \end{aligned}$$

We may generalize this theorem to the sum of any number of functions.

Theorem 6 If $h(x) = f(x) - g(x)$ then

$$h'(x) = f'(x) - g'(x).$$

In words, the derivative of the difference of two functions is the difference of their derivatives.

Example 18 If $f(x) = 7x^2$ and $g(x) = 3x$, so that $h(x) = 7x^2 - 3x$, then

$$\begin{aligned} h'(x) &= f'(x) - g'(x) \\ &= 7 \cdot 2x - 3 \cdot 1 \\ &= 14x - 3. \end{aligned}$$

Theorem 7 If $h(x) = f(x) \cdot g(x)$, then

$$\begin{aligned} h'(x) &\neq f'(x) \cdot g'(x) \\ h'(x) &= f(x) \cdot g'(x) + g(x) \cdot f'(x). \end{aligned}$$

In words, the derivative of the product of two functions is *not* the product of their derivatives; it is the first function times the derivative of the second function plus the second function times the derivative of the first function.

The reason for this unexpected result is found in a general application of the delta process to the expression $f(x) \cdot g(x)$. Additionally, if we assumed that

$$h'(x) = f'(x) \cdot g'(x)$$

for

$$h(x) = x^2, \quad f(x) = x, \quad \text{and} \quad g(x) = x,$$

then $h'(x)$ would have to be $f'(x) \cdot g'(x)$, which is $1 \cdot 1$ or 1. Example 14 says that $h'(x) = 2x$ and $2x \neq 1$ (except when $x = 1/2$).

Example 19 If $f(x) = (3x + 5)$ and $g(x) = (x^2 + 7)$, so that

$$h(x) = (3x + 5) \cdot (x^2 + 7),$$

then

$$f'(x) = 3 \quad \text{and} \quad g'(x) = 2x.$$

Consequently,

$$\begin{aligned} h'(x) &= f(x) \cdot g'(x) + g(x) \cdot f'(x) \\ &= (3x+5) \cdot 2x + (x^2+7) \cdot 3 \\ &= 6x^2 + 10x + 3x^2 + 21 \\ &= 9x^2 + 10x + 21. \end{aligned}$$

If we perform the multiplication of $f(x)$ and $g(x)$ before differentiation, then, as predicted by the theorem,

$$\begin{aligned} h(x) &= (3x+5) \cdot (x^2+7) \\ h(x) &= 3x^3 + 5x^2 + 21x + 35 \quad \text{(Right?)} \\ h'(x) &= 3 \cdot (3x^2) + 5 \cdot (2x) + 21 \cdot 1 + 0 \\ h'(x) &= 9x^2 + 10x + 21. \end{aligned}$$

Theorem 8 If $h(x) = f(x)/g(x)$ and $g(x) \neq 0$, then

$$h'(x) \neq \frac{f'(x)}{g'(x)}$$

$$h'(x) = \frac{g(x) \cdot f'(x) - f(x) \cdot g'(x)}{[g(x)]^2}.$$

In words, the derivative of the quotient of two functions is *not* the quotient of their derivatives; it is the *d*enominator function times the *d*erivative of the *n*umerator function minus the *n*umerator function times the *d*erivative of the *d*enominator function, with this entire difference divided by the *d*enominator function squared. The italic letters provide the mnemonic for the derivative of the quotient of two functions:

$$h'(x) = \frac{ddn - ndd}{d^2}.$$

Example 20 If $f(x) = 2x^2 + x - 15$ and $g(x) = 2x - 5$, so that $h(x) = (2x^2 + x - 15)/(2x - 5)$, then

$$f'(x) = 4x + 1 \quad \text{and} \quad g'(x) = 2.$$

Consequently,

$$\begin{aligned} h'(x) &= \frac{g(x) \cdot f'(x) - f(x) \cdot g'(x)}{[g(x)]^2} \\ &= \frac{(2x-5) \cdot (4x+1) - (2x^2+x-15) \cdot 2}{(2x-5)^2} \qquad \text{(A)} \\ &= \frac{8x^2 - 18x - 5 - 4x^2 - 2x + 30}{(2x-5)^2} \qquad \text{(Right?)} \end{aligned}$$

$$= \frac{4x^2 - 20x + 25}{(2x-5)^2}$$

$$= \frac{4x^2 - 20x + 25}{4x^2 - 20x + 25}$$

$$= 1.$$

The gigantic collapse of equation (A) into 1 should not be too surprising since

$$\begin{aligned} h(x) &= (2x^2 + x - 15)/(2x - 5) \\ &= (2x-5)\cdot(x+3)/(2x-5) \quad (x \neq 5/2) \\ h(x) &= x + 3 \\ h'(x) &= 1 + 0 \\ h'(x) &= 1. \end{aligned}$$

Problem set 9–5

Find the derivatives of these functions by using the theorems of this section.

1. $f(x) = x^3$
2. $f(x) = 5x + 7$
3. $f(x) = -4$
4. $f(x) = \sqrt{719} + 3.14159$
5. $g(x) = 8x^3 - 5x^2 + 4$
6. $g(t) = 31t^3 + 16t^2 - 14t + 7$
7. $h(x) = 3x$
8. $g(t) = 5(x^2 - 7)$
9. $h(u) = 9u \cdot (15u^4 - 7u)$
10. $f(x) = 135x^5 - 63x^2$
11. $f(x) = (x^3 + 7x)\cdot(2x + 5)$
12. $f(t) = 13t^3 \cdot (7 - 5t)$
13. $f(x) = (3x - 5)/x^2$
14. $g(t) = (9t^2 - 5t)/(3 + 2t)$
15. $f(t) = (7t^2)/(3t - 5t^3)$
16. $f(x) = 5/(x + 7)$

Find the slope of the tangent line to the graph of these functions at the indicated points of tangency.

17. $f(x) = -3x^2 + 4x + 6$; $(5, f(5))$, $(1, f(1))$, $(-2, f(-2))$
18. $g(t) = 4t - 3$; $(5, f(5))$, $(2, f(2))$, $(3/4, f(3/4))$
19. $g(x) = 5/(x + 7)$; $(4, f(4))$, $(-8, f(-8))$, $(-6, f(-6))$, $(-7, f(-7))$
20. $f(t) = -t^3 + 3t^2 + 9t - 3$; $(2, f(2))$, $(-1, f(-1))$, $(3, f(3))$

Find the equation of the tangent line to the graph of these functions at the indicated points of tangency.

21. Use the data in Problem 17.
22. Use the data in Problem 18.
23. Use the data in Problem 19.
24. Use the data in Problem 20.
25. In Problem 3 of the first section of this chapter we found that the minimum value of $f(x)$ was 4, occurring at $x = -3$. Find the slope and equation of the tangent line to the graph at $(-3, 4)$. What is $f'(-3)$?
26. Find the derivative of $f(x) = x^3$ by the delta process.
27. Find the derivative of $g(x) = (2x + 5)^3$ by the delta process.

28. As an extension of Theorem 3 of this section we could prove that if

$$g(x) = [f(x)]^n \text{ (} n \text{ an integer), then}$$
$$g'(x) = n \cdot [f(x)]^{n-1} \cdot f'(x).$$

In words, the derivative of a function raised to an integral power is the product of three factors: the power, the function itself raised to one less than that power, and the derivative of the function itself. If we look at Problem 27 we have

$$g(x) = [(2x+5)]^3$$

since $f(x) = (2x+5)$ and $n = 3$.

$$\begin{aligned} g'(x) &= 3 \cdot [(2x+5)]^{3-1} \cdot D_x(2x+5) \\ &= 3 \cdot (2x+5)^2 \cdot 2 \\ &= 6 \cdot (2x+5)^2 \quad \text{or, optionally,} \\ &= 6 \cdot (4x^2+20x+25) \\ &= 24x^2 + 120x + 150. \end{aligned}$$

The symbol $D_x(2x+5)$ ("d sub x of $2x+5$") is a convenient way to indicate the derivative (D) of the function given by the algebraic expression in x. Find the derivatives of these functions, leaving the answers in simplest form.

a. $f(x) = (3x+7)^3$
b. $g(x) = (7x-4)^2$
c. $h(x) = (14x-27)^7$
d. $f(t) = (4t^2-6t+5)$
e. $f(x) = (18x^4-7x^2+9)^8$
f. $g(x) = (3.65x^2-11.61x)^4$

29. Theorem 3 can also be extended to have n represent a fraction. For example, the derivative of

$$f(x) = x^{(3/2)} \quad (x > 0)$$

is

$$\begin{aligned} f'(x) &= (3/2) \cdot x^{(3/2-1)} \cdot D_x(x) \\ &= (3/2) \cdot x^{(1/2)} \cdot 1 \\ &= \frac{3x^{(1/2)}}{2}. \end{aligned}$$

Recall also that n may represent a negative integer or a negative fraction.

a. Differentiate $f(x) = x$ by the delta process.
b. Differentiate $f(x) = x^{(1/2)}$ by Theorem 3.
c. Differentiate $f(x) = x^{-2}$
d. Differentiate $f(x) = x^{-3}$
e. Differentiate $f(x) = 1/(x^3)$ by the quotient theorem.
f. Differentiate $f(x) = x^3 + 1/(x^4)$
g. Differentiate $f(x) = \dfrac{3x^{-2}+5x^2}{8x}$
h. Differentiate $f(x) = \dfrac{x^2-3x+17}{5x^3}$
i. Differentiate $s(t) = \dfrac{6t+7}{t^2}$ by the quotient theorem.
j. Differentiate $s(t) = (6t+7) \cdot t^{-2}$ by the product theorem.

30. If you studied the previous section on velocity you may be assigned this problem. A point moves along a number line in such a way that its distance (s feet) from point 0 t seconds after a stopwatch is started is

$$s(t) = 4t^3 - 7t^2 + 3t - 5.$$

Find its velocity and direction (right or left, according to whether $s'(t) > 0$ or $s'(t) < 0$, respectively) for these values of t along with its distance from 0.

a. $t = 2$ b. $t = 1.5$
c. $t = 3$ d. $t = 5$
e. $t = 0$ f. $t = .4$

31. When we first learned about slopes of lines we saw that whenever the y-coordinate of a point on a line was decreasing as we moved along the line from left to right, its slope was a negative number. Conversely, whenever the y-coordinate of a point on a line was increasing as we moved along the line from left to right, its slope was a positive number. Likewise, if the slope of a line is a negative number, then the y-coordinate of a point on the line decreases as we move it from left to right along the line; and if the slope of a line is a positive number, then the y-coordinate of a point on the line increases as we move it from left to right along the line. Thus, it seems natural to say that a function is increasing at a point $(x, f(x))$ if its tangent line there has a positive slope, and to say that a function is decreasing at a point $(x, f(x))$ if its tangent line there has a negative slope. Since the slope of the tangent line at $(x, f(x))$ is given by $f'(x)$, we conclude that where $f'(x) > 0$ the tangent line is increasing and so is the function itself. Where $f'(x) < 0$ the tangent line is decreasing and so is the function itself. Determine whether the function given by

$$y = f(x) = x^3 - 3x^2 + x + 1$$

is increasing or decreasing at these points.

a. $(-1.5, f(-1.5))$ b. $(-.5, f(-.5))$ c. $(0, f(0))$
d. $(.5, f(.5))$ e. $(1.5, f(1.5))$ f. $(2.5, f(2.5))$

32. Differentiate the function defined by

$$y = f(x) = \frac{x^2 - 3x + 17}{5x^3}.$$

Applications of the derivative

We are finally ready to use our knowledge of differentiation to solve maximum/minimum type problems. We begin with an example.

Finding the extrema of a function

Example 21 Find the maximum value, minimum value, and turning (or plateau) points of the function

$$f(x) = -x^3 + 3x^2 + 9x - 3$$

in the interval $\{x \mid -4 \leqslant x \leqslant 6\}$.

Solution We begin with a graphic approach and evaluate $f(x)$ for those integers in the given domain.

Table 9.13

x	−4	−3	−2	−1	0	1	2	3	4	5	6
$f(x)$	73	24	−1	−8	−3	8	19	24	17	−8	−57

Figure 9.7

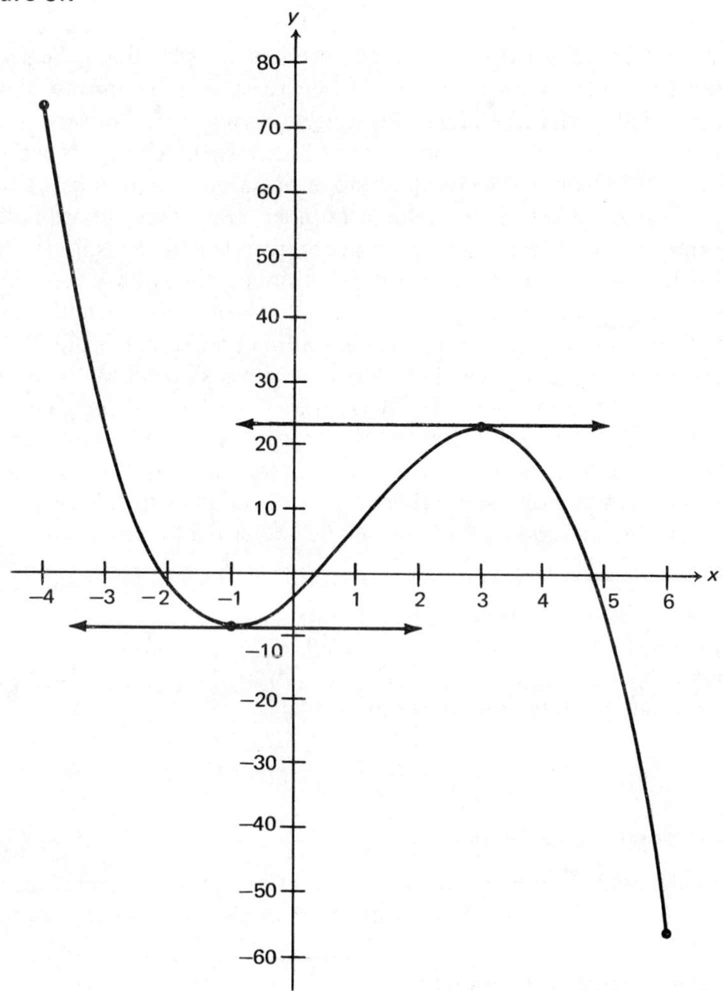

The minimum value of $f(x) = -57$ occurs at $x = 6$, and the maximum value of $f(x) = 73$ occurs at $x = -4$. We turn our attention to picking the exact coordinates of the two plateau points which appear to be $(-1, -8)$ and $(3, 24)$. We examine the next two tables.

Table 9.14

x	−1.2	−1.1	−1	−.9	−.8
$f(x)$	−7.752	−7.939	−8	−7.941	−7.768

Table 9.15

x	2.8	2.9	3	3.1	3.2
$f(x)$	23.768	23.941	24	23.939	23.752

If we used a computer as a magnifying glass to examine $[-1.2, -.8]$ and $[2.8, 3.2]$ in steps of .01, .001, or even .0001, we would become more and more convinced that $(-1, -8)$ and $(3, 24)$ were indeed turning or plateau points. Technically, we say that -8 is a *relative minimum* value of $f(x)$, since in a set of nearby domain points centered around $x = -1$ the corresponding range values are all greater than $f(-1) = -8$ (see Figure 9.8). We also say that 24 is a *relative maximum* value of $f(x)$, since in a set of nearby domain points centered around $x = 3$ the corresponding range values are all less than $f(3) = 24$ (see Figure 9.9).

Figure 9.8

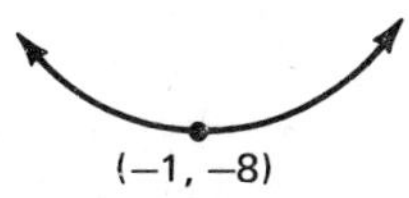

Figure 9.9

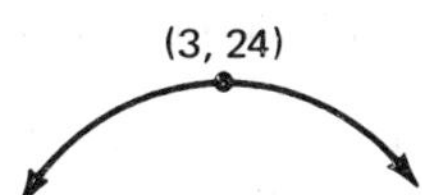

The *absolute minimum* of $f(x)$ in $[-4, 6]$ is $f(6) = -57$, and $f(-4) = 73$ is the *absolute maximum* of $f(x)$ in $[-4, 6]$. If we considered this function over the interval $[2, 6]$, then $f(3) = 24$ is both a relative and an absolute maximum while $f(6) = -57$ is an absolute minimum; there is no relative minimum in this interval. We often use the term *extrema* to include absolute or relative maximum or minimum values of the range elements $f(x)$ of a function. The word "relative" is often interchanged with "local," and "absolute" is often interchanged with "global."

Basic max/min theorem

So far, of course, nothing is new in this example about the way we look for extrema. Suppose we differentiate

$$f(x) = -x^3 + 3x^2 + 9x - 3$$

to obtain

$$f'(x) = -3x^2 + 6x + 9$$

and add a row to our previous tables with the specific values of $f'(x)$.

Table 9.16

x	-1.2	-1.1	-1	$-.9$	$-.8$
$f(x)$	-7.752	-7.939	-8	-7.941	-7.768
$f'(x)$	-2.52	-1.23	0	1.17	2.28

Table 9.17

x	2.8	2.9	3	3.1	3.2
$f(x)$	23.768	23.941	24	23.939	23.752
$f'(x)$	2.28	1.17	0	-1.23	-2.52

Notice that where $f(x)$ *attains a relative minimum* ($x = -1$ in the domain, so that $f(-1) = -8$), *then* $f'(-1) = 0$! *Where* $f(x)$ *attains a relative maximum* ($x = 3$ in the domain so that $f(3) = 24$), *then* $f'(3) = 0$! Graphically, at these plateau points $(-1, -8)$ and $(3, 24)$ we see that a tangent line would be parallel to the x-axis. Any line parallel to the x-axis has a slope of 0, and since $f'(x)$ measures the slope of a tangent line at $(x, f(x))$, $f'(x)$ would be 0 at $x = -1$ and at $x = 3$.

We make a *very* important generalization from Tables 9.16 and 9.17 and from the observation expressed in italicized letters above.

Basic max/min theorem If a differentiable function has any extrema for an interval of its domain $[a, b]$, then they must occur either among those values of x in $[a, b]$ such that $f'(x) = 0$ or at the endpoints a or b.

In other words, the turning points may occur only among those values of x for which the value of the derivative, $f'(x)$, is 0. This fact does *not* mean that whenever $f'(x) = 0$ $f(x)$ will always be a relative minimum or relative maximum, as we will see in a later example and in Problem 37 of this section.

Example 22 Find the maximum and minimum values of

$$f(x) = x^2 - 10x + 2$$

in the interval $[3, 8]$ of its domain.

Solution Since any extrema may occur for those values of x in the domain such that $f'(x)$ is 0, we differentiate, set the derivative equal to 0, and solve the resulting equation for x. A number c in $[a, b]$ such that $f'(c) = 0$ is called a *critical* value of x.

$$\begin{aligned} f(x) &= x^2 - 10x + 2 \\ f'(x) &= 2 \cdot x^{2-1} - 10 \cdot 1 + 0 \\ f'(x) &= 2x - 10 \\ f'(x) &= 0 \\ 2x - 10 &= 0 \\ x &= 5 \end{aligned}$$

The function

$$f(5) = 5^2 - 10 \cdot 5 + 2 = 23$$

may be either a relative minimum or a relative maximum corresponding to this critical number $x = 5$—but which one? If we pick a value of x on either side of $x = 5$, then we have this table.

Table 9.18

x	4.9	5	5.1
$f(x)$	−22.99	−23	−22.99

Since -23 is less than both of its neighbors, we see that $f(5) = -23$ is a relative minimum value of $f(x)$ in $[3, 8]$. To see if -23 is also an absolute minimum in the entire interval we evaluate $f(x)$ at the endpoints $x = 3$ and $x = 8$ of the interval.

Table 9.19

x	3	5	8
$f(x)$	−19	−23	−14

We see, then, that $f(5) = -23$ is both the absolute minimum and a relative minimum in this interval, and $f(8) = -14$ is the absolute maximum in this interval. There is no possibility of any other extrema in this interval of the domain since, according to our Basic Max/min Theorem, all extrema must occur only where $f'(x) = 0$ or at the endpoints of the interval. We always test the endpoints of the interval $[a, b]$ because there may well be an absolute maximum where the derivative is not 0, as illustrated in Example 21. There $f'(-4)$ was not 0, but $f(-4) = 73$ was an absolute maximum.

Thus far the only way we have to determine if a number x in the domain such that $f'(x) = 0$ (a critical value of x) yields a relative maximum or a relative minimum value of $f(x)$ is to pick a pair of neighboring points around x and plug them into the expression for $f(x)$, as we did in the previous example. However, there is a better way than this time-consuming and error-prone method, and we can discuss it by looking closely at our previous Tables 9.16 and 9.17 for

$$f(x) = -x^3 + 3x^2 + 9x - 3.$$

Notice that in $[-1.2, -.8]$, where $f(x)$ attains a relative minimum, $f'(x)$ is an increasing function, and that in $[2.8, 3.2]$, where $f(x)$ is a relative maximum, $f'(x)$ is a decreasing function. Since

$$f'(x) = -3x^2 + 6x + 9$$

is a function in its own right, we know by Problem 31 in the last section that *its* derivative, which indicates whether $f'(x)$ is increasing or decreasing, should be positive in $[-1.2, -.8]$.

$$D_x(f'(x)) = D_x(-3x^2 - 6x + 9)$$
$$D_x(f'(x)) = -6x + 6$$

We say that the expression $(-6x+6)$ is the *second derivative* of the original function given by the expression $(-x^3+3x^2+9x-3)$, and we symbolize this second derivative $D_x(f'(x))$ by y'' or $f''(x)$ ("y prime prime," "f prime prime of x").

Table 9.20

Original function	First derivative	Second derivative
$y = f(x) = (-x^3+3x^2+9x-3)$	$y' = f'(x) = (-3x^2-6x+9)$	$y'' = f''(x) = (-6x+6)$

To see if $f''(x)$ is positive in $[-1.2, -.8]$ we evaluate it there. We also evaluate $f''(x)$ in $[2.8, 3.2]$.

Table 9.21

x	−1.2	−1.1	−1	3.1	−.8
$f(x)$	−7.752	−7.939	−8	−7.941	−7.768
$f'(x)$	−2.52	−1.23	0	1.17	2.28
$f''(x)$	13.2	12.6	12	11.4	10.8

Table 9.22

x	2.8	2.9	3	3.1	3.2
$f(x)$	23.768	23.941	24	23.939	23.752
$f'(x)$	2.28	1.17	0	−1.23	−2.52
$f''(x)$	−10.8	−11.4	−12	−12.6	−13.2

Inspection of the fourth row of Table 9.21, which is centered about a known relative minimum $f(-1) = 8$, shows that $f''(x)$ is a positive number at $x = -1$ ($f''(-1) = 12$). Similarly, inspection of the fourth row of Table 9.22, which is centered about a known relative maximum $f(x) = 24$, shows that $f''(x)$ is a negative number at $x = 3$ ($f''(3) = -12$). The value of $f'(x)$ is decreasing in $[2.8, 3.2]$, so that its derivative $f''(x)$, indicating whether $f'(x)$ is increasing or decreasing, should be negative in $[2.8, 3.2]$, which it is (refer to Problem 31 of Problem set 9–5). The graphs of $f(x)$, $f'(x)$, and $f''(x)$ appear in Figure 9.10.

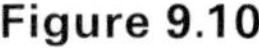

Figure 9.10

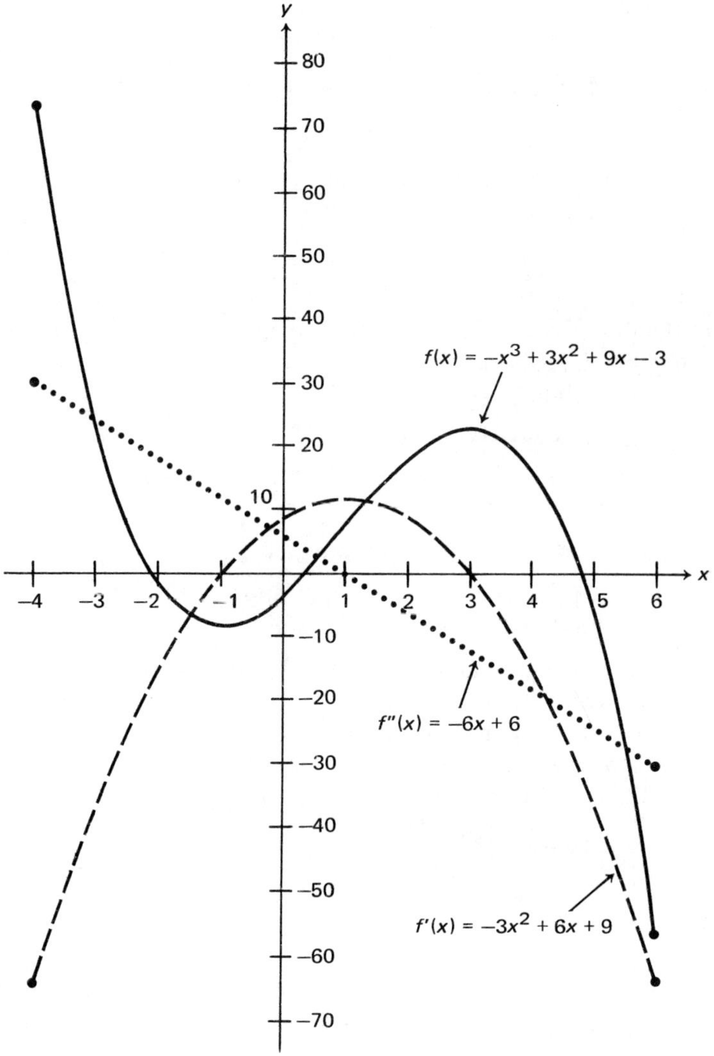

Extended max/min theorem

We summarize our work so far in this entire chapter by making a valid generalization from our discussion of the first and second derivatives of a function.

> Extended max/min theorem If a differentiable function has any extrema in an interval $[a, b]$ of its domain, then they must occur either among those values of x in $[a, b]$ such that $f'(x) = 0$ or at the endpoints a or b. If $f'(x) = 0$ and $f''(x) < 0$, then $f(x)$ is a relative maximum. If $f'(x) = 0$

and $f''(x) > 0$, then $f(x)$ is a relative minimum. If $f'(x) = 0$ and $f''(x) = 0$, then we cannot conclude from this information whether $f(x)$ is a relative minimum, a relative maximum, or neither.

Example 23 Find all extrema of the function

$$f(x) = x^3 - 6x^2 - 5$$

in $[-3, 7]$ and use these points in graphing the function.
Solution We begin by finding the first two derivatives.

$$f'(x) = 3x^2 - 12x$$
$$f''(x) = 6x - 12$$

We set the first derivative equal to 0 to identify the possible values of x where extrema may occur.

$$3x^2 - 12x = 0$$
$$3x \cdot (x - 4) = 0$$
$$3x = 0 \quad \text{or} \quad (x - 4) = 0$$
$$x = 0 \quad \text{or} \quad x = 4$$

Thus, $f(0) = -5$ and $f(4) = -37$ are possible extrema.

$$f''(0) = 6 \cdot 0 - 12 = -12, \quad \text{and} \quad -12 < 0.$$
$$f''(4) = 6 \cdot 4 - 12 = 12, \quad \text{and} \quad 12 > 0.$$

Hence $f(0) = -5$ is a relative maximum because $f''(0) < 0$, and $f(4) = -37$ is a relative minimum because $f''(4) > 0$. The points $(0, f(0)) = (0, -5)$ and $(4, f(4)) = (4, -37)$ are relative extrema or plateau points. We obtain three more points $(-2, -37)$, $(2, -21)$, and $(6, -5)$ to aid in the graphing and also obtain the endpoints $(-3, f(-3))$ and $(7, f(7))$.

Table 9.23

x	−3	−2	0	2	4	6	7
$f(x)$	−86	−37	−5	−21	−37	−5	44

The absolute minimum is $f(-3) = -86$, the absolute maximum is $f(7) = 44$, the relative minimum is $f(4) = -37$, and the relative maximum is $f(0) = -5$.

We accept without proof the statement that an equation such as

$$3x^2 - 12x + 0 = 0$$

has at most two solutions among the real numbers and that an equation such as

$$5x - 13 = 0$$

has only one solution among the real numbers. The derivative of

$$f(x) = ax^3 + bx^2 + cx + d$$

is

$$f'(x) = 3ax^2 + 2bx + c,$$

and when it is set equal to 0 there are at most two values of x such that $f'(x) = 0$ and thus at most two relative extrema. The derivative of

$$f(x) = ax^2 + bx + c$$

is

$$f'(x) = 2ax + b,$$

and when we set it equal to 0 to locate the critical values of x there is just one such number. This paragraph explains why the function in this example has at most two relative extrema. It also explains the instructions preceding Problems 2 through 12 of Problem set 9–1.

We finish this example with its graph.

Figure 9.11

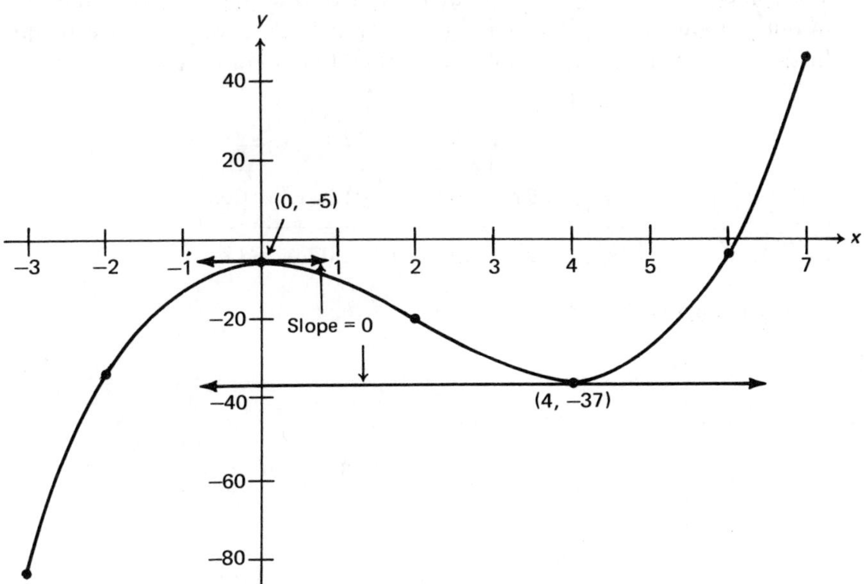

Example 24 Find all extrema of the function defined by

$$y = f(x) = x^3 - 6x^2 + 12x - 7$$

in the interval $[0, 4]$.

Solution As usual, we begin by finding the first two derivatives of function and setting the first one equal to 0. Solving the resulting equation will locate those values of x for which $f(x)$ may possibly be extrema.

$$\begin{aligned} f(x) &= x^3 - 6x^2 + 12x - 7 \\ f'(x) &= 3x^2 - 12x + 12 \\ f''(x) &= 6x - 12 \end{aligned}$$

Setting $f'(x) = 0$ to find the critical values of x, we have

$$3x^2 - 12x + 12 = 0$$
$$3 \cdot (x^2 - 4x + 4) = 0$$
$$x^2 - 4x + 4 = 0$$
$$(x-2)^2 = 0 \qquad \text{(Factoring)}$$
$$x - 2 = 0 \qquad \text{(Taking the square root of both nonnegative numbers)}$$
$$x = 2.$$

We evaluate $f''(2)$ to see if $x = 2$ produces a relative minimum or a relative maximum when $f(2)$ is evaluated.

$$f''(x) = 6x - 12$$
$$f''(2) = 6 \cdot 2 - 12$$
$$f''(2) = 0$$

The last equation, according to our Extended Max/min Theorem, tells us nothing about whether $f(2)$ is an extremum. One way to investigate further is to evaluate and graph the original function for [1.8, 2.2].

Table 9.24

x	1.8	1.9	2	2.1	2.2
$f(x)$	.992	.999	1	1.001	1.008
$f'(x)$	.12	.03	0	.03	.12

Figure 9.12

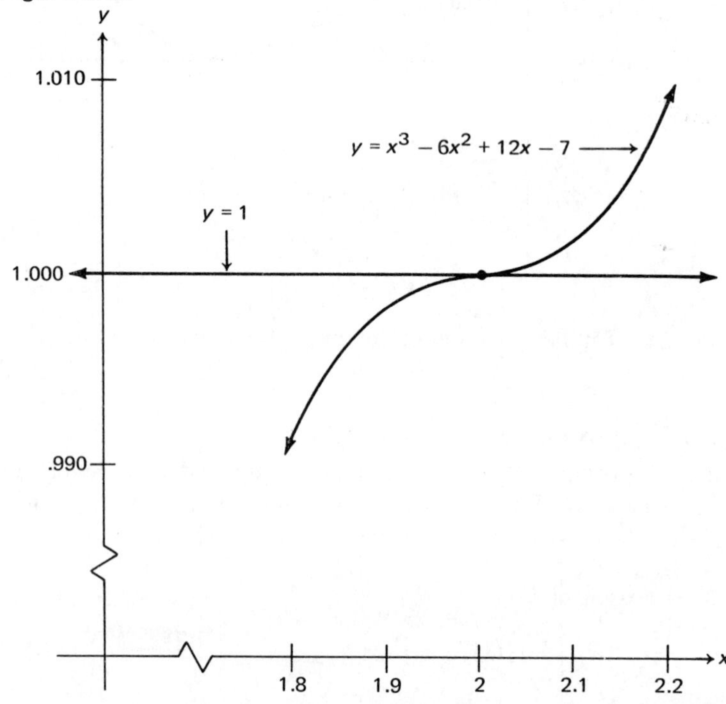

Notice that the point with coordinates $(2, 1)$ is not a turning point; consequently, $f(2) = 1$ is not a relative minimum, even though $f'(2) = 0$. The formal definition of the tangent line to a curve at a point $(x, f(x))$ on the curve is the limiting position of successive approaching secant lines passing through the fixed base point $(x, f(x))$. According to this definition, the line whose equation is $y = 1$ is the tangent line to the graph at $(2, f(2)) = (2, 1)$, even though the line actually cuts the graph at that point.

Since there are no relative extrema ($f(2) = 1$ is the only candidate), all we can do is to evaluate $f(x)$ at the endpoints $x = 0$ and $x = 4$.

$$\begin{aligned} f(0) &= 0^3 - 6 \cdot 0^2 + 12 \cdot 0 - 7 \\ &= -7 \\ f(4) &= 4^3 - 6 \cdot 4^2 + 12 \cdot 4 - 7 \\ &= 9 \end{aligned}$$

Finally, the absolute minimum is $f(0) = -7$, and the absolute maximum is $f(4) = 9$. Here is the complete graph of the function for the interval $[0, 4]$.

Figure 9.13

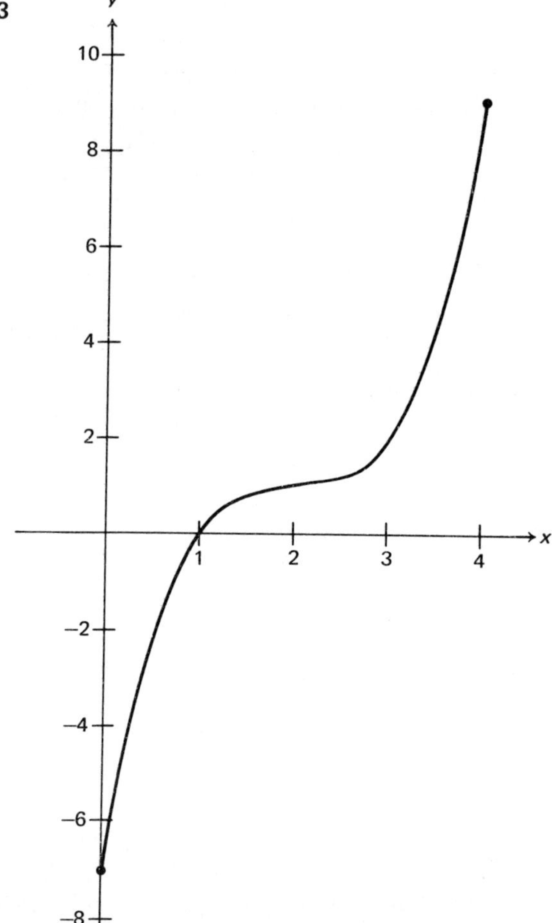

Word problems involving max/min function

Example 25 Continental Leasing Company leases new automobiles in fleets to corporations. They charge \$800 per car per year, but for contracts with more than twelve cars leased the rental fee is discounted by 2% for *each* car in a single contract. How many cars leased to a single corporation in one year will produce maximum income, and what is that income?

Solution We now know the pure mathematics necessary to solve this type of problem. It is just a matter of translating the description of the problem into an algebraic equation involving a function. Basically,

$$\text{total income} = (\text{total number of cars leased}) \cdot (\text{income per car}).$$

We let

$$T = \text{total income}$$

and x = total number of cars that are leased so that if $0 \leqslant x \leqslant 12$, then clearly

$$T = f(x) = x \cdot 800$$

and the maximum value of T is $12 \cdot 800 = \$9{,}600$. If x is greater than 12, then it must also be at most 50 because $50 \cdot 2\% = 100\%$, and a 100% discount would result in zero income. Consequently, if $12 < x \leqslant 50$,

$$\begin{aligned} T = f(x) &= x \cdot (800 - x \cdot .02 \cdot 800) \\ &= x \cdot (800 - 16x) \\ &= 800 - 16x^2. \end{aligned}$$

We find the first and second derivatives of $T = f(x)$,

$$\begin{aligned} T' &= f'(x) = 800 - 32x \\ T'' &= f''(x) = -32 = -32 \cdot x^0, \end{aligned}$$

and continue by finding the critical values of x.

$$\begin{aligned} f'(x) &= 0 \\ 800 - 32x &= 0 \\ 32x &= 800 \\ x &= 800/32 \\ x &= 25 \end{aligned}$$

Since

$$f''(25) = -32 \cdot 25^0 = -32,$$

which is a negative number, a relative maximum value, $f(25)$, exists.

$$\begin{aligned} f(25) &= 800 \cdot 25 - 16 \cdot 25^2 \\ &= 10{,}000 \end{aligned}$$

In conclusion, since $f(13) = 7{,}696$ and $f(50) = 0$, the leasing of 25 cars will yield a maximum of $f(25) = \$10{,}000$ of revenue. Notice that $f(25) > f(12)$; $f(25)$ is the absolute maximum of $f(x)$ over the interval $[0, 50]$. Problem 36 is an interesting variation of this example.

Any max/min function problem is attacked in a similar fashion. Usually, the most difficult part is setting up a function that is truly reflective of the physical situation. We should be sure to begin each of the word problems by clearly identifying the independent variable and the dependent variable and making a diagram wherever possible.

This section is really the capstone of the entire chapter so far. It was necessary to understand, in order, the concepts of maximum and minimum values of the range of a function, average rate of change, secant line, tangent line, limit, instantaneous rate of change, delta process, derivative, and occurrence of extrema. There is no shortcut for learning how to find extrema of most functions that arise from the physical world around us or from the use of calculus in other areas such as economics (see Problem 40).

Problem set 9–6

Find the extrema (both relative and absolute) of these functions in these intervals.

1. $y = f(x) = x^2 - 4x + 7, \ -3 \leqslant x \leqslant 5$
2. $y = f(x) = x^2 + 6x + 13, \ -5 \leqslant x \leqslant 1$
3. $y = f(x) = x^2 - 8x + 14, \ -1 \leqslant x \leqslant 7$
4. $y = f(x) = x^2 - 5.2x + 8.26, \ 1 \leqslant x \leqslant 4$
5. $y = f(x) = x^2 - 7x + 3, \ 4 \leqslant x \leqslant 7$
6. $y = f(x) = 13 + 5x - 7x^2, \ -2 \leqslant x \leqslant 2$
7. $y = f(x) = 18 + 5x - 7x^2, \ -2 \leqslant x \leqslant 2$
8. $y = f(x) = 7 + 36x - 5x^2, \ -3 \leqslant x \leqslant 5$
9. $y = f(x) = 7 + 36x - 5x^2, \ 6 \leqslant x \leqslant 10$
10. $y = f(x) = 13 - 12x - 3x^2, \ -5 \leqslant x \leqslant 0$
11. $y = f(x) = x^3 - 12x, \ -4.1 \leqslant x \leqslant 4.1$
12. $y = f(x) = x^3 - 6x^2 + 12x - 7, \ 0 \leqslant x \leqslant 6$
13. $y = f(x) = x^3 - 9x^2 + 24x - 14, \ 0 \leqslant x \leqslant 6$
14. $y = f(x) = 2x^3 + 3x^2 - 36x + 6, \ -4 \leqslant x \leqslant 4$
15. $y = f(x) = 2x^3 - 3x^2 - 12x - 5, \ -2 \leqslant x \leqslant 4$
16. $y = f(x) = 2x + 7, \ -1 \leqslant x \leqslant 5$
17. $y = f(x) = 5x - 8, \ -2 \leqslant x \leqslant 6$
18. $y = f(x) = 7x^0, \ -3 \leqslant x \leqslant 5$
19. $y = f(x) = (1 + 2x)^2/3x, \ -1 \leqslant x \leqslant 1$ (Careful!)
20. $y = f(x) = (1 + 2x)^2/3x, \ -1 \leqslant x \leqslant -1/8$
21. In Chapter 3 refer to the first example in the section Word Problems with Functions. What value of w, the width of the field, minimizes C, the total cost? What is this cost? What must the length of the field be at this minimizing value of w?
22. In Chapter 3 refer to the third example in the section Word Problems with Functions. How long should the farmer store the grain in order to have the maximum profit, and what is this profit?

23. Refer to Problem 1 in Problem set 3–2. How long should the cooperative association keep the wheat so that its value is a maximum, and what is this maximum value?
24. Refer to Problem 4 in Problem set 3–2. What value of x produces a box of maximum value, and what is this value?
25. Refer to Problem 5 in Problem set 3–2. What value of x produces a box of minimal cost, and what is this minimal cost?
26. Refer to Problem 6 in Problem set 3–2. If 50 feet of fencing is available, then what dimensions of the bed produce a maximum area?
27. Refer to Problem 6 in Problem set 3–2. If 50 feet of fencing is available and he covers the flower bed with topsoil to a depth of d inches that costs c dollars a cubic yard, then find the dimensions of the bed that minimizes the total cost of the fencing and topsoil.
28. Rework the previous problem, this time assuming that k feet of fencing are available.
29. Refer to Problem 7 in Problem set 3–2. Write the bus company's total revenue, R, as a function of x, the number of people in excess of thirty who go. What value of x maximizes R, and what is this value of R?
30. Refer to Problem 8 of Problem set 3–2. Find the dimensions of the paper with minimum total area. What is this total area? Find the dimensions of the paper with minimum area if the pages may have a maximum length of 14 inches and a maximum width of 9 inches. What is this area?
31. Refer to Problem 9 of Problem set 3–2. How much should he charge per car in order to maximize his profit, and what is this profit?
32. What should the diameter of a tin can be if it is to hold 58 cubic inches, the surface area being minimized and the can having an open top?
33. Rework Problem 32, this time giving the can a closed top.
34. Rework Problem 32, this time assessing the volume as c cubic units.
35. Rework Problem 32, this time giving the can a closed top and assessing the volume as c cubic units.
36. Refer to the example of Continental Leasing Company. If *only* those cars in excess of twelve are given a discount of 2% each (so that fifteen cars would cost

$$12 \cdot 800 + 3 \cdot (800 - 6\% \text{ of } 800)$$

dollars), then how many cars leased to a single corporation in one year will produce maximum income, and what is that income?

37. Draw the graph of

$$g(x) = x^3 - 3x^2 + 3x + 2$$

for the interval $[-1, 3]$.

a. Evaluate $g(.9)$ and $g(1.1)$.
b. Evaluate $g(.95)$ and $g(1.05)$.
c. Evaluate $g(1)$.
d. Find $g'(x)$ and evaluate $g'(1)$.

This problem shows that just because $g'(c) = 0$ we are not forced to conclude that $(c, g(c))$ is a relative extremum.

e. Find $g''(1)$. The result, together with the previous parts of this problem, illustrates the last part of our Extended Max/min Theorem.

38. Consider the function defined by

$$y = f(x) = 13 - |2x - 6| \text{ in } [1, 7].$$

For example,

$$f(2) = 13 - |2 \cdot 2 - 6| = 13 - |-2| = 13 - 2 = 11.$$

a. Complete this table.

x	1	2	3	4	5	6	7	2.95	2.99	3.01	3.05
$f(x)$											

b. By examining this table, determine the maximum value of $f(x)$.
c. By examining this table, determine the minimum value of $f(x)$.
d. Make a graph of the function by using the results of part (a).
e. Notice that this is an example of a function that is continuous at a point $(x, f(x))$ and has a relative maximum there, but the derivative of this function does not exist at the "sharp" point $(3, f(3))$.

39. A real estate office handles an apartment complex with 200 units. When the rent of each unit is \$200 per month, all units are occupied. Experience has shown that for each \$10 per month increase in rent, five units become vacant. It is also relevant that the cost of servicing a rented apartment is \$20 a month. What rent should the office charge in order to maximize profit? How much is that maximum profit? How many apartments will be rented at that maximum profit?

40. Let

$$C = f(x) = 10 + 3x + 75/x$$

be the cost function for producing x axles where x must be a minimum order of two axles. There is an overhead of \$10 per order, a cost for labor of \$3 per axle and a one-time set-up charge of \$75 per order for the assembling machine. The set-up charge is spread equally over all the axles produced for an order. The derivative of this cost function,

$$\frac{dC}{dx} = C' = f'(x),$$

is called the *marginal cost function*, and it is defined as the change in total cost (C) per change in unit of product produced (x). We recognize this definition as being equivalent to the concept of instantaneous rate of change. Find the marginal and total cost for these numbers of axles in one order.

a. 3 b. 5 c. 7
d. 8 e. 13 f. 14

g. What order size produces the minimum cost for the company, and what is this minimum cost?

h. A similar definition holds true for *marginal revenue*. That is, if $R = f(x)$ is the revenue produced by the sale of x units of output, then $R' = f'(x) = dR/dx$ is the marginal revenue and represents the additional revenue taken in by the sale of one additional unit of output.

41. A steel company knows that if it charges x dollars a ton it can sell $300 - x$ tons in a single order. There is a charge of \$120 to manufacture each ton and a total cost of \$5,000 which is spread equally among each of the $300 - x$ tons produced in a single batch. How much should the company charge per ton to maximize its revenue, what is that revenue, and how many tons are sold at the maximum revenue?

42. Now use differential calculus to solve the example at the very beginning of this chapter about the salesman and his washers.

43. An important application of calculus is determining the *economic order quantity*, EOQ, of a given product. Suppose a firm estimates that it will use 600 wood-reinforced cardboard boxes each year for shipping aircraft generators. It knows that each box costs 20¢ in storage costs, regardless of how long it is held, and that each order costs \$14 in various paperwork and overhead expenses. The firm assumes that the x boxes ordered in a single lot are used at a constant rate and that replenishment occurs just as the last of the previous lot is used. This means that the average number of boxes in stock is $(x + 0)/2 = x/2$.

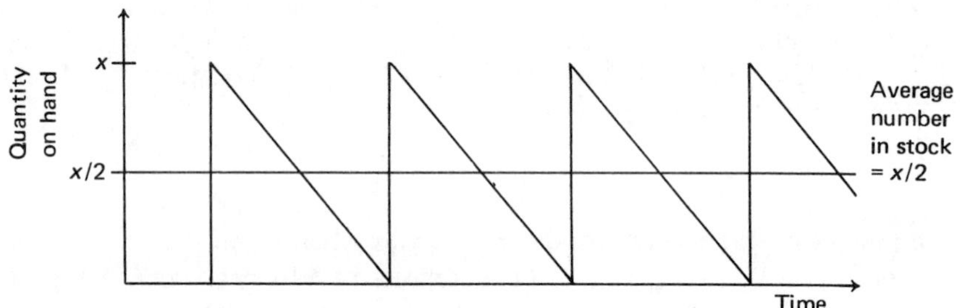

We seek that value of x for which the total annual ordering cost, T, is a minimum. It is apparent from the information that

$$\begin{aligned}&\text{total annual order cost}\\&\quad = \text{annual storage cost} + \text{annual paperwork cost}\end{aligned}$$

or

$$\begin{aligned}T = {} & [(\text{average quantity on hand}) \cdot (\text{storage cost of each box})]\\ & + [(\text{number of orders each year}) \cdot (\text{cost of each order})].\end{aligned}$$

Thus, in our case,

$$T = f(x) = (x/2) \cdot .20 + (600/x) \cdot 14.$$

a. If $x = 6$ boxes are ordered each time, what is the average number of boxes in stock, how many orders are placed each year, and what is the total annual ordering cost?
b. Rework part (a), letting $x = 20$.
c. Rework part (a), letting $x = 60$.
d. Rework part (a), letting $x = 150$.
e. What is the EOQ, that is, what value of x minimizes T?
f. What is this value of T from part (e)?
g. Supposing that each box costs \$1.50, answer parts (e) and (f) again.
h. Suppose that d boxes will be used each year, that each box costs s dollars in storage costs per year, and that each order costs r dollars in paperwork expenses. Suppose also that the usage cycle is similar to the time diagram above. Determine the total annual order cost, T, in terms of d, s, r, and x where x is the number of boxes in each order.
i. What is the EOQ from part (h)?

44. A watermelon grower wishes to ship his product to market as early as possible in the season to obtain the highest possible prices. Right now he can ship 8 tons at a profit of \$5 per ton. By letting the melons grow he estimates he can add 2 tons per week to his selling capacity. On the other hand, he also estimates that the profit per ton declines by 50¢ per week because other growers will also have increased capacity. In how many weeks should he ship his melons in order to obtain the maximum profit, and what is that profit?
45. A manufacturer can sell x items per week at a selling price of

$$p(x) = 200 - .1x$$

cents per item, and it costs

$$c(x) = 50x + 10{,}000$$

cents to produce the batch of x items. How many items should he produce in a single batch to maximize the profit, and what is the maximum profit?
46. Refer to Problem 45. Suppose a tax of 20¢ is placed on each item sold and the price and cost functions change accordingly. How much of the tax increase should the manufacturer pass on to the consumer, and why?
47. Suppose that a company making a certain item knows that its cost function is

$$C = f(x) = \frac{x^3}{3} - 17x^2 - 111x + 50$$

and its revenue function is

$$R = g(x) = x(100 - x).$$

a. Find the quantity of output that minimizes the cost.
b. What is the marginal cost at the value of x which minimizes the cost?

c. Find the quantity of output that maximizes the revenue.
d. What is the marginal revenue at the value of x which maximizes the revenue?
e. Find the quantity of output that maximizes the profit.
f. What is the maximum profit?

48. In a certain truck factory the total cost of producing x trucks per week is

$$C = f(x) = x^2 + 75x + 1000$$

dollars. How many trucks should be produced to maximize the profit if the number produced is limited by a production capacity of fifty per week and the sale price per truck is

$$s = g(x) = \begin{cases} \frac{5}{3}(125 - x) & \text{if } x \leq 25, \\ \frac{500}{3} & \text{if } 25 < x \leq 50? \end{cases}$$

49. Match the graph of each function f_1 through f_{10} with the graph of its derivative. The graphs of the derivatives are symbolized by $g_1, g_2, g_3, \ldots, g_{10}$.

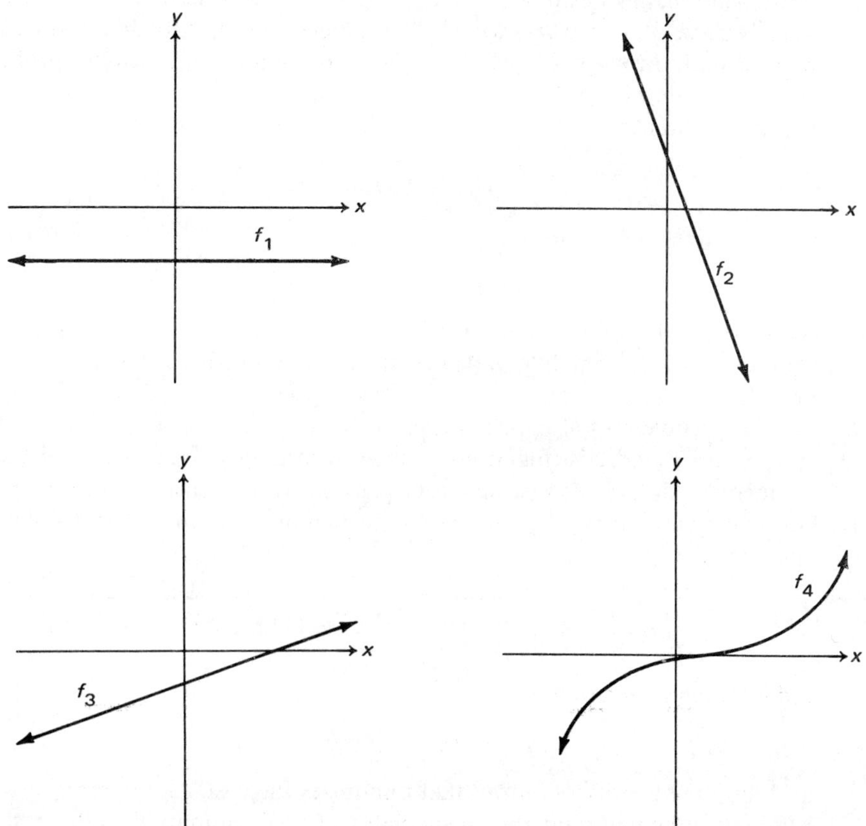

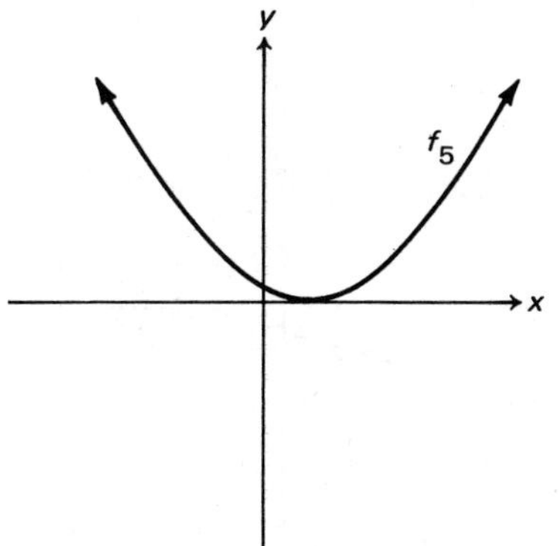
y
x
f_5

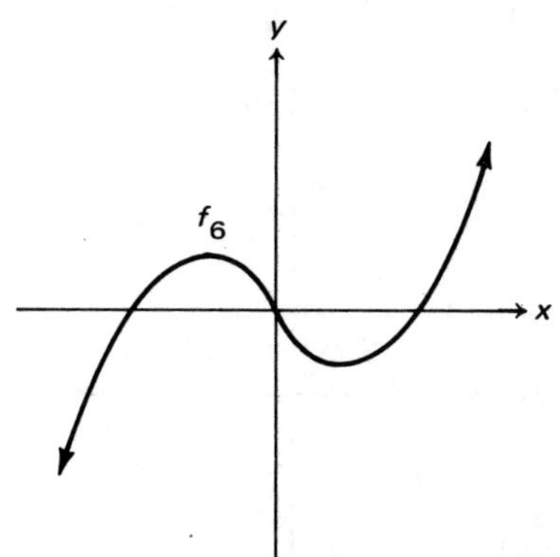
y
x
f_6

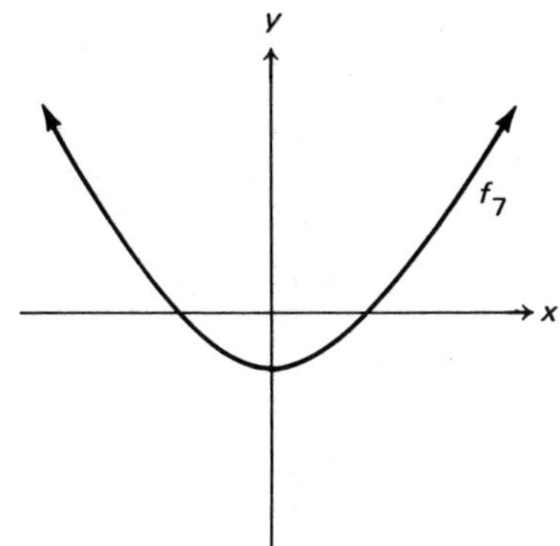
y
x
f_7

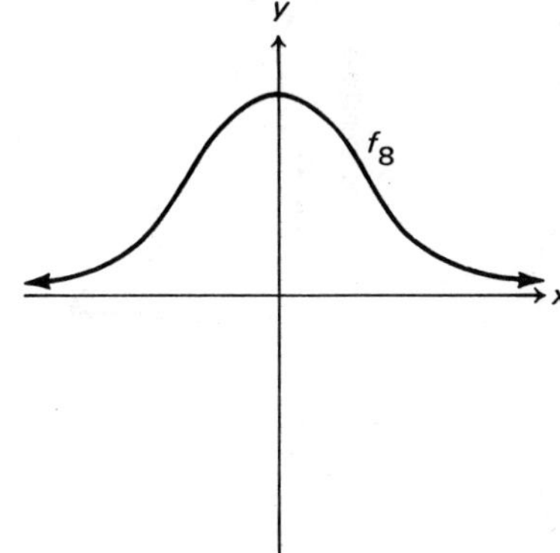
y
x
f_8

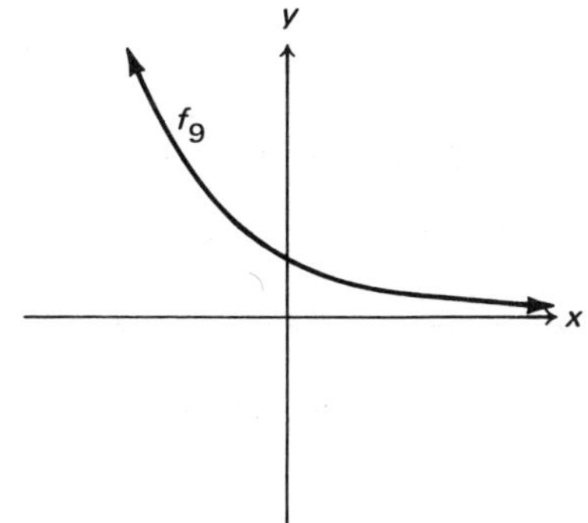
y
x
f_9

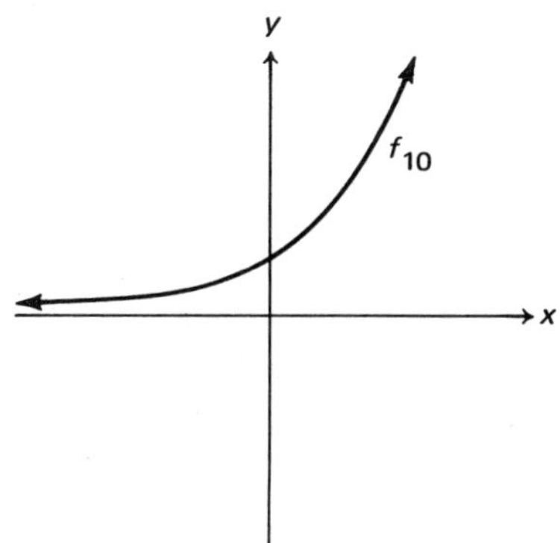
y
x
f_{10}

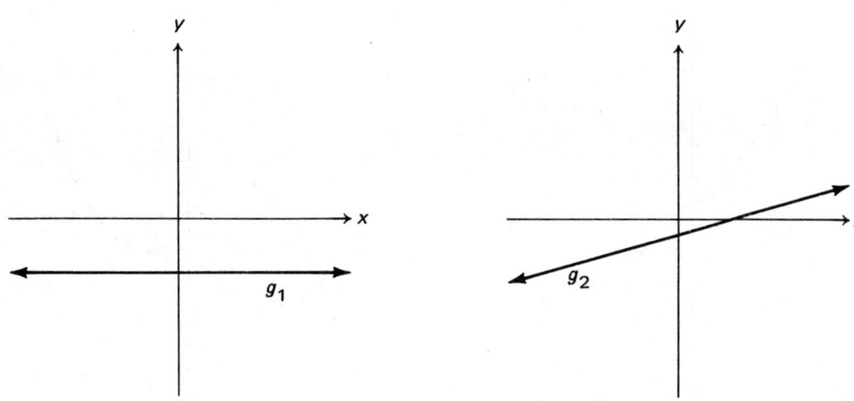
y
x
g_1
y
x
g_2

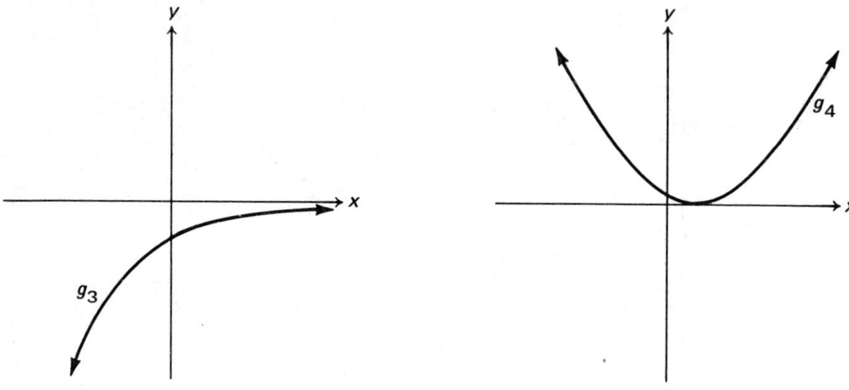
y
x
g_3
y
x
g_4

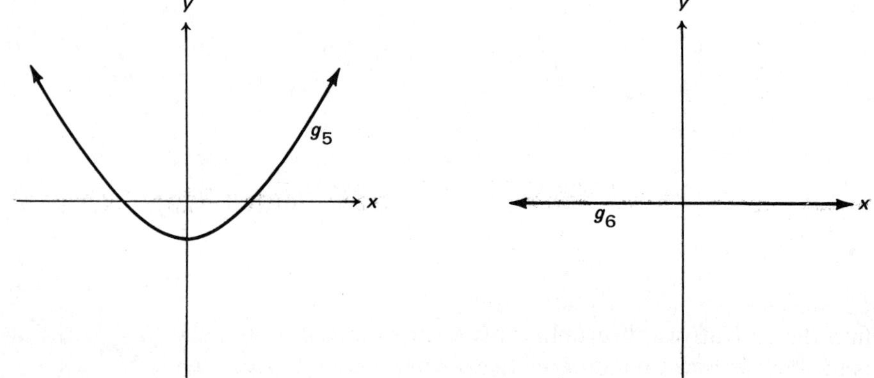
y
x
g_5
y
x
g_6

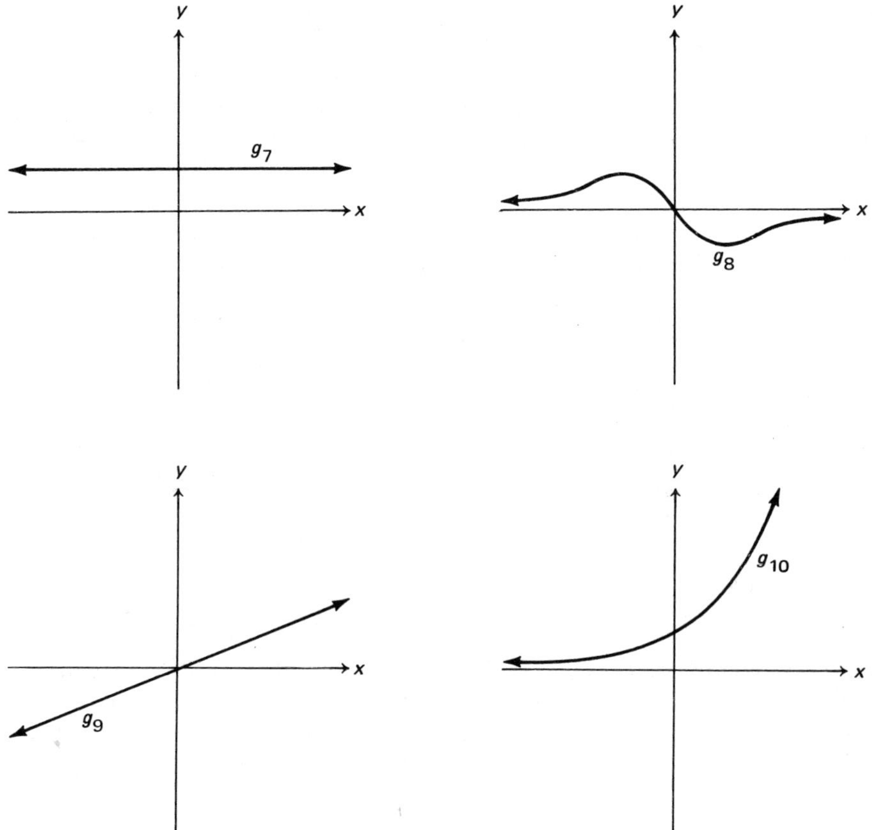

Function	Derivative
f_1	
f_2	g_1
f_3	
f_4	
f_5	
f_6	
f_7	
f_8	
f_9	
f_{10}	

Further applications of the derivative

So far we have learned some basic principles of differentiation which let us find the derivatives of certain kinds of functions. The purpose of this section is to show us how to find derivatives of more complicated functions.

Composite functions

The first type of function we are studying here is known as a *composite function.* Recall from Chapter 3 that we may picture a function g given by $u = g(x)$ as follows.

Figure 9.14

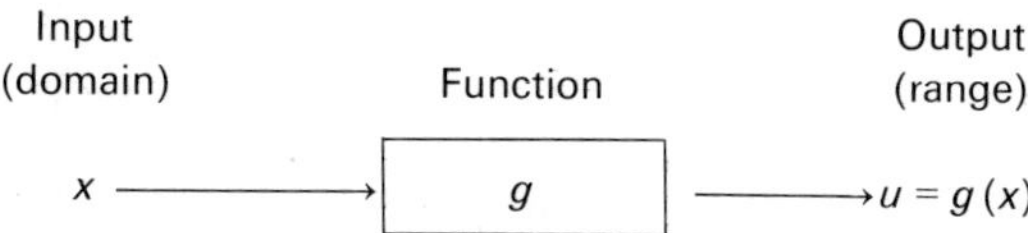

Suppose that we have another function f given by $y = f(u)$.

Figure 9.15

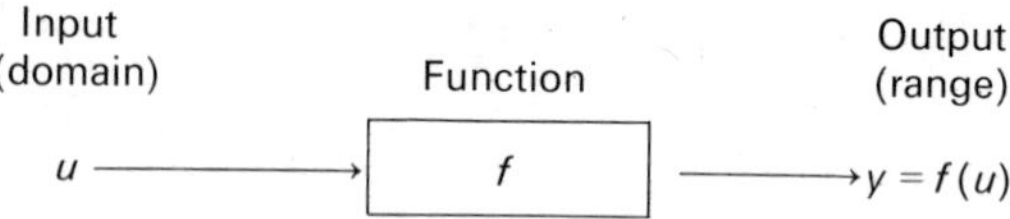

As indicated by the following picture, we put both functions f and g together into one function h in order to feed the outputs (u) from the first function g as inputs into the second function f. The outputs from f are symbolized by y. This chain of functions defines a *composite function* h whose initial inputs are values of x and whose final outputs are values of $y = h(x)$.

Figure 9.16

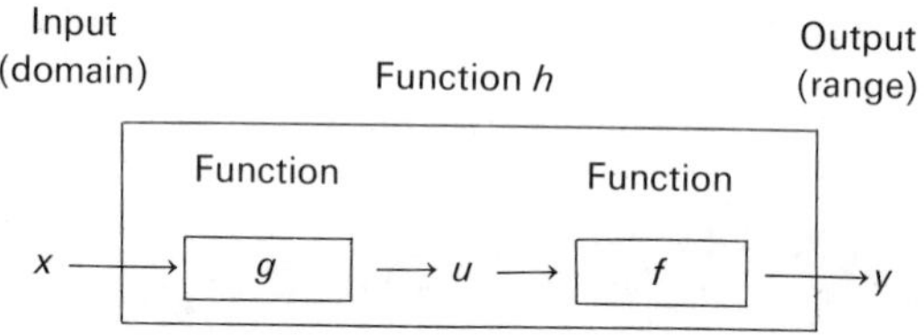

The big FUNCTION h is actually made up, or composed, of the functions g and f placed together as links in a chain. Given an input of x into h, we may write the final output y as

$$y = f(u),$$

or, since $u = g(x)$,

$$y = f(g(x)).$$

Example 26 Consider the function $y = h(x)$ defined by the chain

$$u = g(x) = 2x - 5$$

and

$$y = f(u) = 3u^2 - 7u - 6.$$

If the input of x is 4, then what is the output of $y = h(4)$?

Solution We work first with function g when $x = 4$.

$$u = g(4) = 2 \cdot 4 - 5 = 3$$

The output of $u = g(4) = 3$ now becomes the input to function f.

$$y = f(3) = 3 \cdot 3^2 - 7 \cdot 3 - 6 = 0$$

We conclude from this chain of equations that

$$y = h(4) = 0.$$

It is worthwhile to note that in the previous example we may combine functions g and f into a single general expression for $y = h(x)$ by replacing u with $(2x - 5)$ in the expression

$$y = f(u) = 3u^2 - 7u - 6,$$

as follows.

$$\begin{aligned} y &= f(u) = 3 \cdot (u)^2 - 7 \cdot (u) - 6 \\ y &= h(x) = f(g(x)) = 3 \cdot (2x - 5)^2 - 7 \cdot (2x - 5) - 6 \\ &= 3 \cdot (4x^2 - 20x + 25) - 14x + 35 - 6 \\ &= 12x^2 - 60x + 75 - 14x + 35 - 6 \\ y &= h(x) = 12x^2 - 74x + 104 \end{aligned}$$

Notice that

$$h(4) = 12 \cdot 4^2 - 74 \cdot 4 + 104 = 0,$$

as expected. It is not always possible to obtain

$$y = h(x) = \text{(an expression in } x\text{'s and constants)}$$

from the original functions given by $u = g(x)$ and $y = f(u)$.

In Problems 27 and 28 of Problem set 9–5 we actually had a first function

$$u = (2x + 5)$$

whose output u served as input to the second function

$$y = u^3.$$

The composite function which gives y directly in terms of x was

$$y = (2x + 5)^3.$$

Differentiating a composite function: the chain rule

The problem we wish to consider now is how to find the derivative of a composite function

$$y = h(x) = f(g(x))$$

if we already know how to evaluate the derivatives of the individual functions. In other words, given $g'(x)$ and $f'(u)$, how do we find $h'(x)$?

As an example of two independent but related rates of change, consider a moving car which travels 16 miles per gallon of gasoline and which also uses 4 gallons of gasoline per hour. The product of the two rates,

$$16\,\frac{\text{miles}}{\text{gallon}} \cdot 4\,\frac{\text{gallons}}{\text{hour}},$$

gives the overall rate at which the car travels in miles per hour. Notice that a change of 1 hour of time ultimately causes a change of $4 \cdot 16 = 64$ miles of distance. More generally, if the two independent rates are constant, and $u = g(x)$ changes by 3 units for each unit change of x, and if y changes by 7 units for each unit change of $u = g(x)$, then a change of 1 unit of x would cause a change of $3 \cdot 7 = 21$ units in y. This case of two constant rates suggests what we hope to expect of varying rates. Our expectation is a valid one, and we accept without proof the following general theorem, known as the *chain rule* of derivatives.

Theorem If y is a function of x, $y = h(x)$, through the chain of functions $u = g(x)$, $y = f(u)$, then

$$y' = h'(x) = f'(u) \cdot g'(x)$$

when the appropriate limits exist and when u is in the domain of f.

The chain rule may also be stated as

$$\frac{dy}{dx} = \frac{dy}{du} \cdot \frac{du}{dx}.$$

This equation does *not* represent the multiplication of ordinary fractions; it is an aid for memorizing the chain rule.

It is possible for a composite function to be made up of more than two functions, and here the most natural extension of the chain rule holds. That is, if

$$u = k(x), \quad v = g(u), \quad \text{and} \quad y = f(v),$$

so that $y = h(x)$ is defined by this chain as a function of x, if the appropriate limits exist, if u is in the domain of g, and if v is in the domain of f, then

$$y' = h'(x) = f'(v) \cdot g'(u) \cdot k'(x)$$

or

$$\frac{dy}{dx} = h'(x) = \frac{dy}{dv} \cdot \frac{dv}{du} \cdot \frac{du}{dx}.$$

Example 27 Consider the composite function $y = h(x)$ defined by the chain

$$u = g(x) = (2x - 5)$$

and

$$y = f(u) = 3u^2 - 7u - 6.$$

Find the derivative of the composite function given by $y = h(x)$.

Solution We first find the individual derivatives

$$u' = g'(x) = 2$$

and

$$y = f'(u) = 6u - 7$$

before substituting into the general expression

$$h'(x) = f'(u) \cdot g'(x) \qquad (u = g(x)).$$

Then

$$h'(x) = (6u - 7) \cdot 2 = 12u - 14$$

or, since $u = g(x) = (2x - 5)$,

$$\begin{aligned} h'(x) &= 12 \cdot (2x - 5) - 14 \\ &= 24x - 60 - 14 \\ &= 24x - 74 \end{aligned}$$

as expected.

Differentiating exponential functions

Readers who have not studied logarithms should go on to Problems 1 through 12, inclusive.

Those of us who have a basic knowledge of logarithms will now learn how to differentiate exponential functions. We will soon see that our knowledge of the chain rule is essential for this process.

Theorem If an exponential function is given by an equation of the form

$$y = f(x) = a^x \qquad (a > 0),$$

then its derivative is

$$y' = f'(x) = a^x \cdot \log_e a \tag{B}$$

where e is an irrational number close to 2.718.

This is the same number we met in Chapter Six when we studied the normal distribution curve. Often $\log_e a$ is written as $\ln a$ to show that we are working with the so-called *natural logarithms.*

Example 28 Find the derivative of the function defined by

$$y = f(x) = 2^x$$

and the slope of the tangent line to its graph at $x = 3$.

Solution We use equation (B) to obtain the derivative.

$$y' = f'(x) = 2^x \cdot \log_e 2 \tag{C}$$

We accept without proof the fact that

$$\log_e x \doteq 2.303 \cdot \log_{10} x, \tag{D}$$

so that, for example,

$$\begin{aligned} \log_e 2 &= 2.303 \cdot \log_{10} 2 \\ &= 2.303 \cdot .3010 \qquad \text{(From Table II)} \\ &= .693. \end{aligned}$$

We now rewrite equation (C) as

$$y' = f'(x) = .693 \cdot 2^x$$

and conclude that the slope of the tangent line to the graph at

$$(3, f(3)) = (3, 2^3) = (3, 8)$$

is

$$f'(3) = .693 \cdot 2^3 = 5.544.$$

Example 29 Find the derivative of the function defined by

$$y = f(x) = e^x.$$

Solution We use equation (B).

$$y' = f'(x) = e^x \cdot \log_e e.$$

Logarithms have a certain property which says that whenever b is an allowable base

$$\log_b b = 1 \quad \text{and} \quad \log_b 1 = 0.$$

In our case we conclude

$$\log_e e = 1 \quad \text{and} \quad \log_e 1 = 0.$$

Note also in Table II that

$$\log_{10} 10 = 1 \quad \text{and} \quad \log_{10} 1 = 0.$$

Thus, our solution is

$$\begin{aligned} y' = f'(x) &= e^x \cdot \log_e e \\ &= e^x \cdot 1 \\ &= e^x. \end{aligned}$$

Notice that the derivative of this exponential function is the function itself! This fact that $f(x) = f'(x)$ is one of the reasons that the exponential function with base e, often abbreviated to $\exp(x)$ or just exp, is widely used in theoretical and applied mathematics.

Example 30 Find the derivative of the function defined by

$$y = h(x) = 5^{(2x-3)}.$$

Solution We must realize that this equation actually defines a composite function where

$$u = g(x) = (2x - 3)$$

and

$$y = f(u) = 5^u.$$

The chain rule from earlier in this section is absolutely necessary at this point.

$$\begin{aligned} y' = h'(x) &= f'(u) \cdot g'(x) \quad \text{where} \quad u = (2x-3) \\ &= (5^u \cdot \log_e 5) \cdot 2 \\ &= 2 \cdot 5^{(2x-3)} \cdot 2.303 \cdot \log_{10} 5 \qquad \text{(from eq. (D))} \\ &= 2 \cdot 5^{(2x-3)} \cdot 2.303 \cdot .699 \\ &= 3.220 \cdot 5^{(2x-3)} \end{aligned}$$

The instantaneous rate of change of h per unit of change of x at

$$(1, h(1)) = (1, 5^{(2 \cdot 1 - 3)}) = (1, 5^{-1}) = (1, .2)$$

is

$$\begin{aligned} h'(1) &= 3.220 \cdot 5^{-1} \\ &= 3.220 \cdot .2 \\ &= .644. \end{aligned}$$

We add that the logarithm function defined by

$$y = f(x) = \log_b x \qquad (b > 0,\ b \neq 1)$$

has a derivative, and that derivative is

$$y' = f'(x) = \frac{1}{x} \cdot \log_b e.$$

Applying the chain rule to the composite function

$$y = h(x) = \log_b u \quad (\text{where } u = g(x))$$

yields a derivative defined by

$$y' = h'(x) = \frac{1}{u} \cdot \log_b e \cdot f'(u).$$

We close this chapter by noting that all kinds of unexpected things can happen as a result of applying the limit concept. For example, suppose the independent variable x in the function defined by

$$f(x) = (1 + 1/x)^x$$

increases without bound. Intuitively, the corresponding $f(x)$'s would appear to approach 1 as a limiting value from above. Actually, the limiting value is the now-familiar irrational number e. Our functions have been rather docile and well-behaved, so that our highly intuitive and restricted introduction to differential calculus and some of its applications has been valid. The precise relationship of limits, differentiation, and continuity belong in a solid course in calculus along with a study of not-so-well-behaved functions.

Problem set 9–7

Find the derivatives of the composite function $y = h(x) = f(g(x))$ defined by the following chains of equations in Problems 1 through 10.

1. $y = f(u) = 3u - 7;\ u = g(x) = 4x^2 - 7x + 2$
2. $y = f(u) = 4u^2 - 7u + 16;\ u = g(x) = 5x - 3$
3. $y = f(u) = u^3;\ u = g(x) = 8x + 13$
4. $y = f(u) = u^3;\ u = g(x) = 2x + 5$
5. $y = f(u) = u^{3/2};\ u = g(x) = x$
6. Compare your answers from 4 and 5 with Problems 27 and 28 of Problem set 9–5.
7. $y = f(u) = u;\ u = g(x) = x^{1/3}$
8. $y = f(u) = 3u^4 - u;\ u = g(x) = 2x^2$
9. $y = f(u) = 6u - 11;\ u = k(t) = 4t + 3;\ t = g(x) = 2 - 5x$
10. $y = f(u) = u + 3;\ u = g(x) = 2x - 7$
11. Find the slope of the tangent line to the composite function from Problem 10 at the point $(4, f(4))$.
12. Find the slope of the tangent line to the composite function from Problem 10 at the point $(1, f(1))$.

Find the derivatives of the functions in Problems 13 through 28 and the slope of the tangent line at the indicated points.

13. $y = f(x) = 7^x;\ (3, f(3))$
14. $s = f(t) = 4 \cdot 3^t;\ (2, f(2))$
15. $y = g(x) = 8^x;\ (1/3, g(1/3))$
16. $y = f(x) = e^x;\ (4, f(4))$
17. $y = f(x) = e^x;\ (-4, f(-4))$
18. $y = f(u) = 8^u,\ u = g(x) = 2x - 5;\ (3, f(3))$
19. $y = 8^{2x-5};\ (3, f(3))$
20. $y = e^{2t^2-6t+7};\ (2, f(2))$
21. $y = 3e^{2t^2-6t+7};\ (2, f(2))$
22. $y = 3^x;\ (-2, f(-2))$
23. $y = 3^x;\ (0, f(0))$
24. $y = 3^x;\ (2, f(2))$
25. $y = 3^x;\ (4, f(4))$
26. $y = 7 \cdot 3^x;\ (2, f(2))$
27. $y = 7 \cdot 3^x;\ (4, f(4))$
28. $y = 5 \cdot 9^x;\ (-1/2, f(-1/2))$
29. Differentiate $y = \log_{10} x$.
30. Differentiate $y = \log_7 x$.
31. Differentiate $y = \log_e x$.
32. Differentiate $y = \log_e(x^2 - 7x + 3)$.
33. Differentiate $y = \ln(3x^2 + 5x - 12)$.
34. Assume that the number of carburetors assembled per day x days after

the start of the assembly line is given by

$$y = f(x) = 50 - 50 \cdot e^{-0.2x}.$$

a. How many carburetors are assembled on the first day?
b. The third day?
c. The fifth day?
d. The tenth day?
e. The fifteenth day?
f. Graph the function in the interval [0, 20].
g. What is the slope of the graph at $x = 0$?
h. At $x = 1$?
i. At $x = 3$?
j. At $x = 5$?
k. At $x = 10$?
l. At $x = 15$?
m. At $x = 20$.

35. Work Problem 34 again, letting the equation be

$$y - f(x) - 50 - 50 \cdot e^{-0.8x}.$$

36. Refer to Example 17 in Chapter 3. Find the instantaneous rate of change of the learning curve at these values of x.

x	$f(x)$	Rate of change
0		
1		
2		
3		
4		
5		
6		
7		
8		
9		
10		

Chapter 10

Integral Calculus

The indefinite integral

The previous chapter devoted itself to the study of derivatives—their development and applications. In this brief chapter, we study the problem of "antidifferentiation." In cases where we are presented with the derivative of a function but not with the function itself our job is to see if we can work back from the given derivative to find the function from which it was differentiated. It may even be that our given derivative is the derivative of several different functions, in which case we should like to find them all. For example, if

$$f'(x) = 2x,$$

then we know that the function

$$f(x) = x^2$$

has $f'(x)$ as its derivative. It is the purpose of this chapter to learn more about the processes and applications of going from $f'(x)$ back to $f(x)$ in contrast to going from $f(x)$ to $f'(x)$.

Antiderivatives and antidifferentiation

Definition A function $F(x)$ is an *antiderivative* of a given function $f(x)$ if $F'(x) = f(x)$.

Example 1 $F(x) = x^2$ is an antiderivative of $f(x) = 2x$ because

$$F'(x) = D_x(x^2) = 2x = f(x).$$

Example 2 $F(x) = 3x^5$ is an antiderivative of $f(x) = 15x^4$ because

$$F'(x) = D_x(3x^5) = 15x^4 = f(x).$$

Example 3 $F(x) = 3x^5 + 7$ is an antiderivative of $f(x) = 15x^4$ because

$$F'(x) = D_x(3x^5 + 7) = 15x^4 + 0 = 15x^4 = f(x).$$

Example 4 $F(x) = 3x^5 - 17.1 = 3x^5 + -17.1$ is an antiderivative of $f(x) = 15x^4$ because

$$F'(x) = D_x(3x^5 + -17.1) = 15x^4 + 0 = 15x^4 = f(x).$$

The last three examples show us that there is *not* just one function $F(x)$ that is *the* antiderivative of a given function $f(x)$; there is an infinite *set* of functions of the form

$$F(x) = 3x^5 + c,$$

where c can be any real number constant, whose derivatives are all

$$F'(x) = 15x^4 + 0 = 15x^4 = f(x).$$

Symbolically, this last statement is written

$$\int 15x^4 \, dx = 3x^5 + c$$

for which we say, "The indefinite integral of $15x^4$ dee ex is $3x^5$ plus a constant." The symbol $\int$ is the integral sign, used to indicate that we are to find the antiderivative of the function

$$f(x) = 15x^4,$$

and the symbol dx tells us that the independent variable is x. The letter c represents the *constant of integration.*

In general, we say that

$$\int f(x) \, dx = F(x) + c$$

whenever

$$D_x(F(x) + c) = F'(x) + 0 = F'(x)$$

equals $f(x)$ for all values of x in the intersection of the domains of $F'(x)$ and $f(x)$. The indefinite integral of a function $f(x)$ is called "indefinite" because there are many functions $F(x) + c$ whose derivatives equal $f(x)$. Often the term *antidifferentiation* is used instead of indefinite integration. This term is appropriate because we are going opposite to or against (anti) the process of differentiation.

Example 5 $\int z \cdot x \, dx = zx^2/2 + c$ because if we differentiate

$$F(x) = \frac{zx^2}{2} + c$$

where z represents a specific number or constant and x is the independent variable, then

$$F'(x) = \frac{z}{2} \cdot D_x(x^2) + D_x(c)$$

$$= \frac{z}{2} \cdot 2x + 0$$

$$= zx.$$

Integration formulas

In the previous examples we have been given an antiderivative $F(x)$ and have merely differentiated to confirm that $F'(x) = f(x)$. In almost all cases we are given only $f(x)$ and have to track down

$$\int f(x)\,dx = F(x) + c.$$

Unfortunately, there is no set of formulas that will let us integrate as conveniently and universally as the set of eight shortcut differentiation formulas. The best we can do—and it is sufficient in our limited breadth of applications—is to learn a few special formulas which are illustrated in the following examples. (Be sure to differentiate each answer in Problem Set 10–1 to confirm its correctness.) All the functions we are dealing with are "nice" functions in that they and their indefinite integrals are continuous and defined throughout their domains, except perhaps at a few easily identified points; for example

$$f(x) = 2/(x+3)$$

exists for all values of x except $x = -3$.

Example 6 If $f(x) = x^4$, then

$$\int f(x)\,dx = \int x^4\,dx = \frac{x^5}{5} + c = \frac{x^{4+1}}{4+1} + c;$$

if $f(x) = x^6$, then

$$\int f(x)\,dx = \int x^6\,dx = \frac{x^7}{7} + c = \frac{x^{6+1}}{6+1} + c;$$

if $f(x) = x^{-3}$, then

$$\int f(x)\,dx = \int x^{-3}\,dx = \frac{x^{-2}}{-2} + c = \frac{x^{-3+1}}{-3+1} + c;$$

if $f(x) = \sqrt{x} = x^{1/2}$, then

$$\int f(x)\,dx = \int x^{1/2}\,dx = \frac{2x^{3/2}}{3} + c = \frac{x^{1/2+1}}{1/2+1} + c;$$

and, in general, if $f(x) = x^n$ (n a rational number), then

$$\int f(x)\,dx = \int x^n\,dx = \frac{x^{n+1}}{n+1} + c \qquad (n \neq -1\text{: why?})$$

Example 7 If $f(x) = 5 \cdot x^2$, then

$$\int f(x)\,dx = \int 5 \cdot x^2\,dx = \frac{5x^3}{3} + c = \frac{5x^3}{3} + \text{a constant.}$$

Notice also that

$$5 \cdot \int x^2\,dx = 5 \cdot \left[\frac{x^3}{3} + c\right] = \frac{5x^3}{3} + \text{a constant} = \int 5 \cdot x^2\,dx.$$

We see by examining this example and reading the last equation from right to left that

$$\int 5 \cdot x^2\,dx = 5 \cdot \int x^2\,dx.$$

The generalization holds true for any constant k and any function $f(x)$, so that

$$\int k \cdot f(x)\,dx = k \cdot \int f(x)\,dx$$

In Example 7, $k = 5$ and $f(x) = x^2$.

Finally, we accept without proof the reasonable theorem that the indefinite integral of the sum (or difference) of two or more functions is the sum (or difference) of their individual indefinite integrals provided, of course, that they exist.

Example 8 Find $\int (4x^3 + 7x^2 - 9x + 4)\,dx$.
Solution We first rewrite the problem slightly as

$$\int (4x^3 + 7x^2 - 9x^1 + 4x^0)\,dx,$$

so that by this most recent theorem the answer is

$$\int 4x^3\,dx + \int 7x^2\,dx - \int 9x^1\,dx + \int 4x^0\,dx;$$

or, removing the constants to their proper places, we have

$$4 \cdot \int x^3\,dx + 7 \cdot \int x^2\,dx - 9 \cdot \int x^1\,dx + 4 \cdot \int x^0\,dx,$$

which gives

$$4 \cdot \frac{x^4}{4} + 7 \cdot \frac{x^3}{3} - 9 \cdot \frac{x^2}{2} + 4 \cdot \frac{x^1}{1} + c$$

or, for the final answer,

$$x^4 + \frac{7x^3}{3} - 4 \cdot 5x^2 + 4x + c.$$

The letter c represents the sum of the four constants of integration:

$$c = c_1 + c_2 + c_3 + c_4.$$

Suppose we have a graph such that the slope of the tangent line to the graph at any point on it $(x, f(x))$ is given by the expression $4x - 5$, and we want to know the equation of the graph.

Since the slope of the tangent line is $4x - 5$, we know, from the last chapter, that this expression must be the first derivative of the equation of the graph:

$$f'(x) = 4x - 5.$$

To find the function with $4x - 5$ as its derivative we undo the differentiation that has already taken place by performing the integration process.

$$f'(x) = 4x - 5$$

$$\int f'(x)\,dx = \int (4x - 5)\,dx$$

$$f(x) = \int (4x - 5)\,dx$$

$$f(x) = \int 4x\,dx - \int 5\,dx$$

$$f(x) = 4 \cdot \int x\,dx - 5 \cdot \int x^0\,dx \; (x^0 = 1)$$

$$f(x) = 4 \cdot \frac{x^2}{2} - 5 \cdot \frac{x^1}{1} + c$$

$$f(x) = 2x^2 - 5x + c$$

Here is the partial graph of the *set* of functions of the form (See Fig. 10.1 opposite)

$$f(x) = 2x^2 - 5x + c.$$

Which member, if any, of this family of curves passes through (6, 9)? In order for a curve to pass through a point its coordinates must satisfy the equation of the curve; since (6, 9) must be on the curve

$$f(6) = 9$$

$$2 \cdot 6^2 - 5 \cdot 6 + c = 9$$

$$72 - 30 + c = 9$$

$$42 + c = 9$$

$$c = -33$$

This means that the equation of the particular member of the family is

$$y = f(x) = 2x^2 - 5x - 33.$$

Figure 10.1

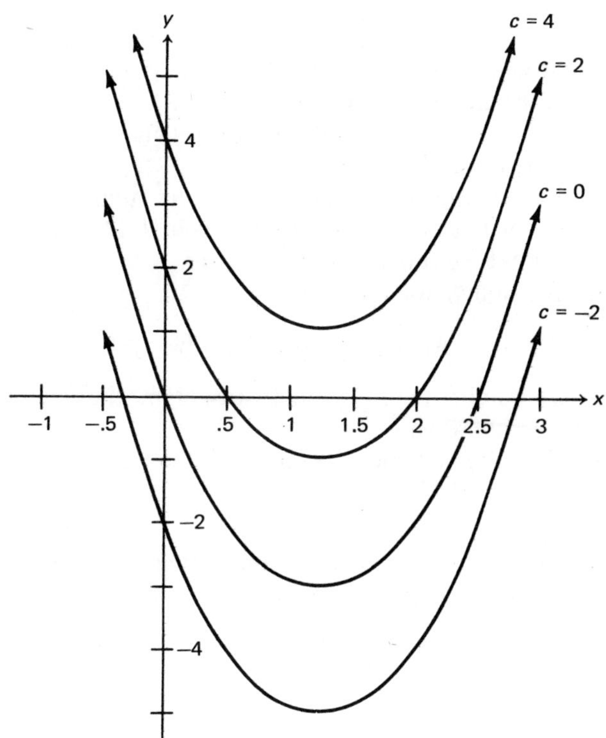

Problem set 10–1

Perform the indicated integrations in Problems 1 through 14. Check each answer by differentiation.

1. $\int 3x\,dx$
2. $\int 7.1x\,dx$
3. $\int 4\,dx$
4. $\int dx = \int 1\,dx = \int x^0\,dx$
5. $\int (3x+7)\,dx$
6. $\int (x^2+13x-7)\,dx$
7. $\int x^{-5}\,dx$
8. $\int 6x^{-4}\,dx$
9. $\int 6/x^{-4}\,dx$
10. $\int (x+2)\cdot(x+3)\,dx$
    (First, multiply the terms.)
11. $\int (x+2)\,dx$
12. $\int (x+3)\,dx$
13. $\int (5x^4-4x^3+3x^2-2x+1)\,dx$
14. $\int (7x^3-2.5x^2+x-13)\,dx$
15. Compare your answers in 10, 11, and 12. In general, what can we say about $\int f(x)\cdot g(x)\,dx$, $\int f(x)\,dx$, and $\int g(x)\,dx$?
16. Find the equation of the graph whose slope function is given by

$$f'(x) = 7x - 2$$

and passes through $(1, 8)$.

17. Find the equation of the graph whose slope function is given by

$$f'(x) = -3x + 1$$

and passes through $(-4, 8)$.

18. If the slope of the graph of a function at any point $(x, f(x))$ on the graph is given by $(3x - 7)$ and the graph passes through $(-2, 5)$, then find the equation which defines the function.

19. The second derivative of an equation defining a function is $y'' = f''(x) = 45x$, and the graph of the function passes through $(1, 13)$ with a slope of 30 at that point. Find the equation of the function.

20. The marginal cost function for item Z is

$$M(w) = 10 - 12w + 15w^2$$

where w is the number of Z's manufactured. Find the total cost function $C = f(w)$ if $C = 136$ when $w = 2$.

21. The marginal revenue function for one unit of item A is

$$M(x) = 100x - 8x^2$$

where x is the number of units of item A that are sold. Find the total revenue function $R = f(x)$ if $R = 600$ when $x = 3$.

The definite integral

In this section we begin with a problem which, though similar to ones we already know how to solve, requires additional pure mathematics. This additional mathematics, learned from making careful observations and a valid generalization based on the one problem, will prove to be valuable to us. This cycle of meeting a temporarily unsolvable specific problem, solving it by an extension of familiar mathematics, generalizing from the problem to obtain a general formula or method, and then applying this generalization to similar specific cases is one we have continually done all through the text. The situation of Karen Jones and her trip to Hawaii at the beginning of the section on annuities in Chapter 4, is a good example of this cycle.

Using area—approximating rectangles

Example 9 What is the area of the region in the Cartesian Plane bounded by the graph of $y = f(x) = x^2$, the vertical line $x = 1$, the vertical line $x = 3$, and the x-axis?

Solution We begin by making a diagram.

Figure 10.2

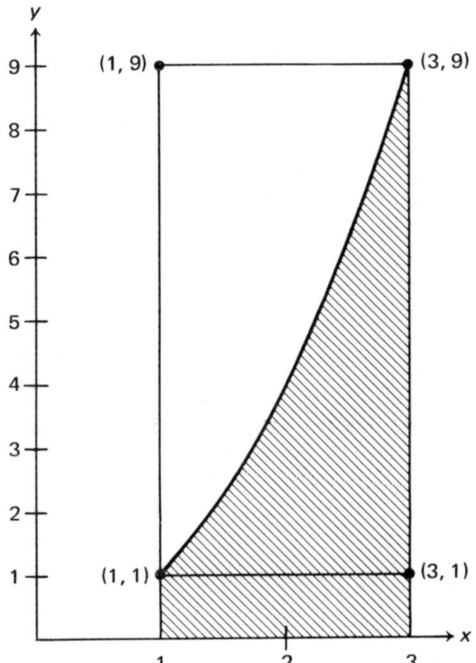

With our present methods of calculating areas we have difficulty in seeing how to solve this problem because the shaded region is not a familiar polygon or a circle. Let us call the area of this shaded region A square units. The area of the region bounded by the graph of $y = 1$ is

$$(\text{width}) \cdot (\text{height}) = 2 \cdot 1 = 2,$$

and the area of the region bounded by the graph of $y = 9$ and the x-axis is

$$(\text{width}) \cdot (\text{height}) = 2 \cdot 9 = 18.$$

Inspection of the diagram tells us that the area of the entire shaded region is between the area of the small rectangle and the area of the large rectangle:

$$2 < A < 18.$$

The key question here is how to find exactly where between 2 and 18 A is located.

Suppose we divided the x-axis between 1 and 3 into four subintervals and erect rectangles based on these subintervals so that their upper *left-hand* corners touch the graph of $f(x) = x^2$.

Figure 10.3

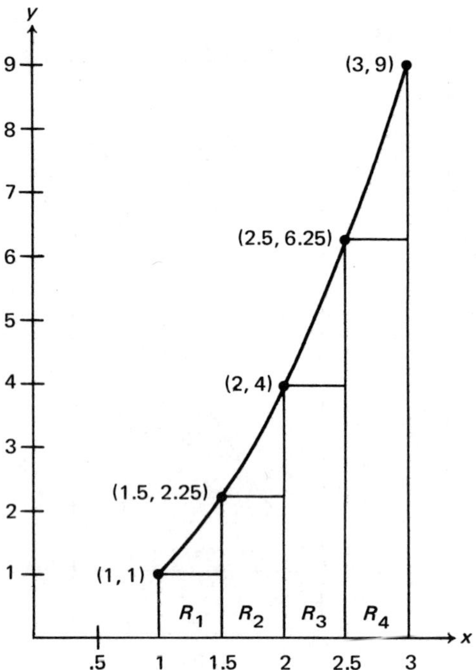

It is fairly obvious that finding the sum of the areas of the four rectangles will give us a better approximation to the true area A than the 2 square units from before, in which we determined the smallest possible value of A. Also, our area here will still be less than A. We see that with $n = 4$ rectangles

$$A_{\text{left}} = R_1 + R_2 + R_3 + R_4$$

$$A_{\text{left}} = .5 \cdot 1 + .5 \cdot 2.25 + .5 \cdot 4 + .5 \cdot 6.25$$

$$A_{\text{left}} = 6.75 \text{ square units}$$

The symbol A_{left} represents the approximation to the area under the curve found by adding the areas of the rectangles, all of which are below the graph. We also could have chosen the symbol $A_{\min}$ since 6.75 is the smallest possible value of A.

Suppose we now divide the x-axis between 1 and 3 into four subintervals and erect rectangles based on these same four subintervals so that their upper *right-hand* corners touch the graph of $f(x) = x^2$ (See Figure 10.4). It is fairly obvious that finding the sum of the areas of the four rectangles will give us a better approximation to the true area A under the curve than the 18 square units from before where we determined the largest

Figure 10.4

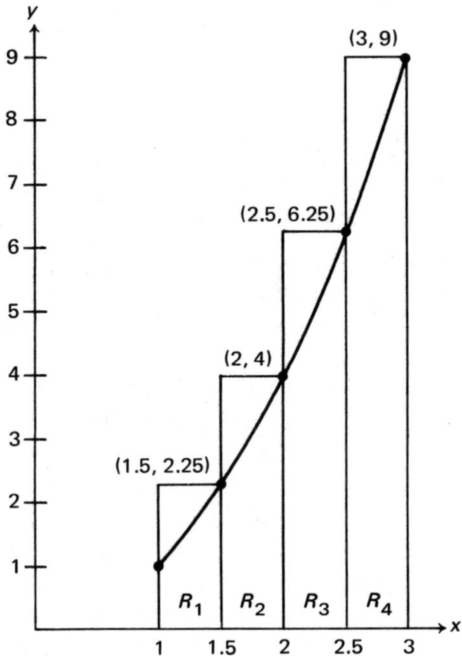

possible value of A. Also, our area here will still be greater than A. We see that with $n = 4$ rectangles

$$A_{\text{right}} = R_1 + R_2 + R_3 + R_4$$

$$A_{\text{right}} = .5 \cdot 2.25 + .5 \cdot 4 + .5 \cdot 6.25 + .5 \cdot 9$$

$$A_{\text{right}} = 10.75 \text{ square units.}$$

The symbol A_{right} represents the approximation to the area under the curve found by adding the areas of the rectangles, all of which are above the graph. We also could have chosen the symbol A_{max} since 10.75 is the largest possible value of A.

We have made some definite progress in our search for the true value of A. Previously, the best we could say about A was that it was between 2 and 18; now we can say that A is between 6.75 and 10.75 square units. It is tempting to take the mean of the last two numbers and conclude that 8.75 is the area, but the mean of 2 and 18 is 10, so the average method does not give us a consistent or correct answer. If we worked with $n = 10$ intervals and $n = 10$ rectangles in exactly the same way that we worked with $n = 4$ intervals, we would obtain

$$A_{\text{left}} = A_{\text{min}} = 7.88 \quad \text{and} \quad A_{\text{right}} = A_{\text{max}} = 9.48.$$

The fly swatter principle

It is revealing to organize our results so far into a table.

Table 10.1

Number of intervals	Length of each interval	Area with rectangles all under curve	Area with rectangles all above curve	Difference
1	2/1 = 2	2	18	16
4	2/4 = .5	6.75	10.75	4
10	2/10 = .2	7.88	9.48	1.6

By going from $n=1$ to $n=4$ to $n=10$ we have reduced the interval in which A must exist from 16 units to 4 units to 1.6 units. We can now use the so-called Fly Swatter Principle to track down A. That is, if n, the number of intervals into which we subdivide $[1, 3]$ becomes larger and larger (and the length of each subinterval in the corresponding subdivisions of $[1, 3]$ become smaller and smaller), we may think of the values of the third column of our table as being a fly swatter approaching a fly (A, the true area) from below. At the same time we may think of the values of the fourth column of our table as being a fly swatter approaching the same fly (A) from above. The fifth column represents the length of the interval the fly may travel on. We can see that as n becomes larger and larger the possible values of A become trapped in smaller and smaller intervals. If n increases without bound, we want to know the limiting values of the third and fourth columns representing A_{left} and A_{right}.

What are these limiting values? After much more detective work, observations of patterns, and advanced work with sequences, it turns out that

$$A_{\text{left}} = \frac{26n^2 - 24n + 4}{3n^2}$$

and

$$A_{\text{right}} = \frac{26n^2 + 24n + 4}{3n^2}$$

where n is the number of equal intervals into which $[1, 3]$ is subdivided. These expressions are applicable only to this specific function and interval. Table 10.2 shows the results after a computer has evaluated these expressions and their difference for increasing values of n. Verify—right now—that replacing n by 4 in these equations does give us 6.75 and 10.75, respectively.

The equation

$$\operatorname*{limit}_{n\to\infty} \frac{1}{n} = 0$$

Table 10.2

NUMBER OF INTERVALS	LENGTH OF EACH INTERVAL	AREA W/ RTGLS UNDER CURVE	AREA W/ RTGLS ABOVE CURVE	DIFFERENCE IN AREA
1	2	2	18	16
2	1	5	13	8
3	.66667	6.14815	11.4815	5.33333
4	.5	6.75	10.75	4
5	.4	7.12	10.32	3.2
6	.33333	7.37037	10.037	2.66667
7	.28571	7.55102	9.83674	2.28572
8	.25	7.6875	9.6875	2
9	.22222	7.79424	9.57202	1.77778
10	.2	7.88	9.48	1.6
12	.16667	8.00926	9.34259	1.33333
15	.13333	8.13926	9.20593	1.06667
20	.1	8.27	9.07	.8
25	.08	8.3488	8.9888	.64
30	.06667	8.40148	8.93482	.53334
50	.04	8.5072	8.8272	.32
60	.03333	8.5337	8.80037	.26667
80	.025	8.56688	8.76687	.19999
100	.02	8.5868	8.7468	.16
150	.01333	8.61339	8.72006	.10667
200	.01	8.6267	8.7067	.08
250	.008	8.63469	8.69869	.064
300	.00667	8.64002	8.69335	.05333
400	.005	8.64668	8.68667	.03999
500	.004	8.65067	8.68267	.032
600	.00333	8.65334	8.68	.02666
700	.00286	8.65524	8.6781	.02286
800	.0025	8.65667	8.67667	.02
900	.00222	8.65778	8.67556	.01778
1000	.002	8.65867	8.67467	.016
1200	.00167	8.66	8.67334	.01334
1400	.00143	8.66095	8.67238	.01143
1600	.00125	8.66167	8.67167	.01
1800	.00111	8.66222	8.67111	.00889
2000	.001	8.66267	8.67067	.008
2500	.0008	8.66347	8.66987	.0064
3000	.00067	8.664	8.66933	.00533
4000	.0005	8.66467	8.66867	.004
5000	.0004	8.66507	8.66827	.0032
6000	.00033	8.66533	8.668	.00267
7000	.00029	8.66553	8.66781	.00228
8000	.00025	8.66567	8.66767	.002
9000	.00022	8.66578	8.66756	.00178
10000	.0002	8.66587	8.66747	.0016

is read "the limit, as n increases without bound, of $1/n$ is 0," and it does agree with common sense since we are dividing an infinitely increasing variable into a fixed number, 1, with ever decreasing quotients.[1]

[1] Strictly speaking, we are now working with the limit of a sequence (whose domain is only positive integers) where the independent variable increases without bound instead of the limit of a function (whose domain is chosen from the real numbers) where the independent variable approaches a fixed number. Our results here are intuitively valid, and we could prove them with the use of the formal definition of the limit of a sequence.

The limiting value of A_{left} as n increases without bound is then

$$\begin{aligned}\lim_{n\to\infty} A_{\text{left}} &= \lim_{n\to\infty}\left(\frac{26n^2-24n+4}{3n^2}\right)\\ &= \lim_{n\to\infty}\left(\frac{26n^2}{3n^2}-\frac{24n}{3n^2}+\frac{4}{3n^2}\right)\\ &= \lim_{n\to\infty}\left(\frac{26}{3}\cdot n^0+\frac{-8}{n}+\frac{4}{3}\cdot\frac{1}{n^2}\right).\end{aligned}$$

We now accept without proof the plausible statement that the limit of the sum of these algebraic expressions is the sum of the individual limits.

$$\begin{aligned}\lim_{n\to\infty} A_{\text{left}} &= \lim_{n\to\infty}\left(\frac{26}{3}\cdot n^0\right)+\lim_{n\to\infty}\left(\frac{-8}{n}\right)+\lim_{n\to\infty}\left(\frac{4}{3}\cdot\frac{1}{n^2}\right)\\ &= \frac{26}{3}+0+0\\ &= \frac{26}{3}\end{aligned}$$

Similar reasoning lets us find the limiting value of A_{right} by evaluating the expression which generates the values in the fourth column of the table as the independent variable n increases without bound.

$$\begin{aligned}\lim_{n\to\infty} A_{\text{right}} &= \lim_{n\to\infty}\left(\frac{26n^2+24n+4}{3n^2}\right)\\ &= \text{limit}\left(\frac{26}{3}\cdot n^0+\frac{8}{n}+\frac{4}{3}\cdot\frac{1}{n^2}\right)\\ &= \lim_{n\to\infty}\left(\frac{26}{3}\cdot n^0\right)+\lim_{n\to\infty}\left(\frac{8}{n}\right)+\lim_{n\to\infty}\left(\frac{4}{3}\cdot\frac{1}{n^2}\right)\\ &= \frac{26}{3}+0+0\\ &= \frac{26}{3}\end{aligned}$$

Both fly swatters smack against each other where

$$A_{\text{left}} = A_{\text{right}} = 26/3,$$

and since A, the true area, is the fly trapped between A_{left} and A_{right}, then A must also take on the value $A = 26/3$ square units. Notice that

$$\lim_{n\to\infty} A_{\text{left}} = \lim_{n\to\infty} A_{\text{right}}.$$

This fact holds true in any similar area-under-the-curve problem.

If we wanted to find the area under the graph of

$$y = f(x) = 9 - x^2$$

in $[0, 2.4]$, we would begin with a graph including, say, eight subintervals of length $2.4/8 = .3$ units each and having the area-approximating rectangles under the curve to make for a neater diagram.

Figure 10.5

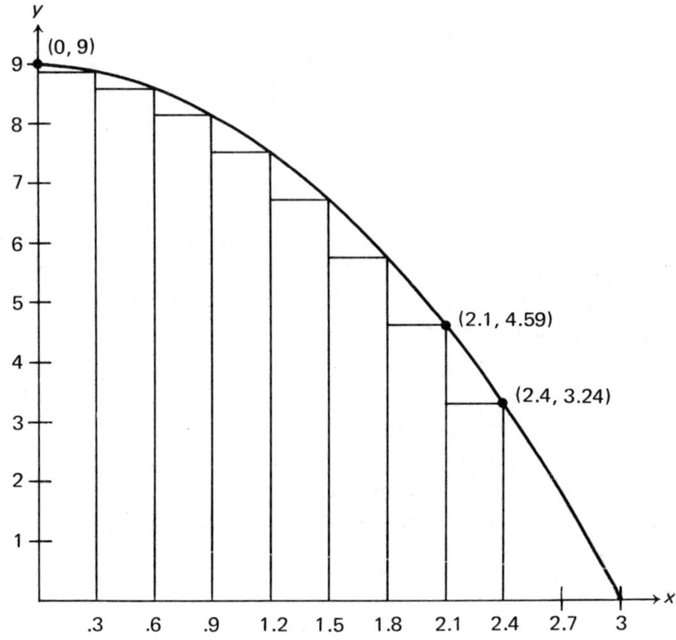

Finding *A* by increasing *n*

We know that the sum of the areas of the eight rectangles is less than the actual area. Intuition says that increasing the number of subdivisions along the x-axis increases the accuracy of the resulting approximations because the rectangles become thinner and thinner, packing in much tighter under the curve. The shaded region in Figure 10.6 overleaf does not have its area included in the estimate of .972 for the entire area under the curve between $x = 2.1$ and $x = 2.4$ (Right on that estimate?). In Figure 10.7, where we are still working with the area under the curve between $x = 2.1$ and $x = 2.4$, the sum of the six thinner rectangles is 1.145125. This numeric result indicates that increasing the number of rectangles increases the accuracy of the resulting approximations to the true area, which certainly agrees with common sense.

Figure 10.6

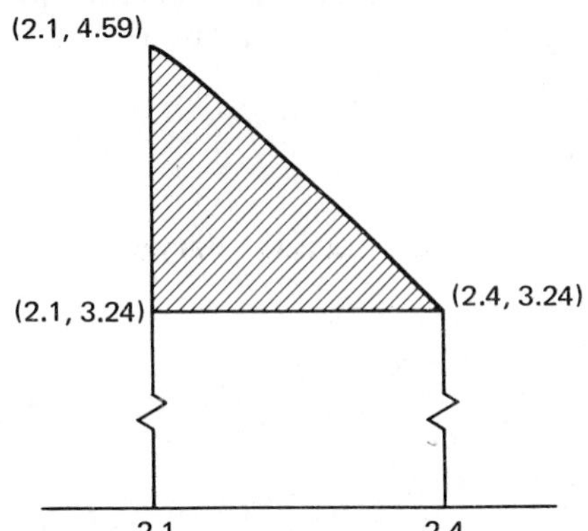

Figure 10.7

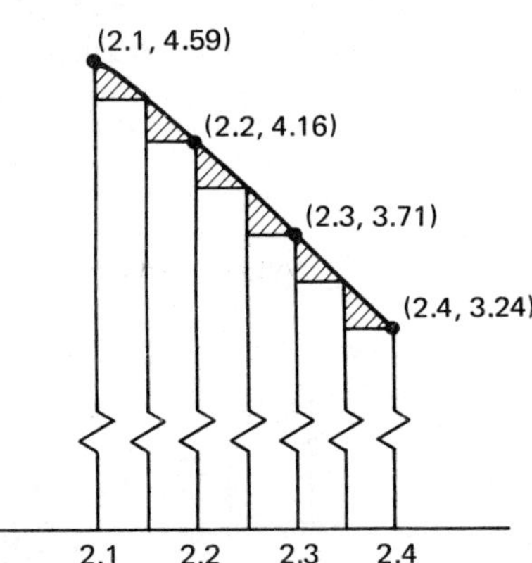

We could gradually increase n for the entire interval $[0, 3]$, and carry out the horrible manual calculations of the areas of the n rectangles to obtain better approximations of A as the values of A_{right} slowly increased. Is there a better and easier way? There certainly is, and we will learn it in the next section. Until then, we will do a few area problems manually to make sure that we know what the easier method of the next section actually represents. We leave this section with a summary.

Let $f(x)$ be a continuous function in $[a, b]$ with $f(x) \geqslant 0$ for all x in $[a, b]$. Then the area bounded by the graph of $y = f(x)$, the vertical linc $x = a$, the vertical line $x = b$, and the x-axis is the *number* obtained by the following process.

1. Divide the x-axis from a to b into n equal subintervals of length $(b-a)/n$ each.
2. Erect rectangles, each based on one of these n subintervals, until they just intersect the curve or at least algebraically determine the heights of these rectangles.
3. Calculate an approximation to the true area A by summing the areas of these n rectangles. The resulting number, symbolized by $A_{\min}$, will be less than A.
4. Determine the limit, as n increases without bound, of the sequence of approximations obtained by repeating steps 1 through 3.
5. This limit (if it exists), symbolized by $\int_a^b f(x)\,dx$, is the area we are looking for.

Everything in the previous paragraph is plausible enough except for the symbol in part 5 which is read "the definite integral from a to b of $f(x)\,dx$." It seems entirely uncalled for to symbolize an area by an integral sign when $\int f(x)\,dx$ represents a set of functions. After all, how can we justify allowing two such diverse ideas as area and antidifferentiation to have such similar symbols?

We will soon find out just why it makes good sense to do this. We also delay the calculation of the area of the region in Figure 10.5, which is symbolized by

$$\int_a^b f(x)\,dx = \int_0^3 (9-x^2)\,dx,$$

until Problem 21 of Problem set 10–3.

Problem set 10–2

1. Verify that if $f(x)=x^2$, $a=1$, $b=3$, and $n=2$, then the resulting approximation to $\int_1^3 f(x)\,dx$ is 5 square units with the approximating rectangles constructed under the curve. Also verify that the resulting approximation is 13 square units with the approximating rectangles constructed above the curve.
2. Carry through the same instructions as in the previous problem, but use $n=5$ rectangles. Compare your answers with $A_{\text{min}}=7.12$ and $A_{\text{max}}=10.32$ from Table 10.2.
3. This problem refers to Problem 2. Show that

$$A_{\text{min}} = \frac{26n^2 - 24n + 4}{3n^2}$$

yields 7.12 for $n=5$ and that

$$A_{\text{max}} = \frac{26n^2 + 24n + 4}{3n^2}$$

yields 10.32 for $n=5$.
4. If $f(x)=2x-1$, then find the approximation to the area of the region under the graph bounded by $x=3$, $x=7$, and the x-axis by dividing $[3,7]$ into eight equal intervals and erecting rectangles with their tops just touching the graph. Symbolize your answer by A_{min}.
5. Rework Problem 4, erecting the rectangles until they are just above the graph; symbolize your answer by A_{max}.
6. By using the triangle area formula $A=(\text{base})\cdot(\text{height})/2$ find the true area of the region mentioned in Problem 4.
7. Find an approximation to the area bounded by the x-axis, the y-axis, the curve $y=x^2/2+1$, and the line $x=2$ by dividing $[0,2]$ into ten equal parts and erecting the rectangles to just touch the curve from below. Make a careful diagram to help with the work.
8. Rework Problem 7, erecting the rectangles to just touch the curve from above.

9. Rework Problem 7, erecting the rectangles until the vertical line from the midpoint of each base intersects the curve; for example,

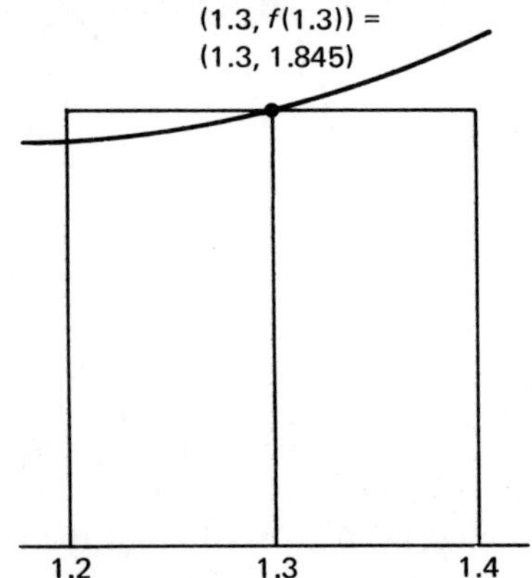

Label your answer A_{mid}.

10. If we worked in the same fashion as Problem 9 with the midpoints of the first problem in this section of the text, then the area of the n approximating rectangles is

$$A_{\text{mid}} = \frac{26n^2 - 2}{3n^2}.$$

a. What is the value of A_{mid} for $n = 2$?
b. What is the value of A_{mid} for $n = 5$?
c. Compare these answers with Problems 1 and 2. Is

$$A_{\text{left}} < A_{\text{mid}} < A_{\text{right}}$$

for both $n = 2$ and $n = 5$?
d. What is the limiting value for A_{mid} as n becomes infinitely large?

Applications of the definite integral

In the last section we learned a cumbersome method for calculating the area under a curve. Actually, the limit process was involved because we calculated the area by letting the number of rectangles under the curve become more numerous and also smaller at their bases. The resulting areas of these rectangular approximations formed a sequence that approached the true area from below (or from above, depending on how we erected the rectangles). Of course, we hope to calculate areas without the brutal arithmetic of our present approximation techniques. The desired method of calculation is found in the following theorem.

Fundamental theorem of calculus

If $f(x)$ is continuous in $[a,b]$ and if we can find an antiderivative $F(x)$ of $f(x)$ in $[a,b]$, then

$$\int_a^b f(x)\,dx = F(b) - F(a).$$

Example 10 In our main example of the last section $f(x) = x^2$ is continuous in $[1,3]$, and we can find an antiderivative of it, which is

$$F(x) = x^3/3. \qquad (\text{check: } F'(x) = D_x(x^3/3) = x^2 = f(x)).$$

According to the Fundamental Theorem

$$\int_1^3 f(x)\,dx = \int_1^3 x^2\,dx$$

becomes, as expected

$$\begin{aligned} F(b) - F(a) &= F(3) - F(1) \\ &= \frac{3^3}{3} - \frac{1^3}{3} \\ &= \frac{27}{3} - \frac{1}{3} \\ &= \frac{26}{3}. \end{aligned}$$

The number

$$\int_a^b f(x)\,dx = F(b) - F(a)$$

is often symbolized by

$$F(x)]_a^b$$

to show that we find an antiderivative of $f(x)$ which is $F(x)$, evaluate $F(x)$ at both $x = b$ and $x = a$, and then subtract $F(a)$ from $F(b)$. Sir Isaac Newton was one of the geniuses of the seventeenth century who first saw how the antiderivative of a function helps to calculate the area under its curve by discovering a relationship between tangent lines and areas, two seemingly unrelated notions. In fact, the integral sign $\int$ is a distorted way of making a capital letter S which stands for "sum."

Example 11 Find the area under the graph of

$$y = f(x) = 9 - x^2$$

in the interval bounded by vertical lines $x = 2.1$, 2.4, and the x-axis.

Solution We first make a quick sketch of the function in the interval as we always do for a definite integral problem.

Figure 10.8

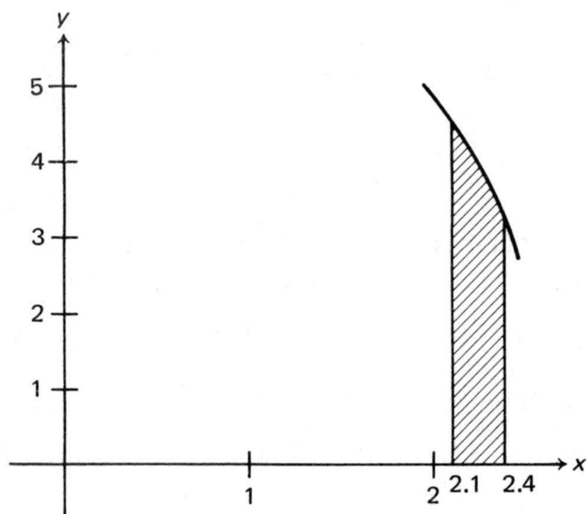

The area, by the Fundamental Theorem, is

$$\int_{2.1}^{2.4} (9 - x^2)\,dx,$$

since

$$\int (9 - x^2)\,dx = 9x - x^3/3 = F(x) \qquad \text{(we delete } +c\text{: see below)} \quad \text{(A)}$$

$$\begin{aligned}\int_{2.1}^{2.4} (9 - x^2)\,dx &= F(2.4) - F(2.1)\\ &= [9 \cdot 2.4 - (2.4)^3/3] - [9 \cdot 2.1 - (2.1)^3/3]\\ &= [21.6 - 4.608] - [18.9 - 3.087]\\ &= 16.992 - 15.813\\ &= 1.179,\end{aligned}$$

which, as expected, is slightly higher than our approximations of .972 and 1.145125 obtained in the last section.

In equation (A) we said that

$$\int (9 - x^2)\,dx = 9x - x^3/3$$

without writing the customary $+c$. If we had said

$$\int (9 - x^2)\,dx = 9x - x^3/3 + 15 = F(x),$$

then

$$F(2.4) = 9 \cdot 2.4 - (2.4)^3/3 + 15$$

and

$$F(2.1) = 9 \cdot 2.1 - (2.1)^3/3 + 15,$$

so that

$$F(2.4) - F(2.1) = 9 \cdot 2.4 - (2.4)^3/3 \cancel{+15} - 9 \cdot 2.1 + (2.1)^3/3 \cancel{-15},$$

which shows that the constant of integration c, regardless of its value, merely cancels out in a definite integral problem, and that is why we ignored it in equation (A). We will continue to do so in all similar problems.

Working with absolute values

Example 12 Find the area under the graph of $y = f(x) = x^2 - 25$ from $x = 2$ to $x = 4$.

Solution

Figure 10.9

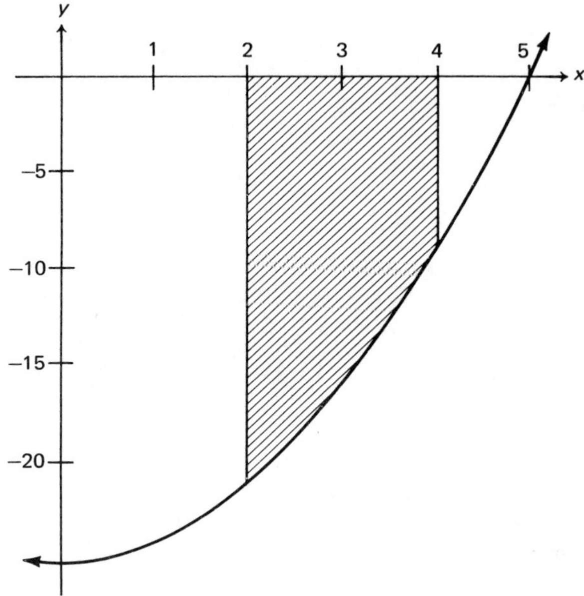

Here "area under" means area between the graph and the x-axis.

$$\text{Area} = \int_2^4 (x^2 - 25)\,dx$$

Here

$$\int (x^2 - 25)\,dx = x^3/3 - 25x = F(x),$$

so that

$$\begin{aligned}\int_2^4 (x^2 - 25)\,dx &= F(4) - F(2)\\ &= (x^3/3 - 25x)\Big]_2^4\\ &= (4^3/3 - 25\cdot 4) - (2^3/3 - 25\cdot 2)\\ &= (64/3 - 300/3) - (8/3 - 150/3)\\ &= -236/3 - (-142/3)\\ &= 142/3 - 236/3\\ &= -94/3\\ &= -31\ 1/3 \text{ square units.}\end{aligned}$$

Wait a minute—how can area be negative? It really is not, so here we take the absolute value of the answer to arrive at the correct answer of 31 1/3 square units.

If we had to find the area of a region such as the shaded one in Figure 10.10,

Figure 10.10

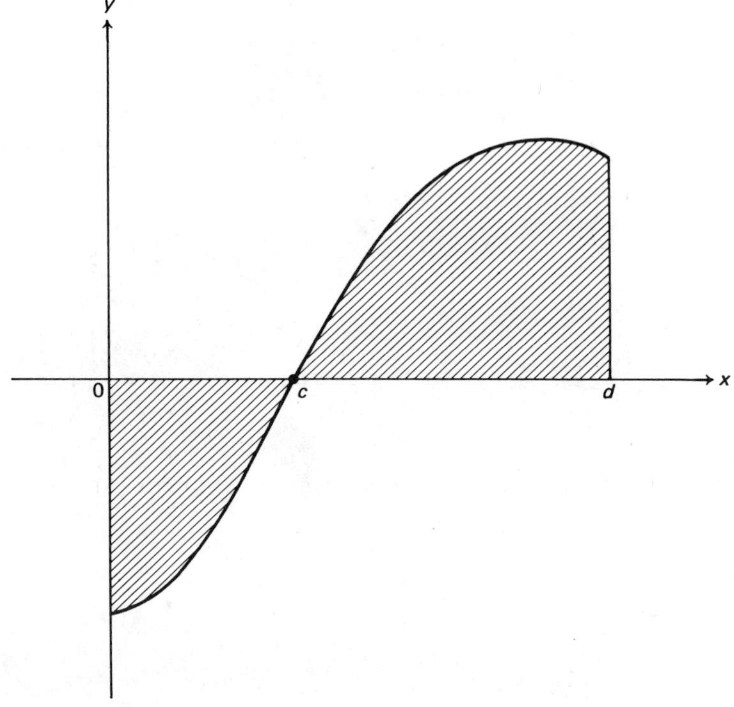

we would add together the areas of the two regions and work with absolute values when necessary.

$$\left|\int_0^c f(x)\,dx\right| + \int_c^d f(x)\,dx$$

This situation illustrates why a quick sketch is so important.

The are certain applications of the definite integral which result in a negative number where such a number is perfectly acceptable. One such application could be the difference between a company's total revenue and its total expenses for a stated period of time. In Figure 10.10 we "desensitized" the definite integral, which we would do if all we wanted to know was how many square feet of carpet it would take to cover the region, whether it was above or below the x-axis.

Problem set 10–3

Evaluate these definite integrals. They all represent the area under the graph of $f(x)$ bounded by the x-axis, the vertical line $x = a$, and the vertical line $x = b$. In all problems make a quick sketch of the function in the interval $[a, b]$.

1. $\int_3^5 x^2\,dx$
2. $\int_{-5}^{-3} x^2\,dx$
3. $\int_{-3}^{-5} x^2\,dx$
4. $\int_{-4}^{+4} (16 - x^2)\,dx$
5. $\int_1^3 (x^2 + 5)\,dx$
6. $\int_2^5 (3x - 6)\,dx$
7. Do Problem 6 without calculus. The results should be equal.
8. $\int_2^5 4 \cdot x^0\,dx$
9. $\int_2^5 4x\,dx$
10. $\int_2^5 4x^2\,dx$
11. $\int_1^4 (x^2 + 2)\,dx$
12. $\int_{-2}^1 (x^3 - 2x^2 - 5x + 6)\,dx$
13. $\int_1^3 (x^3 - 2x^2 - 5x + 6)\,dx$
14. $\int_{-2}^3 (x^3 - 2x^2 - 5x + 6)\,dx$
15. $\int_1^3 (12x - x^3)\,dx$
16. $\int_0^4 (x^2 - 4)\,dx$
17. Can you find the area of this region of the first quadrant?

$$R = \{(x, y) \mid y \leqslant 12x - x^3\} \cap \{(x, y) \mid y \geqslant 8x\}.$$

18. Find the exact area of the region described in Problem 1 of the previous problem set.
19. Find the exact area of the region described in Problem 4 of the previous problem set.
20. Find the exact area of the region described in Problem 7 of the previous problem set.
21. Evaluate $\int_0^3 (9 - x^2)\,dx$.
22. Evaluate $\int_3^6 (9 - x^2)\,dx$.

Further applications of the definite integral

In the first three sections of this chapter we have seen how the number

$$\int_a^b f(x)\,dx$$

is used to calculate the area under a curve. We may broaden the applications of the definite integral to include many situations whose geometric interpretations are those of finding the area under a curve.

Finding the annual sales rate

Example 13 Refer to the first example of Chapter 9 in which the salesman believed that the annual sales rate of the metal washers was a function of time strictly according to the formula

$$S = f(t) = t^3 - t^2 - 10t + 17,$$

where t was between -3 and 5 and S was in thousands of washers per year. How many washers were sold during the seventh year (between $t = 3$ and $t = 4$)?

Solution We begin with the graph of f between $t = 3$ and $t = 4$, after making an appropriate table of values with $S = f(t)$ rounded to three decimal places.

Table 10.3

t	3	3.25	3.5	3.75	4
$S = f(t)$	5	8.266	12.625	18.172	25

Figure 10.11

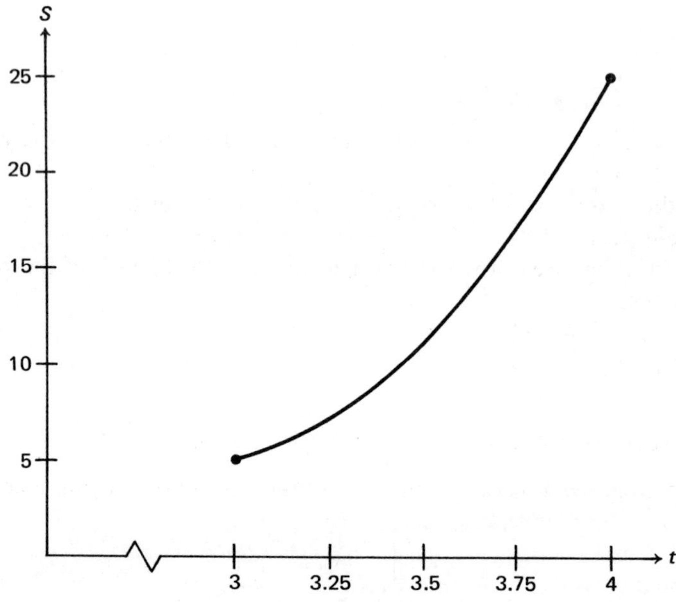

At the beginning of the seventh year the fact that $f(3) = 5$ does *not* mean that during the first minute or hour or day that five (thousand) washers will be sold. It is only the interpretation of $f(3) = 5$ as an *annual rate* that is the correct one. This interpretation means that if the annual rate were to remain constant, then 5,000 washers would be sold during the entire year. Of course, the annual rate does *not* remain constant. The rate at the beginning of the second quarter is 8,266 washers/year; at the beginning of the third quarter 12,625 washers/year; at the beginning of the fourth quarter 18,172 washers/year; and at the end of the fourth quarter (or the beginning of the eighth year), 25,000 washers/year. Suppose now that we return to the graph and construct rectangles as shown in Figure 10.12.

Figure 10.12

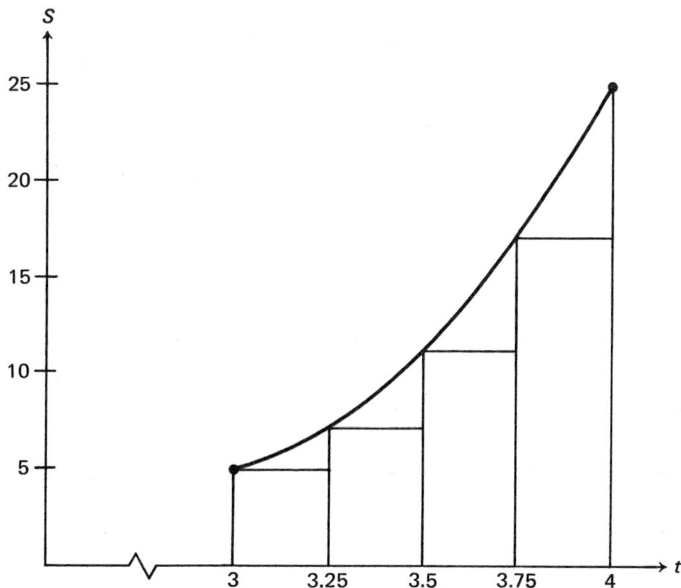

We now examine the first quarter under the assumption that the initial sales rate of 5,000 washers/month remains constant for the entire period. Then the number of washers sold in the quarter is

$$.25 \not{\text{years}} \cdot 5{,}000 \frac{\text{washers}}{\not{\text{year}}},$$

or 1,250 washers. We know the true number of washers sold is somewhat greater than 1,250 because the graph tells us that the sales rate is increasing continuously during the quarter. *Notice that the area of the rectangle based between $t = 3$ and $t = 3.25$ is numerically equal to the estimated number of washers sold between $t = 3$ and $t = 3.25$!*

This fact is the key to solving the original problem. We combine it with the three remaining quarters of the year, according to Table 10.4.

Table 10.4

Quarter	Assumed annual sales rate	Number of washers sold = area of rectangle
First: $t = 3$ to $t = 3.25$	5,000 washer/year	$.25\ \text{years} \cdot 5{,}000\ \dfrac{\text{washers}}{\text{year}} = 1{,}250$
Second: $t = 3.25$ to $t = 3.5$	8,266 washers/year	$.25\ \text{years} \cdot 8{,}266\ \dfrac{\text{washers}}{\text{year}} = 2{,}067$
Third: $t = 3.5$ to $t = 3.75$	12,625 washers/year	$.25\ \text{years} \cdot 12{,}625\ \dfrac{\text{washers}}{\text{year}} = 3{,}156$
Fourth: $t = 3.75$ to $t = 4$	18,172 washers/year	$.25\ \text{years} \cdot 18{,}172\ \dfrac{\text{washers}}{\text{year}} = 4{,}543$

By adding the four numbers in the right-hand column we obtain 11,016 as the estimated number of washers sold. Again this figure is low because of the assumption that the initial rate of sales each quarter remained constant throughout that quarter.

We could continue and obtain a better estimate of the total annual sales by erecting twelve rectangles under the curve. We would assume that the rate of sales for a given month remained constant at its initial level. Summing the twelve products of

$$\left(\frac{1}{12}\text{years}\right) \cdot \left(\text{annual sales in } \frac{\text{washers}}{\text{year}}\right),$$

where the second factor varies each month, would give us the estimated annual sales. This estimate would be better than using four rectangles but still would be lower than the true number of washers sold. However, we are spared the drudgery of the computing task. It should be clear by now that the total area of the rectangles under the curve approximates the total number of washers sold during the year. We can approximate the total sales of washers by making the rectangles thinner and thinner before calculating their area, so that they pack together under the curve more and more closely. Instead of doing this enormous amount of arithmetic we simply use the definite integral to calculate the true area under the curve and hence the true number of washers sold. We symbolize this number by N.

$$N = \int_3^4 (t^3 - t^2 - 10t + 17)\, dt$$

Since

$$\int (t^3 - t^2 - 10t + 17)\, dt = \frac{t^4}{4} - \frac{t^3}{3} - 5t^2 + 17t,$$

then

$$\begin{aligned} N &= \frac{t^4}{4} - \frac{t^3}{3} - 5t^2 + 17t \Big]_3^4 \\ &= \left[\frac{4^4}{4} - \frac{4^3}{3} - 5 \cdot 4^2 + 17 \cdot 4\right] - \left[\frac{3^4}{4} - \frac{3^3}{3} - 5 \cdot 3^2 + 17 \cdot 3\right] \\ &= \frac{92}{3} - \frac{69}{4} \\ &= \frac{161}{12} \\ &= 13.417. \end{aligned}$$

Our conclusion is that 13,417 washers will be sold in that year.

Calculating probability

Example 14 A student is chosen at random from a dormitory whose normally distributed grade point average (mean) is 2.3 and whose standard deviation is .6. What is the probability that his GPA is between 2.7 and 3.2?

Figure 10.13

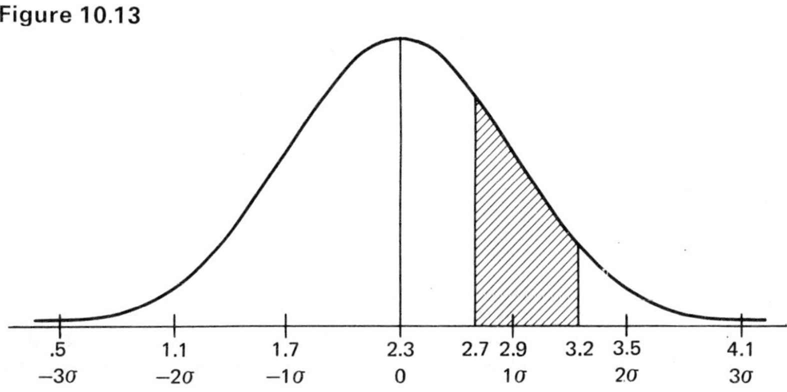

Solution At first glance this problem may not seem to belong in this chapter on integral calculus, especially since we solved problems similar to this one in Chapter 6. Nevertheless, we make a diagram and proceed as usual.

The corresponding z-scores are

$$z_{2.7} = \frac{2.7 - 2.3}{.6} = .67$$

and

$$z_{3.2} = \frac{3.2 - 2.3}{.6} = 1.5.$$

The area between $z = 0$ and $z = 0.67$ is .2486, and the area between $z = 0$ and $z = 1.50$ is .4332, so that the area between $z = .67$ and $z = 1.5$ is

$$.4332 - .2486 = .1846.$$

Therefore,

$$\begin{aligned} &P(GPA \text{ is between } 2.7 \text{ and } 3.2) \\ &\quad = P(z\text{-score is between } .67 \text{ and } 1.5) \\ &\quad = (\text{area under the normal curve between } z = .67 \text{ and } z = 1.5) \\ &\quad = .1846. \end{aligned}$$

The interesting part of this problem is its interpretation of probability as area under the normal curve. Recall from Chapter 6 that the equation of the normal curve is

$$y = f(z) = \frac{1}{\sqrt{2\pi}} \cdot e^{-z^2/2}$$

where e is an irrational number approximated by 2.718. In this example we actually calculate

$$\int_{.67}^{1.5} \frac{1}{\sqrt{2\pi}} \cdot e^{-z^2/2}\, dz.$$

Since the function defined by

$$y = f(z) = e^{-z^2/2}$$

does not have an elementary antiderivative, we use Table XII (which may be calculated using approximation methods similar to finding the area under a curve with rectangles). A computer would be used to do the enormous amounts of arithmetic.

Calculating the mean of $f(x)$ in a definite interval

We close this chapter with another application of the definite integral.

Example 15 Consider the function defined by the equation

$$y = f(x) = x^2/5$$

in the interval $5 \leqslant x \leqslant 15$. What is the mean of all the values of $f(x)$ in that interval?

Solution At first this problem appears unsolvable, for we only know how to find the mean of a finite set of numbers. Obviously, the values of $f(x)$ for $x \in [5, 15]$ are an infinite set. The best we can do is to pick a few typical values of $f(x)$ and find their mean in order to have a rough guess about the mean of all the $f(x)$'s. We obtain this sample of $f(x)$'s from the equally spaced values of x

$$\{6, 8, 10, 12, 14\}$$

in the domain.

Table 10.5

x	6	8	10	12	14
$f(x)$	7.2	12.8	20	28.8	39.2

The mean of these selected values of $f(x)$ is $108.0/5 = 21.6$.

We are still looking for the true mean of all the $f(x)$'s. The following definition is the one we will use to calculate this mean.

Definition If f is a continuous function in $[a, b]$ with $a < b$, then

$$\frac{1}{b-a} \cdot \int_a^b f(x)\,dx$$

is the mean of the values of $f(x)$ on the interval $[a, b]$.

In our case $a = 5$, $b = 15$, and $f(x) = x^2/5$. Upon substitution, then,

$$\begin{aligned} \text{mean} &= \frac{1}{15-5} \cdot \int_5^{15} \frac{x^2}{5}dx \\ &= \frac{1}{10} \cdot \frac{x^3}{15}\bigg]_5^{15} \\ &= \frac{1}{10} \cdot \left[\frac{15^3}{15} - \frac{5^3}{15}\right] \\ &= \frac{1}{10} \cdot \frac{650}{3} \\ &= \frac{65}{3}, \end{aligned}$$

or about 21.7, which is very close to our estimate of 21.6.

At this point we terminate our useful but limited study of the calculus. The development and the interrelationships of such seemingly diverse concepts as the derivative, antiderivative, and definite integral took our most brilliant ancestors hundreds of years to comprehend. The significance of calculus to the scientist and engineer is overwhelming; the advances they have brought to our society because of applied calculus include atomic energy, the space program, and even the existence of computers. The understanding of certain theories of economics is considerably aided by a knowledge of calculus, as several problems in the last chapter and in this chapter have pointed out. To really see how calculus applies directly for the benefit of social and management scientists requires considerably more breadth and depth of accompanying mathematics (for example, probability, statistics, and linear algebra) than we have learned at this point in our mathematical careers.

Problem set 10–4

The first eleven problems refer to Example 13.

1. Find the number of washers sold during the first year (from $t = -3$ to $t = -2$).
2. Find the number of washers sold during the eighth year.
3. Find the number of washers sold during all eight years.
4. Find the number of washers sold during the first quarter of the seventh year.
5. Find the number of washers sold during the second quarter of the seventh year.
6. Find the number of washers sold during the first half of the seventh year without using the results from Problems 4 and 5.
7. Find the number of washers sold during the sixth year.
8. What is the average rate of change of the sales rate of the washers during the sixth year?
9. What is the average rate of change of the sales rate of the washers during the first quarter of the sixth year?
10. What is the instantaneous rate of change of the sales rate of the washers at the beginning of the sixth year?
11. What is the instantaneous rate of change of the sales rate of the washers at the end of the sixth year?
12. A wholesale grocer receives 5,400 cans of frozen processed meat at the beginning of each month, and the depletion of his inventory always follows the same cycle, that is, the inventory slowly diminishes at the beginning of the month, when fresh meat is readily available, but as time passes the demand for frozen meat increases. The function relating inventory to time is

$$I = f(t) = 5{,}400 - 6t^2.$$

The daily cost of storing one can of food is .04 cents, and we assume that each month has thirty days.

a. How many cans are in storage at $t = 0$ days?

b. How many cans are in storage at $t = 1$ day?
c. How many cans are in storage at $t = 15$ days?
d. How many cans are in storage at $t = 30$ days?
e. What is the rate of depletion at $t = 1$ day?
f. What is the rate of depletion at $t = 15$ days?
g. What is the rate of depletion at $t = 30$ days?
h. Construct the graph of the function from $t = 0$ to $t = 70$.
i. What is the total cost of storage for one month?
j. What is the average number of cans stored in one month?

13. The franchisee of a new outlet for a pizza chain expects her daily sales to grow continuously so that she will be able to sell $(20+3t)$ pizzas a day after t days of business. On which day should she expect to sell her thousandth pizza? (Hint: set up the definite integral which gives her total sales after x days of business, set it equal to 1,000, and solve the resulting equation for x.)

14. The XYZ Computer Leasing Corporation wants to set the rental price on its Series 3000 computer to include the maintenance costs of the machine. Furthermore, XYZ does not wish to make any profit on the maintenance charges. If the annual rate of these charges is given by

$$E = f(t) = 400 + 1{,}200t + 3t^2$$

dollars, where t is the age of the computer in years and its useful age is seven years, then find the total maintenance charge for the seven years. What should the annual maintenance charge be if the total charge is spread equally over 7 years?

15. The Mullins Corporation has a new bookkeeping machine which will save money at the rate of

$$S = f(t) = 9.6t - \frac{2t^2}{5}$$

thousands of dollars after t years. The company estimates that the annual expenses for operating the machine and paying the people involved are given by

$$E = g(t) = 10t - 12$$

thousands of dollars after t years.

a. Find the total gross savings after three years.
b. Find the total expenses after three years.
c. Find the net profit on the machine (gross savings minus expenses) after three years.
d. For what value of t is $f(t) = g(t)$? What is your interpretation of this result?

16. Psychologists have determined that the amount of time a human requires to trace a path through a certain maze without error is given by

$$T = f(x) = c \cdot x^{-1/3}$$

for $x \in \{1, 2, 3, 4, \ldots\}$. The letter c is a constant which represents the number of minutes required for the first trial. If c is ten minutes for a certain maze and subject, then what will be the total amount of tracing time required for a subject to complete one hundred trials?

17. Refer to Problem 16. Does it require twice as long for a person to complete one hundred trials as it does for him to complete fifty trials?
18. Refer to 16. If we agree to train a subject whose first trial took eight minutes until his tracing time is no more than one minute, then how long will it take us?
19. A polling organization has determined that the time required by a survey worker to complete one interview depends on the number of interviews previously completed. The time required to complete the nth interview is

$$T = g(n) = 3 \cdot n^{-0.2}.$$

If each interviewer receives \$3 per hour, how much more expensive would it be to have two interviewers each do 250 interviews instead of having one do 500?

Find the average value of $f(x)$ for the following functions over the indicated intervals of the domain.

20. $f(x) = x^2, [0, 1]$
21. $f(x) = x^2, [1, 2]$
22. $f(x) = x^2, [-2, 2]$
23. $f(x) = x^3 - x^2 - 10x + 17, [-3, 5]$
24. $f(x) = 2x + 1, [3, 7]$
25. $f(x) = 7 \cdot x^0, [2, 5]$
26. This problem, involving an exponential function, is for those students who studied the last part of the last section of Chapter Nine on the differentiation of exponential and logarithmic functions. Paramount Leasing Company rents electric typewriters to businesses. It has found that the annual maintenance cost, after t years, increases continuously for a typewriter according to the formula

$$C = f(x) = 25 + \frac{e^x}{10}$$

where C is in dollars. If Paramount leases thirty typewriters to a company for a five-year period and the maintenance charge is to be the same for each year, then what should the charge be?

Answers to Selected Problems

CHAPTER 1

Problem Set 1–1, Pages 6-7

1. Let x = number of ounces of 80% acid solution to be added.

$$x = \frac{48}{7}$$

3. Let x = number of gallons of pure oil to be added.

$$x = \frac{3}{38}$$

5. If you had let x equal the number of gallons of 20% alcohol solution to be added, set up and solved a proper equation, and obtained $x = -5\,1/3$, then you should suspect that something is wrong. A closer look at the problem shows that it is impossible to perform the physical operations themselves. That is, it is impossible to add a 20% mixture to a 30% mixture in order to obtain a 35% mixture as the final blend. The final *negative* answer reflects this physical impossibility.

7. If we restrict ourselves to just one variable, then we can symbolize the two numbers we are looking for by the following two equations

x = number of pounds of 65¢ per pound candy to mix

and

$(60-x)$ = number of pounds of 125¢ per pound candy to mix.

Then

$x = 35$, and $(60-x) = 25$.

9. Let d = the number of dimes, and $(81-d)$ = the number of quarters. Then

$d = 65$, and $(81-d) = 16$.

11. The catch is that while half of one particular *numeral* or symbol for the number twelve is a *numeral* for the number seven, then half of the *number* or idea symbolized by 12 is the *number* symbolized by 6. The catch is that while half of one particular *numeral* or symbol for the number eight is a *numeral* for the number three, then half of the *number* or idea symbolized by 8 is the *number* symbolized by 4.

Problem Set 1–2, Pages 14-19

1. a. 13 c. −15.78 e. 2.8
   g. 6.8 i. 45 k. 5

2. a. $3x+4$ c. $44-6x$ e. $-6x-16$
   g. $0.4x+6$ i. $54-0.6x$ k. x^2-4x-3

3. a. 10 c. 32 e. 25
   g. −111 i. −56 k. 4
   m. −4 n. 4

4. a. 7 c. 1 e. 3
   g. 1/7 i. 40 k. 3/8
   m. $C/2\pi$ o. $\dfrac{b+c}{a}$ q. $\dfrac{A-p}{pr}$

5. a. $x>2$ c. $x>5$ e. $x<1$
   g. $x<-4$ i. $x>-4/3$ k. $x>15$

6. a. 17 c. 16 e. 4
   g. 4.49 i. 6 k. 2

7. a. 8 c. 6.93 e. 8.23
   g. 5 i. 15.84 k. 26.57

8. a. $3a^2b+3bc$ c. x^2+x-12 e. $10x^2-53x-78$
   g. $x^2+14x+49$ i. $x^2-13x+49$ k. $6x+3h-4$

9. a. $x=3,\ x=7$ c. $x=-2,\ x=8$ e. $x=-3,\ x=1/2$
   g. $y=-5,\ y=-1/2$ i. $c=-3,\ c=3$ k. $x=-9,\ x=2$

10. a. $x=3,\ x=7$ c. $x=-8,\ x=0$ e. $x=\dfrac{-3\pm\sqrt{17}}{4}$
    g. $x=\dfrac{-1\pm\sqrt{3}}{4}$ i. $x=-11,\ x=5$ k. $x=1\pm\sqrt{10}$

11. a. −4 c. 11 e. 3
    g. 49 h. undefined symbol i. 23
    k. −17 m. −70 o. 286
    q. 222 s. 2401 u. 36

12.

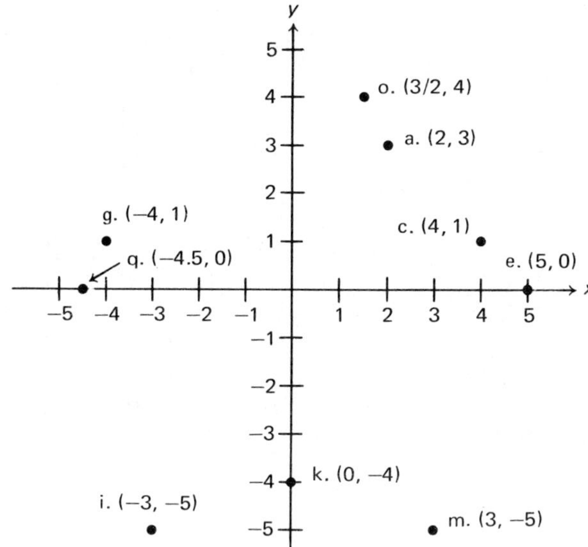

13. a. $x = 1/2$, $y = 21/2$ c. $x = -13$, $y = 6$
 e. $c = 2/21$, $d = -10/7$ g. $x = 4$, $y = 0$
 i. $t = -9.625$, $u = 27.75$ k. $p = 0.608$, $q = -0.3$

CHAPTER 2

Problem Set 2–1, Pages 23-35

1. a. True c. True e. False
 g. True i. True

2. a. $\{13, 14, 15, 16, 17, 18\}$
 c. $\{41, 42, 43, \ldots, 423\}$
 e. $\{7, 9, 11, \ldots, 73\}$
 g. $\{\ \}$
 i. $\{42, 43, 44, \ldots, 422\}$

3. a. $A = \{x \mid x$ is a counting number between 5 and 10, inclusive$\}$
 c. $C = \{x \mid x$ is an even counting number between 2 and 64, inclusive$\}$
 e. $E = \{x \mid x$ is a contiguous state bordered by the Pacific Ocean$\}$
 g. $G = \{x \mid x$ is a counting number greater than 12$\}$

4. a. True c. False e. False
 g. False

5. a. Infinite c. Infinite e. Infinite
   g. Infinite

6. $C = \{4, 5, 7, 8\}$ ↕ ↕ ↕ ↕ $D = \{13, 15, 11, 16\}$ $C = \{4, 5, 7, 8\}$ X ↕ ↕ $D = \{13, 15, 11, 16\}$

8. No. For example, sets $C = \{3, 8\}$ and $D = \{8, 10\}$ are equivalent finite sets but they are not equal.

9. a. 7 belongs to $\{2, 7, 6, 5\}$ but it does not belong to $\{2, 6, 5\}$.
   b. 14 belongs to $\{14, 16, 17, 18, 19\}$ but it does not belong to $\{16, 17\}$.
   c. We can not find an element which belongs to $\{\ \}$ and also does not belong to $\{2, 6, 5\}$ since, by definition, there is no object which belongs to $\{\ \}$.

11. $A = \{3, 7, 11, 15, \ldots, 4n-1, \ldots\}$ ↕ ↕ ↕ ↕ ↕ $B = \{30, 70, 110, 150, \ldots, 40n-10, \ldots\}$

n represents the location (first, second, third, etc.) of an element in either set A or set B. For example, the value of the third element in set B is $40 \cdot 3 - 10 = 110$.

Problem Set 2–2, Pages 32-34

1. a. $\{7\}$ c. $\{7\}$ e. $\{3, 6, 7, 9, 11, 12\}$
   g. $\{\ \}$ i. $\{1, 2, 5, 7, 8\}$ k. $\{3, 4, 6, 9, 10, 11, 12\}$
   m. $\{1, 2, 4, 5, 8, 10\}$ n. $\{4, 10\}$ o. $\{1, 2, 3, 4, 5, 6, 8, 9, 10, 11, 12\}$
   p. $\{4, 10\}$ q. $\{1, 2, 3, 4, 5, 6, 8, 9, 10, 11, 12\}$
   s. $\{1, 2, 10, 11, 12\}$ u. $\{1, 2, 3, 5, 6, 7, 8, 9\}$
   w. $\{5, 7, 8\}$

2. a. True c. False e. True
   g. True i. True k. True
   m. True o. True q. False
   s. True

3. a. True c. True e. True
   g. True i. True k. True
   m. False o. False

a.

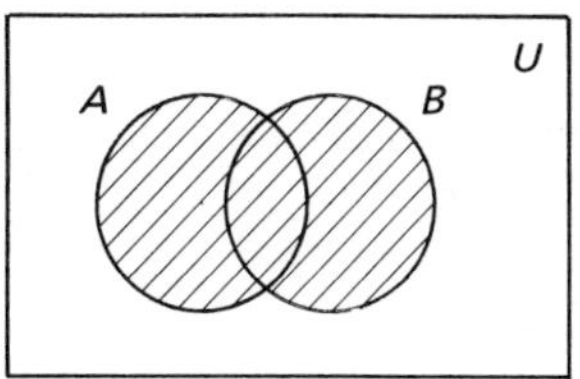

5. a.

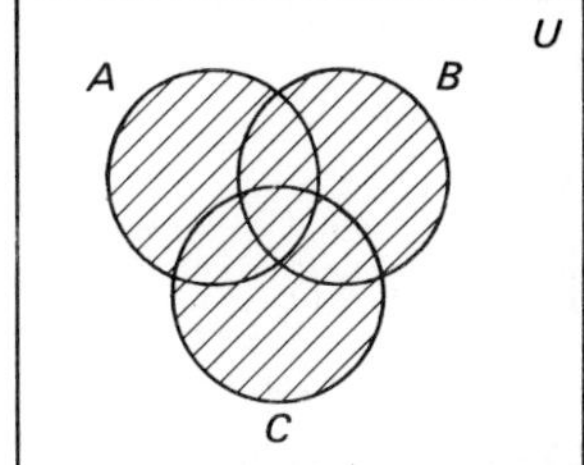

c.

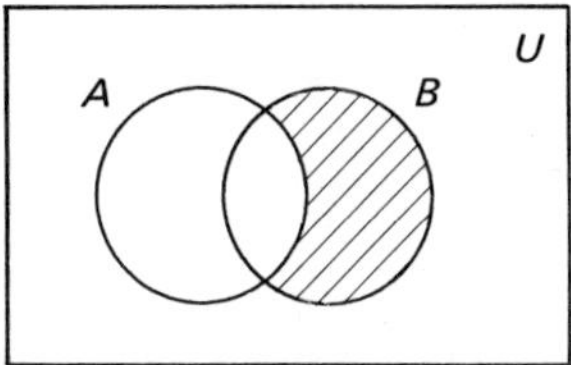

c.

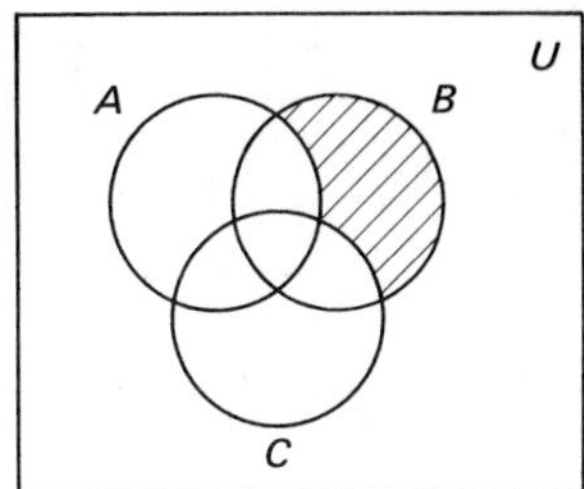

e.

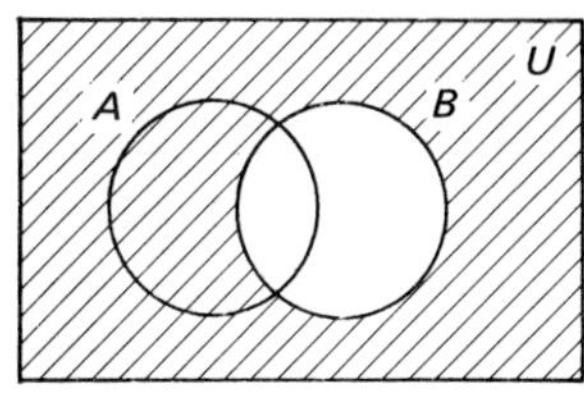

e.

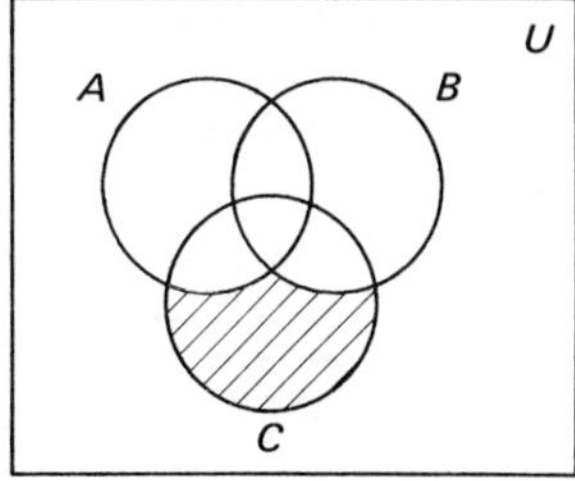

g.

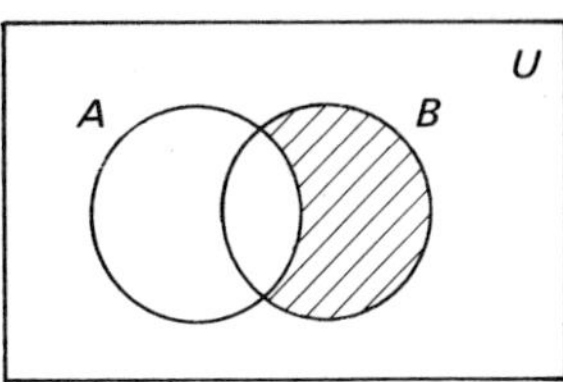

g.

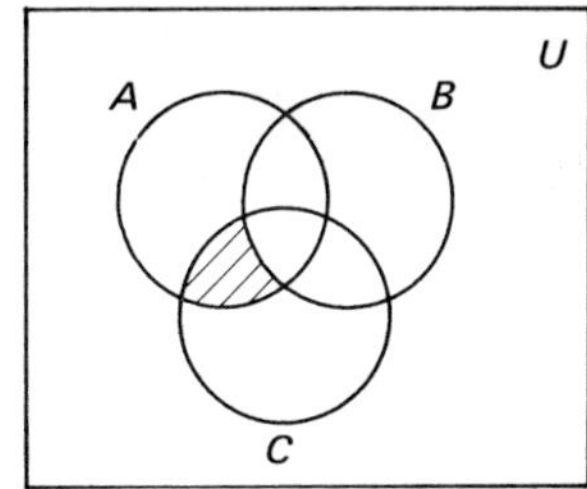

i.

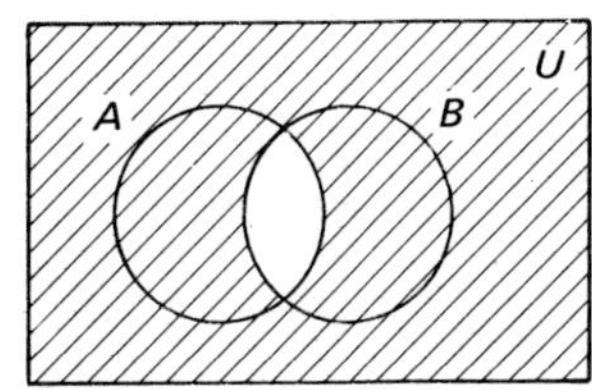

i.

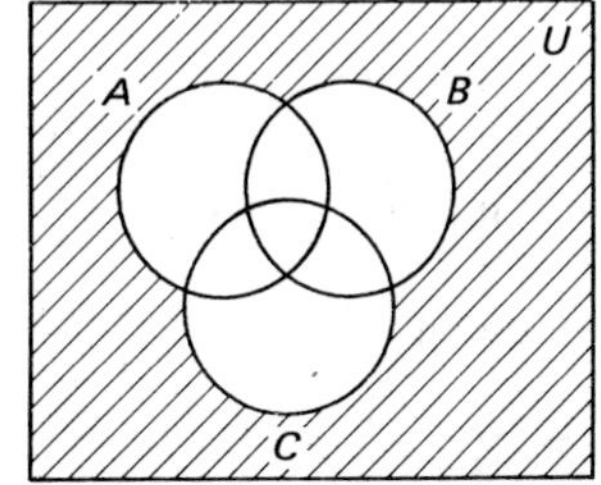

7. Pairs (c) and (g), (d) and (f), (h) and (i)

9. a. 14 b. When A and B have no common elements
 c. 5 d. When A is a subset of B

11. a. $[1, 4] = \{x \mid 1 \leqslant x \leqslant 4\}$

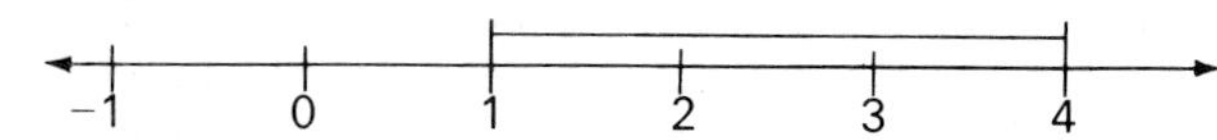

c. $(1, 4] = \{x \mid 1 < x \leqslant 4\}$

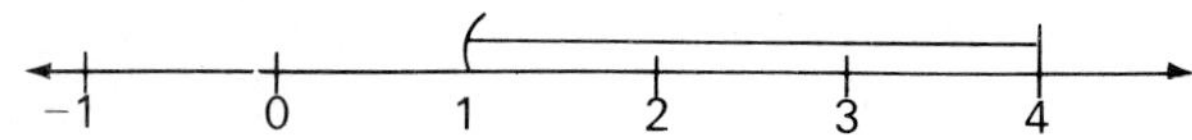

e. $\{x \mid 1 \leqslant x < 4\} = [1, 4)$

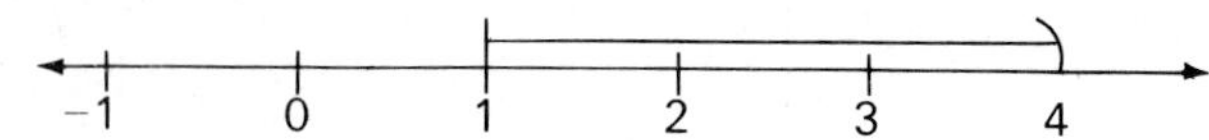

g. $[-2.5, \infty) = \{x \mid x \geqslant -2.5\}$

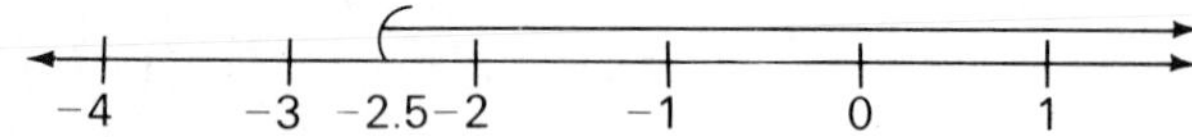

i. $[-3.7, 1.8] = \{x \mid -3.7 \leqslant x \leqslant 1.8\}$

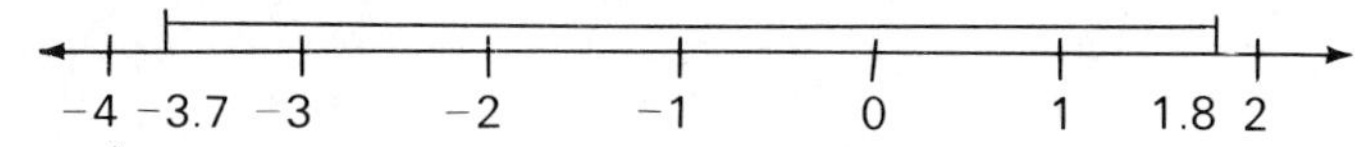

k. $\{x \mid x < 3\} = (-\infty, 3)$

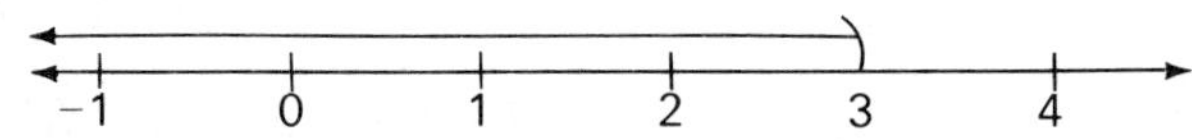

12. a. Yes c. No

Problem Set 2–3, Pages 39-42

1. a. 21 c. 6 e. 6

2. a. $n(C \cap D') = 21$ c. $n(C' \cap D') = 6$, or $n(C \cup D)' = 6$
 e. $n(C' \cap D) = 6$

4. 30; 17

5. a. 45 c. 60

6. a. 4 c. 43 e. 0

8. 3

9. If we perform the arithmetic correctly, the answer is -5 workers. The reason for this nonsense answer is that the original data are inconsistent. It is possible that the person tabulating the original survey sheets made an error.

11. a. 2 c. 4 e. 56

13. a. 38 c. 50 e. 28
g. 6 i. 50 k. $19/30 = 63\%$
m. $6/20 = 30\%$

CHAPTER THREE

Problem Set 3–1, Pages 48-50

1. a. Yes c. Yes e. No
g. Yes

2. a. 25 c. -17 e. $7t-3$
g. $7x+32$ i. $7x-59$ k. 7

3. a. 3 c. 3 e. 3

4. a. 1 c. -6 e. $3x^2+22x+33$
g. $6x+3h+4$

5. a. $x=2$ c. $x=5$, $x=-2$ e. There are no zeroes.

7. $A = f(n) = \dfrac{n \cdot 180 - 360}{n}$;

$n \in \{3, 4, 5, \ldots\}$

8. Let R be the set of all real numbers.
a. $\{t \mid t \in R\}$ c. $\{x \mid x \in R\}$ e. $\{x \mid x \in R\}$
g. $\{t \mid 3t-8 \geqslant 0\} = \{t \mid t \geqslant 8/3\}$

9. a. 16.43 c. 68.43 e. 94.43

Problem Set 3–2, Pages 54-56

1. Let t = number of days the wheat is stored, and let V = the value of the wheat (for each unit of 1000 pounds).

$$\begin{aligned} \text{Value} &= [\text{number of pounds}] \cdot [\text{value per pound}] \\ &= [\text{initial number of pounds} - \text{pounds lost}] \cdot \\ &\quad [\text{initial value per pound} + \text{additional value per pound}] \\ &= [1000 - 5 \cdot t] \cdot [2.5 + 0.08 \cdot t] \\ V = f(t) &= -0.4t^2 + 66.5t + 2500 \end{aligned}$$

The domain is $\{0, 1, 2, 3, \ldots, 200\}$.

2. Let t be the number of hours after midnight, and let d be the distance of the ship from the lighthouse at t hours after midnight. Then $12t$ is the number of miles the ship has traveled in t hours after midnight.

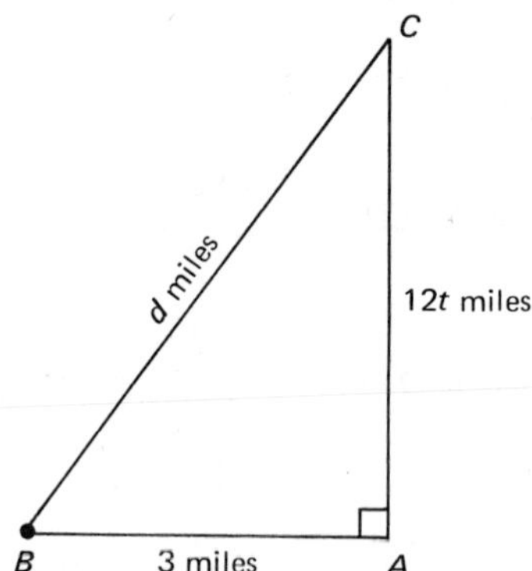

The application of the Pythagorean theorem to this triangle yields

$$d^2 = 3^2 + (12t)^2$$

$$d^2 = 9 + 144t^2$$

$$d = \sqrt{144t^2 + 9} = 3\sqrt{36t^2 + 1}$$

4. Let x represent the length of each side of each square piece cut from the rectangular piece of tin, and let V be the volume of the resulting box.

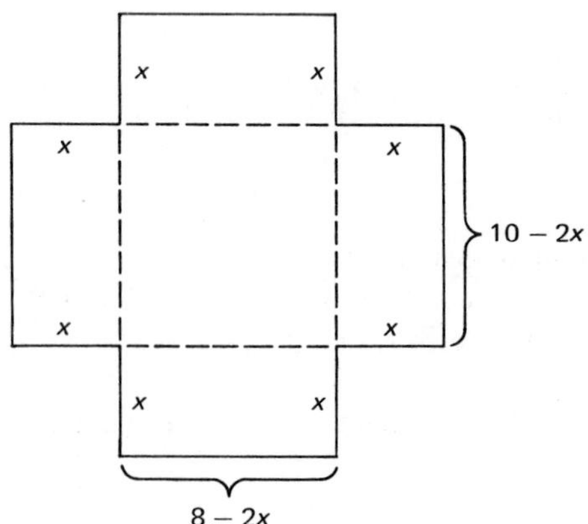

$$\begin{aligned} \text{Volume} &= \text{length} \cdot \text{width} \cdot \text{height} \\ &= (10 - 2x) \cdot (8 - 2x) \cdot x \\ V = f(x) &= 4x^3 - 36x^2 + 80x \end{aligned}$$

The domain is $\{x \mid 0 \leqslant x \leqslant 4\}$.

6\. b. Cost = (cost per cubic yard) · (number of cubic yards)
= 4 · (length in yards) · (width in yards) · (height in yards)

$$= 4 \cdot \frac{30}{3} \cdot \frac{10}{3} \cdot \frac{6}{36}$$

$$= \frac{200}{9}$$

Cost = \$22.22

7\. Let R be the total revenue of the bus company from this trip, and let x be the number of people *above* thirty who go.

Revenue = (number of people who go) · (revenue from each person)

$$= (30+x) \cdot (40-0.4 \cdot x)$$

$$R = f(x) = -0.4x^2+28x+1200$$

The domain is $\{0, 1, 2, 3, \ldots, 50\}$.

9\. a. Let p be the profit per car and let x be the selling price per car. Then $p = x-2$.

b. Let T = the total profit from the sale of the cars. Then

Total profit = (number of cars sold) · (profit per car)

$$= (300-100x) \cdot (x-2)$$

$$T = f(x) = 100 \cdot (-x^2+5x-6)$$

x	2.20	2.30	2.40	2.50	2.60	2.70	2.80
T	16.00	21.00	24.00	25.00	24.00	21.00	16.00

Problem Set 3–3, Pages 64-68

1\. a. $y = 2x+5$

x	−2	0	1
y	1	5	7

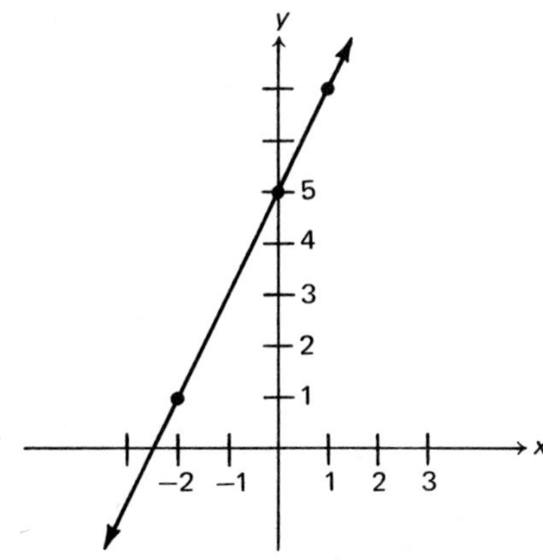

Slope: Choose $(x_1, y_1) = (-2, 1)$ and $(x_2, y_2) = (1, 7)$.

$$\text{Slope} = \frac{y_2 - y_1}{x_2 - x_1} = \frac{7-1}{1--2} = \frac{6}{3} = 2$$

2 is the coefficient of x in the equation $y = 2x+5$.

c. $y = -\frac{5}{4}x + \frac{7}{4}$

x	-4	0	4
y	$\frac{27}{4}$	$\frac{7}{4}$	$-\frac{13}{4}$

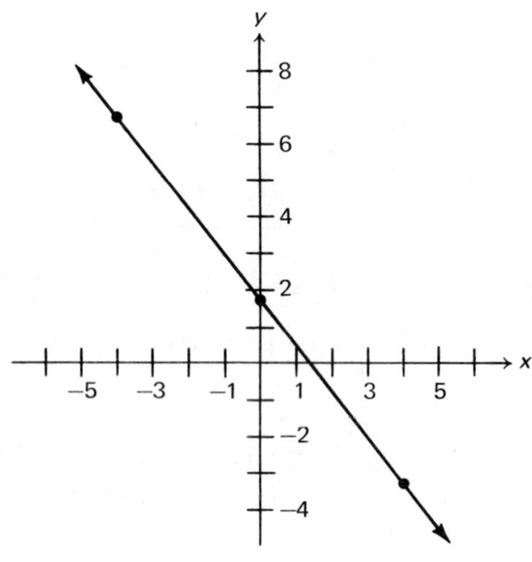

Slope: Choose $(x_1, y_1) = (-4, 27/4)$ and $(x_2, y_2) = (4, -13/4)$

$$\text{Slope} = \frac{y_2 - y_1}{x_2 - x_1} = \frac{-13/4 - 27/4}{4 - -4} = \frac{-40/4}{-8} = -10/8 = -5/4$$

$-5/4$ is the coefficient of x in the equation $y = -5/4x+7/4$.

e. $y = \frac{7}{3}x - \frac{2}{3} = \frac{7}{3}x + \left(-\frac{2}{3}\right)$

x	-6	0	6
y	$-\frac{44}{3}$	$-\frac{2}{3}$	$\frac{40}{3}$

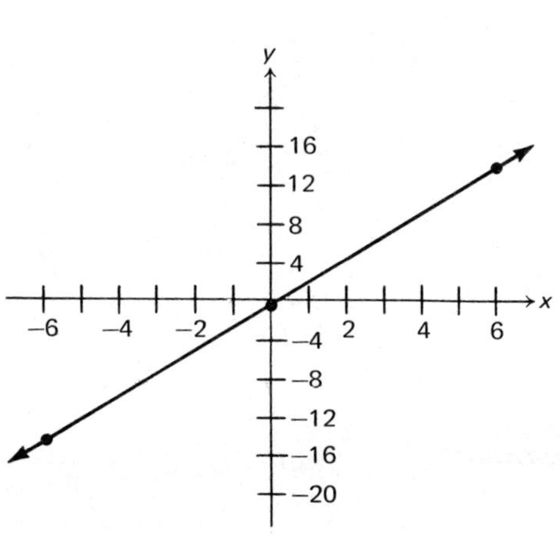

Slope: Choose $(x_1, y_1) = (-6, -44/3)$ and $(x_2, y_2) = (6, 40/3)$

$$\text{Slope} = \frac{y_2 - y_1}{x_2 - x_1} = \frac{40/3 - -44/3}{6 - -6} = \frac{84/3}{12} = \frac{1}{12} \cdot \frac{84}{3} = \frac{7}{3}$$

7/3 is the coefficient of x in the equation $y = 7/3x - 2/3$.

g. $y = 0x - 3 = 0x + (-3)$

x	-2	0	2
y	-3	-3	-3

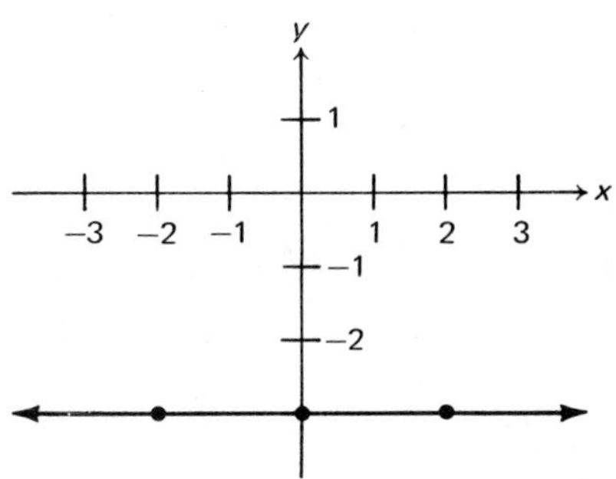

Slope: Choose $(x_1, y_1) = (-2, 3)$ and $(x_2, y_2) = (2, 3)$

$$\text{Slope} = \frac{y_2 - y_1}{x_2 - x_1} = \frac{3 - 3}{2 - -2} = \frac{0}{4} = 0$$

0 is the coefficient of x in the equation $y = 0x - 3$.

i. Attempting to solve $2x + 0y = 6$ for y leads to $y = -2/0x + 6/0$ which is illegal since division by 0 would be involved. However, solving $2x + 0y = 6$ (which is the same as $2x = 6$) for x gives $x = 3$. The graph of $x = 3$ is the set of all points whose x-coordinates equal 3. Clearly, this set is the line which is parallel to the y-axis and three units to the right of it. No restriction is placed on the possible values of y.

x	3
y	any real number

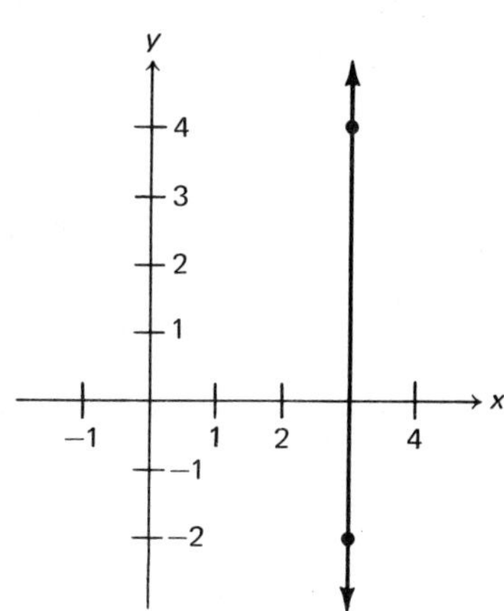

Slope: Choose $(x_1, y_1) = (3, -2)$ and $(x_2, y_2) = (3, 4)$.

$$\text{Slope} = \frac{y_2 - y_1}{x_2 - x_1} = \frac{4 - -2}{3 - 3} = \frac{6}{0} \quad \text{which does not exist.}$$

Hence, we say this line has *no slope* which is in contrast to the line in part (g) above which has *zero slope*.

3. a.

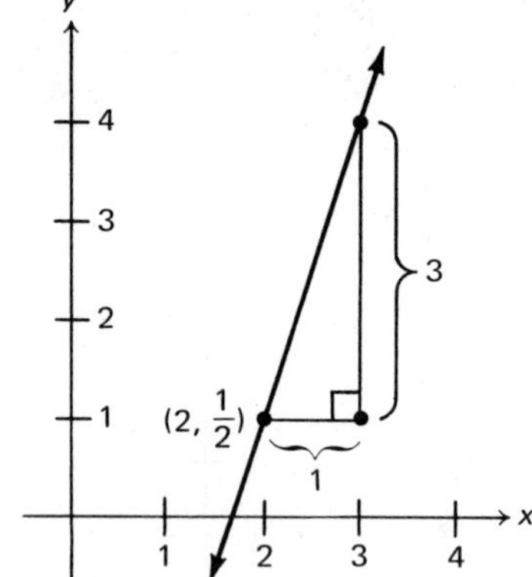

c.

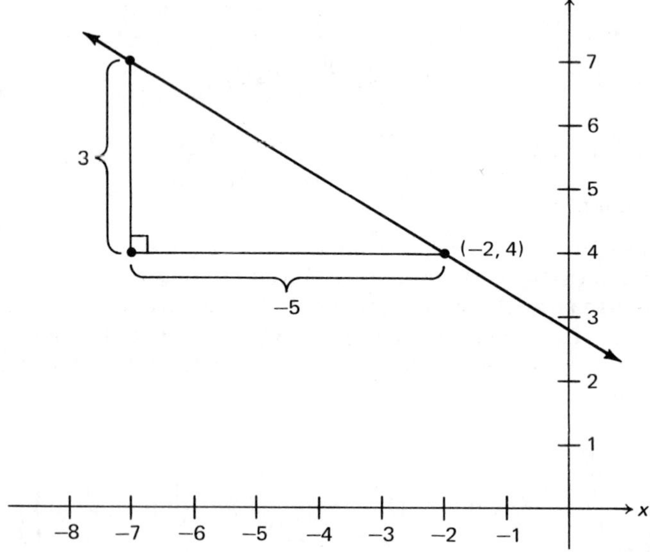

e. The equation $m = 0$ may be written as $m = 0/1$.

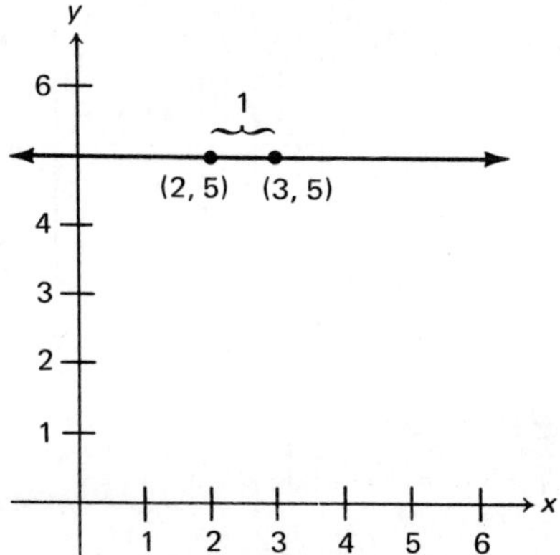

g. The equation $m = 3.5$ may be written as $m = 7/2$.

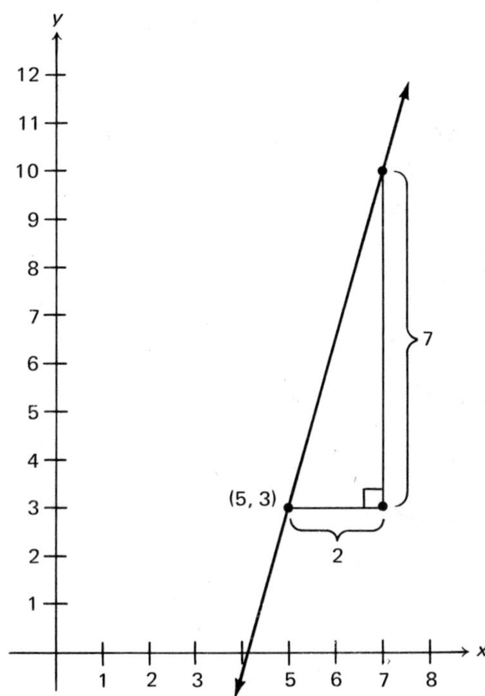

4. Only the quadrilateral in part (d) is a parallelogram. The slopes of its two pairs of parallel sides are 4 and $-7/2$.

5. a.

x	-6	-4	-2	0	2	4	6
y	133	65	21	1	5	33	85

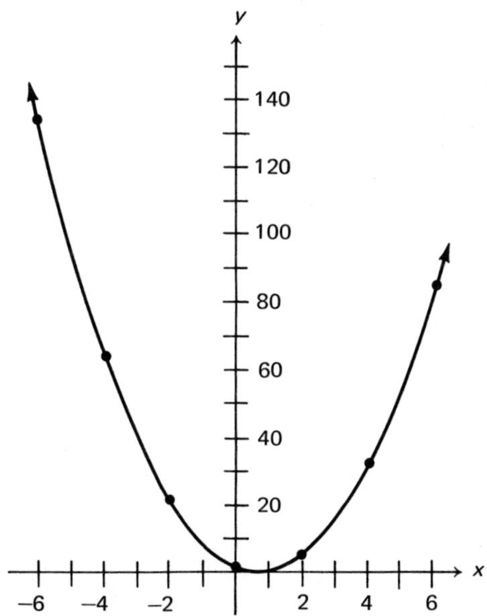

c.

x	−6	−4	−2	0	2	4	6
y	66	32	10	0	2	16	42

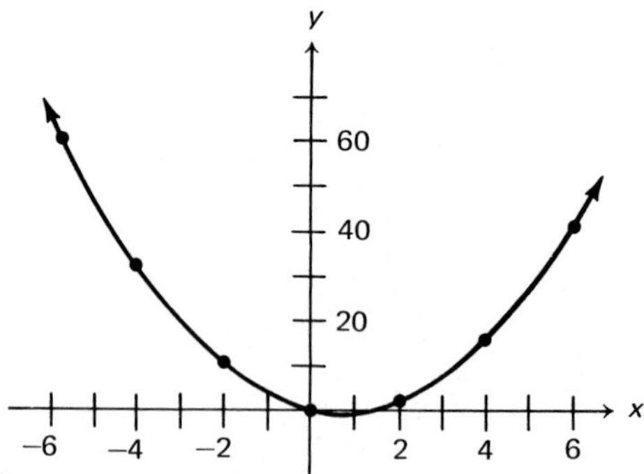

e.

x	−6	−4	−2	0	2	4	6
y	41	21	9	5	9	21	41

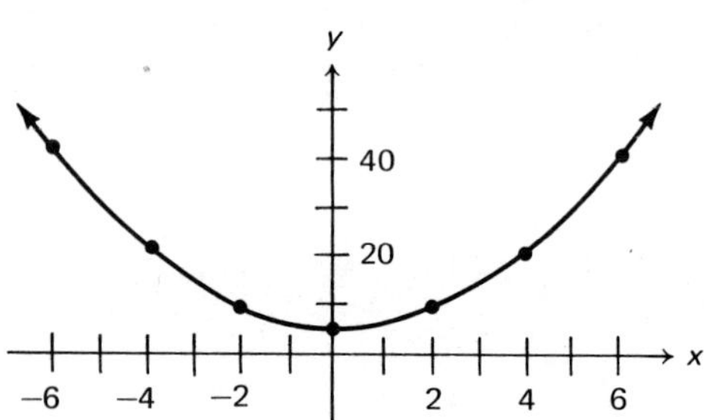

6. a.

x	-6	-4	-2	0	2	4	6
y	-0.4	-0.67	-2	2	0.67	0.4	0.29

x	-1.5	-1	-0.5
y	-4	does not exist	4

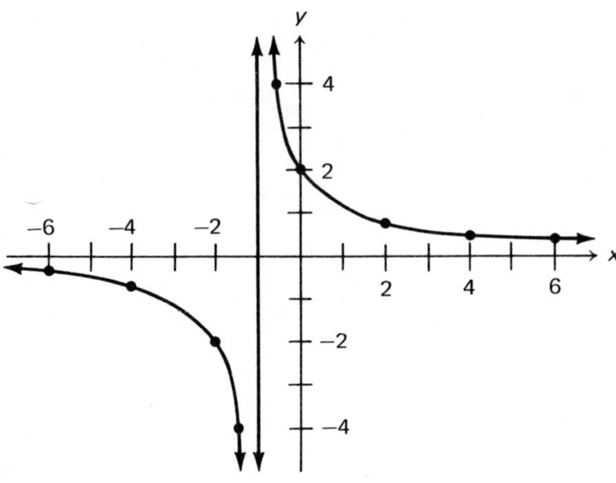

c.

x	−6	−4	−2	0	2	4	6
y	2.67	6	−4	−0.67	0	0.29	0.44

x	−3.5	−3	−2.5
y	11	does not exist	−9

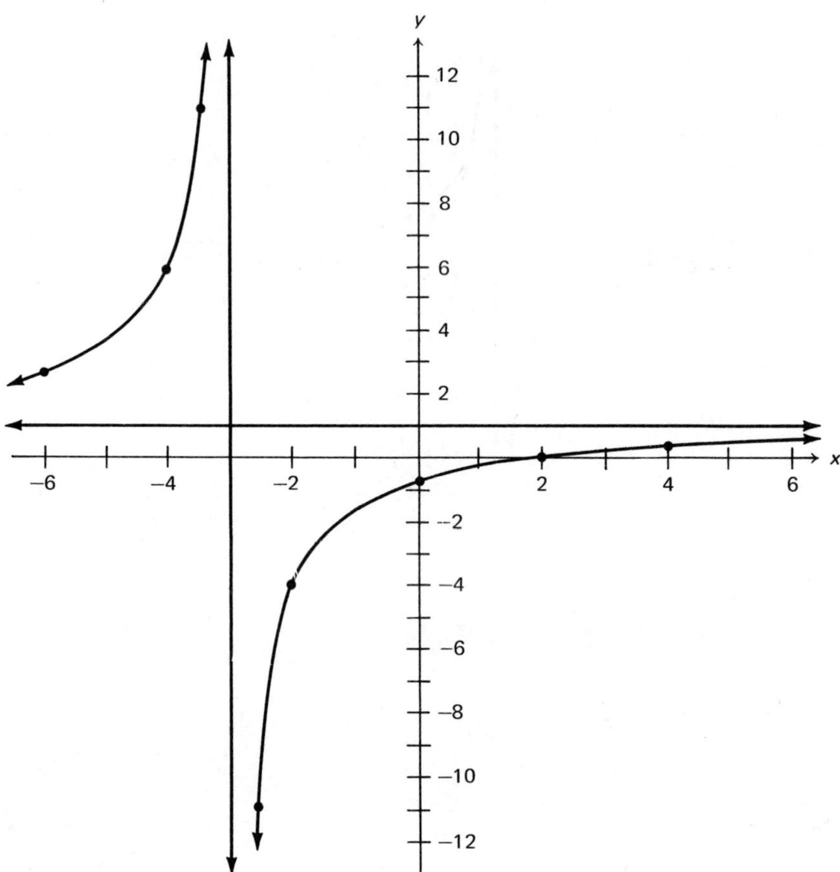

The line whose equation is $y=1$ is an example of a horizontal asymptote. For large negative values of x and for large positive values of x, the expression

$$\frac{x-2}{x+3}=\frac{x}{x+3}-\frac{2}{x+3}$$

represents numbers which are, respectively, slightly above one and slightly below one.

e.

x	−6	−4	−2	0	2	4	6
y	−1	1	3	5	7	9	11

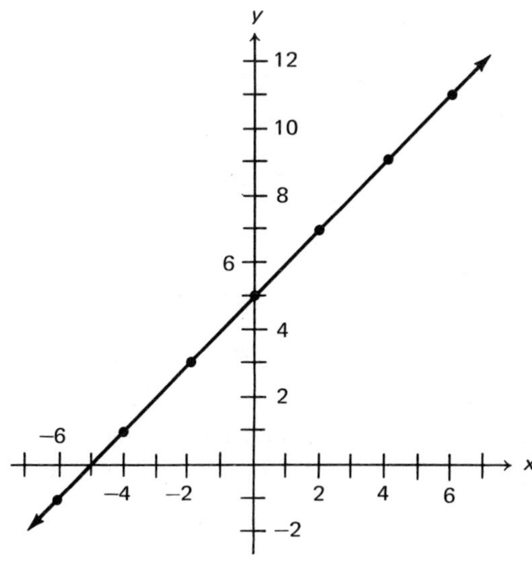

7. a. No c. Yes e. Yes
   g. Yes i. No

9. The x-coordinates of any two separate points on a line parallel to the y-axis are equal to each other. If we apply the slope formula to such a pair of numbers then the denominator is zero, which is not allowed.

11. If we let m be the coefficient of h and b be the constant in the equation $t = f(h) = m \cdot h + b$, then successive substitution of $h = 0$ and $t = 70$, and of $h = 3{,}000$ and $t = 61$, yield the results of $b = 70$ and $m = -0.003$. Therefore, the formula is $t = f(h) = -0.003h + 70$.

h	0	3,000	10,000
t	70	61	40

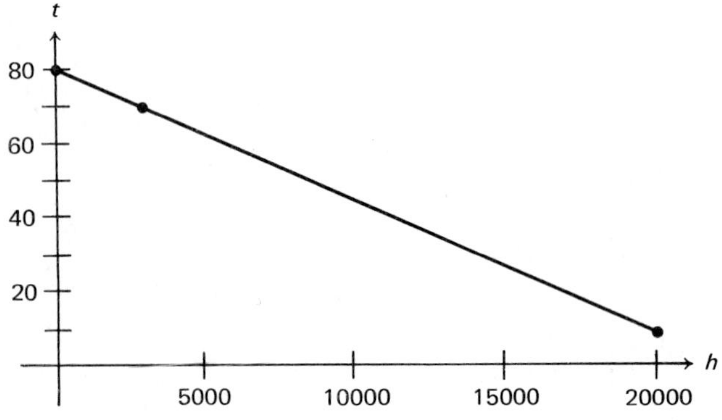

13. Let $(x_1, y_1) = (3, 26/7)$ and $(x_2, y_2) = (-1, 10/7)$. Then

$$m = \frac{y_2 - y_1}{x_2 - x_1} = \frac{10/7 - 26/7}{-1 - 3} = \frac{-16/7}{-4} = \frac{16/7}{4/1} = \frac{16}{7} \cdot \frac{1}{4} = \frac{4}{7}.$$

This example illustrates the fact that calculation of the slope of a line does not depend on which pair of distinct points on the line are chosen for substitution of their coordinates into the slope formula. Notice that, as expected, the arithmetic result of 4/7 is equal to the coefficient of x in the equation $y = 4/7x + 2$.

15. a. $y = 2x + 2$ c. $y = \frac{5}{6}x + \frac{17}{3}$

e. $y = 3x + 30$; the lines in (d) and (e) are parallel.

g. $y = -\frac{9}{7}x - \frac{11}{7}$

Problem Set 3–4, Pages 77-78

1. a. 17, 20, 23 c. 9, 9, 8
   e. 2, 2+3 = 5, 5+3 = 8 (or 2, 5, 8)
   g. 5/16, 5/32, 5/64

3. a. $a_5 = 2$ c. $a_{10} = 5/3$

   e. $a_8 = 243 \cdot \frac{1}{2187} = \frac{1}{9}$

5. a. $S_{14} = 553$ c. $S_{23} = 1127$

6. $a_n = 3+(n-1)\cdot 6$, and $a_7 = 39$

7. $a_n = -187+(n-1)\cdot 16$

9. $d = 43/14, a_8 = 19.5$

11. $a_1 = -11, S_{15} = 150$

13. $a_1 = 1/125, S_3 = 21/125$

15. $a_4 = -16/81, S_4 = 26/81$

17. $a_1 = 5, r = 8$ so that $a_n = 5 \cdot 8^{n-1}$

19. The first 50 odd positive integers are part of the arithmetic progression with $a_1 = 1$ and $d = 2$ so that $S_{50} = 2500$.

21. a. $a_n = 120+0.005 \cdot [15{,}000-(n-1)\cdot 120] = 195.6-0.6n$
    c. The monthly interest payments are part of an arithmetic progression with $a_1 = 75$ and $d = -0.6$. Karl makes $n = 125$ payments, so $S_{125} = \$4{,}725.00$.

23. 156 times

27. The AP with $a_1 = 8$ and $a_6 = 38$ has the equation $a_n = 8+(n-1)\cdot 6$ so that its terms are 8, 14, 20, 26, ... $S_3 = 8+14+20 = 42$, not 34, so our conclusion must be that it is impossible for one AP to meet all three of the specified conditions. The original equations are inconsistent.

Problem Set 3–5, Pages 85-86

1. a. 9 c. −9
   e. −1/9 g. 75
   i. 25/9 k. 5
   m. 64 o. 16

3. Both numbers are equal to 1/16.

5. a. b = 3 c. b = 10 e. $\{b \mid b>0\}$

8.

	x	−3	−2	−1	0	1	2	3
a.	1.5^x	0.296	0.444	0.667	1	1.5	2.25	3.375
c.	$(1/3)^x$	27	9	3	1	0.333	0.111	0.037

a.

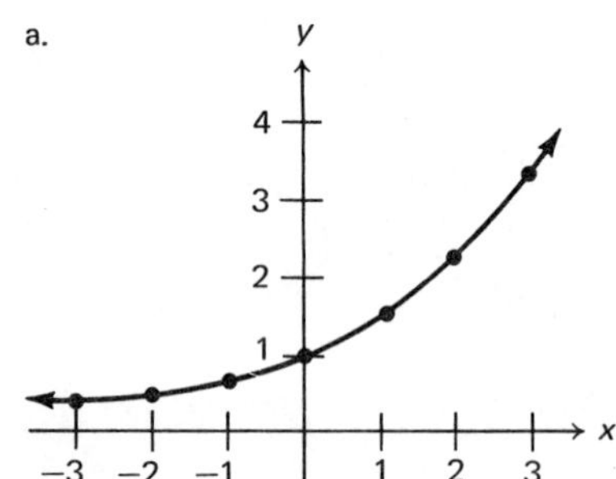

c.

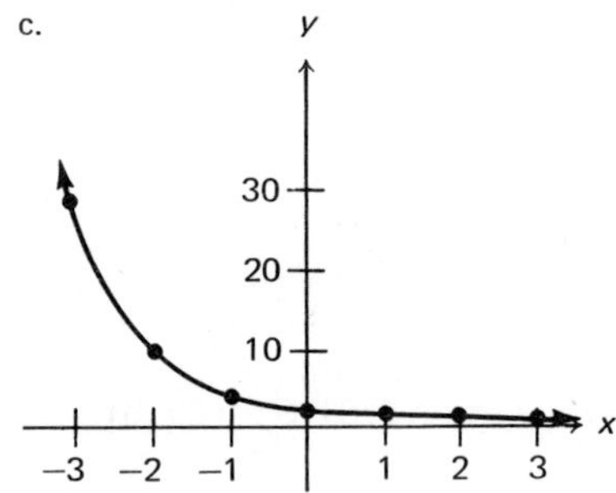

9.

x	1	3	5	10	16
y	33.33	63.41	79.92	95.52	99.26

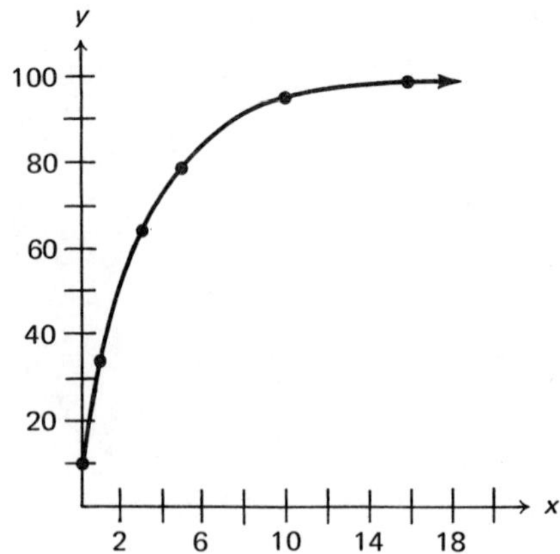

11.

t	0	1	2	3	t
V	4000	$4000 \cdot 0.9$	$(4000 \cdot 0.9) \cdot 0.9 =$ $4000 \cdot (0.9)^2$	$(4000 \cdot (0.9)^2) \cdot 0.9 =$ $4000 \cdot (0.9)^3$	$4000 \cdot$ $(0.9)^t$

13. Let N = the number of grams of Radium 214 present at a point t in time. If a substance loses 20% of itself each minute, this means that the amount present at the end of any minute is 80% of the amount present at the beginning of that minute

t	0	1	2	3	4
N	G	$0.8 \cdot G$	$0.8 \cdot (0.8 \cdot G) =$ $0.64G$	$0.8 \cdot (0.64G) =$ $0.512G$	$0.8 \cdot (0.512G) =$ $0.4096G$

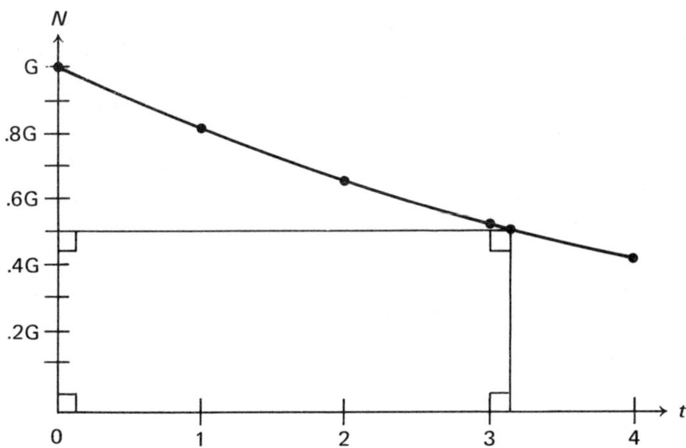

According to this graph, the corresponding value of t for which $0.5G$ grams of Radium 214 remain is about 3.1 minutes.

Problem 3–6, Pages 95-97

1. a. 3.617 billion   c. 5.27 billion
   e. 16.31 billion

3. a. 2025: 352,640,000
   2050: 452,240,000
   c. 2025: 417,600,000
   2050: 576,760,000
   e. 2025: 618,340,000
   2050: 1,014,480,000

4. a. 0.68%   c. 0.91%
   e. 1.59%   g. 2407
   i. 3197   k. 3898

5. $\sqrt{500/300} - 1 = 0.291 = 29.1\%$

9. a. $\sqrt[5]{300/291} - 1 = 0.0061 = 0.61\%$
   c. $\sqrt[5]{373/350} - 1 = 0.0128 = 1.28\%$

CHAPTER 4

Problem Set 4–1, Pages 103-106

1. \$56.00. If we substitute $i = 56$, $p = 700$ (an error), and $t = 1$ into the equation $i = p \cdot r \cdot t$ and solve for r, then $r = 0.08 = 8\%$. The average amount owed is $4242/12 =$ \$353.50, and the substitution of $i = 56$, $p = 353.50$, and $t = 1$ into the equation $i = p \cdot r \cdot t$ yields $r = 15.84\%$.

3. The monthly amounts owed are \$216, 210, 180, 150, 120, 90, 60, and 30. Since their sum is \$1056, the average amount owed is \$132. The substitution of $i = 24$, p = 132, and $t = 8/12$ into the equation $i = p \cdot r \cdot t$ yields $r = 27.27\%$.

5. $r = 624/1222 = 51.06\%$.

7. The average amount owed is $330/8 =$ \$41.25. The substitution of $i = 44$, p = 41.25, and $t = 8/12$ into the equation $i = p \cdot r \cdot t$ yields the solution of $r = 160\%$. The direct ratio formula yields $r = 132.44\%$.

9. \$150.00; \$31.94 (although the thirty-sixth payment would be \$32.10 in order that the total amount repaid equal \$1150.00); 10.20%.

Problem Set 4–2, Pages 112-113

1. The six consecutive compound amounts are \$263.19, \$3,759.10, \$1,593.85, \$1,604.71, \$1,610.33, and \$1,614.17.

3. The ten consecutive present values are \$165.13, \$1,313.19, \$819.54, \$742.47, \$672.97, \$610.27, \$553.68, \$456.39, and \$376.89.

5. \$100 at 4% monthly grows to \$104.07 while \$100 at $4\frac{1}{2}\%$ annually grows to \$104.50. The second way is better by \$0.43.

6. The result of inflation on the purchase price means that in five years the price will be $2000 \cdot (1+3\%)^5$ or \$2,318.54. The present value of this \$2,318.54 is $2{,}318.54 \cdot (1+1.5\%)^{-20}$ or \$1721.44.

8. 40 months.

10. Bank A: \$465.89; Bank B: \$723.52. No, but the interest earned in Bank B (\$423.52) is at least twice the interest earned in Bank A (\$165.89).

Problem Set 4–3, Pages 124-126

1. The ten consecutive amounts are \$3,818.20, \$3,875.27, \$3,933.74, \$3,993.07, \$1,996.54, \$24,609.40, \$26,728.30, \$29,083.20, \$31,696.90, \$15,848.45.

3. \$1,590.42 − \$1,440.00 = \$150.42.

6. $R = 2{,}318.54 \cdot 1/s_{\overline{20|}1.5\%} = 2{,}318.54 \cdot 0.0432452 =$ \$100.27.

7. a. \$181,749 b. \$480,000 c. \$908.75, which means that only \$91.25 of the original deposit is withdrawn.

9. No, since her deposits plus interest will accumulate to only \$2,244.89. Her deposits at the end of each month should be \$96.23.

11. 300 months or 25 years.

12. 2,300 · 0.044777 = \$102.99.

Problem Set 4–4, Pages 132-140

1. The necessary table entries are listed below.

R	*S*	*I*
\$128.94	\$38,682.00	\$18,682.00
120.00	43,200.00	23,200.00
114.12	47,930.40	27,930.40
110.14	52,867.20	32,867.20
141.44	42,432.00	22,432.00
133.14	47,930.40	27,930.40
127.86	53,701.20	33,701.20
124.38	59,702.40	39,702.40
248.76	119,404.80	79,404.80

4. The new monthly payment becomes 30,000 · 0.007072 = \$212.16. The difference in the total amount of interest paid is \$33,648 − \$32,190 = \$1,458.

5. The complete table is listed below.

Month	Principal	Interest	Principal payment	Total payment
1	\$10,000.00	\$50.00	\$21.65	\$71.65
2	9,978.35	49.89	21.76	71.65
3	9,956.59	49.78	21.87	71.65
4	9,934.72	49.67	21.98	71.65
5	9,912.74	49.56	22.09	71.65
6	9,890.65	49.45	22.20	71.65
7	9,868.45	49.34	22.31	71.65
8	9,846.14	49.23	22.42	71.65

6. The missing table entries are listed below.

End of year	Percent of total which is interest	Percent of total paid which reduces amount owed
1	68.9	31.1
2	67.0	33.0
3	65.0	35.0
4	62.8	37.2
5	60.5	39.5
6	58.1	41.9
7	55.5	44.5
8	52.8	47.2
9	49.9	50.1
10	46.8	53.2
11	43.5	56.5
12	40.0	60.0
13	36.3	63.7
14	32.4	67.6
15	28.2	71.8
16	23.8	76.2
17	19.1	80.9
18	14.1	85.9
19	8.8	91.2
20	3.5	96.5

8. $r = 2364.48/38638.8 = 6.12\%$

9. b. The monthly payment is $221.82, the bank's equity after ten years is $23,928.17, and John's equity is $6,071.83. d. The monthly payment is $110.61 and the bank's equity after eight years is $14,638.57, so Karen must pay the current owner $3,361.43. The current owner has paid $10,618.56 of which $7,257.13 has been the interest payments to the bank.

CHAPTER 5

Problem Set 5–1, Pages 149-154

1. a. 1/6 c. 1/3 e. 1/2
   g. 1/2 i. 1/6

3. a. 26/52 = 1/2 c. 2/52 = 1/26 d. 28/52 = 7/13
   e. 1/52 g. 31/52 i. 48/52 = 12/13
   k. 36/52 = 9/13

5. a. False c. True e. False
   g. False

8. a. 3/8 c. 1/8 e. 7/8
   g. 4/8

9. a. $2 = 2^1$ c. $2 \cdot 2 \cdot 2 = 2^3$ e. $2 \cdot 2 \cdot 2 \cdot 2 \cdot 2 = 2^5$

11. $9.8 = 72$

13. a. 1/80 b. 1/45

17. a. No c. Yes e. No

18. If you buy only the forty-fifth ticket, then P(you win) = 1/45. If you buy only the forty-fifth and forty-sixth tickets, then P(you win) = 2/46. $2/46 \neq 2 \cdot 1/45$, so the answer is no.

19. If you buy only the forty-fifth ticket, then P(you win) = 1/50. If you buy only the forty-fifth and forty-sixth tickets, then P(you win) = 2/50. $2/50 = 2 \cdot 1/50$, so the answer is yes.

Problem Set 5–2, Pages 162-166

1. a. 120 c. 20 e. 1
   g. 24 i. 720 k. 504
   m. 1,001,000

2. a. 56 c. 8 e. 10
   g. 385 i. 2600 k. 1
   m. 40 o. 2,598,960 q. 1

3. ${}_{52}C_5 = 2{,}598{,}960$

5. ${}_{12}C_4 = 495$

7. The generalization suggested by the equations

$$ {}_{12}C_4 = {}_{12}C_8 = {}_{12}C_{12-4} $$

is that

$$ {}_nC_r = {}_nC_{n-r} $$

This generalization is proven by the following chain of equations.

$$\begin{aligned}
{}_{n}C_{r} &= \frac{n!}{r!\cdot(n-r)!} \\
&= \frac{n!}{(n-r)!\cdot r!} \\
&= \frac{n!}{(n-r)!\cdot(n-n+r)!} \\
&= \frac{n!}{(n-r)!\cdot(n-(n-r))!} \\
&= {}_{n}C_{n-r}
\end{aligned}$$

9. a. 720 b. 360 c. 1000
   d. 500

11. ${}_{23}C_{7} = 245{,}157$

13. $3\cdot 7\cdot 6\cdot 5\cdot 4\cdot 3\cdot 2\cdot 1 = 15{,}120$

15. $3\cdot {}_{6}P_{6}\cdot 2 = 4{,}320$

17. a. ${}_{17}C_{6} = 12{,}376$
    c. ${}_{10}C_{4}\cdot {}_{7}C_{2} + {}_{10}C_{5}\cdot {}_{7}C_{1} + {}_{10}C_{6}\cdot {}_{7}C_{0} = 4410+1764+210 = 6{,}384$
    e. ${}_{7}C_{3}\cdot {}_{10}C_{3} = 4{,}200$
    g. $12{,}376 = 6{,}384+1{,}792+4{,}200$
    i. $7\cdot 10\cdot {}_{9}C_{3}\cdot {}_{6}C_{1} = 35{,}280$

18. a. 2,520
    c. ${}_{7}C_{3}\cdot {}_{4}C_{2}\cdot {}_{2}C_{2} = 210$
    e. ${}_{11}P_{11} = 11! = 39{,}916{,}800$
    g. If we add the results of the three cases of having the red, blue, and white marbles at the ends, respectively, we have $7{,}560+1{,}260+12{,}600 = 21{,}420$.
    i. 420
    j. ${}_{13}P_{13} = 13! = 6{,}227{,}020{,}800$
    k. ${}_{13}C_{4}\cdot {}_{9}C_{3}\cdot {}_{6}C_{6} = 60{,}060$

19. ${}_{18}C_{7}\cdot {}_{7}C_{4}$

21. $s = f(n) = {}_{n}C_{4}$

22. a. 2^3 b. 3^3 c. 4^3
    d. 5^3 e. n^3

25. a. ${}_{6}C_{3}\cdot {}_{8}C_{2} = 20\cdot 28 = 560$
    c. ${}_{8}C_{3}\cdot {}_{6}C_{2} + {}_{8}C_{4}\cdot {}_{6}C_{1} + {}_{8}C_{5}\cdot {}_{6}C_{0} = 56\cdot 15+70\cdot 6+56\cdot 1 = 1{,}316$
    e. ${}_{8}C_{3}\cdot {}_{8}C_{2} + {}_{8}C_{4}\cdot {}_{8}C_{1} + {}_{8}C_{5}\cdot {}_{8}C_{0} = 56\cdot 28+70\cdot 8+56\cdot 1 = 2{,}184$
    g. ${}_{12}C_{3}\cdot {}_{8}C_{2} + {}_{12}C_{4}\cdot {}_{8}C_{1} + {}_{12}C_{5}\cdot {}_{8}C_{0} = 220\cdot 28+495\cdot 8+792\cdot 1 = 10{,}912$
    i. ${}_{16}C_{3}\cdot {}_{8}C_{2} + {}_{16}C_{4}\cdot {}_{8}C_{1} + {}_{16}C_{5}\cdot {}_{8}C_{0} = 560\cdot 28+1820\cdot 8+4368\cdot 1 = 34608$
    k. ${}_{r}C_{3}\cdot {}_{d}C_{2} + {}_{r}C_{4}\cdot {}_{d}C_{1} + {}_{r}C_{5}\cdot {}_{d}C_{0}$
    l. $\sum_{i=4}^{7} {}_{r}C_{i}\cdot {}_{d}C_{7-i} = {}_{r}C_{4}\cdot {}_{d}C_{3} + {}_{r}C_{5}\cdot {}_{d}C_{2} + {}_{r}C_{6}\cdot {}_{d}C_{1} + {}_{r}C_{7}\cdot {}_{d}C_{0}$
    m. $\sum_{i(c+1)/2}^{c} {}_{r}C_{i}\cdot {}_{d}C_{c-i}$

27. ${}_{25}C_9 \cdot {}_9P_9$

29. $({}_8C_1 \cdot {}_3C_1 \cdot {}_7C_4 \cdot {}_7C_3) \cdot {}_9P_9$

Problem Set 5–3, Pages 175–183

1. a. $26/52 = 1/2$ c. $16/52 = 4/13$

3. The probability of an event is never greater than one. We need to know the probability that he stays in town to work and to have a date so that we can subtract this number from 1.30 to obtain the final answer.

5. a. $0.88 \cdot 0.90 \cdot 0.85 = 0.6732$
   c. $1-(0.12 \cdot 0.10 \cdot 0.15) = 0.9982$

6. a. 1/221 c. 1/102
   e. 1 g. 1/17

9. 61/153

11. $71/200 = 0.355$

14. a. 0.4 c. $1-(0.6)^3 = 0.784$
   e. $1-(0.6)^5 = 0.92224$

15. This problem directly relates to problem 25 of the previous problem set.
   a. $560/{}_{14}C_5 = 560/2002 = 0.280$
   c. $1316/{}_{14}C_5 = 1316/2002 = 0.657$
   e. $2184/{}_{16}C_5 = 2184/4368 = 0.500$
   g. $10912/{}_{20}C_5 = 10912/15504 = 0.704$
   i. $34608/{}_{24}C_5 = 34608/42504 = 0.814$
   k. $({}_rC_3 \cdot {}_dC_2 + {}_rC_4 \cdot {}_dC_1 + {}_rC_5 \cdot {}_dC_0)/{}_{d+r}C_5$
   l. $\sum_{i=4}^{7} {}_rC_i \cdot {}_dC_{7-i}/{}_{d+r}C_7$
   m. $\sum_{i=5}^{9} {}_rC_i \cdot {}_dC_{9-i}/{}_{d+r}C_9$
   n. $\sum_{i=(c+1)/2}^{c} {}_rC_i \cdot {}_dC_{c-i}/{}_{d+r}C_c$

19. a. 456/1341 c. 37/456 e. 189/1341
   g. 60/221 i. 53/221 k. 60/253
   m. 53/359 o. 321/456 q. 970/1341

21. a. $\frac{6}{10} \cdot \frac{4}{11} + \frac{4}{10} \cdot \frac{5}{11} = \frac{44}{110} = \frac{22}{55}$

   c. $\frac{6}{10} \cdot \frac{4}{11} \cdot \frac{3}{10} + \frac{4}{10} \cdot \frac{5}{11} \cdot \frac{4}{11} = \frac{152}{1100} = \frac{38}{275}$

23. a. P(at least one duplication of birthdays)

$= 1 - P$(no duplication of birthdays)

$$= 1 - \frac{28}{28} \cdot \frac{27}{28} \cdot \frac{26}{28} = 1 - \frac{{}_{28}P_3}{28^3}$$

$= 1 - 0.895$

$= 0.105$

c. $P(\text{dup}) = 1 - \frac{28}{28} \cdot \frac{27}{28} \cdot \frac{26}{28} \cdot \frac{25}{28} \cdot \frac{24}{28} = 1 - \frac{{}_{28}P_5}{28^5}$

$= 1 - 0.685 = 0.315$

e. $P(\text{dup}) = 1 - \frac{{}_{28}P_{10}}{28^{10}}$

g. $P(\text{dup}) = 1 - \frac{{}_{28}P_3}{28^3} = 1 - 0.895 = 0.105$

i. $P(\text{dup}) = 1 - \frac{{}_{28}P_5}{28^5} = 1 - 0.685 = 0.315$

k. $P(\text{dup}) = 1 - \frac{{}_{28}P_{14}}{28^{14}}$

m. $P(\text{dup}) = 1 - \frac{{}_{28}P_1}{28^1} = 1 - 1 = 0$

o. $P(\text{dup}) = 1 - \frac{{}_{c}P_n}{c^n}$

Problem Set 5–4, Pages 197-200

1. a. $P(5, \max 1, 0.01) = 0.999$ c. $P(5, \max 1, 0.05) = 0.977$
   e. $P(5, \max 1, 0.20) = 0.737$ g. $P(5, \max 1, 0.40) = 0.337$

3. a. $P(20, \max 1, 0.01) = 0.983$ c. $P(20, \max 1, 0.05) = 0.736$
   e. $P(20, \max 1, 0.20) = 0.069$ g. $P(20, \max 1, 0.40) = 0.001$

5.

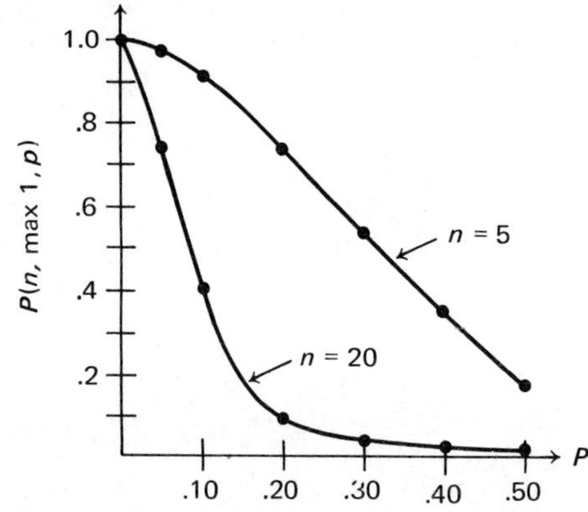

7. a.

n	$P(n, \max 1, 0.10)$	$P(n, \max 2, 0.20)$
5	0.919	0.942
10	0.736	0.678
20	0.392	0.206
30	0.184	0.044

b. As the sample size increases with the other two independent variables held constant, the probability that the sample contains a maximum specified number of defectives decreases.

9. The consecutive values of P(family with b boys and g girls) are listed as follows: 0.1250, 0.3750, 0.3750, 0.1250, 0.0625, 0.2500, 0.3750, 0.2500, 0.0625, 0.5000, 0.0000, 0.3438, 0.3125

11. a. $P(10, 7, 0.6) = P(10, 3, 0.4) = 0.2150$
    c. $P(10, 7, 0.6) + P(10, 8, 0.6) + P(10, 9, 0.6) + P(10, 10, 0.6)$
    $= P(10, 3, 0.4) + P(10, 2, 0.4) + P(10, 1, 0.4) + P(10, 0, 0.4)$
    $= 0.2150 + 0.1209 + 0.0403 + 0.0060$
    $= 0.3822$

13. The probability that Mark will be in the group of six students called is one minus the probability he is not in that group.

P(not in group) = P(not first chosen) · P(not second chosen) · P(not third chosen) · P(not fourth chosen) · P(not fifth chosen) · P(not sixth chosen)

$$= \frac{29}{30} \cdot \frac{28}{29} \cdot \frac{27}{28} \cdot \frac{26}{27} \cdot \frac{25}{26} \cdot \frac{24}{25} = \frac{24}{30} = \frac{4}{5} = 0.8$$

a. 0.8000 c. $P(3, 3, 0.8) = 0.5120$
e. $P(5, 5, 0.8) = 0.3277$

15. a. P(selection of twelve contains a majority believing the defendant guilty)
    = P(selection contains seven believing the defendant guilty) +
    P(selection contains eight believing the defendant guilty) +
    P(selection contains nine believing the defendant guilty) +
    P(selection contains ten believing the defendant guilty) +
    P(selection contains eleven believing the defendant guilty) +
    P(selection contains twelve believing the defendant guilty)
    $= P(12, 7, 0.20) + P(12, 8, 0.20) + P(12, 9, 0.20) + P(12, 10, 0.20) + P(12, 11, 0.20) + P(12, 12, 0.20)$
    $= 0.0033 + 0.0005 + 0.0001 + 0.0000 + 0.0000 + 0.0000$
    $= 0.0036$

Problem Set 5–5, Pages 208-213

1. a. $P(\text{conviction}, n = 12) = \frac{1}{2} \cdot P(12, 6, 0.70) + P(12, 7, 0.70) +$
$P(12, 8, 0.70) + P(12, 9, 0.70) + P(12, 10, 0.70) +$
$P(12, 11, 0.70) + P(12, 12, 0.70)$
$= \frac{1}{2} \cdot P(12, 6, 0.30) + P(12, 5, 0.30) + P(12, 4, 0.30) +$
$P(12, 3, 0.30) + P(12, 2, 0.30) + P(12, 1, 0.30) + P(12, 0, 0.30)$
$= \frac{1}{2} \cdot (0.0792) + 0.1585 + 0.2311 + 0.2397 + 0.1678 + 0.0712 + 0.0138$
$= 0.9217$

3.

Table	Size	f_t	Theoretical number	Actual number
1	6	0.25	1,035	1,059
2	12	0.25	343	324
3	6	0.50	5,000	5,053
4	12	0.50	5,000	4,986
5	6	0.75	8,965	8,948
6	12	0.75	9,657	9,680

Problem Set 5–6, Pages 217-219

1. The consecutive probabilities are 1/4, 1/8, 3/4, 2/11, 3/5, and 5/9.

3.

Odds in favor	Odds on E	$P(E \text{ occurs})$
5 to 1	1 to 5	5/6
8 to 1	1 to 8	8/9
5 to 2	2 to 5	5/7
7 to 3	3 to 7	7/10
1 to 5	5 to 1	1/6
1 to 3	3 to 1	1/4

5. $2/52 \cdot (1.30) + 6/52 \cdot (2.00) = \$0.28$. The maximum entrance fee should be \$0.27 for any expectation of a long-run profit.

7. $(0.2) \cdot 10 + (0.3) \cdot 20 + (0.5) \cdot 30 = 23$ cents

8. We look at this problem from the point of view of Mr. Smith's beneficiaries. Their expectation is their probability of inheriting the policy's benefits times the value of the benefits, or 0.008264 (which is Mr. Smith's probability of dying) $\times$ 10,000 = \$82.64. \$82.64 is the break-even point from the insurance company's point of view.

9. a. 1/4 (240,000) = \$60,000
   c. How much of a gambler are you? You have a 75% chance of losing your life savings on the one play of this game.

11. No. Of the ${}_{10}C_2 = 45$ different ways of choosing two numbers to add, only 20 of these choices result in an even sum so that $P(\text{even sum}) = 20/45 \neq P(\text{odd sum}) = 25/45$. The reason for this inequality lies in the fact that replacement of the cards is not allowed; it is therefore impossible to draw two equal odd numbers or two equal even numbers, both of which have a sum which is an even number.

CHAPTER 6

Problem Set 6–1, Pages 224–225

1. a. 8 19 21 22 22 23 25 26 28 28 30 31 32 33 33 34 35 35 35 36 37 37 37 38 38 40 41 41 42 44 44 45 46 47 48 51 53 54 56 58

b.

Interval	Midpoint, X	Frequency, f
18 up to 25	21.5	6
25 up to 32	28.5	6
32 up to 39	35.5	13
39 up to 46	42.5	7
46 up to 53	49.5	4
53 up to 60	56.5	4

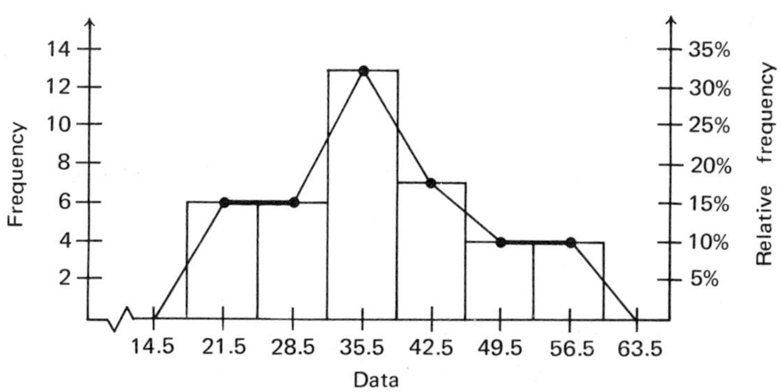

Problem Set 6–2, Pages 235-237

1. $\bar{x} = 39/6 = 6.5$; the median is 7; the mode is 7

3. $\bar{x} = 25/4 = 6.25$; $\sum x^2 = 183$; $\sigma = \sqrt{183/4 - 39.0625} = 2.59$

5. The mean of set U is $38/5 = 7.6$ and the mean of set U' is $76/5 = 15.2$. The mean of U' is two times the mean of U.

7. a. $\sigma = \sqrt{1026} = 32.03$

9.

Interval	X	f	$f \cdot X$	X^2	$f \cdot X^2$
18 up to 25	21.5	6	129	462.25	2773.5
25 up to 32	28.5	6	171	812.25	4873.5
32 up to 39	35.5	13	461.5	1260.25	16383.25
39 up to 46	42.5	7	297.5	1806.25	12643.75
46 up to 53	49.5	4	198	2450.25	9801
53 up to 60	56.5	4	226	3192.25	12769
		40	1483		59244

$\bar{x} = 1483/40 = 37.1$; $\sigma = \sqrt{59244/40 - (37.1)^2} = \sqrt{104.69} = 10.23$

11.

Interval	X	f	$f \cdot X$	X^2	$f \cdot X^2$
18 up to 24	21	6	126	441	2646
24 up to 30	27	4	108	729	2916
30 up to 36	33	9	297	1089	9801
36 up to 42	39	9	351	1521	13689
42 up to 48	45	6	270	2025	12150
48 up to 54	51	3	153	2601	7803
54 up to 60	57	3	171	3249	9747
		40	1476		58752

$\bar{x} = 1476/40 = 36.9$; $\sigma = \sqrt{58752/40 - (36.9)^2} = \sqrt{107.19} = 10.35$

13. a. 87 b. 107; No

15. $\sum f = 120$, $\sum f \cdot X = 324.6$, $\sum f \cdot X^2 = 889.36$, $\bar{x} = 2.705$,
$\sigma = \sqrt{7.4113 - 7.3170} = \sqrt{0.0943} = 0.307$
The estimated median is $2.60 + 21.5 \cdot 0.20/40 = 2.7075$

Problem Set 6–3, Pages 245-248

1. a. 0.1331 c. 0.4049 e. 0.9912
   g. 0.1271 i. 0.2643 k. 0.3478

2. a. $0.7967 \cdot 500 = 398$ c. $0.2749 \cdot 500 = 137$ e. $0.9525 \cdot 500 = 476$
   f. $0.0544 \cdot 500 = 27$

4. a. If the numbers were normally distributed, we would expect to find 95.44% of them, or 24, within two standard deviation units of the mean. In the actual set of numbers, $\bar{x} - 2\sigma = 61.64 - 2\,(8.53) = 44.58$ and $\bar{x} + 2\sigma = 61.64 + 2\,(8.53) = 78.70$. All 25 numbers actually are between 44.58 and 78.70.

   c. We expect to find 15.87% of the 25 numbers, or four, located above one standard deviation unit above the mean. In the actual set of numbers, there are four numbers above $\bar{x} + 1\sigma = 70.17$.

7. 42.07%

11. The fullback is the better player to use since he has exceeded two yards on 95.25% of his runs while the halfback has exceeded two yards on only 89.44% of his runs.

Problem Set 6–4, Page 254

1. $y' = 0.096 \cdot 70 - 4.535 = 2.185$

3. Parts (a), (b), and (c) are all contained in the following table. The least squares linear prediction equation is $y' = 1 \cdot x + 3.4$.

x	y	$y' = 1 \cdot x + 3.4$	$y - y'$	$(y - y')^2$
4	6	7.4	1.4	1.96
5	7	8.4	1.4	1.96
7	16	10.4	5.6	31.36
8	10	11.4	1.4	1.96
11	13	14.4	1.4	1.96
				39.2

5. Parts (a), (b), and (c) are all contained in the following table. The least squares linear prediction equation is $y' = 0 \cdot x + 2$.

x	y	$y' = 0 \cdot x+2$	$y-y'$	$(y-y')^2$
−2	4	2	2	4
−1	1	2	−1	1
0	0	2	2	4
1	1	2	1	1
2	4	2	2	4
				14

Problem Set 6–5, Pages 260-261

1. a. Positive c. Negative
   e. How much do you believe in astrology?

3. +0.49

5. a. +0.93 b. $y' = 1.5 \cdot x+5$

7. a. 0.00 b. $y' = 0 \cdot x+6$

9. According to the short cut formula, $r = 0.904$. $r^2 = (0.904)^2 = 0.817216$, and 0.817216 rounded to three decimal places is 0.817.

Problem Set 6–6, Pages 269-275

1. a. We reject the null hypothesis at the 2% level of significance (the probability that the sample obtained has a χ^2 value of 9.04 or greater is less than 2%), but must accept it at the 1% level.
   c. We reject the null hypothesis at the 0.5% level of significance.
   e. We reject the null hypothesis at the 2% level of significance, but must accept it at the 1% level.

2. a.

Outcome	$(f_o - f_h)^2/f_h$
A	0.90
B	0.90
C	0.08
D	0.40
E	0.10

$df = 4$; $\chi^2 = 2.38$. We accept the null hypothesis at the 5% significance level since $2.38 < 9.49$.

c. Outcome	$(f_o - f_h)^2/f_h$
Q	2.00
R	4.50
S	0.125
T	2.00
U	3.125

$df = 4$; $\chi^2 = 11.75$. We reject the null hypothesis at the 5% significance level since $11.75 > 9.49$. We can even reject the null hypothesis at the 2% level since $11.75 > 11.7$, but must accept it at the 1% level.

3. If we hypothesize that the coin is balanced, the resulting value of χ^2 is 5.12. We therefore reject this hypothesis at the 5% level of significance since $5.12 > 3.84$ with one degree of freedom.

5. We hypothesize that there is no significant difference in performance among the various five brands. Since $df = 5 - 1 = 4$, we will reject this hypothesis at the 5% level if the value of χ^2 from the calculations is greater than 9.49.

Brand	f_o	f_h	$(f_o - f_h)^2/f_h$
H	13	10	0.90
K	13	10	0.90
N	5	10	2.50
T	4	10	3.60
Y	15	10	2.50
			10.40

Since $\chi^2 = 10.40$ is greater than 9.49, we reject the null hypothesis at the 5% level of significance. It is therefore very likely that Japanese motorcycles are better than British motor cycles.

7. The hypothetical frequencies are based on the numbers in Table XI. For example, $P(12, 3, 0.25)$ is 0.2581, or 25.81%. This means that 25.81% of all the juries selected from a pool, of which 25% would believe the prosecution at the trial, should contain three jurors who will be conviction-prone. Since we are choosing 10,000 juries, and we hypothesize that our selections do come from this pool, we expect 25.81% of 10,000, or 2,581 juries, to contain three conviction-prone jurors. According to the second table of problem 3 of Problem Set 5–5, 2,561 juries actually contain three conviction-prone jurors. The remaining 11 pairs of values of f_h and f_o are determined in the same way.

Number of conviction-prone jurors	f_o	f_h	$(f_o - f_h)^2/f_h$
0	301	317	0.81
1	1309	1267	1.39
2	2284	2323	0.65
3	2561	2581	0.15
4	1985	1936	1.24
5	1044	1032	0.14
6	385	401	0.64
7	106	115	0.70
8	23	24	0.04
9	2	4	1.00
above 9	0	0	0.00

$\chi^2 = 6.76$, $df = 12$, and the probability is a little under 90%.

10. This problem is analyzed in the same way as was problem 7 above.

Number of conviction-prone jurors	f_o	f_h	$(f_o - f_h)^2/f_h$
0	3	2	0.50
1	51	44	1.11
2	358	330	2.38
3	1281	1318	1.04
4	2956	2966	0.03
5	3575	3560	0.06
6	1776	1780	0.01

$\chi^2 = 5.13$, $df = 6$, and the probability is a little over 50%

12.

Month	f_o	f_h	$(f_o - f_h)^2/f_h$
January	12	15.5	0.79
February	12	14.5	0.43
March	10	15.5	1.95
April	11	15.0	1.07
May	14	15.5	0.15
June	14	15.0	0.07
July	14	15.5	0.15
August	19	15.5	0.79
September	17	15.0	0.27
October	13	15.5	0.40
November	21	15.0	2.40
December	26	15.5	7.11

Since our answer of $\chi^2 = 15.58$ with $df = 11$ is under the critical value of 19.7, we accept the hypothesis of equal distribution of birthdays among the months for the first 183 birthdays selected. That is, the probability that a sample such as ours has a χ^2 value of 15.58 or larger is under 0.25. Of course, our result is of small comfort to 26 out of each group of 31 men born in December, 1951!

13. a.

Months from birth month	f_o	f_h	$(f_o-f_h)^2/f_h$
−6	24	29	0.86
−5	31	29	0.14
−4	20	29	2.79
−3	23	29	1.24
−2	34	29	0.86
−1	16	29	5.83
0	26	29	0.31
1	36	29	1.69
2	37	29	2.21
3	41	29	4.97
4	26	29	0.31
5	34	29	0.86

$\chi^2 = 22.07$, $df = 11$

b. The probability is between 2% and 5% and is very close to 2%.

15. The value of χ^2 is 4.93 with $df = 3$, so a sample with as much as or slightly more variation than his occurs between 10 and 25 percent of the time. He does not have much reason to doubt the labeling.

16. a. $\chi^2 = 3.20$ with one degree of freedom. The probability is between 5% and 10%.

b. $P(20, 14, 0.50) + P(20, 15, 0.50) + P(20, 16, 0.50) + P(20, 17, 0.50) + P(20, 18, 0.50) + P(20, 19, 0.50) + P(20, 20, 0.50) = 0.0370 + 0.0148 + 0.0046 + 0.0011 + 0.0002 + 0.0000 + 0.0000 = 0.0577$

17. a. $164 + 111 + 27 + 29 + 14 + 5 + 1 = 351$

c. $10{,}000 \cdot 50 = 500{,}000$

Problem Set 6–7, Pages 279-284

A computer was used to calculate the values of χ^2 for each of the answers given below. Hypothetical frequencies were rounded to the nearest 0.1, and each value of $(f_o-f_h)^2/f_h$ was calculated to the nearest 0.01.

1. $df = 2$, $\chi^2 = 6.86$

3. $df = 2$, $\chi^2 = 0.91$

5. $df = 6$, $\chi^2 = 22.79$. We might eliminate the third row from our observed frequencies because of its small values of 3 and 2 which unduly affect the value of χ^2. The corresponding values of $(f_o - f_h)^2/f_h$ are 4.90 and 4.50, which account for 9.40/22.79 = 41% of the total value of χ^2.

7. $df = 1$, $\chi^2 = 12.37$

9. $df = 3$, $\chi^2 = 0.46$

11. $df = 6$, $\chi^2 = 8.41$

13. $df = 4$, $\chi^2 = 73.87$

15. a. $df = 2$, $\chi^2 = 8.38$ c. $df = 4$, $\chi^2 = 15.48$
e. $df = 10$, $\chi^2 = 12.27$

Chapter 7

Problem Set 7–1, Pages 291-292

1. $x = 2$, $y = 2$

3. $x = -6.625$, $y = 5.5$

5. Dependent; Any pair of numbers of the form $(x, (-x+7)/2)$ will satisfy both equations.

7. Inconsistent; No pair of numbers will satisfy both equations.

9. $x = -75/111$, $y = -225/111$

11. Mary is 16 years old and John is 20 years old.

13. 67

15. a. 26 b. $\neq 26$

17. The graphs result in one line whose slope is $-2/5$ and whose y-intercept is $-13/5$.

Problem Set 7–2, Pages 297-299

1. a. $x = 9$, $y = -4$, $z = 2$
c. $x = 7$, $y = 65/3$, $z = -55/3$

e. $x = 0.797385$, $y = 1.12745$, $z = 12.0359$
g. $x = 2$, $y = -4$, $z = 8$

2. a. Inconsistent
   c. $x = 1, y = 2, z = 3, w = 4$
   e. $x = 0, y = 0, z = 0, w = 1$

3. a. $\left[\begin{array}{ccccc|c} 1 & 0 & 0 & 0 & -1 & 1 \\ 0 & 1 & 1 & 1 & 0 & 6 \\ 0 & 0 & 3 & -1 & 1 & 7 \\ 0 & 1 & 0 & 0 & 1 & 3 \\ 1 & 0 & 1 & 1 & -1 & -6 \end{array}\right]$

5. 96 cases of Brand A, 80 cases of Brand B, 48 cases of Brand C

7. No, since fractional numbers of nickels, dimes, and quarters are necessary in order to solve the appropriate system of equations.

9. If we let S, E, N, and W represent the respective enrollments of the four schools, then the four equations arising from the total number of students and the numbers of A's, B's, and C's given are, respectively,

$$S + E + N + W = 1000$$

$$0.1S + 0.15E + 0.25N + 0.15W = 150$$

$$0.25S + 0.25E + 0.15N + 0.20W = 250$$

$$0.5S + 0.55E + 0.35N + 0.4W = 500.$$

Attempted solution of these equations leads to the division of a non-zero number by zero, so the original information is inconsistent and incorrect.

11. $\left[\begin{array}{ccc|c} 1 & 0 & 0 & 2 \\ 0 & 1 & 0 & -4 \\ 0 & 0 & 1 & 8 \end{array}\right]$

$x = 2, y = -4, z = 8$

13. One approach to the problem led to the augmented matrix

$$\left[\begin{array}{ccc|c} 0 & 0 & 0 & -6 \\ -3 & -1 & 0 & -10 \\ 1 & 7 & -4 & 2 \end{array}\right]$$

It leads to the equation $0 \cdot x = -6$ which has no solution, so the original system has no solution and is inconsistent.

Problem Set 7–3, Pages 306-308

1. a. 26
   c. 14
   e. 30
   g. 0
   i. The determinant of a matrix with two identical rows or two identical columns is 0.
   k. This matrix does not have a determinant.

2. a. $x = 7/5, y = -2/5$ c. $x = 12/48 = 1/4, y = -52/48 = -13/12$
   e. The system is inconsistent.

4. a. $x = 15/15 = 1, y = 30/15 = 2, z = 45/15 = 3$
   c. Since $D = 0$ and $D_x = -2$, the system is inconsistent.
   d. Since $D = 0$ and $D_x = 0$, the system is dependent.
   e. $x = -2/3, y = 4/3, z = -7/3$

5. a. $x = 4$

8.
```
10 DATA 3946.1, 74.298, -19.774, 6003.8, -123.45, 98.007
20 READ A, B, K1, C, D, K2
30 LET X = (K1*D-K2*B)/(A*D-C*B)
40 LET Y = (A*K2-C*K1)/(A*D-C*B)
50 PRINT X, Y
60 END
```

9. If we rewrite the first two equations and divide through the second one by 2, we have

$$2x + y = -1$$

$$2x + y = 1.5.$$

There is no pair of numbers (x, y) such that $2 \cdot x + y$ can both equal -1 and 1.5, so the lines which are the graphs of these two equations are parallel (never meet) and the three lines do not meet in one point.

11. 18 men, 6 women, and 22 children

13. $x = 10, y = 6, z = 4$

Problem Set 7–4, Pages 313-317

1. a. 2 c. -5 e. 3
   g. does not exist i. -3 k. 1

3. a. 7 c. -44 e. 82
   g. -16

5. a. $x = 2, y = -1, z = 3$
   c. $u = 3, v = -2, w = -1$

6. a. -5 c. 132

7. a. $x = -2, y = 1, z = -1, t = 3$
   c. This system is dependent.

8. a. $a = 1, b = -3, c = -10$
   c. $a = 1, b = -2, c = -6$
   e. $a = 0, b = -5, c = 5$. Notice that the graph is a line rather than a parabola.

9. $t = f(h) = -0.00003h^2 + 0.007h + 212$; 759

12. a. Three determinants of 2×2 matrices
    b. Four determinants of 3×3 matrices
    c. Five determinants of 4×4 matrices

CHAPTER 8

Problem Set 8–1, Pages 325-326

1.

Part	x intercept	y intercept
a	(1, 0)	(0, −1/3)
c	(2, 0)	(0, 2)
e	(4, 0)	None
g.	(6, 0)	(0, −4)
i.	(14/3, 0)	(0, 4/7)

3. a. True c. False

Problem	Equation	Solid (S) or dotted (D) line	Two points on the line	(0, 0) included?
5	$-2x+3y \geqslant -6$	S	(0, −2), (3, 0)	Yes
7	$x \geqslant 3.6$	S	(3.6, 0), (3.6, 2)	No
9	$y > 2x+1$	D	(0, 1), (−1/2, 0)	No
11	$x-2y \geqslant -4$	S	(0, 2), (−4, 0)	Yes
11	$x+y \geqslant 2$	S	(0, 2), (2, 0)	No
13	$3x+y \geqslant 5$	S	(0, 5), (5/3, 0)	No
13	$2x-3y \leqslant 7$	S	(0, −7/3), (7/2, 0)	Yes
15	$x-3y \geqslant 6$	S	(0, −2), (6, 0)	No
15	$2x-6y \leqslant 8$	S	(0, −4/3), (4, 0)	Yes
17	$x-3y \leqslant 6$	S	(0, −2), (6, 0)	Yes
17	$1.5x+y > 6$	D	(0, 6), (4, 0)	No

19. $x > 1$, $y > 3$, $x < 5$, $y < 7$

21.

Equation	Two points on the line	(0, 0) included?
$x+2y \leqslant 20$	(0, 10) and (20, 0)	Yes
$5x+3y \leqslant 44$	(0, 44/3) and (44/5, 0)	Yes
$3x-2y \leqslant 15$	(0, −15/2), (5, 0), and (10, 15/2)	Yes
$4x+5y \geqslant 40$	(0, 8) and (10, 0)	No
$4x+5y \leqslant 50$	(0, 10) and (12.5, 0)	Yes

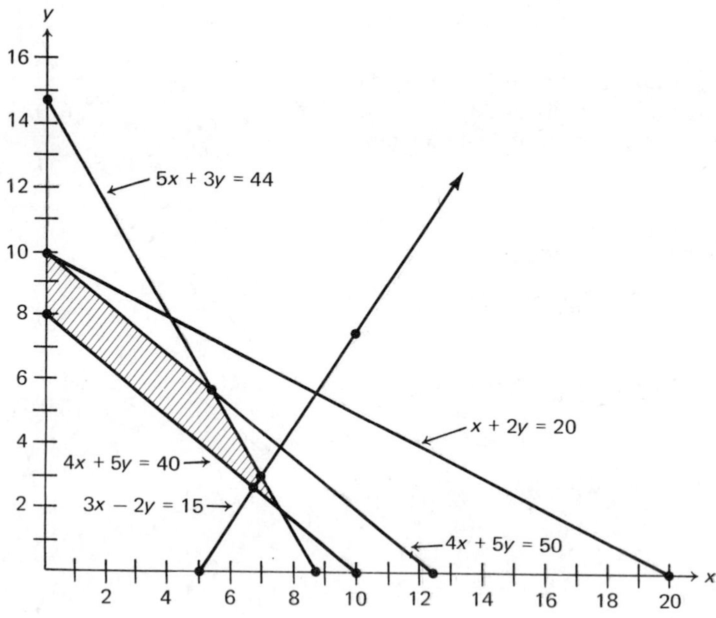

Problem Set 8–2, Pages 329-332

The optimal values of the objective functions are indicated by an asterisk.

1. b.

Corner point	$5x+7y$
(0, 0)	0*
(0, 7)	49*
(1, 5)	40
(8/3, 0)	13 1/3

d.

Corner point	$5x+7y$
(0, 2)	14
(0, 6)	42*
(3, 3)	36
(4, 0)	20
(2, 0)	10*

3. b.

Corner point	$5x-7y$
(0, 0)	0
(0, 7)	−49*
(1, 5)	−30
(8/3, 0)	13 1/3*

d.

Corner point	$5x-7y$
(0, 2)	−14
(0, 6)	−42*
(3, 3)	−6
(4, 0)	20*
(2, 0)	10

5.

Corner point	$4x+5y+2$
(0, 5)	27
(2, 3)	25*
(7, 0)	30

7.

Corner point	$5x+4y$
(0, 5)	20*
(2, 3)	22
(7, 0)	35

9.

Corner point	$25x+35y$
(0, 2)	70
(0, 5)	175*
(7/2, 0)	87.5
(2, 0)	50*

11.

Corner point	$2x-9y$
(0, 0)	0*
(0, 3/2)	−27/2
(7, 23/6)	−41/2*
(7, 7/3)	−7

Problem Set 8–3, Pages 341-344

1. The production of two high backs and three rockers results in a maximum profit of $64.00.

3. The production of 800 barrels of chemical A and 200 barrels of chemical B results in a maximum profit of $1800.00.

5. The use of 2.4 pounds of ingredient A along with 0.8 pounds of ingredient B in one bag of concrete minimizes the cost at 22.4¢.

7. The production of three chairs and two tables results in a maximum profit of $35.00.

9. The production of no chairs and three tables results in a maximum profit of $60.00.

12.

Department	Unused time (minutes) if $x = 1{,}000$ and $y = 300$
Fabricating	$27{,}000-(15 \cdot 1000+40 \cdot 300) = 0$
Assembly	$27{,}000-(12 \cdot 1000+50 \cdot 300) = 0$
Upholstery	$27{,}000-(18.75 \cdot 1000+0 \cdot 300) = 8250$
Linoleum	$27{,}000-(0 \cdot x+56.25 \cdot 300) = 10{,}125$

13. The manufacturer should produce 1440 chairs and 135 desks for a maximum profit of \$41,400.

Department	Unused time (minutes) if $x = 1{,}440$ and $y = 135$
Fabricating	$27{,}000-(15 \cdot 1440+40 \cdot 135) = 0$
Assembly	$27{,}000-(12 \cdot 1440+50 \cdot 135) = 2970$
Upholstery	$27{,}000-(18.75 \cdot 1440+0 \cdot 135) = 0$
Linoleum	$27{,}000-(0 \cdot 1440+56.25 \cdot 135) = 19{,}406.2$

15. The producer should schedule the comic for fifteen minutes, the orchestra for five minutes, and the commercials for ten minutes. The resulting minimum cost would be \$6,500.00

18. He should raise 400 capons, 200 geese, and 400 turkeys which will maximize his profit at \$1850.00

19. The president should choose 40 pounds of compound *R*, 10 pounds of compound *S*, and 50 pounds of compound *T* for the minimum cost of \$5.20. If we convert these amounts to percentages of the total of 100 pounds we have, respectively, 40% of compound *R*, 10% of compound *S*, and 50% of compound *T*. In order to produce 200 pounds of the mixture we would use 40% of 200, or 80 pounds of compound *R*; 10% of 200, or 20 pounds of compound *S*; and 50% of 200, or 100 pounds of compound *T*.

CHAPTER 9

Problem Set 9–1, Pages 349–351

1. a. $f(3) = 3^3 - 3^2 - 10(3) + 17$

$= 5$

c. $f(-1.4) = (-1.4)^3 - (-1.4)^2 - 10(-1.4) + 17$

$= 26.296$

e. $f(2.1) = (2.1)^3 - (2.1)^2 - 10(2.1) + 17$

$= .851$

3. $f(-5) = 25 - 30 + 13 = 8$

$f(-3) = 9 - 18 + 13 = 4$

$f(1) = 1 + 6 + 13 = 20$

Therefore, maximum value $= 20$; minimum value $= 4$

5. $f(1) = 1 - 5.2 + 8.26 = 4.06$

$f(2.6) = 6.76 - 13.52 + 8.26 = 1.5$

$f(4) = 16 - 20.8 + 8.26 = 3.46$

Therefore, maximum value $= 4.06$; minimum value $= 1.5$

7. $f(-5) = 25 - 34 + 6.56 = -2.44$

$f(-3.4) = 11.56 - 23.12 + 6.56 = -5$

$f(-1) = 1 - 6.8 + 6.56 = 0.76$

Therefore, maximum value $= 0.76$; minimum value $= -5$

9. $f(-4.1) = (-4.1)^3 - 12(-4.1) = -19.71$

$f(-2) = (-2)^3 - 12(-2) = -8 + 24 = 16$

$f(2) = 2^3 - 12(2) = 8 - 24 = -16$

$f(4.1) = (4.1)^3 - 12(4.1) = 19.71$

Therefore, maximum value $= 19.71$; minimum value $= -19.71$

11. $f(0) = -14$

$f(2) = 8 - 36 + 48 - 14 = 6$

$f(4) = 64 - 144 + 96 - 14 = 2$

$f(6) = 216 - 324 + 144 - 14 = 22$

Therefore, maximum value $= 22$; minimum value $= -14$

13.
```
10 FOR T = −2 to −1 STEP 0.1
20 LET S = T↑3 − T↑2 − 10*T + 17
30 PRINT T, S
40 NEXT T
50 END
```

15.
```
10 FOR X = −1.6 to −1.4 STEP 0.02
20 LET Y = X↑3 − X↑2 − 10*X + 17
30 PRINT X, Y
40 NEXT X
50 END
```
No difference with output from problem 14

Problem Set 9–2, Pages 357-359

1. $4 \cdot 1 + 7 = 11$

3. $-4/2 + 5 = 3$

5. 8

7. 8

9. $2+3=5$

11. $\lim\limits_{x\to 1/2} \dfrac{2x^2+5x-3}{4x-2} = \lim\limits_{x\to 1/2} \dfrac{(2x-1)(x+3)}{2(2x-1)} = \lim\limits_{x\to 1/2} \dfrac{x+3}{2} = \dfrac{1/2+3}{2} = \dfrac{7}{4}$

13. $3^2=9$

15. $5\cdot 3^2+7\cdot 3 = 45+21 = 66$

17. $2^2+2\cdot 2^3 = 4+16 = 20$

19. $\lim\limits_{x\to -5/3} \dfrac{3x^2-x-10}{3x+5} = \lim\limits_{x\to -5/3} \dfrac{(3x+5)(x-2)}{3x+5} = \lim\limits_{x\to -5/3} x-2 = \dfrac{-5}{3}-2 = -\dfrac{11}{3}$

21. $\lim\limits_{x\to -5/3} 10/(x-3)$ does not exist

x	2.7	2.8	2.9	2.95	2.99	3.01	3.05	3.1	3.2	3.3
$10/(x-3)$	−33.3	−50	−100	−200	−1000	1000	200	100	50	33.3

23.

a.

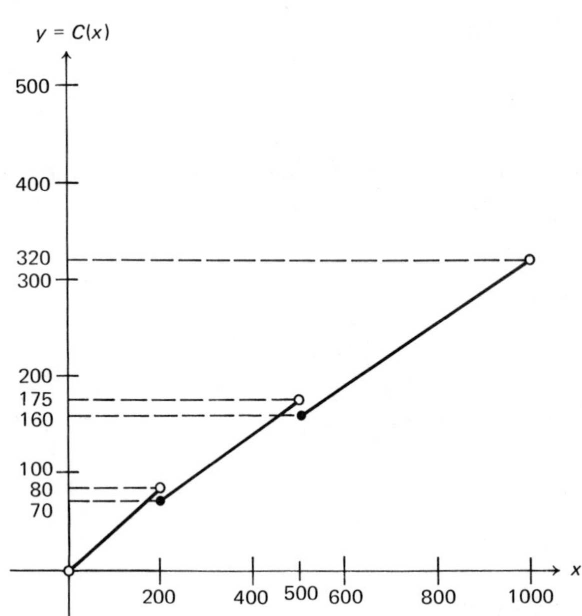

b. $\lim\limits_{x\to 200^+} C(x) = \lim\limits_{x\to 200^+} 0.35x = 70$

c. $\lim\limits_{x\to 200^-} C(x) = \lim\limits_{x\to 200^-} 0.40x = 80$

25. a. $\lim\limits_{x\to 3^-} \dfrac{x-3}{x^2-3x} = \lim\limits_{x\to 3^-} \dfrac{1}{x} = \dfrac{1}{3}$

b. $\lim\limits_{x\to 3^+} \dfrac{x-3}{x^2-3x} = \lim\limits_{x\to 3^+} \dfrac{1}{x} = \dfrac{1}{3}$

c. Since $\lim\limits_{x\to 3^-} f(x) = \lim\limits_{x\to 3^+} f(x) = \dfrac{1}{3}$, $\lim\limits_{x\to 3} f(x) = \dfrac{1}{3}$

26. $\lim\limits_{h\to 0} \dfrac{f(x+h)-f(x)}{h} = \lim\limits_{h\to 0} \dfrac{[3(x+h)^2-5(x+h)]-[3x^2-5x]}{h}$

$= \lim\limits_{h\to 0} \dfrac{3x^2+6xh+3h^2-5x-5h-3x^2+5x}{h}$

$= \lim\limits_{h\to 0} \dfrac{6xh+3h^2-5h}{h} = \lim\limits_{h\to 0} 6x+3h-5$

$= 6x-5$

27. $\lim\limits_{x\to 3} \dfrac{x+3}{x^2-9} = \lim\limits_{x\to 3} \dfrac{1}{x-3}$, which does not exist

Problem Set 9–3, Pages 369–371

1. a. $\dfrac{f(2)-f(1)}{2-1} = \dfrac{12-9}{2-1} = 3$

c. $\dfrac{f(4)-f(3)}{4-3} = \dfrac{12-13}{1} = -1$

e. $\dfrac{f(4)-f(2.5)}{4-2.5} = \dfrac{12-12.75}{1.5} = -0.5$

3.

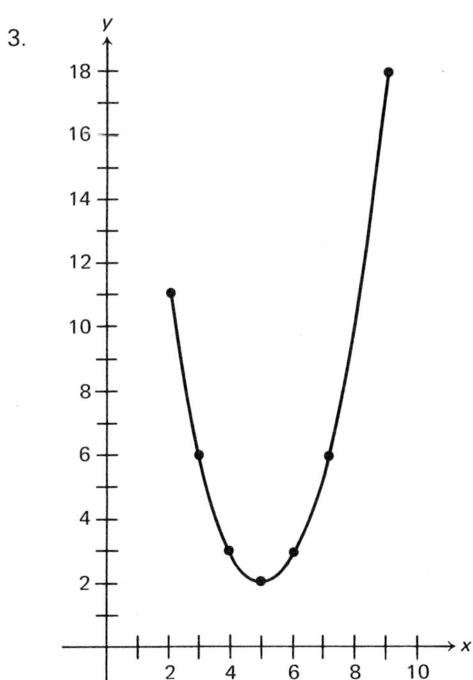

a. $\dfrac{18-3}{9-6} = \dfrac{15}{3} = 5$

c. $\dfrac{8.25-3}{7.5-6} = \dfrac{5.25}{1.5} = 3.5$

e. $\dfrac{4.25-3}{6.5-6} = \dfrac{1.25}{0.5} = 2.5$

g. $\dfrac{3.21-3}{6.1-6} = \dfrac{0.21}{0.1} = 2.1$

i. $\dfrac{3.0201-3}{6.01-6} = \dfrac{0.0201}{0.01} = 2.01$

5. $y' = 2x-10$
   a. At $x = 7$, $y' = 2 \cdot 7-10 = 4$
   c. At $x = 5$, $y' = 2 \cdot 5-10 = 0$
   e. At $x = 3$, $y' = 2 \cdot 3-10 = -4$
   g. At $x = 6.468$, $y' = 2 \cdot 6.468-10 = 2.936$

7. $y' = 10x$
   a. $10 \cdot 2 = 20$
   c. $10 \cdot 0 = 0$
   e. $10 \cdot 0.3 = 3$

9. $y' = 3 = 3 \cdot x^0$
   a. $3 \cdot 1^0 = 3$
   c. $3 \cdot (-2)^0 = 3$
   e. They are all the same line.

Problem Set 9–4, Pages 374-375

1. $s = f(t) = 4+15t-3t^2, \quad 0 \leqslant t \leqslant 10$

a. $$s' = \lim_{\Delta t \to 0} \frac{[4+13(t+\Delta t)-3(t+\Delta t)^2]-[4+15t-3t^2]}{\Delta t}$$

$$= \lim_{\Delta t \to 0} \frac{4+15t+15\Delta t-3t^2-6t\Delta t-3 \cdot (\Delta t)^2-4-15t+3t^2}{\Delta t}$$

$$= \lim_{\Delta t \to 0} \frac{15\Delta t-6t\Delta t-3 \cdot (\Delta t)^2}{\Delta t}$$

$$= \lim_{\Delta t \to 0} 15-6t-3\Delta t$$

$$= 15-6t$$

c. When $t = 1$, $S' = 9$ and $S = 16$. Therefore, the point is 16 feet to the right of the origin, the instantaneous velocity is 9 feet per second, and the point is moving from left to right.

e. When $t = 2.5$, $S' = 0$ and $S = 22.75$. Therefore, the point is 22.75 feet to the right of the origin, the instantaneous velocity is 0 feet per second, and the point is stopped.

g. When $t = 5$, $S' = -15$ and $S = 4$. Therefore, the point is 4 feet to the right of the origin, the instantaneous velocity is -15 feet per second, and the point is moving from right to left.

i. When $t = 9$, $S' = -39$ and $S = -104$. Therefore, the point is 104 feet to the left of the origin, the instantaneous velocity is -39 feet per second, and the point is moving from right to left.

Problem Set 9–5, Pages 383-385

1. $f'(x) = 3x^2$

3. $f'(x) = 0$

5. $g'(x) = 24x^2 - 10x$

7. $h'(x) = 3$

9. $$\begin{aligned} h'(u) &= 9u \cdot (60u^3 - 7) + (15u^4 - 7u) \cdot 9 \\ &= 540u^4 - 63u + 135u^4 - 63u \\ &= 675u^4 - 126u \end{aligned}$$

11. $$\begin{aligned} f'(x) &= (x^3 + 7x)(2) + (2x + 5)(3x^2 + 7) \\ &= 2x^3 + 14x + 6x^3 + 15x^2 + 14x + 35 \\ &= 8x^3 + 15x^2 + 28x + 35 \end{aligned}$$

13. $$f'(x) = \frac{x^2(3) - (3x - 5)(2x)}{x^4} = \frac{3x^2 - 6x^2 + 10x}{x^4} = \frac{-3x + 10}{x^3}$$

15. $$\begin{aligned} f'(t) &= \frac{(3t - 5t^3)(14t) - 7t^2(3 - 15t^2)}{(3t - 5t^3)^2} \\ &= \frac{42t^2 - 70t^4 - 21t^2 + 105t^4}{(3t - 5t^3)^2} \\ &= \frac{35t^4 + 21t^2}{(3t - 5t^3)^2} \\ &= \frac{35t^2 + 21}{(3 - 5t^2)^2} \end{aligned}$$

17. $$\begin{aligned} f'(x) &= -6x + 4 \\ f'(5) &= -30 + 4 = -26 \\ f'(1) &= -6 + 4 = -2 \\ f'(-2) &= 12 + 4 = 16 \end{aligned}$$

19. $g'(x) = \dfrac{-5}{(x+7)^2}$

$g'(4) = \dfrac{-5}{121}$

$g'(-8) = \dfrac{-5}{1} = -5$

$g'(-6) = \dfrac{-5}{1} = -5$

$g'(-7)$ does not exist

21. At $(5, -49)$ the equation is $y+49 = -26(x-5)$, or $y = -26x+81$
At $(1, 7)$ the equation is $y-7 = -2(x-1)$, or $y = -2x+9$
At $(-2, -14)$ the equation is $y-2 = 16(x+14)$, or $y = 16x+18$

25. $f'(x) = 2x+6$
$f'(-3) = 0$

Therefore, the slope of the tangent line at $(-3, 4)$ is 0, and the equation of this horizontal tangent line at $(-3, 4)$ is $y = 4$.

28. a. $f(x) = (3x+7)^3$; $f'(x) = 3(3x+7)^2 \cdot (3) = 9(3x+7)^2$
c. $h(x) = (14x-27)^7$; $h'(x) = 7(14x-27)^6 \cdot (14) = 98(14x-27)^6$
e. $f(x) = (18x^4-7x^2+9)^8$; $f'(x) = 8(18x^4-7x^2+9)^7 \cdot (72x^3-14x) = (18x^4-7x^2+9)^7(576x^3-112x)$

30. $S(t) = 4t^3-7t^2+3t-5$
$v(t) = S'(t) = 12t^2-14t+3$
a. When $t = 2$, $S = 5$ and $v = 23$. Therefore, the velocity is 23 feet per second, the point is 5 feet to the right of the origin, and the point is moving from left to right.
c. When $t = 3$, $S = 40$ and $v = 69$. Therefore, the velocity is 69 feet per second, the point is 40 feet to the right of the origin, and the point is moving from left to right.
e. When $t = 0$, $S = -5$ and $v = 3$. Therefore, the velocity is 3 feet per second, the point is 5 feet to the left of the origin, and the point is moving from left to right.

31. $f'(x) = 3x^2-6x+1$
a. $f'(-1.5) = 16.75$, and f is increasing at $(-1.5, f(-1.5))$.
c. $f'(0) = 1$, and f is increasing at $(0, f(0))$.
e. $f'(1.5) = -1.25$, and f is decreasing at $(1.5, f(1.5))$.

Problem Set 9–6, Pages 397-405

1. $f'(x) = 2x-4$; $f'(x) = 0$ at $x = 2$.
$f''(x) = 2, 2 > 0$. Therefore f has a relative minimum at $(2, 3)$.
$f(-3) = 9+12+7 = 19$
$f(5) = 25-20+7 = 12$

f has an absolute minimum at $(2, 3)$, and f has an absolute maximum at $(-3, 19)$.

3. $f'(x) = 2x-8$; $f'(x) = 0$ at $x = 4$.
   $f''(x) = 2,\ 2 > 0$.
   $f(-1) = 1+8+14 = 23$
   $f(7) = 49-56+14 = 7$
   $f(4) = 16-32+14 = -2$

Therefore, f has a relative and an absolute minimum at $(4, -2)$, and f has an absolute maximum at $(-1, 23)$.

5. $f'(x) = 2x-7$; $f'(x) = 0$ at $x = 7/2$
   $f''(x) = 2,\ 2 > 0$
   $f(7/2) = 49/4 - 49/2 + 3 = -37/4$
   $f(4) = 16-28+3 = 9$
   $f(7) = 49-49+3$

Therefore, f has a relative and absolute minimum at $(7/2, -37/4)$, and f has an absolute maximum at $(4, 9)$.

7. $f'(x) = 5-14x$; $f'(x) = 0$ at $x - 5/14$
   $f''(x) = -14;\ -14 < 0$
   $f(5/14) = 18 + 25/14 - 175/196 = 529/28$
   $f(-2) = 18-10-28 = -20$
   $f(2) = 18+10-28 = 0$

Therefore, f has a relative and absolute maximum at $(5/14, 529/28)$, and f has an absolute minimum at $(-2, -20)$.

9. $f'(x) = 36-10x$; $f'(x) = 0$ at $x = 3.6$
   $f''(x) = -10,\ -10 < 0$.
   $f(3.6) = 71.8$; $f(6) = 43$; $f(10) = -133$

Therefore, f has an absolute maximum at $(6, 43)$, f has an absolute minimum at $(10, -133)$, and f has no relative extrema in the interval.

11. $f'(x) = 3x^2-12$; $f'(x) = 0$ at $x = 2$ or $x = -2$
    $f''(x) = 6x$; $f''(x) > 0$ when $x > 0$; $f''(x) < 0$ when $x < 0$
    $f(-4.1) = -19.721$
    $f(4.1) = 19.721$
    $f(2) = -16$
    $f(-2) = 16$

Therefore, f has an absolute minimum at $(-4.1, -19.721)$, f has an absolute maximum at $(4.1, 19.721)$, f has a relative minimum at $(2, -16)$, and f has a relative maximum at $(-2, 16)$.

13. $f'(x) = 3x^2-18x+24$; $f'(x) = 0$ when $x^2-6x+8 = 0$, at $x = 4$ or $x = 2$.
    $f''(x) = 6x-18 = 6(x-3)$; $f''(x) > 0$ when $x > 3$, $f''(x) < 0$ when $x < 3$.
    $f(0) = -7$; $f(2) = 13$; $f(4) = 9$; $f(6) = 29$

Therefore, f has an absolute minimum at $(0, -7)$, f has an absolute maximum at $(6, 29)$, f has a relative minimum at $(4, 9)$, and f has a relative maximum at $(2, 13)$.

15. $f'(x) = 6x^2 - 6x - 12$; $f'(x) = 0$ when $x^2 - x - 2 = 0$, at $x = -1$ or $x = 2$.
$f''(x) = 12x - 6 = 6(2x - 1)$; $f''(x) > 0$ when $x > 1/2$, $f''(x) < 0$ when $x < 1/2$
$f(-2) = -9$, $f(-1) = 2$, $f(2) = -25$, $f(4) = 27$

Therefore, f has an absolute and relative minimum at $(2, -25)$, f has a relative maximum at $(-1, 2)$, and f has an absolute maximum at $(4, 27)$.

17. $f'(x) = 5$, so $f'(x) > 0$
$f(-2) = -18$, $f(6) = 22$

Therefore, f has an absolute minimum at $(-2, -18)$, f has an absolute maximum at $(6, 22)$, and f has no relative extrema.

19. $f'(x) = -(1 - 4x^2)/3x^2$; $f'(x) = 0$ when $1 - 4x^2 = 0$ at $x = -1/2$ or $x = 1/2$.
$f''(x) = 2/3x^3$; $f''(x) > 0$ when $x > 0$ and $f''(x) < 0$ when $x < 0$.
$f(-1) = -1/3$, $f(-1/2) = 0$, $f(1/2) = 8/3$, $f(1) = 3$

Therefore, f has an absolute minimum at $(-1, -1/3)$, f has an absolute maximum at $(1, 3)$, f has a relative minimum at $(1/2, 8/3)$, and f has a relative maximum at $(-1/2, 0)$.

21.

	l	
w | \$6/yd.	\$8/yd. (top), \$6/yd. (bottom)	\$6/yd. | w
	l	

$$C = 6(2w + l) + 8l$$
$$C = 12w + 14l$$
$$wl = 200$$
$$l = 200/w$$

Total cost $C = f(w) = 12w + 2800/w$, $w \in (0, +\infty)$.

$$f'(w) = 12 - \frac{2800}{w^2} = \frac{12w^2 - 2800}{w^2}; \quad f''(w) = \frac{5600}{w^3}$$

$$f'(w) = 0 \text{ at } w = +\frac{10\sqrt{21}}{3}, \text{ and } f''\left(\frac{10\sqrt{21}}{3}\right) > 0$$

Therefore, f has a minimum when $w = (10\sqrt{21})/3$ yds, and the minimum cost of $f((10\sqrt{21})/3) = 80\sqrt{21}$ occurs when $w = (10\sqrt{21})/3$ yds and $l = (20\sqrt{21})/3$ yards. The respective decimal values are 15.275, 30.451, and 366.606.

23. Let x = number of days wheat is stored. Then number of pounds sold is $1000 - 5x$ for each 1000 pounds stored and the price per thousand pounds is $25 + 0.8x$. Therefore, value $= f(x) = (25 + 0.8x)(1000 - 5x)$, $x \in [0, 200]$

$$f'(x) = (25 + 0.8x)(-5) + (1000 - 5x)(0.8)$$
$$= -125 - 4x + 800 - 4x$$
$$= 675 - 8x$$
$$f''(x) = -8, \ -8 < 0$$
$$f'(x) = 0 \text{ at } x = 84.375$$

Therefore, wheat should be kept $84\frac{3}{8}$ days which produces a maximum value of \$53.4766 per thousand pounds. If it is kept for 84 days, the value is \$53.476, and if it is kept for 85 days, the value is \$53.475. Keeping it for 84 days is the better choice.

24.

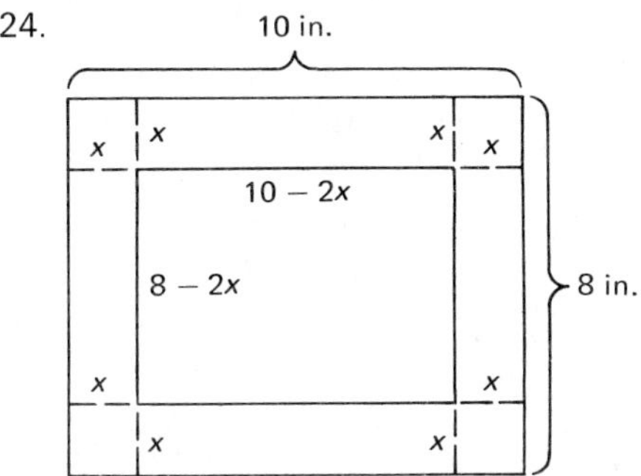

$$
\begin{aligned}
V = f(x) &= x(8-2x)(10-2x) \\
&= 4x^3-36x^2+80x,\ x\in[0,4] \\
f'(x) &= 12x^2-72x+80 \\
f''(x) &= 24x-72 = 24(x-3) \\
f'(x) &= 0 \text{ when } 12x^2-72x+80=0 \text{ at } x = 3\pm\sqrt{21}/3 \\
f''(3-\sqrt{21}/3) &< 0
\end{aligned}
$$

Therefore, f has a maximum value when $x = 3-\sqrt{21}/3$, and the maximum value is $(300-1240\sqrt{21})/9$. The respective decimal values are 1.472 and −598.044.

26.

House

w w

l

$$
\begin{aligned}
A &= w\cdot l \\
l+2w &= 50 \\
l &= 50-2w
\end{aligned}
$$

$$
\begin{aligned}
A = f(w) &= w(50-2w) = 50w-2w^2,\ w\in(0,25) \\
f'(w) &= 50-4w \\
f''(w) &= -4,\ -4<0 \\
f'(w) &= 0 \text{ at } w = 25/2
\end{aligned}
$$

Maximum area is obtained when the width is 25/2 feet and the length is 25 feet.

29. Let x − number of people in excess of 30 who go. Then the fare per person is $40-0.5x$ dollars. Therefore, $R = f(x) = (30+x)(40-0.5x) = 1200+25x-0.5x^2$, $x\in[0,80]$.

$$
\begin{aligned}
f'(x) &= 25-x;\ f''(x) = -1,\ -1<0 \\
f'(x) &= 0 \text{ at } x = 25
\end{aligned}
$$

Therefore, maximum revenue of \$1512.50 results when 25 people in excess of 30 go or when a total of 55 people go.

31. Let x = selling price per car. Then $x(300-100x)$ = total income and $2(300-100x)$ = total costs.

Therefore, profit $= f(x) = x(300-100x) - 2(300-100x),\ x \in [2, 3]$
$= -100x^2+500x-600.$
$f'(x) = -200x+500;\ f''(x) = -200,\ -200<0$
$f'(x) = 0$ at $x = 2.5$

Therefore, maximum profit of $25.00 is obtained when the price per car is $2.50.

32. Surface area $A = \pi \dfrac{d^2}{4} + \pi dh$

Volume $= \pi \dfrac{d^2}{4} h = 58,\ h = \dfrac{232}{\pi d^2}$

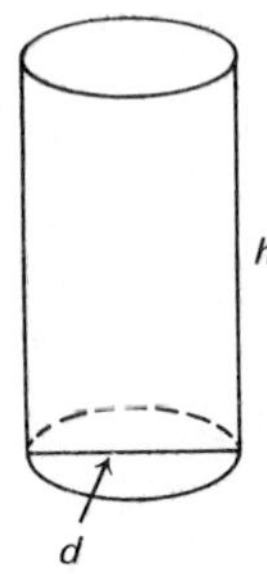

Therefore surface area $= f(d) = \pi \dfrac{d^2}{4} + \pi d\left(\dfrac{232}{\pi d^2}\right)$

$$f(d) = \frac{\pi d^2}{4} + \frac{232}{d},\ d = (0, \infty).$$

$$f'(d) = \frac{\pi d}{2} - \frac{232}{d^2} = \frac{\pi d^3-464}{2d^2},\ f''(d) = \frac{\pi}{2} + \frac{464}{d^3}$$

$$f'(d) = 0 \text{ when } d = \sqrt[3]{\frac{464}{\pi}} = 2\sqrt[3]{\frac{58}{\pi}},\ f''(d)>0 \text{ when } d>0$$

Therefore, f has a minimum value when the diameter is $2\sqrt[3]{58/\pi}$ inches. The respective decimal values of $2 \cdot \sqrt[3]{58/\pi}$ and $f(2 \cdot \sqrt[3]{58/\pi})$ are 5.286 and 65.835.

34. Surface area $A = \dfrac{\pi d^2}{4} + \pi dh$

Volume $= \dfrac{\pi d^2 h}{4} = c,\ h = \dfrac{4c}{\pi d^2}$

Therefore, surface area $= f(d) = \pi d^2/4 + 4c/d,\ d \in (0, \infty).$

$$f'(d) = \frac{\pi d}{2} - \frac{4c}{d^2} = \frac{\pi d^3-8c}{2d^2},\ f''(d) = \frac{\pi}{2} + \frac{8c}{d^3}$$

$$f'(d) = 0 \text{ when } d = 2\sqrt[3]{\frac{c}{\pi}},\ f''(d)>0 \text{ when } d>0$$

Therefore, f has a minimum value when the diameter is $2\sqrt[3]{c/\pi}$ inches.

37. a. $g(0.9) = 2.999$
$g(1.1) = 3.001$
c. $g(1) = 3$
e. $g'(x) = 3x^2-6x+3$
$g''(x) = 6x-6$
$g''(1) = 0$

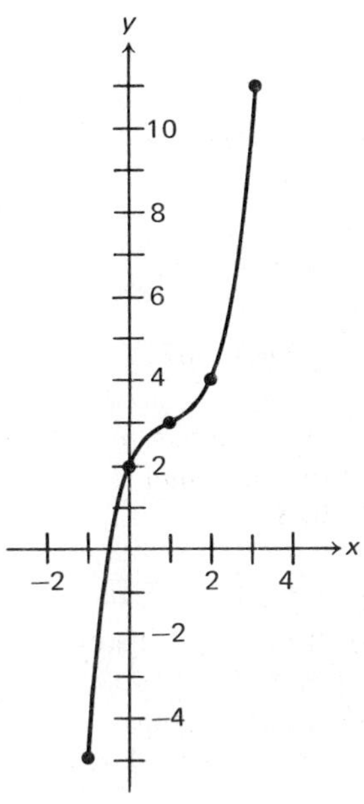

39. Let $x =$ the number of ten-dollar increases. Then rent per apartment is $200+10x$ dollars, the number of apartments rented is $200-5x$, the servicing charges are $20(200-5x)$ dollars, and the income per month is $(200+10x)\cdot(200-5x)$ dollars.

$$\begin{aligned}\text{Profit} = f(x) &= (200+10x)(200-5x)-20(200-5x)\\ &= (180+10x)(200-5x)\\ &= 3600+1100x-50x^2,\ x\in[0,40]\end{aligned}$$

$f'(x) = 1100-100x;\ f''(x) = -100,\ -100<0$
$f'(x) = 0$ at $x = 11$

Therefore, maximum profit of $42,050 is attained when there are eleven ten-dollar increases which results in 145 apartments being rented.

41. Let $x =$ price per ton.

Therefore, revenue $= f(x) = (300-x)x,\ x\in[0,300]$

$f'(x) = 300-2x;\ f''(x) = -2,\ -2<0$
$f'(x) = 0$ at $x = 150$

Therefore, maximum revenue of $22,500 is realized when the price per ton is $150 at which price 150 tons will be sold. Since total profit equals revenue minus cost,

$$\begin{aligned}P = g(x) &= (300-x)\cdot x-(5000+120\cdot(300-x))\\ &= 300x-x^2-5000-36000+120x\\ &= -x^2+420x-41000.\end{aligned}$$

The maximum profit is $3,100 when 210 tons are made and sold.

42. $f'(t) = 3t^2-2t-10,$ $\qquad f''(t) = 6t-2$

$$f'(t) = 0 \text{ at } t = \frac{1 \pm \sqrt{31}}{3} \qquad f''(t) > 0 \text{ when } t > \frac{1}{3}$$

$$f''(t) < 0 \text{ when } t < \frac{1}{3}$$

$$f(-3) = 11,\ f\left(\frac{1-\sqrt{31}}{3}\right) = \frac{94-34\sqrt{31}}{27}$$

$$f\left(\frac{1+\sqrt{31}}{3}\right) = \frac{94+\sqrt{34}}{27},\ f(5) = 67$$

Therefore, maximum value is 67 which occurs when $t = 5$, and minimum value is $(94-34\sqrt{31})/27$ which occurs when $t = (1-\sqrt{31})/3$. The respective decimal values are -3.530 and -1.523.

45. $$\begin{aligned} \text{Profit} = f(x) &= x \cdot P(x) - C(x) \\ &= x(200-0.1x)-50x+10000 \\ &= 200x-0.1x^2-50x+10000 \\ &= -0.1x^2+150x+10000 \end{aligned}$$
$f'(x) = -0.2x+150;\ f''(x) = -0.2,\ -0.2 < 0$
$f'(x) = 0$ at $x = 750$.

Therefore, maximum profit of $562.50 is realized when 750 items are sold.

Problem Set 9–7, Pages 412-413

1. $$\begin{aligned} y = f(u) = f(g(x)) &= f(4x^2-7x+2) \\ &= 3(4x^2-7x+2)-7 \\ &= 12x^2-21x+6-7 \\ &= 12x^2-21x-1 \end{aligned}$$

Therefore, $f'(x) = 24x-21$

3. $y = f(u) = f(g(x)) = f(8x+13) = (8x+13)^3$
$f'(x) = 3(8x+13)^2 \cdot 8 = 24(8x+13)^2$

5. $y = f(u) = f(g(x)) = f(x) = x^{3/2}$
$f'(x) = 3/2x^{1/2}$

7. $y = f(u) = f(g(x)) = f(x^{1/3}) = x^{1/3}$
$$f'(x) = \frac{1}{3}x^{-2/3} = \frac{1}{3\sqrt[3]{x^2}}$$

9. $$\begin{aligned} y = f(u) = f(k(t)) &= f(k(g(x))) = f(k(2-5x)) \\ &= f(4(2-5x)+3) = f(8-20x+3) = f(11-20x) \\ &= 6(11-20x)-11 = 66-120x-11 = -120x+55 \end{aligned}$$
$f'(x) = -120$

10. $y = f(u) = f(g(x)) = f(2x-7) = 2x-7+3 = 2x-4$
    $f'(x) = 2$

11. Since $f(x) = 2x-4$ is a linear function with slope 2, the slope of the tangent line at any point on the graph of f is 2.

12. Since $f(x) = 2x-4$ is a linear function, the equation of the tangent line to the graph of f at any point is $y = 2x-4$.

13. $f(x) = 7^x$, $f'(x) = 7^x \cdot \log_e 7 = 7^x \cdot 2.303 \cdot \log_{10} 7$
    $f'(3) = 7^3\,(1.946) = 343\,(1.946) = 667.478$

Therefore, the slope of the tangent line at $(3, f(3))$ is 667.478.

15. $y = g(x) = 8^x$, $g'(x) = 8^x \cdot \log_e 8 = 8^x \cdot 2.303 \cdot \log_{10} 8$
    $g'(1/3) = 8^{1/3}\,(2.080) = 2\,(2.080) = 4.160$

Therefore, the slope of the tangent line at $(1/3,\ g(1/3))$ is 4.160.

17. $y = f(x) = e^x$, $f'(x) = e^x \cdot \log_e e = e^x \cdot 1 = e^x$
    $f'(-4) = e^{-4} = 0.0183$

Therefore, the slope of the tangent line at $(-4, f(-4))$ is 0.0183

19. $y = 8^{2x-5}$, $y' = 2.8^{(2x-5)} \cdot \log_e 8 = 2 \cdot 8^{(2x-5)} \cdot 2.303 \cdot \log_{10} 8$
    At $x = 3$, $y' = 2\,(8)\,(2.080) = 33.280$.

Therefore, the slope of the tangent line at $(3, f(3))$ is 33.280.

21. $y = f(t) = 3e^{2t^2-6t+7}$,
    $f'(t) = 3e^{2t^2-6t+7}\,(4t-6)$, $f'(2) = 3\,(e^3)\,(2)$
    $f'(2) = 6\,(20.086) = 120.516$

Therefore, the slope of the tangent line at $(2, f(2))$ is 120.516.

23. $y = f(x) = 3^x$, $f'(x) = 3^x \cdot \log_e 3 = 3^x \cdot 2.303 \cdot \log_{10} 3$
    $f'(0) = 1 \cdot 2.303 \cdot \log_{10} 3 = 1.099$

Therefore, the slope of the tangent line at $(0, f(0))$ is 1.099.

25. $y = f(x) = 3^x$, $f'(x) = 3^x \cdot 2.303 \cdot \log_{10} 3$
    $f'(4) = 3^4 \cdot \log_e 3 = 81 \cdot (1.099) = 89.019$

Therefore, the slope of the tangent line at $(4, f(4))$ is 89.019.

27. $y = f(x) = 7 \cdot 3^x$, $f'(x) = 7 \cdot 3^x \cdot \log_e 3$
    $f'(4) = 7 \cdot 3^4 \cdot \log_e 3 = 567 \cdot (1.099) = 623.133$

Therefore, the slope of the tangent line at $(4, f(4))$ is 623.133.

29. $y = f(x) = \log_{10} x$
    $f'(x) = \dfrac{1}{x}\log_{10} e = \dfrac{0.4343}{x}$

31. $y = f(x) = \log_e x$

$$f'(x) = \frac{1}{x} \cdot \log_e e = \frac{1}{x}$$

33. $y = f(x) = \log_e (3x^2+5x-12)$

$$f'(x) = \frac{1}{3x^2+5x-12} \cdot 6x+5 = \frac{6x+5}{3x^2+5x-12}$$

CHAPTER 10

Problem Set 10–1, Pages 419-420

1. $3/2\,x^2+c$

3. $4x+c$

5. $3/2\,x^2+7x+c$

7. $-1/4\,x^{-4}+c$

9. $6/5\,x^5+c$

11. $x^2/2+2x+c$

13. $x^5-x^4+x^3-x^2+x+c$

15. $\int (x+2) \cdot (x+3)\,dx \neq \int (x+2)\,dx \cdot \int (x+3)\,dx$. In general, $\int f(x) \cdot g(x)\,dx \neq \int f(x)\,dx \cdot \int g(x)\,dx$, or in words, the integral of the product of two functions is not equal to the product of the integrals of the two functions.

17. $y = f(x) = -3/2\,x^2+x+36$

19. $y = f(x) = 15/2\,x^3+15/2\,x-2$

21. $R = f(x) = -8/3\,x^3+50x^2+222$

Problem Set 10–2, Pages 429-430

1. The area of the two rectangles is $(1+4) = 5$; the area of the two rectangles is $(4+9) = 13$.

3. $$\frac{26 \cdot 5^2-24 \cdot 5+4}{3 \cdot 5^2} = \frac{534}{75} = 7.12 \text{ and}$$
$$\frac{26 \cdot 5^2+24 \cdot 5+4}{3 \cdot 5^2} = \frac{774}{75} = 10.32$$

5. $(3+3.5+4+4.5+5+5.5+6+6.5) = 38$

6. The area of the triangular region whose vertices are $(0.5, 0)$, $(7, 0)$, and $(7, 13)$ is 42.25 square units. The area of the triangular region whose vertices are $(0.5, 0)$, $(3, 0)$, and $(3, 5)$ is 6.25. Thus, the area of the region specified in problem 4 is $(42.25-6.25) = 36$ square units.

7. $(0.200+0.204+0.216+0.236+0.264+0.300+0.344+0.396+0.456+0.524) =$ 3.14 square units

9. $(0.201+0.209+0.225+0.249+0.281+0.321+0.369+0.425+0.489+0.561) =$ 3.33 square units

10. a. $\dfrac{26 \cdot 2^2-2}{3 \cdot 2^2} = \dfrac{102}{12} = 8.5$

   c. Yes

Problem Set 10-3, Page 435

1. $\dfrac{125}{3} - 9 = \dfrac{98}{3}$

3. $-\dfrac{125}{3} - -9 = -\dfrac{98}{3}$; $\left|-\dfrac{98}{3}\right| = \dfrac{98}{3}$

5. $24 - \dfrac{16}{3} = \dfrac{56}{3}$

7. 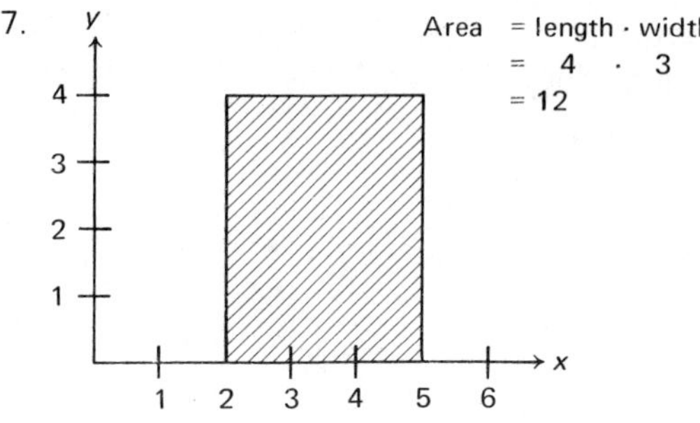

Area $= \text{length} \cdot \text{width}$
$= 4 \cdot 3$
$= 12$

9. $50 - 8 = 42$

11. $\dfrac{88}{3} - \dfrac{7}{3} = 27$

13. $-2.25 - \dfrac{37}{12} = -\dfrac{16}{3}$; $\left|-\dfrac{16}{3}\right| = \dfrac{16}{3}$

15. $33.75 - 5.75 = 28$

17. $\int_0^2 ((12x - x^3) - 8x)\, dx = \int_0^2 (-x^3 + 4x)\, dx = 4 - 0 = 4$

19. $\int_3^7 (2x-1)\, dx = 42 - 6 = 36$

21. $18 - 0 = 18$

Problem Set 10–4, Pages 442-444

1. 15 1/3 – (–66 3/4); 82,083 washers

3. 74.583 – (–66 3/4); 141,333 washers

5. 21.474 – 18.886; 2,588 washers

7. 17.25 – 15.333; 1,917 washers

9. $f(2.25) = 0.828125$, $f(2) = 1$, average rate $= -0.6875$

10. $f'(t) = 3t^2 - 2t - 10$, $f'(2) = -2$

11. $f'(3) = 11$

12. a. 5,400 cans c. 4,650 cans
    e. $f'(1) = -12$; 12 cans per day
    g. $f'(30) = -360$; 360 cans per day
    i. $\int_{10}^{30} 0.04 \cdot (5400 - 6t^2)\, dt = 4{,}320 = \$43.20$

13. $\int_{10}^{x} (20 + 3t)\, dt = 20x + 3/2\, x^2$; the twentieth day

15. a. $\int_0^3 (9.6t - 2t^2/5)\, dt = 39.6$; \$39,600
    c. $\int_0^3 (-2t^2/5 - 0.4t + 12)\, dt = 30.6$; \$30,600

16. $\int_0^{100} 10 \cdot x^{-1/3}\, dx = 323.2$; 323 minutes

18. $8x^{-1/3} = 1$; $x = 512$, or 512 trials

21. 1/3

23. $1/8 \cdot 141.333 = 17.667$

25. 7

Appendix

Table I

x	e^x	e^{-x}
.1	1.10517	.904837
.2	1.2214	.818731
.3	1.34986	.740818
.4	1.49183	.67032
.5	1.64872	.606531
.6	1.82212	.548811
.7	2.01375	.496585
.8	2.22554	.449329
.9	2.4596	.40657
1	2.71828	.367879
1.1	3.00417	.332871
1.2	3.32012	.301194
1.3	3.6693	.272532
1.4	4.0552	.246597
1.5	4.48169	.22313
1.6	4.95303	.201897
1.7	5.47395	.182684
1.8	6.04965	.165299
1.9	6.6859	.149569
2	7.38906	.135335
2.1	8.16617	.122456
2.2	9.02501	.110803
2.3	9.97419	.100259
2.4	11.0232	.090718
2.5	12.1825	.082085
2.6	13.4637	.074274
2.7	14.8797	.067205
2.8	16.4447	.06081
2.9	18.1741	.055023
3	20.0855	.049787
3.1	22.198	.045049
3.2	24.5325	.040762
3.3	27.1126	.036883
3.4	29.9641	.033373
3.5	33.1155	.030197
3.6	36.5983	.027324
3.7	40.4473	.024724
3.8	44.7012	.022371
3.9	49.4025	.020242
4	54.5982	.018316

Table I—*continued*

x	e^x	e^{-x}
4.1	60.3403	.016573
4.2	66.6863	.014996
4.3	73.6998	.013569
4.4	81.4509	.012277
4.5	90.0172	.011109
4.6	99.4844	.010052
4.7	109.947	.009095
4.8	121.51	.00823
4.9	134.29	.007447
5	148.413	.006738
5.1	164.022	.006097
5.2	181.272	.005517
5.3	200.337	.004992
5.4	221.406	.004517
5.5	244.692	.004087
5.6	270.427	.003698
5.7	298.867	.003346
5.8	330.3	.003028
5.9	365.037	.002739
6	403.429	.002479
6.1	445.859	.002243
6.2	492.749	.002029
6.3	544.572	.001836
6.4	601.845	.001662
6.5	665.142	.001503
6.6	735.095	.00136
6.7	812.406	.001231
6.8	897.848	.001114
6.9	992.275	.001008
7	1096.63	.000912
7.1	1211.97	.000825
7.2	1339.43	.000747
7.3	1480.3	.000676
7.4	1635.98	.000611
7.5	1808.04	.000553
7.6	1998.2	.0005
7.7	2208.35	.000453
7.8	2440.61	.00041
7.9	2697.28	.000371
8	2980.96	.000335

Table I—*continued*

x	e^x	e^{-x}
8.1	3294.47	.000304
8.2	3640.95	.000275
8.3	4023.87	.000249
8.4	4447.07	.000225
8.5	4914.77	.000203
8.6	5431.66	.000184
8.7	6002.92	.000167
8.8	6634.25	.000151
8.9	7331.98	.000136
9	8103.09	.000123
9.1	8955.3	.000112
9.2	9897.14	.000101
9.3	10938	.000091
9.4	12088.4	.000083
9.5	13359.7	.000075
9.6	14764.8	.000068
9.7	16317.6	.000061
9.8	18033.7	.000055
9.9	19930.4	.00005
10	22026.5	.000045

Table II

FOUR-PLACE TABLE OF COMMON LOGARITHMS

N	0	1	2	3	4	5	6	7	8	9
10	0000	0043	0086	0128	0170	0212	0253	0294	0334	0374
11	0414	0453	0492	0531	0569	0607	0645	0682	0719	0755
12	0792	0828	0864	0899	0934	0969	1004	1038	1072	1106
13	1139	1173	1206	1239	1271	1303	1335	1367	1399	1430
14	1461	1492	1523	1553	1584	1614	1644	1673	1703	1723
15	1761	1790	1818	1847	1875	1903	1931	1959	1987	2014
16	2041	2068	2095	2122	2148	2175	2201	2227	2253	2279
17	2304	2330	2355	2380	2405	2430	2455	2480	2504	2529
18	2553	2577	2601	2625	2648	2672	2695	2718	2742	2765
19	2788	2810	2833	2856	2878	2900	2923	2945	2967	2989
20	3010	3032	3054	3075	3096	3118	3139	3160	3181	3201
21	3222	3243	3263	3284	3304	3324	3345	3365	3385	3404
22	3424	3444	3464	3483	3502	3522	3541	3560	3579	3598
23	3617	3636	3655	3674	3692	3711	3729	3747	3766	3784
24	3802	3820	3838	3856	3874	3892	3909	3927	3945	3962

Table II—*continued*

N	0	1	2	3	4	5	6	7	8	9
25	3979	3997	4014	4031	4048	4065	4082	4099	4116	4133
26	4150	4166	4183	4200	4216	4232	4249	4265	4281	4298
27	4314	4330	4346	4362	4378	4393	4409	4425	4440	4456
28	4472	4487	4502	4518	4533	4548	4564	4579	4594	4609
29	4624	4639	4654	4669	4683	4698	4713	4728	4742	4757
30	4771	4786	4800	4814	4829	4843	4857	4871	4886	4900
31	4914	4928	4942	4955	4969	4983	4997	5011	5024	5038
32	5051	5065	5079	5092	5105	5119	5132	5145	5159	5172
33	5185	5198	5211	5224	5237	5250	5263	5276	5289	5302
34	5315	5328	5340	5353	5366	5378	5391	5403	5416	5428
35	5441	5453	5465	5478	5490	5502	5514	5527	5539	5551
36	5563	5575	5587	5599	5611	5623	5635	5647	5658	5670
37	5682	5694	5705	5717	5729	5740	5752	5763	5775	5786
38	5798	5809	5821	5832	5843	5855	5866	5877	5888	5899
39	5911	5922	5933	5944	5955	5966	5977	5988	5999	6010
40	6021	6031	6042	6053	6064	6075	6085	6096	6107	6117
41	6128	6138	6149	6160	6170	6180	6191	6201	6212	6222
42	6232	6243	6253	6263	6274	6284	6294	6304	6314	6325
43	6335	6345	6355	6365	6375	6385	6395	6405	6415	6425
44	6435	6444	6454	6464	6474	6484	6493	6503	6513	6522
45	6532	6542	6551	6561	6571	6580	6590	6599	6609	6618
46	6628	6637	6646	6656	6665	6675	6684	6693	6702	6712
47	6721	6730	6739	6749	6758	6767	6776	6785	6794	6803
48	6812	6821	6830	6839	6848	6857	6866	6875	6884	6893
49	6902	6911	6920	6928	6937	6946	6955	6964	6972	6981
50	6990	6998	7007	7016	7024	7033	7042	7050	7059	7067
51	7076	7084	7093	7101	7110	7118	7126	7135	7143	7152
52	7160	7168	7177	7185	7193	7202	7210	7218	7226	7235
53	7243	7251	7259	7267	7275	7284	7292	7300	7308	7316
54	7324	7332	7340	7348	7356	7364	7372	7380	7388	7396
55	7404	7412	7419	7427	7435	7443	7451	7459	7466	7474
56	7482	7490	7497	7505	7513	7520	7528	7536	7543	7551
57	7559	7566	7574	7582	7589	7597	7604	7612	7619	7627
58	7634	7642	7649	7657	7664	7672	7679	7686	7694	7701
59	7709	7716	7723	7731	7738	7745	7752	7760	7767	7774
60	7782	7789	7796	7803	7810	7818	7825	7832	7839	7846
61	7853	7860	7868	7875	7882	7889	7896	7903	7910	7917
62	7924	7931	7938	7945	7952	7959	7966	7973	7980	7987
63	7993	8000	8007	8014	8021	8028	8035	8041	8048	8055
64	8062	8069	8075	8082	8089	8096	8102	8109	8116	8122

Table II—*continued*

N	0	1	2	3	4	5	6	7	8	9
65	8129	8136	8142	8149	8156	8162	8169	8176	8182	8189
66	8195	8202	8209	8215	8222	8228	8235	8241	8248	8254
67	8261	8267	8274	8280	8287	8293	8299	8306	8312	8319
68	8325	8331	8338	8344	8351	8357	8363	8370	8376	8382
69	8388	8395	8401	8407	8414	8420	8426	8432	8439	8445
70	8451	8457	8463	8470	8476	8482	8488	8494	8500	8506
71	8513	8519	8525	8531	8537	8543	8549	8555	8561	8567
72	8573	8579	8585	8591	8597	8603	8609	8615	8621	8627
73	8633	8639	8645	8651	8657	8663	8669	8675	8681	8686
74	8692	8698	8704	8710	8716	8722	8727	8733	8739	8745
75	8751	8756	8762	8768	8774	8779	8785	8791	8797	8802
76	8808	8814	8820	8825	8831	8837	8842	8848	8854	8859
77	8865	8871	8876	8882	8887	8893	8899	8904	8910	8915
78	8921	8927	8932	8938	8943	8949	8954	8960	8965	8971
79	8976	8982	8987	8993	8998	9004	9009	9015	9020	9025
80	9031	9036	9042	9047	9053	9058	9063	9069	9074	9079
81	9085	9090	9096	9101	9106	9112	9117	9122	9128	9133
82	9138	9143	9149	9154	9159	9165	9170	9175	9180	9186
83	9191	9196	9201	9206	9212	9217	9222	9227	9232	9238
84	9243	9248	9253	9258	9263	9269	9274	9279	9284	9289
85	9294	9299	9304	9309	9315	9320	9325	9330	9335	9340
86	9345	9350	9355	9360	9365	9370	9375	9380	9385	9390
87	9395	9400	9405	9410	9415	9420	9425	9430	9435	9440
88	9445	9450	9455	9460	9465	9469	9474	9479	9484	9489
89	9494	9499	9504	9509	9513	9518	9523	9528	9533	9538
90	9542	9547	9552	9557	9562	9566	9571	9576	9581	9586
91	9590	9595	9600	9605	9609	9614	9619	9624	9628	9633
92	9638	9643	9647	9652	9657	9661	9666	9671	9675	9680
93	9685	9689	9694	9699	9703	9708	9713	9717	9722	9727
94	9731	9736	9741	9745	9750	9754	9759	9763	9768	9773
95	9777	9782	9786	9791	9795	9800	9805	9809	9814	9818
96	9823	9827	9832	9836	9841	9845	9850	9854	9859	9863
97	9868	9872	9877	9881	9886	9890	9894	9899	9903	9908
98	9912	9917	9921	9926	9930	9934	9939	9943	9948	9952
99	9956	9961	9965	9969	9974	9978	9983	9987	9991	9996

Table III

N	log (N)
1.001	.000434
1.002	.000868
1.003	.001301
1.004	.001734
1.005	.002166
1.006	.002598
1.007	.003029
1.008	.00346
1.009	.003891
1.01	.004321
1.011	.004751
1.012	.00518
1.013	.005609
1.014	.006038
1.015	.006466
1.016	.006894
1.017	.007321
1.018	.007748
1.019	.008174
1.02	.0086
1.021	.009026
1.022	.009451
1.023	.009876
1.024	.0103
1.025	.010724
1.026	.011147
1.027	.01157
1.028	.011993
1.029	.012415
1.03	.012837
1.031	.013259
1.032	.01368
1.033	.0141
1.034	.01452
1.035	.01494
1.036	.01536
1.037	.015779
1.038	.016197
1.039	.016616
1.04	.017033

Table III—*continued*

N	log (*N*)
1.041	.017451
1.042	.017868
1.043	.018284
1.044	.0187
1.045	.019116
1.046	.019532
1.047	.019947
1.048	.020361
1.049	.020775
1.05	.021189

Table IV

n	$(1.019)^n$
1	1.019
2	1.03836
3	1.05809
4	1.07819
5	1.09868
6	1.11955
7	1.14083
8	1.1625
9	1.18459
10	1.2071
11	1.23003
12	1.2534
13	1.27722
14	1.30148
15	1.32621
16	1.35141
17	1.37709
18	1.40325
19	1.42991
20	1.45708
21	1.48477
22	1.51298
23	1.54172
24	1.57102
25	1.60087
26	1.63128
27	1.66228
28	1.69386
29	1.72604
30	1.75884

Table IV—*continued*

n	$(1.019)^n$
31	1.79226
32	1.82631
33	1.86101
34	1.89637
35	1.9324
36	1.96911
37	2.00653
38	2.04465
39	2.0835
40	2.12309
41	2.16343
42	2.20453
43	2.24642
44	2.2891
45	2.33259
46	2.37691
47	2.42207
48	2.46809
49	2.51499
50	2.56277
51	2.61146
52	2.66108
53	2.71164
54	2.76316
55	2.81566
56	2.86916
57	2.92367
58	2.97922
59	3.03583
60	3.09351
61	3.15229
62	3.21218
63	3.27321
64	3.3354
65	3.39878
66	3.46335
67	3.52916
68	3.59621
69	3.66454
70	3.73417

Table IV—*continued*

n	$(1.019)^n$
71	3.80511
72	3.87741
73	3.95108
74	4.02615
75	4.10265
76	4.1806
77	4.26003
78	4.34097
79	4.42345
80	4.5075
81	4.59314
82	4.68041
83	4.76934
84	4.85995
85	4.95229
86	5.04639
87	5.14227
88	5.23997
89	5.33953
90	5.44098
91	5.54436
92	5.6497
93	5.75705
94	5.86643
95	5.9779
96	6.09148
97	6.20721
98	6.32515
99	6.44533
100	6.56779

Table V

COMPOUND AMOUNT AT END OF n COMPOUNDING PERIODS WITH GROWTH RATE OF r% PER PERIOD: $(1+r)^n$

n	(4/12)%	(4.5/12)%	(5/12)%	(5.5/12)%	(6/12)%
1	1.00333	1.00375	1.00417	1.00458	1.005
2	1.00668	1.00751	1.00835	1.00919	1.01003
3	1.01003	1.01129	1.01255	1.01381	1.01508
4	1.0134	1.01508	1.01677	1.01846	1.02015
5	1.01678	1.01889	1.02101	1.02313	1.02525
6	1.02017	1.02271	1.02526	1.02782	1.03038

Table V—*continued*

n	(4/12)%	(4.5/12)%	(5/12)%	(5.5/12)%	(6/12)%
7	1.02357	1.02655	1.02953	1.03253	1.03553
8	1.02698	1.0304	1.03382	1.03726	1.04071
9	1.0304	1.03426	1.03813	1.04201	1.04591
10	1.03384	1.03814	1.04246	1.04679	1.05114
11	1.03728	1.04203	1.0468	1.05159	1.0564
12	1.04074	1.04594	1.05116	1.05641	1.06168
13	1.04421	1.04986	1.05554	1.06125	1.06699
14	1.04769	1.0538	1.05994	1.06611	1.07232
15	1.05118	1.05775	1.06435	1.071	1.07769
16	1.05469	1.06172	1.06879	1.07591	1.08307
17	1.0582	1.0657	1.07324	1.08084	1.08849
18	1.06173	1.0697	1.07771	1.08579	1.09393
19	1.06527	1.07371	1.08221	1.09077	1.0994
20	1.06882	1.07773	1.08671	1.09577	1.1049
21	1.07238	1.08178	1.09124	1.10079	1.11042
22	1.07596	1.08583	1.09579	1.10584	1.11598
23	1.07955	1.08991	1.10035	1.11091	1.12156
24	1.08314	1.09399	1.10494	1.116	1.12716
25	1.08675	1.09809	1.10954	1.12111	1.1328
26	1.09038	1.10221	1.11417	1.12625	1.13846
27	1.09401	1.10635	1.11881	1.13141	1.14416
28	1.09766	1.1105	1.12347	1.1366	1.14988
29	1.10132	1.11466	1.12815	1.14181	1.15563
30	1.10499	1.11884	1.13285	1.14704	1.16141
31	1.10867	1.12304	1.13757	1.1523	1.16721
32	1.11237	1.12725	1.14231	1.15758	1.17305
33	1.11608	1.13147	1.14707	1.16289	1.17891
34	1.1198	1.13572	1.15185	1.16822	1.18481
35	1.12353	1.13998	1.15665	1.17357	1.19073
36	1.12727	1.14425	1.16147	1.17895	1.19669
37	1.13103	1.14854	1.16631	1.18435	1.20267
38	1.1348	1.15285	1.17117	1.18978	1.20868
39	1.13858	1.15717	1.17605	1.19523	1.21473
40	1.14238	1.16151	1.18095	1.20071	1.2208
41	1.14619	1.16587	1.18587	1.20621	1.22691
42	1.15001	1.17024	1.19081	1.21174	1.23304
43	1.15384	1.17463	1.19577	1.2173	1.23921
44	1.15769	1.17903	1.20075	1.22288	1.2454
45	1.16155	1.18346	1.20576	1.22848	1.25163
46	1.16542	1.18789	1.21078	1.23411	1.25789
47	1.1693	1.19235	1.21583	1.23977	1.26418
48	1.1732	1.19682	1.22089	1.24545	1.2705
49	1.17711	1.20131	1.22598	1.25116	1.27685
50	1.18104	1.20581	1.23109	1.25689	1.28324

Table V—*continued*

n	(4/12)%	(4.5/12)%	(5/12)%	(5.5/12)%	(6/12)%
51	1.18497	1.21034	1.23622	1.26265	1.28965
52	1.18892	1.21487	1.24137	1.26844	1.2961
53	1.19288	1.21943	1.24654	1.27426	1.30258
54	1.19686	1.224	1.25173	1.2801	1.30909
55	1.20085	1.22859	1.25695	1.28596	1.31564
56	1.20485	1.2332	1.26219	1.29186	1.32222
57	1.20887	1.23782	1.26745	1.29778	1.32883
58	1.2129	1.24247	1.27273	1.30373	1.33547
59	1.21694	1.24713	1.27803	1.3097	1.34215
60	1.221	1.2518	1.28335	1.3157	1.34886
61	1.22507	1.2565	1.2887	1.32173	1.35561
62	1.22915	1.26121	1.29407	1.32779	1.36239
63	1.23325	1.26594	1.29946	1.33388	1.3692
64	1.23736	1.27069	1.30488	1.33999	1.37604
65	1.24149	1.27545	1.31031	1.34613	1.38292
66	1.24562	1.28023	1.31577	1.3523	1.38984
67	1.24978	1.28503	1.32126	1.3585	1.39679
68	1.25394	1.28985	1.32676	1.36473	1.40377
69	1.25812	1.29469	1.33229	1.37098	1.41079
70	1.26232	1.29955	1.33784	1.37727	1.41785
71	1.26652	1.30442	1.34341	1.38358	1.42494
72	1.27075	1.30931	1.34901	1.38992	1.43206
73	1.27498	1.31422	1.35463	1.39629	1.43922
74	1.27923	1.31915	1.36028	1.40269	1.44642
75	1.28349	1.3241	1.36595	1.40912	1.45365
76	1.28777	1.32906	1.37164	1.41558	1.46092
77	1.29207	1.33405	1.37735	1.42207	1.46822
78	1.29637	1.33905	1.38309	1.42858	1.47556
79	1.30069	1.34407	1.38885	1.43513	1.48294
80	1.30503	1.34911	1.39464	1.44171	1.49036
81	1.30938	1.35417	1.40045	1.44832	1.49781
82	1.31375	1.35925	1.40629	1.45496	1.5053
83	1.31812	1.36435	1.41214	1.46162	1.51283
84	1.32252	1.36946	1.41803	1.46832	1.52039
85	1.32693	1.3746	1.42394	1.47505	1.52799
86	1.33135	1.37975	1.42987	1.48181	1.53563
87	1.33579	1.38493	1.43583	1.48861	1.54331
88	1.34024	1.39012	1.44181	1.49543	1.55103
89	1.34471	1.39533	1.44782	1.50228	1.55878
90	1.34919	1.40057	1.45385	1.50917	1.56658

Table V—*continued*

n	(4/12)%	(4.5/12)%	(5/12)%	(5.5/12)%	(6/12)%
91	1.35369	1.40582	1.45991	1.51608	1.57441
92	1.3582	1.41109	1.46599	1.52303	1.58228
93	1.36273	1.41638	1.4721	1.53001	1.59019
94	1.36727	1.42169	1.47823	1.53703	1.59815
95	1.37183	1.42703	1.48439	1.54407	1.60614
96	1.3764	1.43238	1.49058	1.55115	1.61417
97	1.38099	1.43775	1.49679	1.55826	1.62224
98	1.38559	1.44314	1.50302	1.5654	1.63035
99	1.39021	1.44855	1.50929	1.57257	1.6385
100	1.39484	1.45398	1.51557	1.57978	1.64669

n	(7/12)%	(8/12)%	(9/12)%	1.0%	1.25%
1	1.00583	1.00667	1.0075	1.01	1.0125
2	1.0117	1.01338	1.01506	1.0201	1.02516
3	1.0176	1.02013	1.02267	1.0303	1.03797
4	1.02354	1.02693	1.03034	1.0406	1.05095
5	1.02951	1.03378	1.03807	1.05101	1.06408
6	1.03552	1.04067	1.04585	1.06152	1.07738
7	1.04155	1.04761	1.0537	1.07213	1.09085
8	1.04763	1.05459	1.0616	1.08286	1.10449
9	1.05374	1.06163	1.06956	1.09368	1.11829
10	1.05989	1.0687	1.07758	1.10462	1.13227
11	1.06607	1.07583	1.08566	1.11567	1.14642
12	1.07229	1.083	1.09381	1.12683	1.16075
13	1.07855	1.09022	1.10201	1.13809	1.17526
14	1.08484	1.09749	1.11028	1.14947	1.18995
15	1.09117	1.1048	1.1186	1.16097	1.20483
16	1.09753	1.11217	1.12699	1.17258	1.21989
17	1.10393	1.11958	1.13544	1.1843	1.23514
18	1.11037	1.12705	1.14396	1.19615	1.25058
19	1.11685	1.13456	1.15254	1.20811	1.26621
20	1.12337	1.14212	1.16118	1.22019	1.28204
21	1.12992	1.14974	1.16989	1.23239	1.29806
22	1.13651	1.1574	1.17867	1.24471	1.31429
23	1.14314	1.16512	1.18751	1.25716	1.33072
24	1.14981	1.17289	1.19641	1.26973	1.34735
25	1.15651	1.18071	1.20539	1.28243	1.36419
26	1.16326	1.18858	1.21443	1.29525	1.38125
27	1.17005	1.1965	1.22353	1.30821	1.39851
28	1.17687	1.20448	1.23271	1.32129	1.41599
29	1.18374	1.21251	1.24196	1.3345	1.43369
30	1.19064	1.22059	1.25127	1.34785	1.45161

Table V—*continued*

n	(7/12)%	(8/12)%	(9/12)%	1.0%	1.25%
31	1.19759	1.22873	1.26065	1.36133	1.46976
32	1.20458	1.23692	1.27011	1.37494	1.48813
33	1.2116	1.24517	1.27964	1.38869	1.50673
34	1.21867	1.25347	1.28923	1.40258	1.52557
35	1.22578	1.26182	1.2989	1.4166	1.54464
36	1.23293	1.27024	1.30864	1.43077	1.56394
37	1.24012	1.2787	1.31846	1.44508	1.58349
38	1.24735	1.28723	1.32835	1.45953	1.60329
39	1.25463	1.29581	1.33831	1.47412	1.62333
40	1.26195	1.30445	1.34835	1.48886	1.64362
41	1.26931	1.31314	1.35846	1.50375	1.66416
42	1.27672	1.3219	1.36865	1.51879	1.68497
43	1.28416	1.33071	1.37891	1.53398	1.70603
44	1.29166	1.33958	1.38925	1.54832	1.72735
45	1.29919	1.34851	1.39967	1.56481	1.74895
46	1.30677	1.3575	1.41017	1.58046	1.77081
47	1.31439	1.36655	1.42075	1.59626	1.79294
48	1.32206	1.37566	1.4314	1.61222	1.81535
49	1.32977	1.38484	1.44214	1.62835	1.83805
50	1.33753	1.39407	1.45296	1.64463	1.86102
51	1.34533	1.40336	1.46385	1.66108	1.88428
52	1.35318	1.41272	1.47483	1.67769	1.90784
53	1.36107	1.42214	1.48589	1.69446	1.93169
54	1.36901	1.48162	1.49704	1.71141	1.95583
55	1.377	1.44116	1.50826	1.72852	1.98028
56	1.38503	1.45077	1.51958	1.74581	2.00503
57	1.39311	1.46044	1.53097	1.76326	2.0301
58	1.40124	1.47018	1.54245	1.7809	2.05547
59	1.40941	1.47998	1.55402	1.79871	2.08117
60	1.41763	1.48984	1.56568	1.81669	2.10718
61	1.4259	1.49977	1.57742	1.83486	2.13352
62	1.43422	1.50977	1.58925	1.85321	2.16019
63	1.44259	1.51984	1.60117	1.87174	2.18719
64	1.451	1.52997	1.61318	1.89046	2.21453
65	1.45947	1.54017	1.62528	1.90936	2.24221
66	1.46798	1.55044	1.63747	1.92846	2.27024
67	1.47654	1.56078	1.64975	1.94774	2.29862
68	1.48516	1.57118	1.66212	1.96722	2.32735
69	1.49382	1.58165	1.67459	1.98689	2.35644
70	1.50253	1.5922	1.68715	2.00676	2.3859

Table V—*continued*

n	(7/12)%	(8/12)%	(9/12)%	1.0%	1.25%
71	1.5113	1.60281	1.6998	2.02683	2.41572
72	1.52011	1.6135	1.71255	2.04709	2.44592
73	1.52898	1.62426	1.72539	2.06757	2.47649
74	1.5379	1.63508	1.73833	2.08824	2.50745
75	1.54687	1.64598	1.75137	2.10912	2.53879
76	1.5559	1.65696	1.76451	2.13022	2.57053
77	1.56497	1.668	1.77774	2.15152	2.60266
78	1.5741	1.67912	1.79107	2.17303	2.63519
79	1.58328	1.69032	1.80451	2.19476	2.66813
80	1.59252	1.70159	1.81804	2.21671	2.70148
81	1.60181	1.71293	1.83168	2.23888	2.73525
82	1.61115	1.72435	1.84541	2.26127	2.76944
83	1.62055	1.73585	1.85925	2.28388	2.80406
84	1.63	1.74742	1.8732	2.30672	2.83911
85	1.63951	1.75907	1.88725	2.32978	2.8746
86	1.64908	1.77079	1.9014	2.35308	2.91053
87	1.6587	1.7826	1.91566	2.37661	2.94691
88	1.66837	1.79448	1.93003	2.40038	2.98375
89	1.67811	1.80645	1.9445	2.42438	3.02105
90	1.68789	1.81849	1.95909	2.44863	3.05881
91	1.69774	1.83061	1.97378	2.47311	3.09705
92	1.70764	1.84282	1.98858	2.49784	3.13576
93	1.71761	1.8551	2.0035	2.52282	3.17496
94	1.72762	1.86747	2.01852	2.54805	3.21464
95	1.7377	1.87992	2.03366	2.57353	3.25483
96	1.74784	1.89245	2.04891	2.59927	3.29551
97	1.75804	1.90507	2.06428	2.62526	3.33671
98	1.76829	1.91777	2.07976	2.65151	3.37841
99	1.77861	1.93055	2.09536	2.67803	3.42064
100	1.78898	1.94342	2.11108	2.70481	3.4634

n	1.5%	2.0%	2.5%	3.0%	3.5%
1	1.015	1.02	1.025	1.03	1.035
2	1.03023	1.0404	1.05063	1.0609	1.07123
3	1.04568	1.06121	1.07689	1.09273	1.10872
4	1.06136	1.08243	1.10381	1.12551	1.14752
5	1.07728	1.10408	1.13141	1.15927	1.18769
6	1.09344	1.12616	1.15969	1.19405	1.22926
7	1.10985	1.14869	1.18869	1.22987	1.27228
8	1.12649	1.17166	1.2184	1.26677	1.31681
9	1.14339	1.19509	1.24886	1.30477	1.3529
10	1.16054	1.21899	1.28009	1.34392	1.4106

Table V—*continued*

n	1.5%	2.0%	2.5%	3.0%	3.5%
11	1.17795	1.24337	1.31209	1.38423	1.45997
12	1.19562	1.26824	1.34489	1.42576	1.51107
13	1.21355	1.29361	1.37851	1.46853	1.56396
14	1.23176	1.31948	1.41298	1.51259	1.6187
15	1.25023	1.34587	1.4483	1.55797	1.67535
16	1.26899	1.37279	1.48451	1.60471	1.73399
17	1.28802	1.40024	1.52162	1.65285	1.79468
18	1.30734	1.42825	1.55966	1.70243	1.85749
19	1.32695	1.45681	1.59865	1.75351	1.9225
20	1.34686	1.48595	1.63862	1.80611	1.98979
21	1.36706	1.51567	1.67958	1.86029	2.05944
22	1.38757	1.54598	1.72157	1.9161	2.13152
23	1.40838	1.5769	1.76461	1.97359	2.20612
24	1.42951	1.60844	1.80873	2.03279	2.28333
25	1.45095	1.64061	1.85395	2.09378	2.36325
26	1.47272	1.67342	1.9003	2.15659	2.44596
27	1.49481	1.70689	1.9478	2.22129	2.53157
28	1.51723	1.74103	1.9965	2.28793	2.62018
29	1.53999	1.77585	2.04641	2.35657	2.71188
30	1.56309	1.81136	2.09757	2.42726	2.8068
31	1.58653	1.84759	2.15001	2.50008	2.90504
32	1.61033	1.88454	2.20376	2.57508	3.00672
33	1.63449	1.92223	2.25886	2.65234	3.11195
34	1.659	1.96068	2.31533	2.73191	3.22087
35	1.68389	1.99989	2.37321	2.81386	3.3336
36	1.70915	2.03989	2.43254	2.89828	3.45028
37	1.73479	2.08069	2.49336	2.98523	3.57104
38	1.76081	2.1223	2.55569	3.07478	3.69602
39	1.78722	2.16475	2.61958	3.16703	3.82538
40	1.81403	2.20804	2.68507	3.26204	3.95927
41	1.84124	2.2522	2.7522	3.3599	4.09785
42	1.86886	2.29725	2.821	3.4607	4.24127
43	1.89689	2.34319	2.89153	3.56452	4.38972
44	1.92535	2.39006	2.96382	3.67145	4.54336
45	1.95423	2.43786	3.03791	3.7816	4.70238
46	1.98354	2.48662	3.11386	3.89504	4.86696
47	2.01329	2.53635	3.19171	4.0119	5.0373
48	2.04349	2.58708	3.2715	4.13225	5.21361
49	2.07415	2.63882	3.35329	4.25622	5.39609
50	2.10526	2.69159	3.43712	4.38391	5.58495

Table V—*continued*

n	1.5%	2.0%	2.5%	3.0%	3.5%
51	2.13684	2.74543	3.52305	4.51542	5.78042
52	2.16889	2.80033	3.61113	4.65089	5.98274
53	2.20142	2.85634	3.7014	4.79041	6.19214
54	2.23445	2.91347	3.79394	4.93413	6.40886
55	2.26796	2.97174	3.88879	5.08215	6.63317
56	2.30198	3.03117	3.98601	5.23462	6.86533
57	2.33651	3.09179	4.08566	5.39165	7.10562
58	2.37156	3.15363	4.1878	5.5534	7.35432
59	2.40713	3.2167	4.2925	5.72	7.61172
60	2.44324	3.28104	4.39981	5.89161	7.87813
61	2.47989	3.34666	4.5098	6.06835	8.15386
62	2.51709	3.41359	4.62255	6.2504	8.43925
63	2.55485	3.48186	4.73811	6.43792	8.73462
64	2.59317	3.5515	4.85657	6.63105	9.04034
65	2.63207	3.62253	4.97798	6.82999	9.35675
66	2.67155	3.69498	5.10243	7.03488	9.68424
67	2.71162	3.76888	5.22999	7.24593	10.0232
68	2.7523	3.84426	5.36074	7.46331	10.374
69	2.79358	3.92115	5.49476	7.68721	10.7371
70	2.83549	3.99957	5.63213	7.91782	11.1129
71	2.87802	4.07956	5.77293	8.15536	11.5018
72	2.92119	4.16115	5.91726	8.40002	11.9044
73	2.96501	4.24438	6.06519	8.65202	12.3211
74	3.00948	4.32926	6.21682	8.91158	12.7523
75	3.05463	4.41585	6.37224	9.17893	13.1986
76	3.10045	4.50417	6.53155	9.4543	13.6606
77	3.14695	4.59425	6.69484	9.73793	14.1387
78	3.19416	4.68613	6.86221	10.0301	14.6336
79	3.24207	4.77986	7.03376	10.331	15.1457
80	3.2907	4.87545	7.20961	10.6409	15.6758
81	3.34006	4.97296	7.38985	10.9601	16.2245
82	3.39016	5.07242	7.5746	11.2889	16.7924
83	3.44102	5.17387	7.76396	11.6276	17.3801
84	3.49263	5.27735	7.95806	11.9764	17.9884
85	3.54502	5.3829	8.15701	12.3357	18.618
86	3.5982	5.49055	8.36094	12.7058	19.2696
87	3.65217	5.60036	8.56996	13.087	19.9441
88	3.70696	5.71237	8.78421	13.4796	20.6421
89	3.76256	5.82662	9.00382	13.884	21.3646
90	3.819	5.94315	9.22892	14.3005	22.1123

Table V—*continued*

n	1.5%	2.0%	2.5%	3.0%	3.5%
91	3.87629	6.06202	9.45964	14.7295	22.8863
92	3.93443	6.18326	9.69613	15.1714	23.6873
93	3.99345	6.30692	9.93854	15.6265	24.5163
94	4.05335	6.43306	10.187	16.0953	25.3744
95	4.11415	6.56172	10.4417	16.5782	26.2625
96	4.17586	6.69296	10.7027	17.0755	27.1817
97	4.2385	6.82682	10.9703	17.5878	28.1331
98	4.30208	6.96335	11.2445	18.1154	29.1177
99	4.36661	7.10262	11.5257	18.6589	30.1369
100	4.43211	7.24467	11.8138	19.2186	31.1917

n	4.0%	4.5%	5.0%	5.5%	6.0%
1	1.04	1.045	1.05	1.055	1.06
2	1.0816	1.09203	1.1025	1.11303	1.1236
3	1.12486	1.14117	1.15762	1.17424	1.19102
4	1.16986	1.19252	1.21551	1.23883	1.26248
5	1.21665	1.24618	1.27628	1.30696	1.33822
6	1.26532	1.30226	1.3401	1.37884	1.41852
7	1.31593	1.36086	1.4071	1.45468	1.50363
8	1.36857	1.4221	1.47746	1.53469	1.59385
9	1.42331	1.4861	1.55133	1.6191	1.68948
10	1.48024	1.55297	1.62889	1.70815	1.79085
11	1.53945	1.62285	1.71034	1.80209	1.8983
12	1.60103	1.69588	1.79586	1.90121	2.01219
13	1.66507	1.7722	1.88565	2.00578	2.13293
14	1.73168	1.85195	1.97993	2.11609	2.2609
15	1.80094	1.93528	2.07893	2.23248	2.39655
16	1.87298	2.02237	2.18287	2.35527	2.54035
17	1.9479	2.11338	2.29202	2.48481	2.69277
18	2.02582	2.20848	2.40662	2.62147	2.85433
19	2.10685	2.30786	2.52695	2.76565	3.02559
20	2.19112	2.41172	2.6533	2.91776	3.20713
21	2.27877	2.52024	2.78596	3.07824	3.39956
22	2.36992	2.63365	2.92526	3.24754	3.60353
23	2.46471	2.75217	3.07152	3.42616	3.81974
24	2.5633	2.87602	3.2251	3.6146	4.04892
25	2.66584	3.00544	3.38635	3.8134	4.29186
26	2.77247	3.14068	3.55567	4.02314	4.54937
27	2.88337	3.28201	3.73345	4.24441	4.82233
28	2.9987	3.4297	3.92013	4.47785	5.11167
29	3.11865	3.58404	4.11613	4.72414	5.41837
30	3.2434	3.74532	4.32194	4.98396	5.74347

Table V—*continued*

n	4.0%	4.5%	5.0%	5.5%	6.0%
31	3.37313	3.91386	4.53804	5.25808	6.08808
32	3.50806	4.08998	4.76494	5.54728	6.45337
33	3.64838	4.27403	5.00318	5.85238	6.84057
34	3.79431	4.46636	5.25334	6.17426	7.251
35	3.94609	4.66735	5.51601	6.51384	7.68606
36	4.10393	4.87738	5.79181	6.87211	8.14722
37	4.26809	5.09687	6.0814	7.25007	8.63605
38	4.43881	5.32622	6.38547	7.64883	9.15422
39	4.61636	5.5659	6.70474	8.06951	9.70347
40	4.80102	5.81637	7.03998	8.51334	10.2857
41	4.99306	6.07811	7.39198	8.98157	10.9028
42	5.19278	6.35162	7.76158	9.47556	11.557
43	5.40049	6.63744	8.14966	9.99671	12.2504
44	5.61651	6.93613	8.55714	10.5465	12.9854
45	5.84117	7.24826	8.985	11.1266	13.7645
46	6.07482	7.57443	9.43425	11.7386	14.5904
47	6.31781	7.91528	9.90596	12.3842	15.4658
48	6.57052	8.27146	10.4013	13.0653	16.3938
49	6.83334	8.64368	10.9213	13.7839	17.3774
50	7.10668	9.03265	11.4674	14.542	18.4201
51	7.39094	9.43912	12.0408	15.3418	19.5253
52	7.68658	9.86388	12.6428	16.1856	20.6968
53	7.99405	10.3078	13.2749	17.0758	21.9386
54	8.31381	10.7716	13.9387	18.015	23.2549
55	8.64636	11.2563	14.6356	19.0058	24.6502
56	8.99221	11.7629	15.3674	20.0512	26.1292
57	9.3519	12.2922	16.1358	21.154	27.6969
58	9.72598	12.8453	16.9425	22.3175	29.3588
59	10.115	13.4234	17.7897	23.5449	31.1203
60	10.5196	14.0274	18.6792	24.8399	32.9875
61	10.9404	14.6587	19.6131	26.2061	34.9667
62	11.378	15.3183	20.5938	27.6474	37.0647
63	11.8331	16.0076	21.6235	29.168	39.2886
64	12.3065	16.728	22.7046	30.7723	41.6459
65	12.7987	17.4807	23.8399	32.4648	44.1447
66	13.3107	18.2674	25.0319	34.2503	46.7934
67	13.8431	19.0894	26.2834	36.1341	49.601
68	14.3968	19.9484	27.5976	38.1215	52.577
69	14.9727	20.8461	28.9775	40.2182	55.7316
70	15.5716	21.7842	30.4264	42.4302	59.0755

Table V—*continued*

n	4.0%	4.5%	5.0%	5.5%	6.0%
71	16.1945	22.7645	31.9477	44.7638	62.62
72	16.8422	23.7889	33.5451	47.2258	66.3772
73	17.5159	24.8594	35.2223	49.8232	70.3599
74	18.2166	25.978	36.9834	52.5635	74.5815
75	18.9452	27.147	38.8326	55.4545	79.0563
76	19.703	28.3687	40.7742	58.5045	83.7997
77	20.4912	29.6453	42.8129	61.7223	88.8277
78	21.3108	30.9793	44.9536	65.117	94.1573
79	22.1632	32.3734	47.2013	68.6985	99.8067
80	23.0498	33.8302	49.5613	72.4769	105.795
81	23.9718	35.3525	52.0394	76.4631	112.143
82	24.9306	36.9434	54.6414	80.6686	118.871
83	25.9279	38.6058	57.3734	85.1054	126.004
84	26.965	40.3431	60.2421	89.7862	133.564
85	28.0436	42.1585	63.2542	94.7244	141.578
86	29.1653	44.0557	66.4169	99.9343	150.072
87	30.3319	46.0382	69.7378	105.431	159.077
88	31.5452	48.1099	73.2247	111.229	168.621
89	32.807	50.2748	76.8859	117.347	178.738
90	34.1193	52.5372	80.7302	123.801	189.463
91	35.4841	54.9014	84.7667	130.61	200.831
92	36.9034	57.3719	89.005	137.794	212.88
93	38.3796	59.9537	93.4553	145.372	225.653
94	39.9147	62.6516	98.128	153.368	239.192
95	41.5113	65.4709	103.034	161.803	253.544
96	43.1718	68.4171	108.186	170.702	268.756
97	44.8987	71.4959	113.595	180.091	284.882
98	46.6946	74.7132	119.275	189.996	301.975
99	48.5624	78.0753	125.239	200.446	320.093
100	50.5049	81.5887	131.501	211.47	339.299

Table VI

PRESENT VALUE OF $1 AT END OF n COMPOUNDING PERIODS WITH GROWTH RATE OF r% PER PERIOD: $(1+r)^{-n}$

n	(4/12)%	(4.5/12)%	(5/12)%	(5.5/12)%	(6/12)%
1	.996678	.996264	.995851	.995438	.995025
2	.993367	.992542	.991719	.990896	.990074
3	.990066	.988834	.987604	.986375	.985148
4	.986777	.985139	.983506	.981875	.980247
5	.983499	.981459	.979425	.977395	.97587
6	.980231	.977792	.975361	.972936	.970517

Table VI—*continued*

n	(4/12)%	(4.5/12)%	(5/12)%	(5.5/12)%	(6/12)%
7	.976975	.974139	.971314	.968497	.965689
8	.973729	.970499	.967284	.964078	.960884
9	.970494	.966874	.96327	.95968	.956103
10	.967269	.963261	.959273	.955301	.951347
11	.964056	.959662	.955293	.950943	.946613
12	.960853	.956077	.951329	.946604	.941904
13	.957661	.952505	.947382	.942285	.937217
14	.954479	.948946	.943451	.937986	.932555
15	.951308	.945401	.939536	.933706	.927915
16	.948148	.941869	.935638	.929447	.923298
17	.944997	.93835	.931755	.925206	.918705
18	.941858	.934844	.927889	.920985	.914134
19	.938729	.931352	.924039	.916783	.909586
20	.93561	.927872	.920205	.9126	.90506
21	.932502	.924406	.916387	.908437	.900557
22	.929404	.920952	.912584	.904292	.896077
23	.926316	.917511	.908798	.900166	.891619
24	.923238	.914083	.905027	.896059	.887183
25	.920171	.910668	.901272	.891971	.882769
26	.917114	.907266	.897532	.887902	.878376
27	.914067	.903876	.893808	.88385	.874006
28	.91103	.900499	.890099	.879818	.869658
29	.908004	.897135	.886406	.875804	.865331
30	.904987	.893783	.882728	.871808	.861026
31	.90198	.890444	.879065	.86783	.856742
32	.898984	.887117	.875418	.863871	.852479
33	.895997	.883803	.871785	.85993	.848238
34	.89302	.880501	.868168	.856007	.844018
35	.890054	.877211	.864566	.852101	.839819
36	.887096	.873934	.860978	.848213	.83564
37	.884149	.870669	.857406	.844343	.831483
38	.881212	.867416	.853848	.840491	.827346
39	.878284	.864175	.850305	.836656	.82323
40	.875366	.860947	.846777	.832839	.819134
41	.872458	.85773	.843264	.82904	.815058
42	.869559	.854525	.839764	.825257	.811003
43	.866671	.851333	.83628	.821492	.806969
44	.863791	.848152	.83281	.817744	.802953
45	.860921	.844983	.829355	.814013	.798958
46	.858061	.841827	.825913	.810299	.794983
47	.855211	.838681	.822486	.806602	.791028
48	.852369	.835548	.819074	.802922	.787093
49	.849537	.832426	.815675	.799259	.783177
50	.846715	.829316	.81229	.795613	.77928

Table VI—*continued*

n	(4/12)%	(4.5/12)%	(5/12)%	(5.5/12)%	(6/12)%
51	.843902	.826218	.80892	.791982	.775403
52	.841098	.823131	.805564	.788369	.771545
53	.838304	.820056	.802221	.784772	.767706
54	.835519	.816992	.798892	.781192	.763887
55	.832743	.81394	.795577	.777628	.760086
56	.829976	.810899	.792276	.77408	.756305
57	.827219	.807869	.788989	.770548	.752542
58	.824471	.804851	.785715	.767033	.748798
59	.821732	.801844	.782455	.763533	.745072
60	.819002	.798848	.779209	.76005	.741365
61	.816281	.795864	.775975	.756582	.737677
62	.813569	.79289	.772755	.75313	.734007
63	.810866	.789928	.769549	.749694	.730355
64	.808172	.786977	.766356	.746274	.726721
65	.805487	.784036	.763176	.742869	.723105
66	.802811	.781107	.760009	.739479	.719508
67	.800144	.778189	.756856	.736106	.715928
68	.797485	.775281	.753715	.732747	.712366
69	.794836	.772385	.750588	.729404	.708822
70	.792195	.769499	.747474	.726076	.705295
71	.789563	.766624	.744372	.722763	.701786
72	.78694	.76376	.741283	.719466	.698295
73	.784326	.760907	.738208	.716183	.694821
74	.78172	.758064	.735144	.712916	.691364
75	.779123	.755232	.732094	.709663	.687924
76	.776534	.75241	.729057	.706426	.684501
77	.773954	.749599	.726031	.703203	.681096
78	.771383	.746799	.723019	.699994	.677707
79	.76882	.744009	.720019	.6968	.674335
80	.766266	.741229	.717031	.693622	.67098
81	.76372	.73846	.714056	.690457	.667642
82	.761183	.735701	.711093	.687307	.66432
83	.758654	.732952	.708143	.684171	.661015
84	.756134	.730213	.705204	.68105	.657726
85	.753622	.727485	.702278	.677942	.654454
86	.751118	.724767	.699364	.674849	.651198
87	.748622	.72206	.696463	.67177	.647958
88	.746135	.719362	.693573	.668705	.644734
89	.743656	.716675	.690695	.665654	.641526
90	741186	.713997	.687829	.662618	.638335

Table VI—*continued*

n	(4/12)%	(4.5/12)%	(5/12)%	(5.5/12)%	(6/12)%
91	.738723	.711329	.684975	.659594	.635159
92	.736269	.708672	.682133	.656585	.631998
93	.733823	.706024	.679302	.653589	.628854
94	.731385	.703386	.676484	.650607	.625726
95	.728955	.700759	.673677	.647639	.622613
96	.726533	.69814	.670881	.644684	.619515
97	.72412	.695532	.668098	.641743	.616432
98	.721714	.692934	.665326	.638815	.613365
99	.719316	.690345	.662565	.6359	.610314
100	.716926	.687765	.659816	.632999	.607277

n	(7/12)%	(8/12)%	(9/12)%	1.0%	1.25%
1	.994201	.993378	.992556	.990099	.987654
2	.988434	.986799	.985167	.980296	.975461
3	.982702	.980264	.977833	.97059	.963418
4	.977003	.973772	.970554	.96098	.951524
5	.971337	.967323	.96333	.951466	.939777
6	.965703	.960917	.956158	.942046	.928175
7	.960103	.954554	.94904	.932718	.916716
8	.954534	.948232	.941976	.923484	.905399
9	.948999	.941952	.934964	.91434	.894221
10	.943495	.935714	.928004	.905287	.883181
11	.938023	.929518	.921095	.896324	.872277
12	.932583	.923362	.914239	.88745	.861509
13	.927174	.917247	.907433	.878663	.850873
14	.921797	.911173	.900678	.869963	.840368
15	.916451	.905138	.893973	.86135	.829993
16	.911136	.899144	.887318	.852822	.819747
17	.905852	.893189	.880713	.844378	.809626
18	.900598	.887274	.874157	.836018	.799631
19	.895375	.881398	.867649	.82774	.789759
20	.890182	.875561	.861191	.819545	.780009
21	.885019	.869763	.85478	.811431	.770379
22	.879887	.864003	.848417	.803397	.760868
23	.874784	.858281	.842101	.795442	.751475
24	.86971	.852597	.835832	.787567	.742197
25	.864667	.846951	.82961	.779769	.733034
26	.859652	.841342	.823434	.772049	.723985
27	.854666	.83577	.817305	.764405	.715046
28	.849709	.830235	.811221	.756836	.706219
29	.844782	.824737	.805182	.749343	.6975
30	.839882	.819275	.799188	.741924	.688889

Table VI—*continued*

n	(7/12)%	(8/12)%	(9/12)%	1.0%	1.25%
31	.835011	.81385	.793239	.734578	.680384
32	.830168	.80846	.787334	.727305	.671984
33	.825354	.803106	.781473	.720104	.663688
34	.820567	.797787	.775655	.712974	.655495
35	.815808	.792504	.769881	.705915	.647402
36	.811077	.787256	.76415	.698926	.639409
37	.806373	.782042	.758462	.692006	.631515
38	.801696	.776863	.752815	.685154	.623719
39	.797047	.771718	.747211	.678371	.616019
40	.792424	.766607	.741649	.671654	.608414
41	.787829	.76153	.736128	.665004	.600902
42	.783259	.756488	.730648	.65842	.593484
43	.778717	.751478	.725209	.651901	.586157
44	.774201	.746501	.719811	.645446	.57892
45	.769711	.741557	.714452	.639056	.571773
46	.765247	.736646	.709134	.632729	.564714
47	.760808	.731768	.703855	.626464	.557742
48	.756396	.726922	.698615	.620261	.550857
49	.752009	.722108	.693415	.61412	.544056
50	.747648	.717326	.688253	.60804	.537339
51	.743312	.712575	.683129	.60202	.530706
52	.739001	.707856	.678044	.596059	.524154
53	.734715	.703168	.672997	.590157	.517683
54	.730454	.698512	.667987	.584314	.511291
55	.726218	.693886	.663014	.578529	.504979
56	.722006	.68929	.658079	.572801	.498745
57	.717819	.684725	.65318	.56713	.492588
58	.713656	.680191	.648318	.561515	.486506
59	.709517	.675687	.643491	.555955	.4805
60	.705402	.671212	.638701	.550451	.474568
61	.701311	.666767	.633946	.545001	.468709
62	.697244	.662351	.629227	.539605	.462923
63	.6932	.657965	.624543	.534262	.457207
64	.689179	.653607	.619894	.528972	.451563
65	.685182	.649279	.615279	.523735	.445988
66	.681209	.644979	.610699	.518549	.440482
67	.677258	.640708	.606153	.513415	.435044
68	.67333	.636465	.601641	.508332	.429673
69	.669425	.63225	.597162	.503299	.424368
70	.665543	.628063	.592717	.498316	.419129

Table VI—*continued*

n	(7/12)%	(8/12)%	(9/12)%	1.0%	1.25%
71	.661683	.623903	.588304	.493382	.413955
72	.657845	.619771	.583925	.488497	.408844
73	.65403	.615667	.579578	.483661	.403797
74	.650237	.61159	.575264	.478872	.398812
75	.646466	.607539	.570981	.474131	.393888
76	.642717	.603516	.566731	.469436	.389025
77	.638989	.599519	.562512	.464788	.384223
78	.635283	.595549	.558325	.460186	.379479
79	.631599	.591605	.554169	.45563	.374794
80	.627936	.587687	.550043	.451119	.370167
81	.624294	.583795	.545949	.446653	.365597
82	.620673	.579929	.541884	.44223	.361084
83	.617074	.576088	.537851	.437852	.356626
84	.613495	.572273	.533847	.433517	.352223
85	.609937	.568483	.529873	.429224	.347875
86	.6064	.564719	.525928	.424975	.34358
87	.602883	.560979	.522013	.420767	.339338
88	.599386	.557264	.518127	.416601	.335149
89	.59591	.553573	.51427	.412476	.331011
90	.592454	.549907	.510442	.408392	.326925
91	.589018	.546265	.506642	.404349	.322888
92	.585602	.542648	.502871	.400345	.318902
93	.582206	.539054	.499127	.396382	.314965
94	.578829	.535484	.495412	.392457	.311077
95	.575472	.531938	.491724	.388571	.307236
96	.572135	.528415	.488063	.384724	.303443
97	.568817	.524916	.48443	.380915	.299697
98	.565518	.521439	.480824	.377144	.295997
99	.562235	.517986	.477245	.373409	.292343
100	.558977	.514556	.473692	.369712	.288734

n	1.5%	2.0%	2.5%	3.0%	3.5%
1	.985222	.980392	.97561	.970874	.966184
2	.970662	.961169	.951814	.942596	.933511
3	.956317	.942322	.928599	.915142	.901943
4	.942184	.923845	.90595	.888487	.871442
5	.92826	.905731	.883854	.862609	.841973
6	.914542	.887971	.862297	.837484	.8135
7	.901026	.87056	.841265	.813092	.785991
8	.88771	.85349	.820746	.789409	.759411
9	.874591	.836755	.800728	.766417	.73373
10	.861666	.820348	.781198	.744094	.708918

Table VI—*continued*

n	1.5%	2.0%	2.5%	3.0%	3.5%
11	.848932	.804263	.762144	.722421	.684945
12	.836386	.788493	.743555	.70138	.661783
13	.824025	.773032	.72542	.680951	.639403
14	.811848	.757875	.707727	.661118	.617781
15	.79985	.743015	.690465	.641862	.59689
16	.788029	.728445	.673624	.623167	.576705
17	.776383	.714162	.657194	.605016	.557203
18	.76491	.700159	.641165	.587395	.53836
19	.753605	.68643	.625527	.570286	.520155
20	.742468	.672971	.61027	.553676	.502565
21	.731496	.659775	.595385	.537549	.48557
22	.720685	.646839	.580864	.521893	.46915
23	.710035	.634156	.566696	.506692	.453285
24	.699542	.621721	.552875	.491934	.437956
25	.689203	.60953	.53939	.477606	.423146
26	.679018	.597579	.526234	.463695	.408837
27	.668983	.585862	.513399	.450189	.395011
28	.659097	.574374	.500877	.437077	.381653
29	.649356	.563112	.48866	.424346	.368747
30	.63976	.55207	.476742	.411987	.356278
31	.630305	.541246	.465114	.399987	.344229
32	.62099	.530633	.45377	.388337	.332589
33	.611813	.520228	.442702	.377026	.321342
34	.602771	.510028	.431904	.366045	.310475
35	.593863	.500027	.42137	.355383	.299976
36	.585087	.490222	.411093	.345032	.289832
37	.57644	.48061	.401066	.334983	.280031
38	.567921	.471187	.391284	.325226	.270561
39	.559528	.461948	.38174	.315754	.261412
40	.551259	.45289	.37243	.306557	.252572
41	.543112	.44401	.363346	.297628	.244031
42	.535086	.435303	.354484	.288959	.235778
43	.527178	.426768	.345838	.280543	.227805
44	.519387	.4184	.337403	.272372	.220102
45	.511712	.410196	.329173	.264439	.212658
46	.504149	.402153	.321145	.256736	.205467
47	.496699	.394268	.313312	.249259	.198519
48	.489358	.386537	.30567	.241999	.191806
49	.482126	.378958	.298215	.23495	.18532
50	.475001	.371527	.290941	.228107	.179053

Table VI—*continued*

n	1.5%	2.0%	2.5%	3.0%	3.5%
51	.467981	.364242	.283845	.221463	.172998
52	.461065	.3571	.276922	.215013	.167148
53	.454252	.350098	.270168	.20875	.161495
54	.447538	.343234	.263578	.20267	.156034
55	.440924	.336504	.25715	.196767	.150758
56	.434408	.329905	.250878	.191036	.145659
57	.427988	.323437	.244759	.185472	.140734
58	.421663	.317095	.238789	.18007	.135975
59	.415432	.310877	.232965	.174825	.131376
60	.409292	.304782	.227283	.169733	.126934
61	.403244	.298805	.221739	.164789	.122641
62	.397284	.292947	.216331	.15999	.118494
63	.391413	.287203	.211055	.15533	.114487
64	.385629	.281571	.205907	.150806	.110615
65	.37993	.27605	.200885	.146413	.106875
66	.374315	.270637	.195985	.142149	.103261
67	.368783	.265331	.191205	.138008	.099769
68	.363333	.260128	.186541	.133989	.096395
69	.357963	.255027	.181992	.130086	.093135
70	.352673	.250027	.177553	.126297	.089986
71	.347461	.245125	.173222	.122619	.086943
72	.342326	.240318	.168997	.119047	.084003
73	.337267	.235606	.164875	.11558	.081162
74	.332283	.230986	.160854	.112214	.078417
75	.327372	.226457	.156931	.108945	.075765
76	.322534	.222017	.153103	.105772	.073203
77	.317768	.217663	.149369	.102691	.070728
78	.313072	.213396	.145726	.0997	.068336
79	.308445	.209211	.142171	.096796	.066025
80	.303887	.205109	.138704	.093977	.063792
81	.299396	.201087	.135321	.09124	.061635
82	.294971	.197144	.13202	.088582	.059551
83	.290612	.193279	.1288	.086002	.057537
84	.286317	.189489	.125659	.083497	.055591
85	.282086	.185774	.122594	.081065	.053712
86	.277917	.182131	.119604	.078704	.051895
87	.27381	.17856	.116687	.076412	.05014
88	.269763	.175059	.113841	.074186	.048445
89	.265776	.171626	.111064	.072026	.046806
90	.261849	.168261	.108355	.069928	.045224

Table VI—*continued*

n	1.5%	2.0%	2.5%	3.0%	3.5%
91	.257979	.164962	.105712	.067891	.043694
92	.254166	.161727	.103134	.065914	.042217
93	.25041	.158556	.100618	.063994	.040789
94	.24671	.155447	.098164	.06213	.03941
95	.243064	.152399	.09577	.06032	.038077
96	.239471	.149411	.093434	.058563	.036789
97	.235932	.146481	.091155	.056858	.035545
98	.232446	.143609	.088932	.055202	.034343
99	.22901	.140793	.086763	.053594	.033182
100	.225626	.138032	.084647	.052033	.03206

n	4.0%	4.5%	5.0%	5.5%	6.0%
1	.961539	.956938	.952381	.947867	.943396
2	.924556	.91573	.90703	.898452	.889997
3	.888997	.876297	.863838	.851614	.83962
4	.854804	.838561	.822703	.807216	.792094
5	.821927	.802451	.783526	.765134	.747259
6	.790315	.767896	.746216	.725246	.704961
7	.759918	.734828	.710682	.687437	.665058
8	.73069	.703185	.67684	.651599	.627413
9	.702587	.672904	.644609	.617629	.591899
10	.675564	.643928	.613913	.58543	.558395
11	.649581	.616199	.58468	.55491	.526788
12	.624597	.589664	.556838	.525981	.49697
13	.600574	.564272	.530322	.49856	.46884
14	.577475	.539973	.505068	.472569	.442302
15	.555265	.51672	.481017	.447933	.417266
16	.533908	.494469	.458112	.424581	.393647
17	.513373	.473176	.436297	.402446	.371365
18	.493628	.4528	.415521	.381465	.350344
19	.474643	.433302	.395734	.361579	.330514
20	.456387	.414643	.37689	.342728	.311805
21	.438834	.396787	.358943	.324861	.294156
22	.421956	.379701	.34185	.307925	.277506
23	.405727	.36335	.325572	.291872	.261798
24	.390122	.347703	.310068	.276656	.246979
25	.375117	.33273	.295303	.262233	.232999
26	.360689	.318402	.281241	.248562	.219811
27	.346817	.304691	.267849	.235604	.207369
28	.333478	.291571	.255094	.223321	.195631
29	.320652	.279015	.242946	.211679	.184557
30	.308319	.267	.231378	.200644	.174111

Table VI—*continued*

n	4.0%	4.5%	5.0%	5.5%	6.0%
31	.29646	.255502	.22036	.190183	.164255
32	.285058	.2445	.209866	.180269	.154958
33	.274094	.233971	.199873	.170871	.146187
34	.263552	.223896	.190355	.161963	.137912
35	.253416	.214254	.18129	.153519	.130106
36	.243669	.205028	.172658	.145516	.122741
37	.234297	.196199	.164436	.13793	.115794
38	.225286	.18775	.156606	.130739	.109239
39	.216621	.179665	.149148	.123923	.103056
40	.208289	.171929	.142046	.117463	.097223
41	.200278	.164525	.135282	.111339	.091719
42	.192575	.15744	.12884	.105535	.086528
43	.185168	.15066	.122705	.100033	.08163
44	.178046	.144173	.116861	.094818	.077009
45	.171199	.137964	.111297	.089875	.07265
46	.164614	.132023	.105997	.085189	.068538
47	.158283	.126338	.100949	.080748	.064659
48	.152195	.120898	.096142	.076539	.060999
49	.146341	.115691	.091564	.072548	.057546
50	.140713	.11071	.087204	.068766	.054289
51	.135301	.105942	.083051	.065181	.051216
52	.130097	.10138	.079096	.061783	.048317
53	.125093	.097014	.07533	.058562	.045582
54	.120282	.092837	.071743	.055509	.043002
55	.115656	.088839	.068327	.052615	.040568
56	.111207	.085013	.065073	.049872	.038271
57	.10693	.081353	.061974	.047272	.036105
58	.102817	.077849	.059023	.044808	.034061
59	.098863	.074497	.056212	.042472	.032133
60	.09506	.071289	.053536	.040258	.030315
61	.091404	.068219	.050986	.038159	.028599
62	.087889	.065281	.048558	.03617	.02698
63	.084508	.06247	.046246	.034284	.025453
64	.081258	.05978	.044044	.032497	.024012
65	.078133	.057206	.041947	.030803	.022653
66	.075128	.054742	.039949	.029197	.021371
67	.072238	.052385	.038047	.027675	.020161
68	.06946	.050129	.036235	.026232	.01902
69	.066788	.047971	.03451	.024864	.017943
70	.064219	.045905	.032866	.023568	.016927

Table VI—*continued*

n	4.0%	4.5%	5.0%	5.5%	6.0%
71	.061749	.043928	.031301	.022339	.015969
72	.059375	.042036	.029811	.021175	.015065
73	.057091	.040226	.028391	.020071	.014213
74	.054895	.038494	.027039	.019025	.013408
75	.052784	.036836	.025752	.018033	.012649
76	.050754	.03525	.024525	.017093	.011933
77	.048802	.033732	.023357	.016202	.011258
78	.046925	.03228	.022245	.015357	.010621
79	.04512	.03089	.021186	.014556	.010019
80	.043384	.029559	.020177	.013798	.009452
81	.041716	.028287	.019216	.013078	.008917
82	.040111	.027068	.018301	.012396	.008412
83	.038569	.025903	.01743	.01175	.007936
84	.037085	.024787	.0166	.011138	.007487
85	.035659	.02372	.015809	.010557	.007063
86	.034287	.022699	.015056	.010007	.006663
87	.032969	.021721	.014339	.009485	.006286
88	.031701	.020786	.013657	.00899	.00593
89	.030481	.019891	.013006	.008522	.005595
90	.029309	.019034	.012387	.008077	.005278
91	.028182	.018214	.011797	.007656	.004979
92	.027098	.01743	.011235	.007257	.004697
93	.026056	.01668	.0107	.006879	.004432
94	.025053	.015961	.010191	.00652	.004181
95	.02409	.015274	.009705	.00618	.003944
96	.023163	.014616	.009243	.005858	.003721
97	.022272	.013987	.008803	.005553	.00351
98	.021416	.013385	.008384	.005263	.003312
99	.020592	.012808	.007985	.004989	.003124
100	.0198	.012257	.007605	.004729	.002947

Table VII

$$s_{\overline{n}|i} = \frac{(1+i)^n - 1}{i}$$

n	(4/12)%	(4.5/12)%	(5/12)%	(5.5/12)%
12	12.2226	12.2509	12.2787	12.3072
24	24.9432	25.0647	25.1855	25.3086
36	38.182	38.4671	38.7527	39.0433
48	51.9602	52.4852	53.014	53.5528
60	66.2996	67.1474	68.0049	68.8808

Table VII—*continued*

n	(4/12)%	(4.5/12)%	(5/12)%	(5.5/12)%
72	81.2234	82.483	83.7628	85.0734
84	96.7553	98.5234	100.327	102.179
96	112.92	115.301	117.738	120.25
108	129.743	132.848	136.041	139.34
120	147.252	151.203	155.279	159.508
132	165.473	170.4	175.502	180.812
144	184.438	190.479	196.76	203.319
156	204.175	211.481	219.105	227.095
168	224.716	233.447	242.593	252.212
180	246.094	256.423	267.283	278.745
192	268.343	280.454	293.236	306.776
204	291.498	305.589	320.517	336.388
216	315.597	331.88	349.194	367.67
228	340.678	359.377	379.338	400.717
240	366.78	388.138	411.024	435.627
252	393.946	418.221	444.331	472.507
264	422.218	449.685	479.342	511.468
276	451.643	482.595	516.144	552.626
288	482.266	517.017	554.83	596.105
300	514.138	553.02	595.494	642.037
312	547.307	590.677	638.239	690.56
324	581.828	630.064	683.17	741.821
336	617.755	671.261	730.401	795.973
348	655.146	714.35	780.047	853.179
360	694.061	759.419	832.234	913.612
372	734.561	806.558	887.091	977.454
384	776.711	855.863	944.754	1044.9
396	820.578	907.433	1005.37	1116.14
408	866.233	961.373	1069.08	1191.41
420	913.748	1017.79	1136.06	1270.92
432	963.198	1076.8	1206.46	1354.92
444	1014.66	1138.52	1280.46	1443.66
456	1068.23	1203.07	1358.25	1537.4
468	1123.97	1270.6	1440.02	1636.43
480	1181.98	1341.22	1525.97	1741.04

Table VII—*continued*

n	(6/12)%	(6.5/12)%	(7/12)%	(7.5/12)%
12	12.336	12.3639	12.3928	12.421
24	25.4328	25.5559	25.6814	25.6063
36	39.3374	39.6314	39.9307	40.2306
48	54.0998	54.6495	55.2101	55.7748
60	69.7726	70.6734	71.5941	72.5257
72	86.412	87.7705	89.1624	90.5769
84	104.078	106.013	108.001	110.03
96	122.833	125.476	128.201	130.992
108	142.746	146.243	149.862	153.582
120	163.886	168.402	173.088	177.926
132	186.331	192.044	197.994	204.16
144	210.159	217.269	224.699	232.43
156	235.458	244.184	253.336	262.895
168	262.317	272.901	284.043	295.725
180	290.832	303.542	316.969	331.103
192	321.107	336.234	352.276	369.228
204	353.249	371.116	390.135	410.313
216	387.373	408.334	430.731	454.587
228	423.602	448.045	474.262	502.298
240	462.066	490.415	520.94	553.713
252	502.902	535.623	570.992	609.12
264	546.256	583.858	624.662	668.827
276	592.285	635.324	682.213	733.17
288	641.153	690.236	743.923	802.508
300	693.036	748.826	810.095	877.229
312	748.118	811.34	881.051	957.75
324	806.597	878.041	957.135	1044.52
336	868.684	949.208	1038.72	1138.03
348	934.601	1025.14	1126.2	1238.8
360	1004.58	1106.16	1220.01	1347.39
372	1078.88	1192.61	1320.6	1464.41
384	1157.76	1284.84	1428.46	1590.52
396	1241.51	1383.25	1544.12	1726.41
408	1330.42	1488.26	1668.14	1872.85
420	1424.82	1600.29	1801.12	2030.67
432	1525.03	1719.83	1943.72	2200.73
444	1631.43	1847.37	2096.62	2384.
456	1744.4	1983.46	2260.58	2581.49
468	1864.33	2128.65	2436.4	2794.32
480	1991.65	2283.58	2624.92	3023.67

Table VII—*continued*

n	(8/12)%	(8.5/12)%	(9/12)%	(9.5/12)%
12	12.4499	12.4788	12.5075	12.5366
24	25.9331	26.0606	26.1883	26.3175
36	40.5353	40.8429	41.1525	41.4661
48	56.3496	56.9317	57.5204	58.1181
60	73.4765	74.4428	75.4237	76.4228
72	92.0248	93.5016	95.0065	96.5443
84	112.113	114.245	116.426	118.663
96	133.868	136.822	139.855	142.976
108	157.428	161.395	165.482	169.703
120	182.945	188.139	193.513	199.083
132	210.579	217.248	224.173	231.378
144	240.507	248.93	257.71	266.878
156	272.918	283.412	294.392	305.902
168	308.02	320.942	334.516	348.799
180	346.035	361.789	378.403	395.953
192	387.206	406.246	426.407	447.787
204	431.794	454.634	478.914	504.766
216	480.082	507.298	536.347	567.4
228	532.378	564.618	599.167	636.25
240	589.015	627.004	667.881	711.933
252	650.352	694.904	743.04	795.128
264	716.781	768.806	825.249	886.58
276	788.723	849.241	915.171	987.108
288	866.636	936.785	1013.53	1097.61
300	951.016	1032.07	1121.11	1219.09
312	1042.4	1135.77	1238.78	1352.61
324	1141.37	1248.64	1367.5	1499.4
336	1248.55	1371.49	1508.29	1660.74
348	1364.63	1505.2	1662.28	1838.11
360	1490.34	1650.72	1830.72	2033.07
372	1626.49	1809.11	2014.96	2247.39
384	1773.93	1981.5	2216.49	2482.97
396	1933.62	2169.12	2436.91	2741.94
408	2106.56	2373.34	2678.02	3026.61
420	2293.85	2595.6	2941.74	3339.53
432	2496.69	2837.5	3230.21	3683.51
444	2716.36	3100.79	3545.73	4061.63
456	2954.26	3387.35	3890.85	4477.28
468	3211.92	3699.25	4268.34	4934.18
480	3490.95	4038.71	4681.25	5436.42

Table VIII

$\frac{1}{(s_{\overline{n}|i})}$

n	(4/12)%	(4.5/12)%	(5/12)%	(5.5/12)%
12	.08182	.081631	.081447	.081258
24	.040096	.039901	.03971	.039517
36	.026195	.026001	.025809	.025617
48	.01925	.019057	.018867	.018678
60	.015088	.014897	.014709	.014522
72	.012316	.012128	.011943	.011759
84	.01034	.010154	.009972	.009791
96	.00886	.008677	.008498	.00832
108	.007712	.007532	.007355	.007181
120	.006796	.006618	.006445	.006274
132	.006048	.005873	.005702	.005535
144	.005426	.005254	.005087	.004923
156	.004902	.004733	.004569	.004408
168	.004455	.004288	.004127	.003969
180	.004068	.003904	.003746	.003592
192	.003731	.00357	.003415	.003264
204	.003435	.003277	.003124	.002977
216	.003173	.003018	.002868	.002724
228	.00294	.002787	.002641	.0025
240	.002731	.002581	.002437	.0023
252	.002543	.002396	.002255	.002121
264	.002373	.002228	.002091	.00196
276	.002219	·002077	.001942	.001814
288	.002078	.001939	.001807	.001682
300	.00195	.001813	.001684	.001562
312	.001832	.001697	.001571	.001453
324	.001723	.001592	.001468	.001353
336	.001623	.001494	.001374	.001261
348	.001531	.001404	.001286	.001177
360	.001445	.001321	.001206	.001099
372	.001366	.001244	.001132	.001028
384	.001292	.001173	.001063	.000962
396	.001223	.001107	.000999	.0009
408	.001159	.001045	.00094	.000844
420	.001099	.000987	.000885	.000791
432	.001043	.000933	.000833	.000743
444	.00099	.000883	.000785	.000697
456	.000941	.000836	.000741	.000655
468	.000894	.000792	.000699	.000616
480	.000851	.00075	.00066	.000579

Table VIII—*continued*

n	(6/12)%	(6.5/12)%	(7/12)%	(7.5/12)%
12	.081068	.080885	.080697	.080513
24	.039324	.039134	.038943	.038755
36	.025426	.025237	.025048	.024861
48	.018489	.018303	.018117	.017934
60	.014337	.014154	.013972	.013793
72	.011577	.011398	.01122	.011045
84	.009613	.009437	.009264	.009093
96	.008146	.007974	.007805	.007639
108	.00701	.006842	.006677	.006516
120	.006106	.005943	.005782	.005625
132	.005371	.005212	.005055	.004903
144	.004763	.004607	.004455	.004307
156	.004252	.0041	.003952	.003808
168	.003817	.003669	.003525	.003386
180	.003443	.003299	.003159	.003025
192	.003119	.002979	.002843	.002713
204	.002835	.002699	.002568	.002442
216	.002586	.002453	.002326	.002204
228	.002365	.002236	.002113	.001995
240	.002169	.002044	.001924	.00181
252	.001993	.001871	.001756	.001646
264	.001835	.001717	.001605	.0015
276	.001693	.001579	.00147	.001368
288	.001564	.001453	.001349	.001251
300	.001447	.00134	.001239	.001144
312	.001341	.001237	.00114	.001049
324	.001244	.001143	.001049	.000962
336	.001156	.001058	.000967	.000883
348	.001074	.00098	.000892	.000812
360	.001	.000909	.000824	.000747
372	.000931	.000843	.000762	.000687
384	.000868	.000783	.000705	.000633
396	.00081	.000727	.000652	.000584
408	.000756	.000676	.000604	.000538
420	.000706	.000629	.00056	.000497
432	.00066	.000586	.000519	.000459
444	.000617	.000546	.000481	.000424
456	.000578	.000509	.000447	.000392
468	.000541	.000474	.000415	.000362
480	.000507	.000442	.000385	.000335

Table VIII—*continued*

n	(8/12)%	(8.5/12)%	(9/12)%	(9.5/12)%
12	.080322	.080136	.079952	.079766
24	.038561	.038372	.038185	.037998
36	.02467	.024484	.0243	.024116
48	.017746	.017565	.017385	.017206
60	.01361	.013433	.013258	.013085
72	.010867	.010695	.010526	.010358
84	.00892	.008753	.008589	.008427
96	.00747	.007309	.00715	.006994
108	.006352	.006196	.006043	.005893
120	.005466	.005315	.005168	.005023
132	.004749	.004603	.004461	.004322
144	.004158	.004017	.00388	.003747
156	.003664	.003528	.003397	.003269
168	.003247	.003116	.002989	.002867
180	.00289	.002764	.002643	.002526
192	.002583	.002462	.002345	.002233
204	.002316	.0022	.002088	.001981
216	.002083	.001971	.001864	.001762
228	.001878	.001771	.001669	.001572
240	.001698	.001595	.001497	.001405
252	.001538	.001439	.001346	.001258
264	.001395	.001301	.001212	.001128
276	.001268	.001178	.001093	.001013
288	.001154	.001067	.000987	.000911
300	.001052	.000969	.000892	.00082
312	.000959	.00088	.000807	.000739
324	.000876	.000801	.000731	.000667
336	.000801	.000729	.000663	.000602
348	.000733	.000664	.000602	.000544
360	.000671	.000606	.000546	.000492
372	.000615	.000553	.000496	.000445
384	.000564	.000505	.000451	.000403
396	.000517	.000461	.00041	.000365
408	.000475	.000421	.000373	.00033
420	.000436	.000385	.00034	.000299
432	.000401	.000352	.00031	.000271
444	.000368	.000322	.000282	.000246
456	.000338	.000295	.000257	.000223
468	.000311	.00027	.000234	.000203
480	.000286	.000248	.000214	.000184

Table IX

$$a_{\overline{n}|i} = \frac{1-(1+i)^{-n}}{i}$$

n	(4/12)%	(4.5/12)%	(5/12)%	(5.5/12)%
12	11.7441	11.7128	11.681	11.65
24	23.0285	22.9112	22.7936	22.678
36	33.8711	33.6177	33.3652	33.1171
48	44.2893	43.8539	43.4224	42.9988
60	54.2995	53.6406	52.99	52.3528
72	63.918	62.9973	62.092	61.2074
84	73.1599	71.9431	70.751	69.5892
96	82.04	80.496	78.9885	77.5234
108	90.5725	88.6731	86.8251	85.034
120	98.771	96.4912	94.2802	92.1436
132	106.648	103.966	101.373	98.8735
144	114.218	111.112	108.12	105.244
156	121.49	117.945	114.538	111.274
168	128.479	124.477	120.645	116.983
180	135.193	130.722	126.454	122.387
192	141.645	136.693	131.98	127.502
204	147.844	142.402	137.238	132.344
216	153.8	147.86	142.239	136.927
228	159.524	153.079	146.997	141.266
240	165.023	158.068	151.524	145.373
252	170.307	162.838	155.83	149.26
265	175.384	167.398	159.927	152.94
276	180.262	171.758	163.824	156.424
288	184.95	175.927	167.532	159.722
300	189.454	179.913	171.059	162.843
312	193.781	183.723	174.414	165.798
324	197.939	187.366	177.606	168.595
336	201.935	190.849	180.643	171.243
348	205.774	194.18	183.532	173.749
360	209.462	197.363	186.28	176.122
372	213.007	200.407	188.895	178.368
384	216.412	203.318	191.382	180.493
396	219.684	206.1	193.749	182.506
408	222.828	208.76	196.	184.411
420	225.85	211.304	198.141	186.214
432	228.752	213.736	200.179	187.921
444	231.541	216.06	202.117	189.537
456	234.221	218.283	203.961	191.066
468	236.796	220.408	205.715	192.514
480	239.271	222.44	207.383	193.885

Table IX—*continued*

n	(6/12)%	(6.5/12)%	(7/12)%	(7.5/12)%
12	11.6193	11.5879	11.5573	11.5262
24	22.5635	22.4484	22.3354	22.2221
36	32.872	32.6273	32.3868	32.1474
48	42.5815	42.1672	41.7607	41.3578
60	51.727	51.1084	50.5026	49.9046
72	60.3411	59.4883	58.6551	57.8358
84	68.4547	67.3423	66.258	65.1956
96	76.0971	74.7032	73.3483	72.0252
108	83.2954	81.6022	79.9606	78.3628
120	90.0755	88.0681	86.1272	84.2438
132	96.4617	94.1281	91.878	89.7012
144	102.477	99.8078	97.2411	94.7655
156	108.143	105.131	102.243	99.4649
168	113.479	110.12	106.907	103.826
180	118.506	114.796	111.257	107.872
192	123.24	119.178	115.313	111.628
204	127.7	123.286	119.097	115.112
216	131.9	127.135	122.625	118.346
228	135.857	130.743	125.915	121.347
240	139.583	134.125	128.983	124.131
252	143.093	137.294	131.845	126.715
264	146.399	140.264	134.513	129.113
276	149.513	143.048	137.002	131.338
288	152.446	145.657	139.323	133.403
300	155.209	148.102	141.488	135.319
312	157.811	150.394	143.506	137.097
324	160.262	152.542	145.389	138.747
336	162.571	154.555	147.144	140.278
348	164.745	156.442	148.781	141.699
360	166.793	158.21	150.308	143.017
372	168.723	159.868	151.732	144.24
384	170.54	161.421	153.06	145.376
396	172.251	162.877	154.298	146.429
408	173.863	164.241	155.453	147.407
420	175.382	165.52	156.53	148.314
432	176.812	166.719	157.535	149.156
444	178.159	167.842	158.471	149.937
456	179.428	168.895	159.345	150.662
468	180.623	169.882	160.16	151.335
480	181.749	170.807	160.919	151.959

Table IX—*continued*

n	(8/12)%	(8.5/12)%	(9/12)%	(9.5/12)%
12	11.4958	11.4653	11.4348	11.4047
24	22.1105	21.9995	21.889	21.7797
36	31.9117	31.6782	31.4467	31.218
48	40.9617	40.5709	40.1846	39.8042
60	49.3183	48.7413	48.1732	47.6151
72	57.0343	56.2482	55.4767	54.7207
84	64.159	63.1455	62.1538	61.1849
96	70.7377	69.4826	68.2582	67.0654
108	76.8123	75.3051	73.8392	72.4149
120	82.4212	80.6547	78.9415	77.2815
132	87.6003	85.5698	83.6062	81.7087
144	92.3825	90.0858	87.8709	85.7361
156	96.7982	94.235	91.7698	89.4
168	100.876	98.0472	95.3343	92.733
180	104.64	101.55	98.5932	95.7651
192	108.117	104.760	101.573	98.5234
204	111.326	107.725	104.296	101.033
216	114.29	110.442	106.787	103.315
228	117.027	112.938	109.063	105.392
240	119.554	115.231	111.145	107.281
252	121.887	117.338	113.048	109.
264	124.042	119.274	114.787	110.563
276	126.031	121.053	116.378	111.986
288	127.868	122.687	117.832	113.279
300	129.564	124.189	119.161	114.456
312	131.13	125.568	120.377	115.527
324	132.577	126.836	121.488	116.501
336	133.912	128.001	122.504	117.387
348	135.145	129.071	123.433	118.193
360	136.283	130.054	124.282	118.927
372	137.335	130.957	125.058	119.594
384	138.305	131.787	125.768	120.201
396	139.201	132.55	126.417	120.753
408	140.029	133.25	127.01	121.255
420	140.793	133.894	127.552	121.712
432	141.499	134.485	128.048	122.128
444	142.15	135.029	128.501	122.506
456	142.752	135.528	128.916	122.85
468	143.307	135.987	129.294	123.163
480	143.82	136.408	129.641	123.447

Table X

$\frac{1}{a_{\overline{n}|i}}$

n	(4/12)%	(4.5/12)%	(5/12)%	(5.5/12)%
12	.085154	.085381	.085613	.085841
24	.043429	.043651	.043877	.0441
36	.029528	.029751	.029976	.0302
48	.022583	.022807	.023034	.023261
60	.018421	.018647	.018876	.019106
72	.01565	.015878	.01611	.016342
84	.013673	.013904	.014139	.014375
96	.012194	.012427	.012665	.012904
108	.011045	.011282	.011522	.011765
120	.010129	.010368	.010611	.010857
132	.009381	.009623	.009869	.010118
144	.00876	.009004	.009254	.009506
156	.008236	.008483	.008735	.008991
168	.007788	.008038	.008293	.008553
180	.007401	.007654	.007913	.008175
192	.007064	.00732	.007581	.007848
204	.006768	.007027	.007291	.007561
216	.006506	.006768	.007035	.007308
228	.006273	.006537	.006807	.007083
240	.006064	.006331	.006604	.006883
252	.005876	.006146	.006422	.006704
264	.005706	.005978	.006257	.006543
276	.005552	.005827	.006109	.006397
288	.005411	.005689	.005974	.006265
300	.005283	.005563	.00585	.006145
312	.005165	.005447	.005738	.006036
324	.005057	.005342	.005635	.005936
336	.004957	.005244	.00554	.005844
348	.004864	.005154	.005453	.00576
360	.004779	.005071	.005373	.005682
372	.004699	.004994	.005298	.005611
384	.004625	.004923	.00523	.005545
396	.004556	.004857	.005166	.005484
408	.004492	.004795	.005107	.005427
420	.004432	.004737	.005051	.005375
432	.004376	.004683	.005	.005326
444	.004323	.004633	.004952	.005281
456	.004274	.004586	.004907	.005238
468	.004228	.004542	.004866	.005199
480	.004184	.0045	.004826	.005162

Table X—*continued*

n	(6/12)%	(6.5/12)%	(7/12)%	(7.5/12)%
12	.086068	.086302	.08653	.086763
24	.044324	.044551	.044777	.045005
36	.030426	.030654	.030881	.031111
48	.023489	.02372	.02395	.024184
60	.019337	.019571	.019805	.020043
72	.016577	.016815	.017053	.017295
84	.014613	.014854	.015097	.015343
96	.013146	.013391	.013638	.013889
108	.01201	.012259	.012511	.012766
120	.011106	.011359	.011615	.011875
132	.010371	.010628	.010889	.011153
144	.009763	.010024	.010288	.010557
156	.009252	.009516	.009785	.010058
168	.008817	.009085	.009358	.009636
180	.008443	.008716	.008993	.009275
192	.008119	.008395	.008677	.008963
204	.007835	.008116	.008401	.008692
216	.007586	.00787	.008159	.008454
228	.007365	.007653	.007946	.008245
240	.007169	.00746	.007757	.00806
252	.006993	.007288	.007589	.007896
264	.006835	.007134	.007439	.00775
276	.006693	.006995	.007304	.007618
288	.006564	.00687	.007182	.007501
300	.006447	.006757	.007072	.007394
312	.006341	.006654	.006973	.007299
324	.006244	.00656	.006883	.007212
336	.006156	.006475	.006801	.007133
348	.006074	.006397	.006726	.007062
360	.006	.006325	.006657	.006997
372	.005931	.00626	.006595	.006937
384	.005868	.006199	.006538	.006883
396	.00581	.006144	.006485	.006834
408	.005756	.006093	.006437	.006788
420	.005706	.006046	.006393	.006747
432	.00566	.006003	.006352	.006709
444	.005617	.005962	.006315	.006674
456	.005578	.005925	.00628	.006642
468	.005541	.005891	.006248	.006612
480	.005507	.005859	.006219	.006585

Table X—*continued*

n	(8/12)%	(8.5/12)%	(9/12)%	(9.5/12)%
12	.ø86989	.ø87219	.ø87452	.ø87683
24	.ø45227	.ø45456	.ø45685	.ø45914
36	.ø31337	.ø31567	.ø318	.ø32ø33
48	.ø24413	.ø24648	.ø24885	.ø25123
6ø	.ø2ø276	.ø2ø516	.ø2ø758	.ø21øø2
72	.ø17533	.ø17778	.ø18ø26	.ø18275
84	.ø15586	.ø15836	.ø16ø89	.ø16344
96	.ø14137	.ø14392	.ø1465	.ø14911
1ø8	.ø13ø19	.ø13279	.ø13543	.ø138ø9
12ø	.ø12133	.ø12399	.ø12668	.ø1294
132	.ø11415	.ø11686	.ø11961	.ø12239
144	.ø1ø825	.ø111ø1	.ø1138	.ø11664
156	.ø1ø331	.ø1ø612	.ø1ø897	.ø11186
168	.øø9913	.ø1ø199	.ø1ø489	.ø1ø784
18ø	.øø9557	.øø9847	.ø1ø143	.ø1ø442
192	.øø9249	.øø9545	.øø9845	.ø1ø15
2ø4	.øø8983	.øø9283	.øø9588	.øø9898
216	.øø875	.øø9ø55	.øø9364	.øø9679
228	.øø8545	.øø8854	.øø9169	.øø9488
24ø	.øø8364	.øø8678	.øø8997	.øø9321
252	.øø82ø4	.øø8522	.øø8846	.øø9174
264	.øø8ø62	.øø8384	.øø8712	.øø9ø45
276	.øø7935	.øø8261	.øø8593	.øø893
288	.øø7821	.øø8151	.øø8487	.øø8828
3øø	.øø7718	.øø8ø52	.øø8392	.øø8737
312	.øø7626	.øø7964	.øø83ø7	.øø8656
324	.øø7543	.øø7884	.øø8231	.øø8584
336	.øø7468	.øø7812	.øø8163	.øø8519
348	.øø7399	.øø7748	.øø81ø2	.øø8461
36ø	.øø7338	.øø7689	.øø8ø46	.øø84ø9
372	.øø7281	.øø7636	.øø7996	.øø8362
384	.øø723	.øø7588	.øø7951	.øø8319
396	.øø7184	.øø7544	.øø791	.øø8281
4ø8	.øø7141	.øø75ø5	.øø7873	.øø8247
42ø	.øø71ø3	.øø7469	.øø784	.øø8216
432	.øø7ø67	.øø7436	.øø781	.øø8188
444	.øø7ø35	.øø74ø6	.øø7782	.øø8163
456	.øø7øø5	.øø7379	.øø7757	.øø814
468	.øø6978	.øø7354	.øø7734	.øø8119
48ø	.øø6953	.øø7331	.øø7714	.øø81ø1

Table XI

$P(n, r, p) = {}_nC_r \cdot p^r \cdot (1-p)^{n-r}$

		p						
n	*r*	.05	.10	.20	.25	.30	.40	.50
1	0	.9500	.9000	.8000	.7500	.7000	.6000	.5000
	1	.0500	.1000	.2000	.2500	.3000	.4000	.5000
2	0	.9025	.8100	.6400	.5625	.4900	.3600	.2500
	1	.0950	.1800	.3200	.3750	.4200	.4800	.5000
	2	.0025	.0100	.0400	.0625	.0900	.1600	.2500
3	0	.8574	.7290	.5120	.4219	.3430	.2160	.1250
	1	.1354	.2430	.3840	.4219	.4410	.4320	.3750
	2	.0071	.0270	.0960	.1406	.1890	.2880	.3750
	3	.0001	.0010	.0080	.0156	.0270	.0640	.1250
4	0	.8145	.6561	.4096	.3164	.2401	.1296	.0625
	1	.1715	.2916	.4096	.4219	.4116	.3456	.2500
	2	.0135	.0486	.1536	.2109	.2646	.3456	.3750
	3	.0005	.0036	.0256	.0469	.0756	.1536	.2500
	4	.0000	.0001	.0016	.0039	.0081	.0256	.0625
5	0	.7738	.5905	.3277	.2373	.1681	.0778	.0312
	1	.2036	.3280	.4096	.3955	.3602	.2592	.1562
	2	.0214	.0729	.2048	.2637	.3087	.3456	.3125
	3	.0011	.0081	.0512	.0879	.1323	.2304	.3125
	4	.0000	.0004	.0064	.0146	.0284	.0768	.1562
	5	.0000	.0000	.0003	.0010	.0024	.0102	.0312
6	0	.7351	.5314	.2621	.1780	.1176	.0467	.0156
	1	.2321	.3543	.3932	.3560	.3025	.1866	.0938
	2	.0305	.0984	.2458	.2966	.3241	.3110	.2344
	3	.0021	.0146	.0819	.1318	.1852	.2765	.3125
	4	.0001	.0012	.0154	.0330	.0595	.1382	.2344
	5	.0000	.0001	.0015	.0044	.0102	.0369	.0938
	6	.0000	.0000	.0001	.0002	.0007	.0041	.0156
7	0	.6983	.4783	.2097	.1335	.0824	.0280	.0078
	1	.2573	.3720	.3670	.3115	.2471	.1306	.0547
	2	.0406	.1240	.2753	.3115	.3177	.2613	.1641
	3	.0036	.0230	.1147	.1730	.2269	.2903	.2734
	4	.0002	.0026	.0287	.0577	.0972	.1935	.2734
	5	.0000	.0002	.0043	.0115	.0250	.0774	.1641
	6	.0000	.0000	.0004	.0013	.0036	.0172	.0547
	7	.0000	.0000	.0000	.0001	.0002	.0016	.0078

Table XI—*continued*

n	r	.05	.10	.20	.25	.30	.40	.50
8	0	.6634	.4305	.1678	.1001	.0576	.0168	.0039
	1	.2793	.3826	.3355	.2670	.1977	.0896	.0312
	2	.0515	.1488	.2936	.3115	.2965	.2090	.1094
	3	.0054	.0331	.1468	.2076	.2541	.2787	.2187
	4	.0004	.0046	.0459	.0865	.1361	.2322	.2734
	5	.0000	.0004	.0092	.0231	.0467	.1239	.2187
	6	.0000	.0000	.0011	.0038	.0100	.0413	.1004
	7	.0000	.0000	.0001	.0004	.0012	.0079	.0312
	8	.0000	.0000	.0000	.0000	.0001	.0007	.0039
9	0	.6302	.3874	.1342	.0751	.0404	.0101	.0020
	1	.2985	.3874	.3020	.2253	.1556	.0605	.0176
	2	.0629	.1722	.3020	.3003	.2668	.1612	.0703
	3	.0077	.0446	.1762	.2336	.2668	.2508	.1641
	4	.0006	.0074	.0661	.1168	.1715	.2508	.2461
	5	.0000	.0008	.0165	.0389	.0735	.1672	.2461
	6	.0000	.0001	.0028	.0087	.0210	.0743	.1641
	7	.0000	.0000	.0003	.0012	.0039	.0212	.0703
	8	.0000	.0000	.0000	.0001	.0004	.0035	.0176
	9	.0000	.0000	.0000	.0000	.0000	.0003	.0020
10	0	.5987	.3487	.1074	.0563	.0282	.0060	.0010
	1	.3151	.3874	.2684	.1877	.1211	.0403	.0098
	2	.0746	.1937	.3020	.2816	.2335	.1209	.0439
	3	.0105	.0574	.2013	.2503	.2668	.2150	.1172
	4	.0010	.0112	.0881	.1460	.2001	.2508	.2051
	5	.0001	.0015	.0264	.0584	.1029	.2007	.2461
	6	.0000	.0001	.0055	.0162	.0368	.1115	.2051
	7	.0000	.0000	.0008	.0031	.0090	.0425	.1172
	8	.0000	.0000	.0001	.0004	.0014	.0106	.0439
	9	.0000	.0000	.0000	.0000	.0001	.0016	.0098
	10	.0000	.0000	.0000	.0000	.0000	.0001	.0010
12	0	.5404	.2824	.0687	.0317	.0138	.0022	.0002
	1	.3413	.3766	.2062	.1267	.0712	.0174	.0029
	2	.0988	.2301	.2835	.2323	.1678	.0639	.0161
	3	.0173	.0852	.2362	.2581	.2397	.1419	.0537
	4	.0021	.0213	.1329	.1936	.2311	.2128	.1208
	5	.0002	.0038	.0532	.1032	.1585	.2270	.1934
	6	.0000	.0005	.0155	.0401	.0792	.1766	.2256
	7	.0000	.0000	.0033	.0115	.0291	.1009	.1934
	8	.0000	.0000	.0005	.0024	.0078	.0420	.1208
	9	.0000	.0000	.0001	.0004	.0015	.0125	.0537

Table XI—*continued*

		p						
n	*r*	.05	.10	.20	.25	.30	.40	.50
12	10	.0000	.0000	.0000	.0000	.0002	.0025	.0161
	11	.0000	.0000	.0000	.0000	.0000	.0003	.0029
	12	.0000	.0000	.0000	.0000	.0000	.0000	.0002
15	0	.4633	.2059	.0352	.0134	.0047	.0005	.0000
	1	.3658	.3432	.1319	.0668	.0305	.0047	.0005
	2	.1348	.2669	.2309	.1559	.0916	.0219	.0032
	3	.0307	.1285	.2501	.2252	.1700	.0634	.0139
	4	.0049	.0428	.1876	.2252	.2186	.1268	.0417
	5	.0006	.0105	.1032	.1651	.2061	.1859	.0916
	6	.0000	.0019	.0430	.0917	.1472	.2066	.1527
	7	.0000	.0003	.0138	.0393	.0811	.1771	.1964
	8	.0000	.0000	.0035	.0131	.0348	.1181	.1964
	9	.0000	.0000	.0007	.0034	.0116	.0612	.1527
	10	.0000	.0000	.0001	.0007	.0030	.0245	.0916
	11	.0000	.0000	.0000	.0001	.0006	.0074	.0417
	12	.0000	.0000	.0000	.0000	.0001	.0016	.0139
	13	.0000	.0000	.0000	.0000	.0000	.0003	.0032
	14	.0000	.0000	.0000	.0000	.0000	.0000	.0005
	15	.0000	.0000	.0000	.0000	.0000	.0000	.0000
20	0	.3585	.1216	.0115	.0032	.0008	.0000	.0000
	1	.3774	.2702	.0576	.0211	.0068	.0005	.0000
	2	.1887	.2852	.1369	.0669	.0278	.0031	.0002
	3	.0596	.1901	.2054	.1339	.0716	.0123	.0011
	4	.0133	.0898	.2182	.1897	.1304	.0350	.0046
	5	.0022	.0319	.1746	.2023	.1789	.0746	.0148
	6	.0003	.0089	.1091	.1686	.1916	.1244	.0370
	7	.0000	.0020	.0545	.1124	.1643	.1659	.0739
	8	.0000	.0004	.0222	.0609	.1144	.1797	.1201
	9	.0000	.0001	.0074	.0271	.0654	.1597	.1602
	10	.0000	.0000	.0020	.0099	.0308	.1171	.1762
	11	.0000	.0000	.0005	.0030	.0120	.0710	.1602
	12	.0000	.0000	.0001	.0008	.0039	.0355	.1201
	13	.0000	.0000	.0000	.0002	.0010	.0146	.0739
	14	.0000	.0000	.0000	.0000	.0002	.0049	.0370

Table XI—*continued*

n	r	p .05	.10	.20	.25	.30	.40	.50
20	15	.0000	.0000	.0000	.0000	.0000	.0013	.0148
	16	.0000	.0000	.0000	.0000	.0000	.0003	.0046
	17	.0000	.0000	.0000	.0000	.0000	.0000	.0011
	18	.0000	.0000	.0000	.0000	.0000	.0000	.0002
	19	.0000	.0000	.0000	.0000	.0000	.0000	.0000
	20	.0000	.0000	.0000	.0000	.0000	.0000	.0000

Table XII

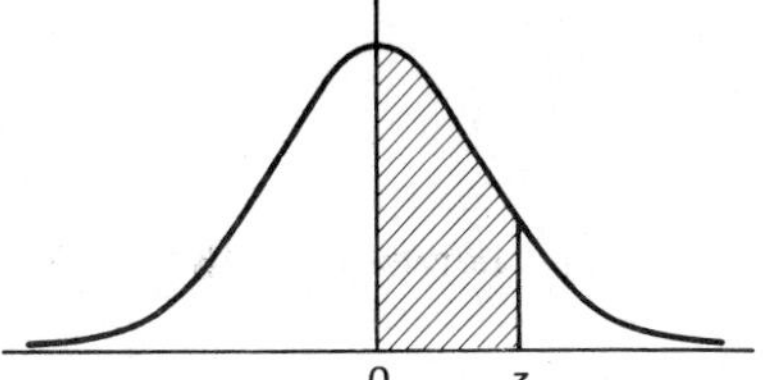

AREAS UNDER THE STANDARD NORMAL CURVE FROM 0 TO z

z	0	1	2	3	4	5	6	7	8	9
0.0	.0000	.0040	.0080	.0120	.0160	.0199	.0239	.0279	.0319	.0359
0.1	.0398	.0438	.0478	.0517	.0557	.0596	.0636	.0675	.0714	.0754
0.2	.0793	.0832	.0871	.0910	.0948	.0987	.1026	.1064	.1103	.1141
0.3	.1179	.1217	.1255	.1293	.1331	.1368	.1406	.1443	.1480	.1517
0.4	.1554	.1591	.1628	.1664	.1700	.1736	.1772	.1808	.1844	.1879
0.5	.1915	.1950	.1985	.2019	.2054	.2088	.2123	.2157	.2190	.2224
0.6	.2258	.2291	.2324	.2357	.2389	.2422	.2454	.2486	.2518	.2549
0.7	.2580	.2612	.2642	.2673	.2704	.2734	.2764	.2794	.2823	.2852
0.8	.2881	.2910	.2939	.2967	.2996	.3023	.3051	.3078	.3106	.3133
0.9	.3159	.3186	.3212	.3238	.3264	.3289	.3315	.3340	.3365	.3389
1.0	.3413	.3438	.3461	.3485	.3508	.3531	.3554	.3577	.3599	.3621
1.1	.3643	.3665	.3686	.3708	.3729	.3749	.3770	.3790	.3810	.3830
1.2	.3849	.3869	.3888	.3907	.3925	.3944	.3962	.3980	.3997	.4015
1.3	.4032	.4049	.4066	.4082	.4099	.4115	.4131	.4147	.4162	.4177
1.4	.4192	.4207	.4222	.4236	.4251	.4265	.4279	.4292	.4306	.4319
1.5	.4332	.4345	.4357	.4370	.4382	.4394	.4406	.4418	.4429	.4441
1.6	.4452	.4463	.4474	.4484	.4495	.4505	.4515	.4525	.4535	.4545
1.7	.4554	.4564	.4573	.4582	.4591	.4599	.4608	.4616	.4625	.4633
1.8	.4641	.4649	.4656	.4664	.4671	.4678	.4686	.4693	.4699	.4706
1.9	.4713	.4719	.4726	.4732	.4738	.4744	.4750	.4756	.4761	.4767
2.0	.4772	.4778	.4783	.4788	.4793	.4798	.4803	.4808	.4812	.4817
2.1	.4821	.4826	.4830	.4834	.4838	.4842	.4846	.4850	.4854	.4857
2.2	.4861	.4864	.4868	.4871	.4875	.4878	.4881	.4884	.4887	.4890
2.3	.4893	.4896	.4898	.4901	.4904	.4906	.4909	.4911	.4913	.4916
2.4	.4918	.4920	.4922	.4925	.4927	.4929	.4931	.4932	.4934	.4936

Table XII—*continued*

z	0	1	2	3	4	5	6	7	8	9
2.5	.4938	.4940	.4941	.4943	.4945	.4946	.4948	.4949	.4951	.4952
2.6	.4953	.4955	.4956	.4957	.4959	.4960	.4961	.4962	.4963	.4964
2.7	.4965	.4966	.4967	.4968	.4969	.4970	.4971	.4972	.4973	.4974
2.8	.4974	.4975	.4976	.4977	.4977	.4978	.4979	.4979	.4980	.4981
2.9	.4981	.4982	.4982	.4983	.4984	.4984	.4985	.4985	.4986	.4986
3.0	.4987	.4987	.4987	.4988	.4988	.4989	.4989	.4989	.4990	.4990
3.1	.4990	.4991	.4991	.4991	.4992	.4992	.4992	.4992	.4993	.4993
3.2	.4993	.4993	.4994	.4994	.4994	.4994	.4994	.4995	.4995	.4995
3.3	.4995	.4995	.4995	.4996	.4996	.4996	.4996	.4996	.4996	.4997
3.4	.4997	.4997	.4997	.4997	.4997	.4997	.4997	.4997	.4997	.4998
3.5	.4998	.4998	.4998	.4998	.4998	.4998	.4998	.4998	.4998	.4998
3.6	.4998	.4998	.4999	.4999	.4999	.4999	.4999	.4999	.4999	.4999
3.7	.4999	.4999	.4999	.4999	.4999	.4999	.4999	.4999	.4999	.4999
3.8	.4999	.4999	.4999	.4999	.4999	.4999	.4999	.4999	.4999	.4999
3.9	.5000	.5000	.5000	.5000	.5000	.5000	.5000	.5000	.5000	.5000

Index